ASTRONOMY AND ASTROPHYSICS LIBRARY

Springer
New York
Berlin
Heidelberg
Hong Kong
London
Milan
Paris
Tokyo

ASTRONOMY AND ASTROPHYSICS LIBRARY

The Stars By E. L. Schatzman and F. Praderie

Modern Astrometry 2nd Edition
By J. Kovalevsky

The Physics and Dynamics of Planetary Nebulae By G. A. Gurzadyan

Galaxies and Cosmology By F. Combes, P. Boissé, A. Mazure and A. Blanchard

Observational Astrophysics 2nd Edition
By P. Léna, F. Lebrun and F. Mignard

Physics of Planetary Rings Celestial Mechanics of Continuous Media
By A. M. Fridman and N. N. Gorkavyi

Tools of Radio Astronomy 4th Edition
By K. Rohlfs and T. L. Wilson

Astrophysical Formulae 3rd Edition (2 volumes)
Volume I: Radiation, Gas Processes and High Energy Astrophysics
Volume II: Space, Time, Matter and Cosmology
By K. R. Lang

Tools of Radio Astronomy Problems and Solutions By T. L. Wilson and S. Hüttemeister

Galaxy Formation By M. S. Longair

Astrophysical Concepts 2nd Edition
By M. Harwit

Astrometry of Fundamental Catalogues The Evolution from Optical to Radio Reference Frames
By H. G. Walter and O. J. Sovers

Compact Stars. Nuclear Physics, Particle Physics and General Relativity 2nd Edition
By N. K. Glendenning

The Sun from Space By K. R. Lang

Stellar Physics (2 volumes)
Volume 1: Fundamental Concepts and Stellar Equilibrium
By G. S. Bisnovatyi-Kogan

Stellar Physics (2 volumes)
Volume 2: Stellar Evolution and Stability
By G. S. Bisnovatyi-Kogan

Theory of Orbits (2 volumes)
Volume 1: Integrable Systems and Non-perturbative Methods
Volume 2: Perturbative and Geometrical Methods
By D. Boccaletti and G. Pucacco

Black Hole Gravitohydromagnetics
By B. Punsly

Stellar Structure and Evolution
By R. Kippenhahn and A. Weigert

Gravitational Lenses By P. Schneider, J. Ehlers and E. E. Falco

Reflecting Telescope Optics (2 volumes)
Volume I: Basic Design Theory and its Historical Development. 2nd Edition
Volume II: Manufacture, Testing, Alignment, Modern Techniques
By R. N. Wilson

Interplanetary Dust
By E. Grün, B. Å. S. Gustafson, S. Dermott and H. Fechtig (Eds.)

The Universe in Gamma Rays
By V. Schönfelder

Astrophysics. A Primer By W. Kundt

Cosmic Ray Astrophysics
By R. Schlickeiser

Astrophysics of the Diffuse Universe
By M. A. Dopita and R. S. Sutherland

The Sun An Introduction. 2nd Edition
By M. Stix

Order and Chaos in Dynamical Astronomy
By G. J. Contopoulos

Astronomical Image and Data Analysis
By J.-L. Starck and F. Murtagh

Continued after Index

Series homepage – http://www.springer.de/phys/books/aal

Carl J. Hansen Steven D. Kawaler
Virginia Trimble

Stellar Interiors

Physical Principles, Structure, and Evolution

Second Edition

With 119 Illustrations and a CD-ROM

Springer

Carl J. Hansen, Department of Astrophysics and Planetary Sciences, University of Colorado, Boulder, CO 80309-0440, USA. Email: chansen@jila.colorado.edu

Steven D. Kawaler, Departments of Physics and Astronomy, Iowa State University, Ames, IA 50011, USA. Email: sdk@iastate.edu

Virginia Trimble, Department of Astronomy, University of Maryland, College Park, MD 20740, and Department of Physics, University of California, Irvine, Irvine, CA 92717, USA

Cover illustration: Calligraphy by Sara Compton.

Library of Congress Cataloging-in-Publication Data
Hansen, Carl J.
Stellar interiors : physical principles, structure, and evolution.—2nd ed. / C.J. Hansen, S.D. Kawaler, and V. Trimble.
p. cm. — (Astronomy and astrophysics library)
Includes bibliographical references and index.
ISBN 0-387-20089-4 (hc : alk. paper)
1. Stars—Structure. 2. Stars—Evolution. 3. Astrophysics. I. Kawaler, Steven D. II. Trimble, Virginia. III. Title. IV. Series.
QB808.H36 2004
523.8′6—dc22 2003056810

ISBN 0-387-20089-4 Printed on acid-free paper.

Printed in the United States of America. (MVY)

9 8 7 6 5 4 3 2 1 SPIN 10939086

Springer-Verlag is a part of *Springer Science+Business Media*

springeronline.com

Preface

The first edition of this text appeared in 1994. Shortly after the third printing, our editor suggested that we attempt a second edition because new developments in stellar structure and evolution had made our original work outdated. We (the original authors, CJH and SDK) reluctantly agreed but with reservations due to the effort involved. Our initial reluctance disappeared when we were able to convince (cajole, twist the arm of, etc.) our new coauthor-colleague Virginia Trimble to join us. (Welcome Virginia!) We (i.e., all three of us) hope that you agree that the present edition is a great improvement compared to the 1994 effort.

Our objectives in this edition are the same ones we set forth in 1994:

> What you will find is a text designed for our target audience: the typical senior undergraduate or beginning graduate student in astronomy or astrophysics who wishes an overview of stellar structure and evolution with just enough detail to understand the general picture. She or he can go on from there to more specialized texts or directly to the research literature depending on talent and interests. To this end, this text presents the basic physical principles without chasing all the (interesting!) details.

For those of you familiar with the first edition, you will find that some things have not been changed substantially ($F = ma$ is still $F = ma$), while others definitely have. For example, Chapter 2 has been completely rewritten. In many respects this chapter is the key to the text because it gives an extensive overview of the subject. The next eight chapters rely on the student's having absorbed large parts of Chapter 2, though complete understanding is not necessary. Many students may wish to start with Chapter 2, although we recommend at least a once-through of Chapter 1, which contains some fundamental material. And, in response to many requests, there is substantially more observational material.

We have also attempted to improve on the graphics and have included more than we did in the first edition. In addition, the instructor will find many more "Exercises" at the end of chapters. They are a mixed bag (easy, moderate, difficult) but we hope they illuminate much of what we have to say. (Chapter 2 has more than its share; and, in fact, Chapters 1 and 2, plus exercises, could be the basis of a mini-course.)

Also new is the inclusion on the inside back cover of a CD-ROM containing computer programs that make decent "zero-age main sequence" stellar models and analyse those models for "pulsations" (radial and nonradial), and stellar evolution codes everyone can play with. All are in `FORTRAN` and should work on most computer platforms. Some of these codes are of our doing and we thank Andy Odell and Dean Pesnell (Nomad Research) for their generous contributions. As an additional bonus we have included portions of a colorful and informative *Stellar Evolution Tutorial* put together by John Lattanzio and his colleagues (as part of a commercial enterprise called Cantanout Ltd.). See the `README` files on the CD-ROM for more information on the programs and tutorial.

Acknowledgments: We wish to thank our many past and present senior colleagues and students for numerous reprints, corrections, suggestions, comments, problems (i.e., exercises), book loans, help with computer glitches, and PostScript figure files. They made our task much easier and enjoyable. Blame the typos, mistakes, and confusion on us. In particular, for the second edition, we thank Dave Arnett, Mitch Begelman, David Branch, Nic Brummell, Joe Cassinelli, Maurice Clement, Peter Conti, Ethan Hansen, Henny Lamers, Michael McCarthy, Cole Miller, Sean O'Brien, Dean Richardson, Dimitar Sasselov, Ted Snow, Peter Stetson, Pat Thaddeus, Juri Toomre, Don Vandenberg, Craig Wheeler, Matt Wood, and Ellen Zweibel. VT gives personal thanks to those people from whom she first learned that stellar structure and evolution is an exciting topic—namely, (the late) Thornton Leigh Page, J. Beverly Oke, and Bohdan Paczyński. She also recognizes the past encouragement and support of UCLA, CalTech, and the Stony Brook Summer School. CJH and SDK wish to thank their families and especially their wives Camille and Leslie: may they not become computer widows *yet* again. Finally, we send many kudos to our editors at Springer-Verlag.

The text was set in LaTeX 2_ε by the authors.

Carl J. Hansen — University of Colorado at Boulder
Steven D. Kawaler — Iowa State University at Ames
Virginia Trimble — University of California at Irvine, University of Maryland at College Park

Contents

1 Preliminaries

"Cosmologists have, nonetheless, made real progress in recent years. This is because what makes things baffling is their degree of complexity, not their sheer size —and a star is simpler than an insect."

— Martin Rees in Scientific American (Dec. 1999)

If you want an important insight into what makes stars work, go out and look at them for a few nights. You will find that they appear to do nothing much at all except shine steadily. This is certainly true from a historical perspective: taking the sun as an example, from fossil evidence, we can extend this period of "inactivity" to roughly three billion years. The reason for this relative tranquility is that stars are, on the whole, very stable objects in which self-gravitational forces are delicately balanced by steep internal pressure gradients. The latter require high temperatures. In the deep interior of a star these temperatures are measured in (at least) millions of degrees Kelvin and, in most instances, are sufficiently high to initiate the thermonuclear fusion of light nuclei. The power so produced then laboriously works its way out through the remaining bulk of the star and finally gives rise to the radiation we see streaming off the surface. The vast majority of stars spend most of their active lives in such an equilibrium state, converting hydrogen into helium, and it is only this gradual transmutation of elements by the fusion process that eventually causes their structure to change in some marked way.

This chapter will introduce some concepts and physical processes that, when tied together, will enable us to paint a preliminary picture of the stellar interior and to make estimates of the magnitudes of various quantities such as pressure, temperature, and lifetimes. Later chapters will expand on these concepts and processes and bring us up to date on some modern developments in stellar structure and evolution. If, in reading this chapter, you begin to get lost, we suggest you review Appendix A on some properties of stars and nomenclature. Chapter 2 also contains similar material in narrative form. For those of you who have no background in the subject at all, a good first-year undergraduate text on astronomy for the nonscientist may be in order, and several excellent texts are available. Some portions of the material we present will also make considerable demands on your understanding of physical processes. We assume that you have a decent background in undergraduate physics. If not, you will have to catch up and review that material.

An alternative route some of you may wish to take is to start with Chapter 2, which discusses stars and their evolution. If you choose this route, however, you may have to return to this chapter for some elementary material. The later chapters go into more detail than you will need for now.[1] Either way, go for it!

1.1 Hydrostatic Equilibrium

We first consider the theoretician's dream: a spherically symmetric, nonrotating, nonmagnetic, single, etc., star on which there are no net forces acting and, hence, no net accelerations. There may be internal motions, such as those associated with convection, but these are assumed to average out overall. We wish to find a relation that expresses this equilibrium. First assume that the stellar material is so constituted that internal stresses are isotropic and thus reduce to ordinary pressures, and define the following quantities, which will be used throughout this text (and see Appendix B for a more complete listing of symbols, including the values of physical and astronomical constants):

radius: r is the radial distance measured from the stellar center (cm)

total stellar radius: $\mathcal{R}$

mass density: $\rho(r)$ is the mass density at r (g cm^{-3})

temperature: $T(r)$ is the temperature at r (deg K)

pressure: $P(r)$ is the pressure at r (dyne cm^{-2} = erg cm^{-3})

mass: $\mathcal{M}_r$ is the mass contained within a sphere of radius r (g)

total stellar mass: $\mathcal{M} = \mathcal{M}_\mathcal{R}$

luminosity: $\mathcal{L}_r$, the rate of energy flow through a sphere at r (erg s^{-1})

total stellar luminosity: $\mathcal{L} = \mathcal{L}_\mathcal{R}$

local gravity: $g(r)$, local acceleration due to gravity (cm s^{-2})

gravitational constant $G = 6.6726 \times 10^{-8}$ g^{-1} cm^3 s^{-2}

solar mass: $\mathcal{M}_\odot = 1.9891 \times 10^{33}$ g

solar luminosity: $\mathcal{L}_\odot = 3.847 \times 10^{33}$ erg s^{-1}

solar radius: $\mathcal{R}_\odot = 6.96 \times 10^{10}$ cm

Note that the above are expressed in cgs units. There is really no good reason for this, but cgs seem to be the units of choice for most researchers dealing with stars. MKS (SI) units could just as well be used instead (and are actually preferred by those dealing with magnetic fields in astrophysics). We will use solar units (e.g., $\mathcal{M}/\mathcal{M}_\odot$ or $\mathcal{R}/\mathcal{R}_\odot$) when appropriate.

---

[1] In any case, we strongly suggest you attempt as many of the exercises found near the end of the chapters as possible. Note also that references are given at the end of each chapter. (Since some of the journal abbreviations we use may seem obscure, Appendix C lists them along with the full journal name.

First we investigate the balance of forces within a star in equilibrium. From elementary physics, the local gravity at spherical radius r is

$$g(r) = \frac{G\mathcal{M}_r}{r^2} = 2.74 \times 10^4 \left(\frac{\mathcal{M}_r}{\mathcal{M}_\odot}\right)\left(\frac{r}{\mathcal{R}_\odot}\right)^{-2} \quad \text{cm s}^{-2} \tag{1.1}$$

and

$$\mathcal{M}_{r+dr} - \mathcal{M}_r = d\mathcal{M}_r = 4\pi r^2 \rho(r)\, dr \tag{1.2}$$

is the mass contained within a spherical shell of infinitesimal thickness dr at r. The integral of (1.2) yields the mass within r,

$$\mathcal{M}_r = \int_0^r 4\pi r^2 \rho\, dr \,. \tag{1.3}$$

Either (1.2) or (1.3) will be referred to as the *mass equation* or the *equation of mass conservation.*

Now consider a 1–cm^2 element of area on the surface of the shell at r. There is an inwardly directed gravitational force on a volume 1 cm^2 $\times$ dr of

$$\rho g\, dr = \rho \frac{G\mathcal{M}_r}{r^2}\, dr \,. \tag{1.4}$$

To counterbalance this force we must rely on an imbalance of pressure forces; that is, the pressure $P(r)$ pushing outward against the inner side of the shell must be greater than the pressure acting inward on the outer face. The net pressure outward is $P(r) - P(r+dr) = -(dP/dr)\, dr$. Adding the gravitational and differential pressure forces then yields

$$\rho \ddot{r} = -\frac{dP}{dr} - \frac{G\mathcal{M}_r}{r^2}\rho \tag{1.5}$$

as the equation of motion, where $\ddot{r}$ is the local acceleration d^2r/dt^2.

By hypothesis, all net forces are zero, with $\ddot{r} = 0$, and we obtain the *equation of hydrostatic* (or mechanical) *equilibrium*:

$$\frac{dP}{dr} = -\frac{G\mathcal{M}_r}{r^2}\rho = -g\rho \,. \tag{1.6}$$

Since $g, \rho \geq 0$, then $dP/dr \leq 0$, and the pressure must decrease outward everywhere. If this condition is violated anywhere within the star, then hydrostatic equilibrium is impossible and local accelerations must occur.

We can obtain the hydrostatic equation in yet another way and, at the same time, introduce some new concepts.

1.2 An Energy Principle

The preceding was a local approach to mechanical equilibrium because only local quantities at r were involved (although a gradient did appear). What we shall do now is take a global view wherein equilibrium is posed as an integral constraint on the structure of the entire star.

Imagine that the equilibrium star is only one of an infinity of possible configurations and the trick is to find the right one. (The wrong ones will not be in equilibrium and just won't do.) Each configuration will be specified by an integral function so constructed that the equilibrium star is represented by a stationary point in the series of possible functions. This begins to sound like a problem in classical mechanics and the calculus of variations—and it is. (We'll ease into the mathematics.) The function in question is the total stellar energy, and so let's see what it is.

The total gravitational potential energy, Ω, of a self-gravitating body is defined as the *negative* of the total amount of energy required to disperse all mass elements of the body to infinity. The zero point of the potential is taken as the final state after dispersal. In other words, Ω is the energy required to assemble the star, in its current configuration, by collecting material from the outside universe. Thus Ω represents (negative) work done on, or by, the system and it must be accounted for when determining the total energy of the star.

We can get to the dispersed state by successively peeling off spherical shells from our spherical star. Suppose we have already done so down to an interior mass of $\mathcal{M}_r + d\mathcal{M}_r$ and we are just about to remove the next shell, which has a mass $d\mathcal{M}_r$. To move this shell outward from some radius r' to $r' + dr'$ requires $(G\mathcal{M}_r/r'^2)\, d\mathcal{M}_r\, dr'$ units of work. To go from r to infinity then gives a contribution to Ω of (remembering the minus sign for Ω)

$$d\Omega = -\int_r^\infty \frac{G\mathcal{M}_r}{r'^2}\, d\mathcal{M}_r\, dr' = -\frac{G\mathcal{M}_r}{r}\, d\mathcal{M}_r .$$

To disperse the whole star requires that we do this for all $d\mathcal{M}_r$ or,

$$\Omega = -\int_0^\mathcal{M} \frac{G\mathcal{M}_r}{r}\, d\mathcal{M}_r \, . \tag{1.7}$$

The potential energy thus has the units of $G\mathcal{M}^2/\mathcal{R}$ and we shall often write it in the form

$$\Omega = -q\,\frac{G\mathcal{M}^2}{\mathcal{R}} \, . \tag{1.8}$$

For a uniform density sphere, with ρ constant, it is easy to show that the pure number q is equal to 3/5. (This should be familiar from electrostatics, where the energy required to disperse a uniformly charged sphere to infinity is $-3e^2/5\mathcal{R}$.) Because density almost always decreases outward for equilibrium stars, the value of 3/5 is, for all practical purposes, a lower limit with $q \geq 3/5$.

For the sun, $G\mathcal{M}_\odot{}^2/\mathcal{R}_\odot \approx 3.8 \times 10^{48}$ erg. If we divide this figure by the present solar luminosity, $\mathcal{L}_\odot$, we find a characteristic time (the Kelvin–Helmholtz time scale) of about 3×10^7 years. More will be said about this time scale later on.

If we neglect gross mass motions or phenomena such as turbulence, then the total energy of the star is Ω plus the total internal energy arising from microscopic processes. Let E be the local specific internal energy in units of ergs per gram of material. It is to be multiplied by ρ if you want energy per unit volume. (Thus E will sometimes have the units of erg cm^{-3} but you will either be forewarned by a statement or the appearance of those units.) The total energy, W, is then the sum of Ω and the mass integral of E,

$$W = \int_{\mathcal{M}} E \, d\mathcal{M}_r + \Omega = U + \Omega \tag{1.9}$$

which also defines the total internal energy

$$U = \int_{\mathcal{M}} E \, d\mathcal{M}_r \ . \tag{1.10}$$

The statement now is that the equilibrium state of the star corresponds to a stationary point with respect to W. This means that W for the star in hydrostatic equilibrium is an extremum (a maximum or minimum) relative to all other possible configurations the star could have (with the possible exception of other extrema). What we are going to do to test this idea is to perturb the star away from its original state in an *adiabatic* but otherwise arbitrary and infinitesimal fashion. The adiabatic part can be satisfied if the perturbation is performed sufficiently rapidly that heat transfer between mass elements does not take place (as in an adiabatic sound wave). We shall show later that energy redistribution in normal stars takes place on time scales longer than mechanical response times. On the other hand, we also require that the perturbation be sufficiently slow that kinetic energies of mass motions can be ignored.

If δ represents either a local or global perturbation operator (think of it as taking a differential), then the stellar hydrostatic equilibrium state is that for which

$$(\delta W)_{\text{ad}} = 0$$

where the "ad" subscript denotes "adiabatic." Thus if arbitrary, but small, adiabatic changes result in no change in W, then the initial stellar state is in hydrostatic equilibrium. To show this, we have to look how U and Ω change when ρ, T, etc., are varied adiabatically. We thus have to look at the pieces of

$$(\delta W)_{\text{ad}} = (\delta U)_{\text{ad}} + (\delta \Omega)_{\text{ad}} \ .$$

A perturbation δ causes U to change by δU with

$$U \longrightarrow U + \delta U = U + \delta \int_{\mathcal{M}} E \, d\mathcal{M}_r = U + \int_{\mathcal{M}} \delta E \, d\mathcal{M}_r \ .$$

The last step follows because we choose to consider the change in specific internal energy of a particular mass element $d\mathcal{M}_r$. (This is a *Lagrangian* description of the perturbation about which more will be said in Chap. 8.) Now consider δE. We label each mass element of $d\mathcal{M}_r$ worth of matter and see what happens to it (and E) when its position r, and ρ, and T are changed.

For an infinitesimal and reversible change (it would be nice to be able to put the star back together again), the combined first and second laws of thermodynamics state that

$$dQ = dE + P\,dV_\rho = T\,dS\ . \tag{1.11}$$

Here dQ is the heat added to the system, dE is the increase in internal specific energy, and $P\,dV_\rho$ is the work done by the system on its surroundings if the "volume" changes by dV_ρ. This volume is the *specific volume*, with

$$V_\rho = 1/\rho \tag{1.12}$$

and is that associated with a given gram of material. It has the units of $\text{cm}^3\ \text{g}^{-1}$. (The symbol V will be reserved for ordinary volume with units of cm^3.) The entropy S, and Q, are also mass-specific quantities. If we replace the differentials in the preceding by δs, then the requirement of adiabaticity ($\delta S = 0$) immediately yields $(\delta E)_{\text{ad}} = -P\,\delta V_\rho$. Thus,

$$(\delta U)_{\text{ad}} = -\int_{\mathcal{M}} P\,\delta V_\rho\,d\mathcal{M}_r\ .$$

What is δV_ρ? From the definition of the specific volume (1.12) and the mass equation (1.2),

$$V_\rho = \frac{1}{\rho} = \frac{4\pi r^2\,dr}{d\mathcal{M}_r} = \frac{d(4\pi r^3/3)}{d\mathcal{M}_r}\ . \tag{1.13}$$

To make life easy, we restrict all perturbations to those that maintain spherical symmetry. Thus if the mass parcel $d\mathcal{M}_r$ moves at all, it moves only in the radial direction to a new position $r + \delta r$. Perturbing V_ρ in (1.13) is then equivalent to perturbing r or

$$V_\rho \longrightarrow V_\rho + \delta V_\rho = \frac{d[4\pi(r+\delta r)^3/3]}{d\mathcal{M}_r} = V_\rho + \frac{d(4\pi r^2 \delta r)}{d\mathcal{M}_r} \tag{1.14}$$

to first order in δr, where we assume that $|\delta r/r| \ll 1$. (Later we will call this sort of thing "linearization.") The variation in total internal energy is then

$$(\delta U)_{\text{ad}} = -\int_{\mathcal{M}} P\,\frac{d(4\pi r^2\,\delta r)}{d\mathcal{M}_r}\,d\mathcal{M}_r\ . \tag{1.15}$$

We now introduce two boundary conditions. The first is obvious: we don't allow the center of our spherically symmetric star to move. This amounts to requiring that $\delta r(\mathcal{M}_r = 0) = 0$. The second is called the "zero boundary condition on pressure" and it requires that the pressure at the surface vanish.

Thus, $P_S = P(\mathcal{M}_r = \mathcal{M}) = 0$. This last is perfectly reasonable in this context because, in our idealized star, the surface is presumably where the mass runs out and we implicitly assume that no external pressures have been applied. (Later, in Chap. 4, we will have to worry quite a bit more about this "surface." It is more subtle than may appear.) Now integrate (1.15) by parts, apply the boundary conditions to the resulting constant term, and find

$$(\delta U)_{\text{ad}} = \int_{\mathcal{M}} \frac{dP}{d\mathcal{M}_r} 4\pi r^2 \, \delta r \, d\mathcal{M}_r \, .$$

The corresponding analysis for $(\delta\Omega)_{\text{ad}}$ yields

$$\Omega \longrightarrow \Omega + \delta\Omega = -\int_{\mathcal{M}} \frac{G\mathcal{M}_r}{r + \delta r} \, d\mathcal{M}_r = \Omega + \int_{\mathcal{M}} \frac{G\mathcal{M}_r}{r^2} \, \delta r \, d\mathcal{M}_r$$

to first order in δr after expansion of the denominator in the first integral.

Putting it all together, we find

$$(\delta W)_{\text{ad}} = \int_{\mathcal{M}} \left[\frac{dP}{d\mathcal{M}_r} 4\pi r^2 + \frac{G\mathcal{M}_r}{r^2} \right] \delta r \, d\mathcal{M}_r \, .$$

The aim is now to see what happens when this expression is set to zero. Is hydrostatic equilibrium regained? This is an exercise from the calculus of variations (as in Goldstein, 1981). If δr is indeed arbitrary (subject to restrictions of symmetry), then the only way $(\delta W)_{\text{ad}}$ can vanish is for the integrand to vanish identically; that is, we must have

$$\frac{dP}{d\mathcal{M}_r} = -\frac{G\mathcal{M}_r}{4\pi r^4} \, . \tag{1.16}$$

The equation of hydrostatic equilibrium (1.6) follows immediately after the mass equation (1.2) is used to convert the differential from $d\mathcal{M}_r$ to dr. The version (1.16) is Lagrangian (the independent variable is $d\mathcal{M}_r$) and, after introducing acceleration in the appropriate place, is often used in one-dimensional hydrodynamical studies of stars (as in Chap. 7). Note that (1.16) is *necessary* for an extremum in W but it does not give us the structure directly nor does it tell us whether more than one extremum exists or, for that matter, whether any exist.

In this regard you may profit from considering the significance of $(\delta^2 W)_{\text{ad}}$, which is the second variation of W. As discussed by Chiu (1968, §2.12), the sign of the second variation determines whether the equilibrium configuration is mechanically stable or unstable to small perturbations. This is like asking whether a pencil balanced on its point is "stable." You are invited to play with this idea in Ex. 1.11 near the end of this chapter.

1.3 The Virial Theorem and Its Applications

We now derive the virial theorem and, from it, obtain some interesting and useful relations between various global stellar quantities such as W and Ω.

This will be primarily an exercise in classical mechanics at first, but the utility of the virial theorem in making simple estimates of temperature, density, and the like will soon be apparent. In addition, the theorem will be applied to yield estimates for some important stellar time scales. Most texts on stellar interiors contain some discussion of this topic. We shall follow Clayton (1968, Chap. 2). A specialized reference is Collins (1978).

Consider the scalar product $\sum_i \mathbf{p}_i \bullet \mathbf{r}_i$ where $\mathbf{p}_i$ is the vector momentum of a free particle of mass m_i located at position $\mathbf{r}_i$, and the sum is over all particles comprising the star. If the mechanics are nonrelativistic, then recognize that

$$\frac{d}{dt}\sum_i \mathbf{p}_i \bullet \mathbf{r}_i = \frac{d}{dt}\sum_i m_i \dot{\mathbf{r}}_i \bullet \mathbf{r}_i = \frac{1}{2}\frac{d}{dt}\sum_i \frac{d}{dt}(m_i r_i^2) = \frac{1}{2}\frac{d^2 I}{dt^2}$$

where I is the moment of inertia, $I = \sum_i m_i r_i^2$. On the other hand, the derivative of the original sum yields

$$\frac{d}{dt}\sum_i \mathbf{p}_i \bullet \mathbf{r}_i = \sum_i \frac{d\mathbf{p}_i}{dt} \bullet \mathbf{r}_i + \sum_i \mathbf{p}_i \bullet \frac{d\mathbf{r}_i}{dt} \, .$$

The last term is just $\sum_i m_i v_i^2$ (v_i is the velocity of particle i) and is equal to twice the total kinetic energy, K, of all the free particles in the star. Furthermore, take note of Newton's law,

$$\frac{d\mathbf{p}_i}{dt} = \mathbf{F}_i$$

where $\mathbf{F}_i$ is the force applied to particle i, which we will take as the force of gravity. Putting this together, we have

$$\tfrac{1}{2}\frac{d^2 I}{dt^2} = 2K + \sum_i \mathbf{F}_i \bullet \mathbf{r}_i \, . \tag{1.17}$$

The last term is the virial of Clausius, but to make any use of it all of the $\mathbf{F}_i \bullet \mathbf{r}_i$ must be specified.

That term in (1.17) is the mutual gravitational interaction of all the particles in the star. (And, remember, we are still ignoring magnetic fields, etc. These make their own kinds of contributions.) To treat gravity, let $\mathbf{F}_{ij}$ be the gravitational force on particle i due to the presence of particle j. Because such forces are equal and opposite, $\mathbf{F}_{ij} = -\mathbf{F}_{ji}$. You may verify by direct construction (with, say, three particles) that

$$\sum_i \mathbf{F}_i \bullet \mathbf{r}_i = \sum_{\substack{i,j \\ i<j}} (\mathbf{F}_{ij} \bullet \mathbf{r}_i + \mathbf{F}_{ji} \bullet \mathbf{r}_j)$$

where the sum is to be taken over all i and j provided that $i < j$. Hereafter, this convention will be assumed and the limits on the sum will not be given.

From elementary physics, the Newtonian gravitational force is

$$\mathbf{F}_{ij} = -\frac{Gm_i m_j}{r_{ij}^3}(\mathbf{r}_i - \mathbf{r}_j)$$

where r_{ij} is the interparticle distance $r_{ij} = |\mathbf{r}_i - \mathbf{r}_j|$. The gravitational contribution to the virial is then

$$\sum \mathbf{F}_{ij} \bullet (\mathbf{r}_i - \mathbf{r}_j) = -\frac{\sum Gm_i m_j}{r_{ij}} = \text{Virial}$$

using the equal and opposite expression to obtain the first term. It should be apparent that the last sum (with minus sign) is just the negative of the work required for dispersal to infinity; that is, we have recovered Ω. Thus we have

$$\text{Virial} = \Omega \,.$$

Combining this with (1.17), we obtain

$$\tfrac{1}{2}\frac{d^2 I}{dt^2} = 2K + \Omega \tag{1.18}$$

as the "virial theorem," which we will often refer to as just the "virial."

Note that this expression refers to quantities derived from sums (or integrals) over the *whole* star. If we had chosen instead to consider only a portion of the star—as, say, defined by a sphere of radius $r_S \leq \mathcal{R}$ and volume V_S—then I, K, and Ω would refer only to that portion. However, the spherical shell containing material within radii $r_S < r \leq \mathcal{R}$ would contribute an additional term to the right-hand side of (1.18) given by $-3P_S V_S$, where P_S is the pressure at the surface r_S. If $r_S \to \mathcal{R}$ and $P_S \to 0$ (as in a zero boundary condition on pressure), then (1.18) is unchanged because we have just encompassed the whole star and no external pressures act at $\mathcal{R}$. (For a derivation of this additional term see, for example, Cox 1968, §17.2, or Clayton, 1968, pp. 134–135, and you can try it yourself in Ex. 1.8.) We will not have occasion to use this term, but its possible presence should be kept in mind.

We now interpret what the energy K represents. For example, is it U or, if not, how does it differ? We had

$$2K = \sum_i m_i v_i^2 = \sum_i \mathbf{p}_i \bullet \mathbf{v}_i \,. \tag{1.19}$$

The scalar product of $\mathbf{p}$ and $\mathbf{v}$ measures the rate of momentum transfer and, hence, from the kinetic theory of gases, must be related to the pressure. In the continuum limit of an isotropic gas, pressure is given by[2]

[2] This may seem to come out of the blue but, as long as the "gas" is perfect and isotropic, it also applies to a radiation "gas" and other situations. Equation (1.20) is the compact way of expressing derivations of ideal gas and radiation pressures as given, for example, in §2–1 of Clayton (1968). You should be able to construct this yourself, realizing that the factor of 1/3 comes from averaging over angle in the isotropic gas.

$$P = \tfrac{1}{3} \int_p n(\mathbf{p})\, \mathbf{p} \bullet \mathbf{v}\, d^3\mathbf{p} \tag{1.20}$$

where $n(\mathbf{p})$ is the number density of particles with momentum $\mathbf{p}$ and the integration is over all momenta. The units of $n(\mathbf{p})$ are number cm^{-3} p^{-3}. Since the sum in (1.19) includes all particles, it should be clear that (1.20) need only be integrated over total volume V (in cm^3) to obtain an expression for K—namely,

$$2K = 3 \int_V P\, dV. \tag{1.21}$$

Furthermore, since $d\mathcal{M}_r = \rho\, d(\frac{4}{3}\pi r^3) = \rho\, dV$, we find

$$2K = 3 \int_{\mathcal{M}} \frac{P}{\rho}\, d\mathcal{M}_r \tag{1.22}$$

and the virial theorem becomes

$$\tfrac{1}{2}\frac{d^2 I}{dt^2} = \int_{\mathcal{M}} \frac{3P}{\rho}\, d\mathcal{M}_r + \Omega\,. \tag{1.23}$$

We now apply this to stars by looking into some possible choices for the equation of state.

1.3.1 Application: Global Energetics

Consider a simple, but useful, relation between pressure and internal energy of the form

$$P = (\gamma - 1)\rho E \tag{1.24}$$

where γ is a constant and E is still in erg g^{-1}. This is usually called a "γ–law equation of state" and is not just of academic interest. For example (and as we will show later), for a monatomic ideal gas $\gamma = c_P/c_V = 5/3$ where c_P and c_V are, respectively, the specific heats at constant pressure and volume. In this instance $P = \frac{2}{3}\rho E$. For radiation or a completely relativistic Fermi gas $\gamma = 4/3$. Since $2K = 3(\gamma - 1)\int E\, d\mathcal{M}_r$—from combining (1.22) and (1.24)—then $K = \frac{3}{2}(\gamma - 1)U$. Thus $K = U$ only if $\gamma = 5/3$; that is, the total kinetic energy is the same as the total internal energy only under certain circumstances. Note that a γ of 5/3 does not necessarily mean the gas is ideal and monatomic.

The virial theorem is now

$$\tfrac{1}{2}\frac{d^2 I}{dt^2} = 3(\gamma - 1)U + \Omega\,. \tag{1.25}$$

If we let $W = U + \Omega$, as in (1.9), then the theorem becomes

$$\tfrac{1}{2}\frac{d^2 I}{dt^2} = 3(\gamma - 1)W - (3\gamma - 4)\Omega\,. \tag{1.26}$$

For hydrostatic equilibrium d^2I/dt^2 must be zero and W is related to Ω by

$$W = \frac{3\gamma - 4}{3(\gamma - 1)}\,\Omega \tag{1.27}$$

which shows explicitly the relation between W and Ω for hydrostatic stars with the γ–law equation of state.

Since the energy W is that which is available to do useful work, a dynamically stable star should have $W < 0$. Otherwise, the star would have enough energy, at least in principle, completely to disperse all or part of itself. Equation (1.27) then implies that a star in hydrostatic equilibrium should have a γ that exceeds 4/3. However, and as we shall find later on, even this condition does not always guarantee safety. The star could contain a potentially explosive fuel which, if ignited, could also cause W to exceed zero for a time. In addition, we do not necessarily expect the total energy to remain absolutely and forever constant. After all, stars do shine and lose energy in doing so.

We shall now explore some consequences of energy losses due to radiation where the energy source is gravitational energy released by contraction.

1.3.2 Application: The Kelvin–Helmholtz Time Scale

Barring bizarre circumstances, a star derives its energy to shine from three sources: internal energy, thermonuclear fuel, and gravitational contraction. One or more sources are used at one time or another. Here we briefly examine the last source. A more complete treatment will be deferred to Chapter 6, where stellar energy sources are discussed in more depth.

Suppose a star contracts very gradually while maintaining sphericity and hydrostatic equilibrium at all times. (Realize that contraction cannot occur without some acceleration unless all mass elements are just coasting. What we mean here is that hydrostatic equilibrium is to be maintained *almost* exactly.) As the star contracts, Ω and, possibly, W change. Denote these changes by $\Delta\Omega$ and ΔW. If γ remains constant during contraction, then (1.27) implies

$$\Delta W = \frac{3\gamma - 4}{3(\gamma - 1)}\,\Delta\Omega\,. \tag{1.28}$$

Because we cannot follow the star's progress exactly (at least at this stage in the text) we use some dimensional arguments to estimate what Ω and W do upon contraction.

Let $\mathcal{R}$ be the total stellar radius (or some other representative radius) and $\Delta\mathcal{R}$ be its change through some stage in the contraction. We assume $\gamma > 4/3$. From (1.8), $\Omega \propto -G\mathcal{M}^2/\mathcal{R}$, which implies that $\Delta\Omega \propto (G\mathcal{M}^2/\mathcal{R}^2)\Delta\mathcal{R}$ for constant q. Since $\Delta\mathcal{R} < 0$ for contraction, then $\Delta\Omega$ is also negative in these circumstances and the star sinks deeper into its own potential well. This means that energy has been liberated in some form. The virial result (1.28) also implies that $\Delta W < 0$ and thus the system as a whole has lost energy. What exactly is the energy budget here? Well, part of what has been made available goes into internal energy. This may be seen from (1.25) (with d^2I/dt^2 set to zero for equilibrium), which becomes

$$\Delta U = -\frac{1}{3(\gamma - 1)} \Delta\Omega \tag{1.29}$$

and yields $\delta U > 0$ for contraction. Of the $|\Delta\Omega|$ units of energy made available, ΔU is used to "heat" up the star. The rest is lost from the system. At this stage in our discourse, it is simplest to assume that this energy has been radiated from the stellar surface during the contraction; that is, power has been expended in the form of luminosity. Note that if $\gamma = 5/3$ (as for an ideal monatomic gas), then $\Delta U = -\Delta W = -\Delta\Omega/2$ and the split between internal energy and time-integrated luminosity is equal. Note also that if an increase in temperature is associated with the increase in U, then the star has an overall specific heat that is negative: a loss of total energy means an increase in temperature. This phenomenon is an important self-regulating mechanism for normal stars. Finally, if $\gamma = 4/3$, then $\Delta W = 0$ and all the energy goes into increasing U and the star need not radiate at all.

Suppose we now extend the above analysis and hypothesize that contraction is *solely* responsible for maintaining stellar luminosities. For an ideal gas star with $\gamma = 5/3$, $\Delta W = \Delta\Omega/2 = (q/2)\left(G\mathcal{M}^2/\mathcal{R}^2\right)\Delta\mathcal{R}$. If we equate $-dW/dt$ to the luminosity $\mathcal{L}$ (as a power output), then

$$\mathcal{L} = -\frac{dW}{dt} = -\frac{q}{2}\frac{G\mathcal{M}^2}{\mathcal{R}}\left(\frac{d\mathcal{R}/dt}{\mathcal{R}}\right) . \tag{1.30}$$

It is clear that if $\mathcal{L}$ is kept constant, then this equation defines a characteristic e-folding time for radius decrease of

$$t_{\rm KH} \approx \frac{q}{2}\frac{G\mathcal{M}^2}{\mathcal{L}\mathcal{R}} \tag{1.31}$$

where the "KH" subscript stands for the originators of the idea, Baron W.T. Kelvin and H.L.F. Helmholtz. Choosing a representative value of q of 3/2 (which is about right for the sun),

$$t_{\rm KH} \approx 2 \times 10^7 \left(\frac{\mathcal{M}}{\mathcal{M}_\odot}\right)^2 \left(\frac{\mathcal{L}}{\mathcal{L}_\odot}\right)^{-1} \left(\frac{\mathcal{R}}{\mathcal{R}_\odot}\right)^{-1} \text{ years.} \tag{1.32}$$

We know that a figure of 2×10^7 years for radius changes for the sun cannot be correct from fossil evidence: terrestrial life is the same now (except for relatively inconsequential developments) as it was many millions of years ago. Any major structural change in the sun would have had profound consequences for life and there is no sign of such consequences. However, we will find that most stars do depend on (or, more accurately, are forced into) gravitational contraction at some stage of evolution, and the corresponding time scales can be comparatively very short.

1.3.3 Application: A Dynamic Time Scale

Consider a star in hydrostatic equilibrium composed purely of an ideal gas so that $W = \Omega/2 \approx -G\mathcal{M}^2/\mathcal{R}$. If, by some magic, an internal process were to take place instantaneously whereby $\gamma \rightarrow 4/3$ but W did not change significantly, then $d^2I/dt^2 \approx -G\mathcal{M}^2/\mathcal{R}$ from (1.26). By dimensional arguments, $I \approx \mathcal{M}\mathcal{R}^2$, so we define a time scale $t_{\rm dyn}$ by $d^2I/dt^2 \approx I/t_{\rm dyn}^2 \approx \mathcal{M}\mathcal{R}^2/t_{\rm dyn}$. Equating the two expressions for d^2I/dt^2 yields $t_{\rm dyn}^2 \approx \mathcal{R}^3/G\mathcal{M}$ or

$$t_{\rm dyn} \approx \frac{1}{[G\langle\rho\rangle]^{1/2}} \tag{1.33}$$

where $\langle\rho\rangle \approx \mathcal{M}/\mathcal{R}^3$ is approximately the average density. The dynamic time scale $t_{\rm dyn}$ is then a measure of the e-folding time for changes in radius as the star makes dynamic adjustments in structure. (In this example, d^2I/dt^2 is negative and the star collapses.) For the sun $t_{\rm dyn}$ is about an hour, which is many orders of magnitude shorter than $t_{\rm KH}$.

Expression (1.33) is a form of the "period–mean density relation" and it will come up again when we discuss variable stars.

1.3.4 Application: Estimates of Stellar Temperatures

We can squeeze even more out of the virial theorem. Consider a star of uniform density and temperature composed of a monatomic ideal gas. The internal energy density is

$$E = \tfrac{3}{2}nkT = \tfrac{3}{2}\rho\frac{N_{\rm A}kT}{\mu} \ \ {\rm erg\ cm^{-3}} \tag{1.34}$$

as we shall show in Chapter 3 (although it is an elementary result). Here n is the *number density* of free particles (in number cm^{-3}); k and $N_{\rm A}$, respectively, are Boltzmann's and Avogadro's constants; and μ is the mean molecular weight (usually in amu) per ion or atom of the stellar mixture. The quantity μ will be discussed in much more detail shortly (in §1.4.1), but, for now, regard it as that thing which makes

$$n = \frac{\rho N_{\rm A}}{\mu} .$$

For a typical stellar mixture of elements it is of order unity. In the language of μ, the ideal gas pressure is

$$P = nkT = \frac{\rho N_{\rm A}kT}{\mu} . \tag{1.35}$$

Multiplying E by the stellar volume V yields U and, since $\rho V = \mathcal{M}$, we find $U = \frac{3}{2}\mathcal{M}N_{\rm A}kT/\mu$. On the other hand, $U = -\Omega/2$ from the virial theorem (1.25) for the $\gamma = 5/3$ gas, and $\Omega = -\frac{3}{5}G\mathcal{M}^2/\mathcal{R}$ for the constant-density sphere. Equate the two forms for U; solve for T in terms of ρ, $\mathcal{M}$, and μ; eliminate $\mathcal{R}$ by way of the density, and find

$$T = 4.09 \times 10^6 \, \mu \left(\frac{\mathcal{M}}{\mathcal{M}_\odot} \right)^{2/3} \rho^{1/3} \text{ K} . \tag{1.36}$$

Before discussing the numerical results obtainable from this expression, it is worthwhile deriving the main components from another perspective.

The Lagrangian expression for the equation of hydrostatic equilibrium (1.16) is useful in this regard. In dimensional form it states that P is proportional to $G\mathcal{M}^2/\mathcal{R}^4$. But P also varies as $\mathcal{M}T/\mathcal{R}^3\mu$ after density has been eliminated in the ideal gas law, $P = nkT$. After equating the two versions of P we find (1.36) (but not the constant). The point is that if $\mathcal{R}$ is made smaller, for example, then ρ increases as $1/\mathcal{R}^3$ and, consequently, so would the ideal gas pressure were T to stay constant. This dependence of P on $\mathcal{R}$ is not strong enough, however, because P must also increase as $1/\mathcal{R}^4$ for hydrostatic equilibrium independent of the temperature. Thus the ideal gas equation of state and hydrostatic equilibrium demand that T must increase as $1/\mathcal{R} \propto \rho^{1/3}$.

Figure 1.1 shows (1.36) plotted as $\log T$ versus $\log \rho$ for $\mu = 1$ with $\mathcal{M}$ ranging between 0.3 and 100 $\mathcal{M}_\odot$. As a typical star, consider the present-day sun, which has an average density of $\langle \rho \rangle \approx 1.4 \text{ g cm}^{-3}$ and a central density of approximately 80 g cm^{-3}. If "average" may be identified with the quantities in (1.36), then an average temperature for the sun is a few million degrees. Even though it doesn't make a lot of sense to talk about an average temperature for a star, we note that the central temperature for the present-day sun is $T_c \approx 15 \times 10^6$ K, which close to the number just found. As we shall see later, a temperature greater than about 10^6 K is just what is needed to initiate hydrogenic nuclear fusion in stars. *A star thus produces energy by nuclear fusion because hydrostatic equilibrium requires high temperatures.*

Figure 1.1 has other lines on it that partition the $\log \rho$–$\log T$ plane into regions where equations of state other than the ideal gas law dominate. The "degeneracy" boundary defines that region where Fermi–Dirac degenerate electrons begin to play a major role (and see Chap. 3). Above a line corresponding to about 25 $\mathcal{M}_\odot$, radiation pressure (with a γ of 4/3 and $P = \frac{1}{3}aT^4$) becomes important. The areas beginning at $\rho \approx 10^6 \text{ g cm}^{-3}$ and $T \approx m_e c^2/k \approx 5 \times 10^9$ K ($m_e c^2$ is the electron rest mass energy) are regions where relativistic effects come in. All of these domains have their own peculiarities, which can greatly modify the simple picture built up thus far. But we shall have to wait for Chapter 3 to see what they are.

1.3.5 Application: Another Dynamic Time Scale

We have already found one dynamic time scale associated with readjustments of the moment of inertia when hydrostatic equilibrium is seriously thrown out

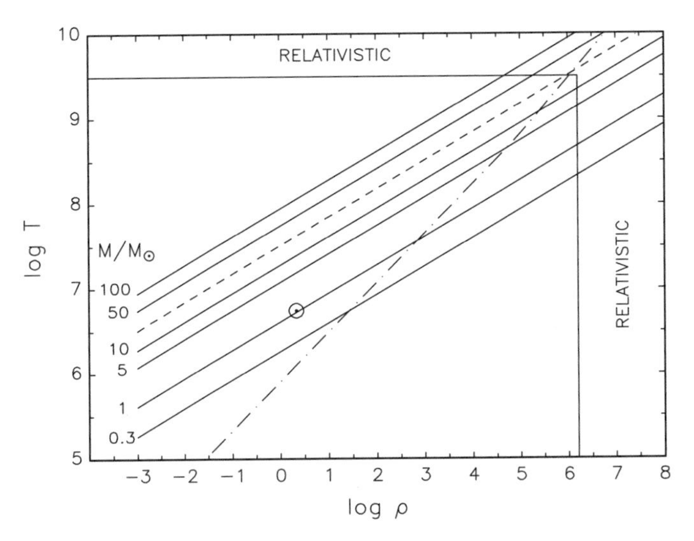

Fig. 1.1. The ideal gas virial result for temperature versus density for various masses (in solar units) from Eq. (1.36). Radiation pressure dominates above the dashed line and degenerate electrons must be considered below the dotted-dashed line. Regions where relativistic effects are important are indicated. The location of a constant-density "sun" is shown by the $\odot$.

of kilter. Now consider perturbations in structure induced by small-amplitude adiabatic sound waves and, specifically, compute how long it takes an adiabatic sound wave to travel from, say, the center of a star to the surface and back to the center again. (A discussion of mechanisms that might make such waves, or cause them to be reflected, we postpone until Chap. 8.) If the stellar sound speed is v_s, taken as constant for now, and Π is the "period" for one complete traversal, then

$$\Pi = \frac{2\mathcal{R}}{v_s} . \tag{1.37}$$

From elementary physics, the square of the local adiabatic sound speed is given by

$$v_s^2 = \left(\frac{dP}{d\rho}\right)_{\rm ad} = \Gamma_1 \frac{P}{\rho} \tag{1.38}$$

where Γ_1, our first "adiabatic exponent," is

$$\Gamma_1 = \left(\frac{d\ \ln P}{d\ \ln \rho}\right)_{\rm ad} = \frac{\rho}{P}\left(\frac{dP}{d\rho}\right)_{\rm ad} . \tag{1.39}$$

We shall find later that Γ_1, which measures how pressure changes in response to changes in density under adiabatic conditions, is of order unity (and, of course, it is dimensionless). For an ideal monotonic gas (1.38) yields the elementary result $v_s \propto \sqrt{T}$.

If hydrostatic equilibrium is closely maintained while the weak sound wave passes through the star, then version (1.23) of the virial theorem yields

$$-\Omega = 3\int_{\mathcal{M}} \frac{P}{\rho}\, d\mathcal{M}_r = 3\int_{\mathcal{M}} \frac{v_s^2}{\Gamma_1}\, d\mathcal{M}_r \approx \frac{3v_s^2}{\Gamma_1}\mathcal{M}$$

where the sound velocity and Γ_1 appearing on the right-hand side represent suitable averages of those quantities. Since $|\Omega| \approx G\mathcal{M}^2/\mathcal{R}$ we then obtain an estimate for the period of $\Pi \approx \left(\mathcal{R}^3/G\mathcal{M}\right)^{-1/2}$. Constants of order unity (such as Γ_1) have been set to unity. After eliminating mass and radius in favor of density we find

$$\Pi \approx \frac{1}{\left[G\langle\rho\rangle\right]^{1/2}} \approx \frac{.04}{\left[\langle\rho\rangle/\langle\rho_\odot\rangle\right]^{1/2}} \quad \text{days} \tag{1.40}$$

where $\langle\rho_\odot\rangle = 1.41$ g cm^{-3}. The final factor of 0.04 days comes from taking care with some of those quantities of order unity and inserting information that will be dealt with in Chapter 8.

Expression (1.40) is the same as $t_{\rm dyn}$ of (1.33) and rightly so because they both describe mechanical phenomena involving the whole star. A more careful analysis of how standing sound waves behave, however, introduces an additional factor of $(3\Gamma_1 - 4)^{1/2}$ in the denominator of (1.40). Again, a "gamma" of 4/3 will do curious things—as is obvious if $\Gamma_1 < 4/3$. We will postpone this discussion until it is time to examine variable stars.

1.4 The Constant-Density Model

We are now going to construct a stellar model by insisting that density be everywhere constant. Of course, in real life, we can't do this—the run of density is determined by many factors—but the model does have some utility. The constant density "model" of §1.3.4 was somewhat of a fudge. There we claimed that the star was in hydrostatic equilibrium, at constant temperature, and the ideal gas law was responsible for the pressure. A little thought, but not very much, should convince you that those conditions are contradictory. They imply that the pressure must be constant and yet hydrostatic equilibrium is still satisfied. We will make amends now.

If we set $\rho = \rho_c =$ constant, with "c" meaning center, then the mass equation (1.2) yields $\mathcal{M}_r = \frac{4}{3}\pi r^3 \rho_c$. This last expression is true up to the surface where $r = \mathcal{R}$ and $\mathcal{M}_r = \mathcal{M}$. Thus, after some trivial algebra,

$$\mathcal{M}_r = \frac{r^3}{\mathcal{R}^3}\mathcal{M}\,.$$

This is now used in the Lagrangian form of the hydrostatic equilibrium equation (1.16) to rid ourselves of r. The pressure gradient is then

$$\frac{dP}{d\mathcal{M}_r} = -\frac{G\mathcal{M}}{4\pi\mathcal{R}^4}\left(\frac{\mathcal{M}_r}{\mathcal{M}}\right)^{-1/3}.$$

Integrate this using the zero pressure boundary condition at $\mathcal{R}$ to find

$$P = P_{\rm c}\left[1 - \left(\frac{\mathcal{M}_r}{\mathcal{M}}\right)^{2/3}\right] = P_{\rm c}\left[1 - \left(\frac{r}{\mathcal{R}}\right)^2\right] \tag{1.41}$$

where $P_{\rm c}$ is the central pressure (at $\mathcal{M}_r = 0$) with

$$P_{\rm c} = \frac{3}{8\pi}\frac{G\mathcal{M}^2}{\mathcal{R}^4} = 1.34 \times 10^{15}\left(\frac{\mathcal{M}}{\mathcal{M}_\odot}\right)^2\left(\frac{\mathcal{R}}{\mathcal{R}_\odot}\right)^{-4} \text{ dyne cm}^{-2}. \tag{1.42}$$

The numerical value for $P_{\rm c}$ can be shown to be a *lower limit* for central pressures in hydrostatic objects if it is assumed that ρ always decreases outward. This assumption is correct except for some very unusual circumstances (which may, in any case, signal an incipient instability in structure). That $P_{\rm c}$ is a lower limit seems reasonable because stronger concentrations of mass toward the center than that of constant density imply stronger gravitational fields which, in turn, require higher pressures to maintain equilibrium. (See, for example, the "linear star model" of Stein 1966, which we include as Ex. 1.3. You might also try Ex. 1.2. It explores another lower limit on $P_{\rm c}$.)

A simple exercise for the reader is to verify that the above expressions for pressure and mass distribution satisfy the equilibrium version of the virial theorem (1.23) with $d^2I/dt^2 = 0$ and $\Omega = -\frac{3}{5}G\mathcal{M}^2/\mathcal{R}$.

To find a temperature distribution we have to specify an equation of state and we again choose the monatomic ideal gas as a useful example with $P = nkT$. But, before we reach our objective, we should first figure out how to compute n or, equivalently, the mean molecular weight μ.

1.4.1 Calculation of Molecular Weights

Assume that the gas is composed of a mixture of neutral atoms, ions (in various stages of ionization), and electrons but, overall, the gas is electrically neutral. These are the free particles composing n. First collect the ions and neutral atoms together into nuclear isotopic species, calling all of them "ions" for now, and denote a specific species by an index i. Thus, for example, assign some particular index to all the ions of ^{4}He. Each nucleus of index i has an integer nuclear charge Z_i and a nuclear mass number, in amu (atomic mass units), of A_i. For ^{4}He, $Z_i = 2$ and $A_i = 4$. (The atomic mass of ^{4}He is not exactly 4, but this is close enough.) Furthermore, let X_i be the *fraction by mass* of species i in the mixture such that $\sum_i X_i = 1$. Thus, for example, if 70% of the mass of a sample of matter were composed of species i, then $X_i = 0.7$. The ion number density, in units of cm^{-3}, of a given species i is then

$$n_{\mathrm{I},i} = \frac{(\text{mass/unit volume) of } i}{(\text{mass of 1 ion) of } i} = \frac{\rho X_i N_{\mathrm{A}}}{A_i} \tag{1.43}$$

where Avogadro's number $N_{\mathrm{A}} = 6.022142 \times 10^{23}$ mole^{-1}. Just to be sure we understand where this comes from, recall that "amu" is so defined that an atom of carbon isotope ^{12}C has a mass of exactly 12 amu, N_{A} is the number of ^{12}C atoms in 12 g of ^{12}C, and a "mole" is the amount of substance in a system that contains as many atoms as there are atoms in 12 g of ^{12}C. You can take it from there.

The total for all ions is

$$n_{\mathrm{I}} = \sum_i n_{\mathrm{I},i} = \rho N_{\mathrm{A}} \sum_i \frac{X_i}{A_i} . \tag{1.44}$$

Now define μ_{I} as the "total mean molecular weight of ions" such that

$$n_{\mathrm{I}} = \frac{\rho N_{\mathrm{A}}}{\mu_{\mathrm{I}}} \tag{1.45}$$

or

$$\mu_{\mathrm{I}} = \left[\sum_i \frac{X_i}{A_i} \right]^{-1} . \tag{1.46}$$

The ion mean molecular weight is then a sort of mean mass of an "average" ion in the mixture and it contains all the information needed to find the number density of ions.

The electrons are a bit more difficult to treat. To find out how many free electrons there are we must have prior knowledge of the states of ionization for all species. This information is difficult to come by and we will defer until later a discussion of how it is obtained. For now we assume that some good soul has done the work for us and has supplied us with the quantities y_i that contain what we want. These y_i are defined such that the number density of free electrons associated with nuclear species i is given by

$$n_{\mathrm{e},i} = y_i \, Z_i \, n_{\mathrm{I},i} = \rho N_{\mathrm{A}} \left(\frac{X_i}{A_i} \right) y_i Z_i . \tag{1.47}$$

Thus, out of the Z_i electrons that a particular ion of species i could possibly contribute to the free electron sea, only the fraction y_i are, on average, actually free. We call y_i the "ionization fraction." A value $y_i = 1$ then means that the species is completely ionized, whereas $y_i = 0$ implies complete neutrality. The total electron number density is therefore

$$n_{\mathrm{e}} = \sum_i n_{\mathrm{e},i} = \rho N_{\mathrm{A}} \sum_i \left(\frac{X_i}{A_i} \right) y_i Z_i = \frac{\rho N_{\mathrm{A}}}{\mu_{\mathrm{e}}} \tag{1.48}$$

which also defines μ_{e}, the "mean molecular weight per free electron." (Note that in no way are we assigning a "weight" to the electron in this sense.) Thus

$$\mu_{\rm e} = \left[\sum_i \frac{Z_i X_i y_i}{A_i} \right]^{-1} . \tag{1.49}$$

If you look carefully at the way $\mu_{\rm e}$ is constructed you will realize that it is the ratio of the total number of nucleons (protons plus neutrons) contained in all nuclei to the total number of free electrons in any sample of the material.

Finally, from the definition of n as the sum of $n_{\rm I}$ and $n_{\rm e}$, we easily find that the total mean molecular weight is

$$\mu = \left[\frac{1}{\mu_{\rm I}} + \frac{1}{\mu_{\rm e}} \right]^{-1} \tag{1.50}$$

with

$$n = n_{\rm I} + n_{\rm e} = \frac{\rho N_{\rm A}}{\mu} . \tag{1.51}$$

For relatively unevolved stars, in which nuclear transformations have not progressed to any great extent, the major nuclear constituents are hydrogen (^{1}H) and helium (^{4}He). We shall refer to their mass fractions (X_i) as, respectively, X and Y. All else shall collectively be called "metals" (or, sometimes, "heavies") and their mass fraction is denoted by Z (not to be confused with ion charge). A typical value of Z might be, at most, a few percent. Obviously

$$X + Y + Z = 1 . \tag{1.52}$$

A catalogue of the relative abundances of metals seen on the surfaces of most stars, including the sun, reveals that the dominant heavy elements are carbon, nitrogen, oxygen, and neon. Elements heavier than those, up to nickel, contribute a little, and past there we find only traces. For the most part, the isotopes of the major heavy elements fall along the "valley of beta-stability" in which $Z_i/A_i \approx 1/2$. The same value of charge to mass number also applies to ^{4}He.

An example of the metal abundances, X_i, seen in the solar atmosphere is shown in Fig. 1.2 for the elements from carbon ($Z_i = 6$) to nickel ($Z_i = 28$). The abscissa is the average mass number for the element using relative isotopic abundances observed for the earth. The set is normalized so that $\sum X_i = Z = 0.02$, which is close to the metal mass fraction for the solar atmosphere. Note that oxygen is the most abundant (by mass), followed by carbon, neon, etc. Assuming that standard versions of the Big Bang are correct, these (and the other metals) are not the result of element production in the very early universe. Of course we wouldn't mention this at all in this text were it not that stars were (and are) responsible—as we shall eventually see. (For an expanded version of Fig. 1.2 see Fig. 2.19 and the discussion in §2.8.1.)

In the deep stellar interior, hydrogen, helium, and most of the metals are completely ionized ($y_i = 1$). If, in addition, metals compose only a minor

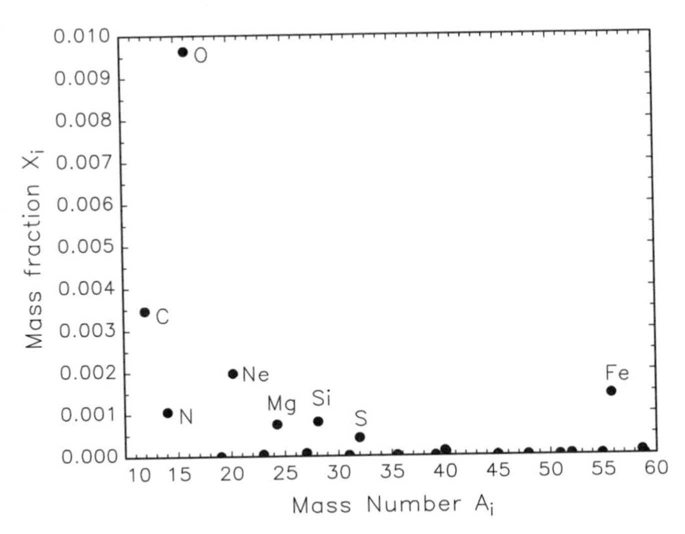

Fig. 1.2. Shown are metal abundances for the solar atmosphere versus average elemental mass number: from material reviewed in Grevesse and Noels (1993).

fraction of the total, with $Z \ll 1$, then you can use the results of the above analysis to find the following convenient approximation for μ_e:

$$\mu_e \approx \frac{2}{1+X} \, . \tag{1.53}$$

Note, however, that in detailed modeling of stars, this is to be used with great caution; ionization might not be complete (or elements may even be completely neutral) and abundances may be quite strange.

The ion mean molecular weight can be similarly approximated under the same conditions as above with the additional observation that Z is small compared to an average A ($A = \langle A_i \rangle \approx 14$ or so). The result is

$$\mu_I \approx \frac{4}{1+3X} \, . \tag{1.54}$$

Using (1.53) and (1.54), an approximation for the total mean molecular weight is then

$$\mu \approx \frac{4}{3+5X} \, . \tag{1.55}$$

For a star just beginning the longest active period of its natural life—a "zero-age main sequence" star (ZAMS)—typical abundances are $X \approx 0.7$, $Y \approx 0.3$, and $Z \approx 0.03$ (or somewhat less for stars formed earlier on in galactic history). These correspond to $\mu_I \approx 1.3$, $\mu_e \approx 1.2$, and $\mu \approx 0.6$.

Now that we have reasonable approximations for the molecular weights and are assured that typical values are near unity, we return to the constant-density model.

1.4.2 The Temperature Distribution

Taking $P = \rho N_{\mathrm{A}} kT/\mu$, and using the central pressure of (1.42), the central temperature for the constant-density model becomes

$$T_{\mathrm{c}} = \frac{1}{2}\frac{G\mathcal{M}}{\mathcal{R}}\frac{\mu}{N_{\mathrm{A}}k} = 1.15 \times 10^{7}\, \mu \left(\frac{\mathcal{M}}{\mathcal{M}_{\odot}}\right)\left(\frac{\mathcal{R}}{\mathcal{R}_{\odot}}\right)^{-1} \mathrm{K}. \tag{1.56}$$

The temperature distribution with respect to r and $\mathcal{M}_r$ is of the same form as that of the pressure in (1.41) with P_{c} replaced by T_{c}.

For solar values of mass and radius, this central temperature is remarkably close to that of the present-day sun found from sophisticated solar models and is an improvement over the virial "average" estimate of (1.36). The constant-density model result for T_{c} is higher than that from the virial because we have found the detailed run of pressure in the model and not just some average pressure.

As noted earlier, however, we cannot just assume a density distribution and expect such a stellar model to satisfy all the equations of stellar structure. We now discuss some of these additional constraints and equations.

1.5 Energy Generation and Transport

One goal of the effort in fusion energy research is to heat up a plasma containing potential thermonuclear fuel to temperatures exceeding about a million degrees and then physically contain it for a sufficiently long period of time. Most stars do that as a matter of course. They have the temperatures, containment mechanism (gravity), fuel, and time, and can fuse together light elements into heavier ones and, by doing so, release energy. We shall not discuss here precisely what kinds of thermonuclear burning take place in stars (see Chap. 6) but we shall extend our notion of equilibrium to include energy generation and how it is balanced by the leakage of energy through the star. In particular, suppose some sort of nuclear burning is taking place within a given localized gram of material. If the energy generated in that gram is not transferred elsewhere, then a nonequilibrium condition holds and the material heats up. If, on the other hand, we succeed in somehow removing energy as fast as it is liberated, and no faster, then we say the material is in "thermal balance." (Note that this term is not universally used by all authors in this context.) The sample of material is, of course, not strictly in equilibrium because, in the case of fusion, the composition is changing with time as more massive nuclear species are produced—but usually very slowly. We shall return to that problem later.

To express thermal balance quantitatively, consider a spherically symmetric shell of mass $d\mathcal{M}_r$ and thickness dr. Within that shell denote the power generated per gram as ε (erg g^{-1} s^{-1}). We shall refer to it as the "energy generation rate." The total power generated in the shell is $4\pi r^2 \rho\varepsilon\, dr = \varepsilon\, d\mathcal{M}_r$.

To balance the power generated, we must have a net flux of energy leaving the shell. If $\mathcal{F}(r)$ is the flux (in units of erg cm^{-2} s^{-1}), with positive values implying a radially directed outward flow, then $\mathcal{L}_r = 4\pi r^2 \mathcal{F}(r)$ is the total power, or luminosity, in erg s^{-1}, entering (or leaving) the shell's inner face, and $\mathcal{L}_{r+dr} = 4\pi r^2 \mathcal{F}(r + dr)$ is the luminosity leaving through the outer face at $r + dr$. The difference of these two terms is the net loss or gain of power for the shell. For thermal balance that difference must equal the total power generated within the shell. That is,

$$\mathcal{L}_{r+dr} - \mathcal{L}_r = d\mathcal{L}_r = 4\pi r^2 \rho \varepsilon \, dr$$

which yields the differential equation

$$\frac{d\mathcal{L}_r}{dr} = 4\pi r^2 \rho \varepsilon \ . \tag{1.57}$$

We will refer to this (or a more general version of it) as the "energy equation." Its Lagrangian form is

$$\frac{d\mathcal{L}_r}{d\mathcal{M}_r} = \varepsilon \tag{1.58}$$

by way of the mass equation. Note that we have used total differentials here. If the "equilibrium state" were also a function of time, then partials would appear instead. Note also that other energy sources, such as gravitational contraction, are being completely ignored at the moment.

Since, for now, we are only considering $\varepsilon \geq 0$, then $\mathcal{L}_r$ must either be constant (in regions where $\varepsilon = 0$) or increase monotonically with r or $\mathcal{M}_r$. We will demonstrate later that ε is usually a strong function of temperature and, because temperature is expected to decrease outward in a star, ε should be largest in the inner stellar regions provided that fuel is present. Thus $\mathcal{L}_r$ should increase rapidly from the center, starting from zero, and then level out to its surface value of $\mathcal{L}$. There are exceptions to these statements for highly evolved stars, but they will suffice for now.

Future discussions will make extensive use of a power law expression for ε of the form

$$\varepsilon = \varepsilon_0 \rho^\lambda T^\nu \tag{1.59}$$

where ε_0, λ, and ν are constants over some sufficiently restricted range of T, ρ, and composition. As important examples, consider briefly the two ways that stars burn hydrogen (^{1}H) into helium (^{4}He). These are the proton-proton (pp) chains, and the carbon-nitrogen-oxygen (CNO) cycles. The first is, for the most part, a simple sequence of nuclear reactions, starting with one involving two protons, that gradually add protons to intermediate reaction products to eventually produce helium. The second cycle uses C, N, and O as catalysts to achieve the same end. For typical hydrogen-burning temperatures and densities ($T \gtrsim$ a few million degrees, ρ of order 1 to 100 g cm^{-3}), the temperature and density exponents ν and λ are given in Table 1.1. We also give the exponents for the "triple-alpha" reaction, which effectively combines three ^{4}He nuclei (alpha particles) to make one nucleus of ^{12}C at temperatures exceeding 10^8 K. The constant term ε_0 need not concern us for the present, and the derivation of all these numbers will be given in Chapter 6.

Table 1.1. Temperature and Density Exponents

Energy generation mode for ε	λ	ν
pp-chains	1	≈ 4
CNO-cycles	1	≈ 15
Triple-α	2	≈ 40

On the hydrogen-burning main sequence (discussed in Chap. 2), the pp-chains dominate for stars of mass less than about one solar mass, but the CNO cycles take over for more massive stars. This sensitivity to mass reflects the combined factors of the general tendency of temperatures to increase with mass (see Fig. 1.1) and the relative values of the temperature exponents, ν, for the two modes of energy generation.

The total energy released in the conversion of hydrogen to helium is approximately 6×10^{18} ergs for every gram of hydrogen consumed. To get an idea of what this might represent, a simple calculation will easily convince you that the sun, with its present-day hydrogen content of roughly 70% by mass, could continue to shine for almost 10^{11} years at its present luminosity just by burning all its available hydrogen.

What about the other factor in thermal balance? What determines $\mathcal{L}_r$? As we shall see, there are three major modes of energy transport: **radiation** (photon) transfer, **convection** of hotter and cooler mass elements, and heat **conduction**, with the first two being most important for most stars. (White dwarfs depend heavily on the last mode, but those stars are in a class by themselves.)

For those of us concerned primarily with the interiors of stars, it is fortunate that the transfer of energy by means of radiation is easily described. Except for the very outermost stellar layer, the energy flux carried by radiation obeys a Fick's law of diffusion; that is, the flow is driven by a gradient of a quantity having something to do with the radiation field. The form is

$$\mathcal{F}(r) = -\mathcal{D}\,\frac{d(aT^4)}{dr}$$

where aT^4 is the radiation energy density and $\mathcal{D}$ is a diffusion coefficient. We shall show in Chapter 4 that the important part of $\mathcal{D}$ is the "opacity," κ, which, by its name alone, lets you know how the flow of radiation is hindered by the medium through which it passes. We suspect that $\mathcal{D}$ should be inversely proportional to κ. Without further ado, multiply $\mathcal{F}(r)$ by $4\pi r^2$

to obtain a luminosity, put in the relevant factors in $\mathcal{D}$ to be derived later, and find

$$\mathcal{L}_r = -\frac{4\pi r^2 c}{3\kappa\rho}\frac{d\,aT^4}{dr} . \tag{1.60}$$

The alternative Lagrangian form is

$$\mathcal{L}_r = -\frac{(4\pi r^2)^2 c}{3\kappa}\frac{d\,aT^4}{d\mathcal{M}_r} . \tag{1.61}$$

We shall have ample opportunity to use both of these forms.

The calculation of opacities is no easy matter and there is a whole industry set up for just that purpose. Chapter 4 will discuss what goes into them but, for now, we write a generic opacity in the power law form

$$\kappa = \kappa_0 \rho^n T^{-s} \;\; \text{cm}^2\ \text{g}^{-1}. \tag{1.62}$$

As in the case of ε, the coefficients and powers, κ_0, n, and s, are constants. Important examples are electron Thomson scattering opacity ($n = s = 0$), which is important for completely ionized stellar regions, and Kramers' opacity ($n = 1$, $s = 3.5$), which is characteristic of radiative processes involving atoms.

The luminosity carried by the transport of hot or colder material, which we call convection, is a good deal more difficult to treat. We shall give a simple prescription in the section below along with simple ideas that tie together what has been discussed thus far.

1.6 Stellar Dimensional Analysis

Some texts on stellar evolution (and see especially Cox, 1968, Chap. 22) discuss the topics of "homology" and "homologous stars." These terms describe sequences of simple spherical stellar models in complete equilibrium where one model is related to any of the others by a simple change in scale. More specifically, assume that the models all have the same constituent physics (equation of state, opacity, etc., as given by power laws), the same uniform composition, and that $\mathcal{M}_r$ and r are related as follows. If one of the stars in the homologous collection is chosen as a reference star—call it star 0 and refer to it by a zero subscript—then these relations must apply in order that the stars be homologous to one another:

$$r = \frac{\mathcal{R}}{\mathcal{R}_0} r_0 \qquad \text{and} \tag{1.63}$$

$$\mathcal{M}_r = \frac{\mathcal{M}}{\mathcal{M}_0}\mathcal{M}_{r,0} \tag{1.64}$$

where those quantities not subscripted with a zero refer to any other star in the collection. These relations mean that the stars have the same relative mass distribution such that radius and the mass interior to that radius are related by simple ratios to the corresponding quantities in the reference star. We may also replace these equations by their derivatives keeping $\mathcal{R}$, $\mathcal{R}_0$, $\mathcal{M}$, and $\mathcal{M}_0$ all constant.

A consequence of the above is that the mass equations (1.2) for two stars may be divided one by the other to give a relation between the densities at equivalent mass points:

$$\rho = \rho_0 \frac{d\mathcal{M}_r}{d\mathcal{M}_{r,0}} \frac{1}{dr/dr_0} \left(\frac{r}{r_0}\right)^{-2} = \rho_0 \left(\frac{\mathcal{M}}{\mathcal{M}_0}\right) \left(\frac{\mathcal{R}}{\mathcal{R}_0}\right)^{-3} . \tag{1.65}$$

For stars of constant, but differing, densities this is obvious. It would also be an obvious result in a comparison of *average* densities between *any* two stars. However, (1.65) is true in general only for homologous stars.

What follows is a simplified treatment of homologous stars using a form of dimensional analysis. We shall follow the scheme of Carson (1986), and the results obtained will turn out to be identical to those obtained from standard homology arguments. They will also be very useful for estimating how various stellar quantities such as mass, radius, etc., are related. Again, however, the results are not to be used blindly.

We start by writing the Lagrangian version of the equation of hydrostatic equilibrium (1.16) in a form that emphasizes the dependence of pressure on mass and radius. Fundamental constants, such as G, could be retained but, at the end, it would be apparent that they were not needed. We have

$$P \propto \frac{\mathcal{M}^2}{\mathcal{R}^4} \tag{1.66}$$

where $\mathcal{M}$ and $\mathcal{R}$ are chosen to represent mass and radius variables as in the spirit of (1.65). The pressure is specified in power law form in the same way as was done for the energy generation rate and opacity. Thus, we write

$$P = P_0 \rho^{\chi_\rho} T^{\chi_T} . \tag{1.67}$$

The constants P_0 (which will not be needed), χ_ρ, and χ_T, are assumed to be the same for all stars in the collection. Note that (1.67) may also be written in logarithmic differential form as

$$d \ln P = \chi_\rho \, d \ln \rho + \chi_T \, d \ln T . \tag{1.68}$$

If (1.66) and (1.67) are equated, we then arrive at a relation between $\mathcal{R}$, ρ, T, and $\mathcal{M}$, which we also written in logarithmic differential form: namely,

$$4 \, d \ln \mathcal{R} + \chi_\rho \, d \ln \rho + \chi_T \, d \ln T = 2 \, d \ln \mathcal{M} . \tag{1.69}$$

The plan is to treat the energy equation (1.58), the power law form of the energy generation rate (1.59), the diffusive radiative transfer equation

(1.61), and the power law opacity (1.62) in the same way as we just did for the pressure. The aim will be to construct separate $\mathcal{R}$, ρ, T, and $\mathcal{L}$ versus $\mathcal{M}$ relations as

$$\mathcal{R} \propto \mathcal{M}^{\alpha_{\mathcal{R}}} \tag{1.70}$$
$$\rho \propto \mathcal{M}^{\alpha_{\rho}} \tag{1.71}$$
$$T \propto \mathcal{M}^{\alpha_{T}} \tag{1.72}$$
$$\mathcal{L} \propto \mathcal{M}^{\alpha_{\mathcal{L}}} \tag{1.73}$$

where the exponents α are to be determined. We have the requisite number of equations to do this. For example, (1.70–1.73) may be inserted into (1.69) to yield one relation between the αs:

$$4\alpha_{\mathcal{R}} + \chi_\rho \alpha_\rho + \chi_T \alpha_T = 2$$

where a common factor of $d \ln \mathcal{M}$ has been divided out. If this sort of thing is done for, in order, the mass equation, the equation of hydrostatic equilibrium (just done), the energy equation, and, finally, the transfer equation, we then obtain the matrix equation

$$\begin{pmatrix} 3 & 1 & 0 & 0 \\ 4 & \chi_\rho & 0 & \chi_T \\ 0 & \lambda & -1 & \nu \\ 4 & -n & -1 & 4+s \end{pmatrix} \begin{pmatrix} \alpha_{\mathcal{R}} \\ \alpha_\rho \\ \alpha_{\mathcal{L}} \\ \alpha_T \end{pmatrix} = \begin{pmatrix} 1 \\ 2 \\ -1 \\ 1 \end{pmatrix} . \tag{1.74}$$

The determinant of the matrix on the left-hand side of (1.74) is

$$\mathrm{D}_{\mathrm{rad}} = (3\chi_\rho - 4)(\nu - s - 4) - \chi_T(3\lambda + 3n + 4) \tag{1.75}$$

where the "rad" subscript reminds us that energy transfer is by radiation in this case. We assume here that $\mathrm{D}_{\mathrm{rad}}$ is not zero but it could be for some particular combination of temperature and density exponents. The latter circumstance leads to some strange situations, which we defer to Ex. 7.1.

The solutions to (1.74) are then (adapted from Carson, 1986, with the correction of a minor typographical error in $\alpha_{\mathcal{R}}$):

$$\alpha_{\mathcal{R}} = \frac{1}{3}\left[1 - 2(\chi_T + \nu - s - 4)/\mathrm{D}_{\mathrm{rad}}\right] \tag{1.76}$$
$$\alpha_\rho = 2(\chi_T + \nu - s - 4)/\mathrm{D}_{\mathrm{rad}} \tag{1.77}$$
$$\alpha_{\mathcal{L}} = 1 + \left[2\lambda(\chi_T + \nu - s - 4) - 2\nu(\chi_\rho + \lambda + n)\right]/\mathrm{D}_{\mathrm{rad}} \tag{1.78}$$
$$\alpha_T = -2(\chi_\rho + \lambda + n)/\mathrm{D}_{\mathrm{rad}} \tag{1.79}$$

where these are to be used in (1.70–1.73) in the situation where radiation is assumed to carry all the luminosity (or where you suspect radiation transfer seems to dominate).

If energy transport is primarily by means of convection, then the above analysis must be modified, and we include that analysis for completeness (although, as we shall see, the results are of limited use). We shall have to wait until Chapter 5 to explore convection in detail, but it will have to suffice for now to state that vigorous and efficient convection implies that the dependence of temperature on density as a function of radius is adiabatic. Specifically this means that

$$T(r) \propto \rho(r)^{\Gamma_3 - 1} \tag{1.80}$$

where $(\Gamma_3 - 1)$ is the adiabatic thermodynamic derivative

$$\Gamma_3 - 1 = \left(\frac{d \ln T}{d \ln \rho} \right)_{\text{ad}} \tag{1.81}$$

similar to Γ_1 of (1.39). Γ_3 is also of order unity and we shall see much more of these Γs later. This relation replaces the radiative transfer equation of the preceding analysis and means that the last row in the matrix of (1.74) is replaced by $(0,\ \Gamma_3 - 1,\ 0,\ -1)$ and the last element of the right-hand side constant column vector is now zero. A simple calculation yields the determinant for the new system

$$\mathrm{D}_{\text{conv}} = (3\chi_\rho - 4) + 3\chi_T(\Gamma_3 - 1) \tag{1.82}$$

and the new exponents α are

$$\alpha_{\mathcal{R}} = \left(1 - 2/\mathrm{D}_{\text{conv}}\right)/3 \tag{1.83}$$

$$\alpha_\rho = 2/\mathrm{D}_{\text{conv}} \tag{1.84}$$

$$\alpha_{\mathcal{L}} = 1 + 2[\nu(\Gamma_3 - 1) + \lambda]/\mathrm{D}_{\text{conv}} \tag{1.85}$$

$$\alpha_T = 2(\Gamma_3 - 1)/\mathrm{D}_{\text{conv}} \tag{1.86}$$

for efficient convective transport.

How well does this analysis work? The stars that we think we know the most about are located on the hydrogen main sequence. For the most part these stars are nearly homogeneous in composition and their masses, luminosities, and radii are relatively well determined. Figure 1.3, constructed primarily from data given in Allen (1973, §100, and see Table 3–6 in Mihalas and Binney 1981) illustrates the observed relation between these three quantities.

From our previous discussion we expect that stars on the upper (more massive and luminous part of the) main sequence should have higher central temperatures just because they are more massive. The appropriate opacity law to use in this case is electron scattering for which $n = s = 0$. Similarly, the energy is generated primarily by the CNO cycles and thus, from Table 1.1, $\lambda = 1$ and $\nu \approx 15$. Although, as we shall show, the inner regions of these stars are convective, radiative transport of energy still dominates in the outer regions from which the power finally escapes. Finally, although radiation pressure is important, the pressure is mostly determined by the ideal gas

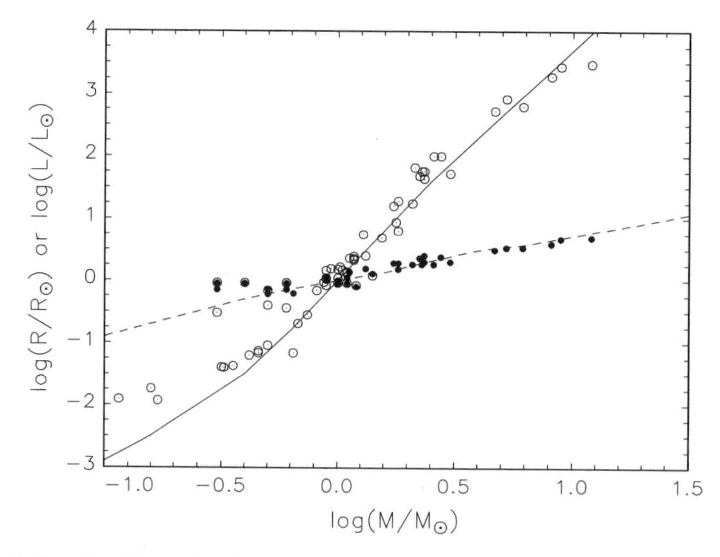

Fig. 1.3. Luminosity and radius versus mass for main sequence stars. All quantities are in solar units. The solid (dashed) line is that for luminosity (radius) adapted from material in Allen (1973). Open (filled) dots are luminosity (radius) for components of binaries, from Harris et al. (1963) and Böhm (1989).

law for which $\chi_\rho = \chi_T = 1$. If these stars represent, roughly, a homologous sequence, then the preceding analysis should give values of $\alpha_\mathcal{R}$ and $\alpha_\mathcal{L}$ that reproduce the slopes in Fig. 1.3. Using equations (1.75) through (1.79) and the exponents just quoted, find that $\alpha_\mathcal{R} = 0.78$, and $\alpha_\mathcal{L} = 3.0$. A fit to the slopes in Fig. 1.3 for stars with masses greater than a few solar masses yields

$$\frac{\mathcal{R}}{\mathcal{R}_\odot} \approx \left(\frac{\mathcal{M}}{\mathcal{M}_\odot}\right)^{0.75} \quad \text{and} \tag{1.87}$$

$$\frac{\mathcal{L}}{\mathcal{L}_\odot} \approx \left(\frac{\mathcal{M}}{\mathcal{M}_\odot}\right)^{3.5} \tag{1.88}$$

where the sun is not only used for normalization of the various quantities, but it appears as the reference star in the homologous set of stars. Obviously the homology relations have done fairly well. In addition, $\alpha_T = 0.22$ and $\alpha_\rho = -1.33$ so that temperature should increase with mass on the upper main sequence whereas density should decrease. We state now, without further proof, that this is indeed what happens (and see §2.15). Stellar models show that central (as a homologous point) temperatures and densities do just this and the exponents are just about what we find.

The lower (less massive) main sequence is more difficult to treat. The pp-chains ($\lambda = 1$, $\nu = 4$) dominate the energy generation rate and Kramers' opacity (with $n = 1$, $s = 3.5$) operates through much of the star but, and especially for very low mass stars, convection is important. This may seem to be no problem because we have derived the homology relations for convection, but the trouble is that the structure of these stars may be almost solely determined by what happens at the very outermost radiative surface (see Chap. 7). But, being intrepid, let us see what happens if we combine the above exponents with an ideal gas law, assume radiative transfer, and try to duplicate stars of around a solar mass. One result is that $\alpha_{\mathcal{L}} \approx 5.5$. We compare this to some results from the astrometric satellite Hipparcos[3] reported by Martin and Mignard (1998), Martin et al. (1998), and Lebreton (2001). The dashed line in Fig. 1.4 is an eyeball fit to the data with $\alpha_{\mathcal{L}} = 3.9$. Even though the slope may be moved around a bit (and we confess that we did use an approximation to the bolometric correction to Hipparcos magnitudes), it would stretch the imagination to claim that $\alpha_{\mathcal{L}} \approx 5.5$ is a good fit for main sequence stars of around a solar mass or less.

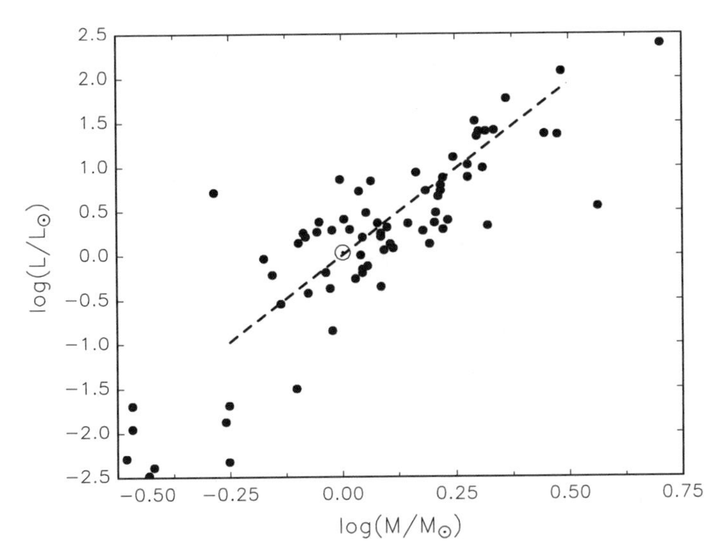

Fig. 1.4. Shown are luminosities versus masses for main sequence stars in binary systems derived from Hipparcos data as reported by Martin and Mignard (1998) and Martin et al. (1998). The dashed line corresponds to a mass–luminosity relation $\mathcal{L} \propto \mathcal{M}^{3.9}$ and the location of the sun is indicated by $\odot$.

[3] This seemingly prosaic space mission was launched in 1989 and was designed to measure precise positions of stars in the sky. After an operational lifetime of nearly four years, the observations have now yielded results of fundamental importance to astronomy, including new insights into stellar evolution and cosmology. Prosaic indeed! A good overview is given by Kovalevsky (1998).

1.7 Evolutionary Lifetimes on the Main Sequence

It is a fact of life that stars spend most of their active life on the main sequence converting hydrogen to helium. Another fact is that when approximately 10% of a star's original hydrogen is converted to helium, the star undergoes structural transformations that cause its luminosity and/or radius to change enough that it can no longer be called a main sequence star. (Why this happens is a later subject.) Thus the main sequence lifetime is geared to the rate at which fusion reactions take place. To estimate that time, $t_{\rm nuc}$, all we have to do is calculate how much energy is released by burning 10% of the star's available hydrogen and compare it to the main sequence luminosity. From the figures quoted before for the energy release per gram in hydrogen burning, it is evident that

$$t_{\rm nuc} \approx \frac{0.1 \times 0.7 \times \mathcal{M} \times 6 \times 10^{18}}{\mathcal{L}} \ {\rm s} \tag{1.89}$$

or, after converting to years and solar units,

$$t_{\rm nuc} \approx 10^{10} \left(\frac{\mathcal{M}}{\mathcal{M}_\odot}\right) \left(\frac{\mathcal{L}}{\mathcal{L}_\odot}\right)^{-1} \ \text{years}\,. \tag{1.90}$$

Note that a factor of 0.7 appears in (1.89). This is the typical value of the hydrogen mass fraction X given previously.

To eliminate the luminosity in (1.90), use the mass–luminosity relation (1.88) and find, for upper main sequence stars,

$$t_{\rm nuc} \approx 10^{10} \left(\frac{\mathcal{M}}{\mathcal{M}_\odot}\right)^{-2.9} \ \text{years}\,. \tag{1.91}$$

The main sequence lifetime of the sun is thus expected to be around 10^{10} years (if we accept, roughly, the luminosity slope in Fig. 1.4). This is to be compared to the present age of the sun of 4.6×10^9 years as a "middle-aged star." More massive stars have shorter lifetimes because they are so profligate in using up their fuel to maintain their high luminosities. Stars on the lower main sequence with masses not much less than the sun have lifetimes that exceed present estimates for the age of the galaxy and universe.

This simple theoretical result explains why the main sequence for clusters—all of whose stars are assumed to have been formed at nearly the same time—terminates at the "turnoff point" leaving only the lower mass stars, and why rough estimates may be made for the ages of those clusters (although more is involved than what we have implied, as discussed in §2.3).

1.8 The Hertzsprung–Russell Diagram

Before we go on in the next chapter to describe real stars, it is essential that we introduce the *Hertzsprung–Russell diagram*—or, more simply, the

HR diagram—which we shall use extensively. This two-dimensional diagram is the astronomer's way of characterizing important observational properties of stars. The vertical axis is a measure of the power output of a star while the abscissa tells us the color or, equivalently, the temperature of the visible surface. The units used for the axes depend on context and who is presenting them. An observer will usually express power in magnitudes of one sort or the other. A theoretician usually prefers luminosity (and the conversion from magnitude to luminosity is sometimes no easy matter). Similarly, the observer will indicate color as a difference in magnitudes between two spectral bands but the theoretician uses *effective temperature*, $T_{\rm eff}$, which is a theoretical construct. The relation between luminosity, total stellar radius, and $T_{\rm eff}$ is

$$\mathcal{L} = 4\pi\sigma \mathcal{R}^2 T_{\rm eff}^4 \tag{1.92}$$

where Stefan–Boltzmann's constant $\sigma = 5.6704 \times 10^{-5}$ erg cm^{-2} K^{-4} s^{-1}. There are some subtleties to what is meant by radius and effective temperature but, in the simplest definition, $\mathcal{R}$ is the radius of the visible surface (photosphere) and $T_{\rm eff}$ is the temperature on that surface. Thus (1.92) is the blackbody radiant luminosity emitted from the surface of a sphere of radius $\mathcal{R}$ whose surface temperature is $T_{\rm eff}$. The effective temperature of the sun is $T_{\rm eff}(\odot) = 5,780$ K. In solar units for $\mathcal{L}$ and $\mathcal{R}$, (1.92) becomes

$$\frac{\mathcal{L}}{\mathcal{L}_\odot} = 8.97 \times 10^{-16} \left(\frac{\mathcal{R}}{\mathcal{R}_\odot}\right)^2 T_{\rm eff}^4 \,. \tag{1.93}$$

We shall usually use the $\mathcal{L}$–$T_{\rm eff}$ version of the HR diagram. One major convenience in doing so is that it is very easy to place straight lines of constant radius on such a diagram if $\mathcal{L}$ and $T_{\rm eff}$ are expressed as logarithms. Note, however, that the effective temperature scale runs from right to left with the highest temperatures appearing on the left (for historical reasons). Note also, the HR diagram gives no further information than $\mathcal{L}$, $T_{\rm eff}$, and $\mathcal{R}$. It says nothing (at least directly) about stellar mass, composition, or state of evolution.

An example of an HR diagram is shown in Fig. 1.5 from the review article by Iben (1991). It shows typical ranges of stellar luminosities and effective temperatures and three lines of constant radius that can be deduced from (1.92–1.93). Nearby and bright stars are also indicated (from data listed in Allen, 1973). It is clear that most of these stars lie along a relatively well-defined locus called the "main sequence." Others are collectively called "giants" (because of their large size) while a small number have radii of about $10^{-2}\,\mathcal{R}_\odot$ and these are the "white dwarfs." There are other kinds of stars than those shown in the figure and part of the task of the next chapter will be to explore possible evolutionary relationships between these diverse objects.

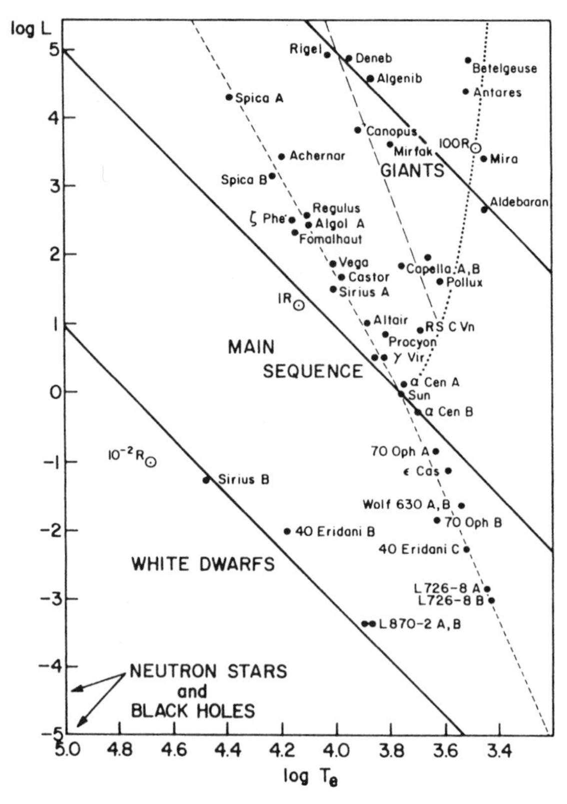

Fig. 1.5. An illustrative Hertzsprung–Russell diagram showing nearby and bright stars as seen from the earth. Reproduced with permission from Iben (1991).

1.9 Summary Remarks

This chapter has discussed a diverse set of topics all tied to an underlying theme; namely, assuming, for the most part, that stars are spherically symmetric and mechanically static, then the application of relatively simple physics allows us to derive their overall characteristics. Thus, because stars are massive they must somehow contrive to build up immense internal pressures to support themselves against collapsing under their own weight. Stars achieve this by way of high internal temperatures and densities. These statements are summarized in the viral theorem (plus the accompanying mass and hydrostatic equilibrium equations and an appropriate equation of state), which gives reasonable estimates of internal pressures and temperatures for most stars. But, because most stars are made up chiefly of material left over from the earliest stages of our evolving universe, hydrogen is a prime fuel for thermonuclear reactions and we now have an energy source. That energy must leak out, by a variety of means, to the surface, thus making stars shine. If we tie these last pieces of the puzzle together and describe thermal balance and, for example, radiative diffusion, the resulting four equations (1.3, 1.6, 1.58, and 1.60, or their variants) plus constituent equations of state, energy generation and opacity form a complete (for now) system that may be solved to construct stellar models. We have not done the latter in this chapter but, using dimensional analysis, we were able to find scaling laws relating models to one another with some success when compared to the real world.

In the next chapter we get down to business by giving an overview of stellar evolution and some of the kinds of stars evolution produces. Later chapters will be devoted to elaborating on the equations governing stellar structure and evolution, and developing the necessary input physics and techniques required to find their solutions.

1.10 Exercises

There's no problem so big or complicated
that it can't be run away from.

—Graffito, London (1979)

Exercise 1.1. This is a little exercise in some items that this book does *not* cover but which are essential to an understanding of stars. It has to do with spectral classification of stars in the *UBV* photometric system and some other matters. We recommend that you browse through the second and third chapters of Mihalas and Binney (1981). Appendix A also contains some information. Most of what you need for this exercise may also be found in Allen (1973, Chap. 10) or Cox (1999). Note that you will have to look up numbers in tables and these tables are not always entirely consistent: it's still not an exact science. Some of the answers you get for the following questions will therefore be estimates, but they will be good ones. In any case, you are told that a star has been observed with a *UBV* color index of B–$V = 1.6$ and that interstellar reddening is negligible. In addition, the parallax of the star is $\pi = 0.25$ seconds of arc, and its apparent visual magnitude is $m_V = 9.8$. Detailed spectroscopy also reveals that the star has all the characteristics of a main sequence star (luminosity class V).

1. What is the spectral class of the star?
2. What is the distance to the star (in parsecs), its distance modulus, its absolute magnitude (M_V), bolometric correction ($B.C.$), bolometric magnitude (M_{bol}), and luminosity (in $\mathcal{L}_\odot$)?
3. What is its effective temperature (T_{eff}) and radius (in $\mathcal{R}_\odot$)?
4. Estimate the mass of the star (in $\mathcal{M}_\odot$).

Exercise 1.2. We stated, without proof, that the central pressure of the constant density star was a lower limit (§1.4); that is, central pressures must exceed $P_c = 3G\mathcal{M}^2/(8\pi\mathcal{R}^4)$. The proof of that statement requires a bit more work than we wish to attempt now. There is, however, a weaker lower limit on P_c. To get at this consider the function

$$f(r) = P(r) + \frac{G\mathcal{M}_r^2}{8\pi r^4} \ .$$

1. Show that $f(r)$ decreases outward with increasing r. (Hint: Differentiate $f(r)$ with respect to r and use the equation of hydrostatic equilibrium to show $df/dr < 0$.)
2. Assuming zero pressure at $\mathcal{R}$, demonstrate (almost immediately) that

$$P_c > \frac{G\mathcal{M}^2}{8\pi\mathcal{R}^4}$$

which is less stringent than that given by (1.42). Note that you must show $\mathcal{M}_r^2/r^4$ goes to zero as $r \to 0$.

Exercise 1.3. A useful (albeit not terribly realistic) model for a homogeneous composition star may be obtained by assuming that the density is a linear function of radius. (See Stein, 1966.) Thus assume that

$$\rho(r) = \rho_c \left[1 - r/\mathcal{R}\right]$$

where ρ_c is the central density and $\mathcal{R}$ is the total radius where zero boundary conditions, $P(\mathcal{R})=T(\mathcal{R})=0$, apply.

1. Find an expression for the central density in terms of $\mathcal{R}$ and $\mathcal{M}$. (You will have to use the mass equation.)
2. Use the equation of hydrostatic equilibrium and zero boundary conditions to find pressure as a function of radius. Your answer will be of the form $P(r) = P_c \times$(polynomial in $r/\mathcal{R}$). What is P_c in terms of $\mathcal{M}$ and $\mathcal{R}$? (It should be proportional to $G\mathcal{M}^2/\mathcal{R}^4$.) Express P_c numerically with $\mathcal{M}$ and $\mathcal{R}$ in solar units.
3. In this model, what is the central temperature, T_c? (Assume an ideal gas.) Compare this result to that obtained for the constant-density model. Why is the central pressure higher for the linear model whereas the central temperature is lower?
4. Verify that the virial theorem is satisfied and write down an explicit expression for Ω (i.e., what is q of Eq. 1.8?).

Exercise 1.4. We shall discuss completely degenerate electron equations of state in Chapter 3, but we can use them now without explaining what they are. If the electrons are nonrelativistic, then the power law exponents for pressure of equation (1.67) can be shown to be $\chi_\rho = 5/3$ and $\chi_T = 0$. Use this information to find the exponent $\alpha_\mathcal{R}$ in $\mathcal{R} \propto \mathcal{M}^{\alpha_\mathcal{R}}$ of (1.70). You will find that it does not matter whether the star is fully convective or fully radiative; you get the same answer from the homology relations.

Exercise 1.5. We used dimensional analysis earlier on (in §1.6) to derive some estimates for the mass–luminosity slope of hydrogen main sequence stars. These stars burn hydrogen into helium. Suppose, by some means, all the hydrogen is converted to ^{4}He and that nucleus begins to combine to form ^{12}C by way of the triple-α reaction. We could, in principle, then imagine equilibrium stars composed of pure ^{4}He that form a "helium–burning main sequence."

1. Assuming an ideal gas, radiative transfer with electron scattering, and using the values of λ and ν from Table 1.1 for the triple-α reaction, find $\alpha_{\mathcal{L}}$ of (1.73) for the helium main sequence.
2. Such a main sequence may not exist in nature but that doesn't stop theoreticians from constructing them on the computer. The following pairs of mass–luminosity results, in the form $[\mathcal{M}/\mathcal{M}_\odot,\ \log_{10}\mathcal{L}/\mathcal{L}_\odot]$, for three helium main sequence models are from
   ▹ Hansen, C.J., & Spangenberg, W.H. 1971, ApJ, 168, 71:
   [4.0, 4.24], [2.0, 3.42], and [1.0, 2.52]. Use these results to estimate $\alpha_{\mathcal{L}}$ and compare to part (1). (Recall that an expression such as 1.73 may be written in differential form as in 1.68.)
3. Using the computer data of part 2 and your $\alpha_{\mathcal{L}}$ result, normalize the mass–luminosity relation and find the constant C in

$$\frac{\mathcal{L}}{\mathcal{L}_\odot} = C\left(\frac{\mathcal{M}}{\mathcal{M}_\odot}\right)^{\alpha_{\mathcal{L}}}.$$

. Suppose stars on the helium main sequence evolve off their main sequence in much the same way as do stars on the hydrogen main sequence. That is, after 10% of a star's helium is converted to ^{12}C they radically change their structure. Call the time it takes to do that the "main sequence lifetime" or $t_{\rm ms}$. If conversion of a gram of ^{4}He to ^{12}C releases 6×10^{17} ergs, then what are D and δ in

$$t_{\rm ms} = D\left(\frac{\mathcal{M}}{\mathcal{M}_\odot}\right)^{-\delta} \text{ years?}$$

We suggest you follow the arguments of §1.7.

Exercise 1.6. (This problem is due to Ellen Zweibel.) It appears that some stars (besides the sun) are orbited by planets. Those extra-solar planets discovered thus far seem to have masses comparable to, or greater than, that of Jupiter and they orbit the parent star close in. Suppose one of these planets is captured by, and accreted onto, the parent. The way we imagine this to take place is that the planet's orbit is circular and just grazes the star before accretion. Once accretion has taken place and the planet is completely assimilated into the star, we expect the combined body to have a radius different than that of the original star. To get a common nomenclature, let $\mathcal{M}$ be the mass of the original star, m the mass of the planet and assume $m << \mathcal{M}$. The radius of the parent star is $\mathcal{R}$ and what we wish to find is how much that changes in terms of the small quantity $m/\mathcal{M}$. Call the change $\Delta\mathcal{R}$.

1. As the planet is accreted the total energy of the star increases by some amount ΔW. Let this amount be due solely to the orbital energy of the planet just as it is accreted (i.e., neglect any chemical or gravitational energy addition from the planet). What is ΔW in terms of $\mathcal{M}$, m, and $\mathcal{R}$ (and G)?
2. If the gravitational potential energy Ω takes the form (1.8), then find $\Delta\Omega$ to first order in the small quantities $m/\mathcal{M}$ and $\Delta\mathcal{R}/\mathcal{R}$. (Note that part of the change in Ω is due to the added mass m and part due to $\Delta\mathcal{R}$.) Assume q of (1.8) remains constant.
3. Use the virial theorem in the form (1.28) to solve finally for $\Delta\mathcal{R}/\mathcal{R}$.
4. If $\gamma = 5/3$ and $q = 3/2$, what would $\Delta\mathcal{R}/\mathcal{R}$ be for the sun if it swallowed Jupiter?

Exercise 1.7. We shall have little to do with general relativity (GR) in this text but here we briefly explore the "Tolman–Oppenheimer–Volkoff" (TOV) GR equation of hydrostatic equilibrium for spherical stars. It is the daunting expression

$$\frac{dP}{dr} = -\frac{G\left[\rho(r) + P(r)/c^2\right]\left[\mathcal{M}(r) + 4\pi r^3 P(r)/c^2\right]}{r\left[r - 2G\mathcal{M}(r)/c^2\right]}$$

where ρ is now the mass–energy density but we still have $\mathcal{M}_r = \int 4\pi r^2 \rho \, dr$. Let's see what this gives for a "star" of constant ρ. For those of you who wish to delve into this further see

▹ Shapiro S.L., & Teukolsky, S.A. 1983, *Black Holes, White Dwarfs, and Neutron Stars* (New York: Wiley & Sons)

or, with even more stuff,

▹ Misner, C.W., Thorne, K.S., & Wheeler, J.A. 1973, *Gravitation* (San Francisco: Freeman).

1. Show that
$$P(r) = \rho(r)c^2 \left[\frac{(1 - 2\mathcal{M}Gr^2/R^3c^2)^{1/2} - (1 - 2\mathcal{M}G/Rc^2)^{1/2}}{3(1 - 2G\mathcal{M}/Rc^2)^{1/2} - (1 - 2\mathcal{M}Gr^2/R^3c^2)^{1/2}}\right]$$
satisfies the TOV equation for ρ constant.
2. Show that this solution reduces to the constant density star solution of §1.4 in the Newtonian limit $c \to \infty$ and that the central pressure of (1.41–1.42) is retrieved.
3. Define the parameter $\alpha = 2G\mathcal{M}/\mathcal{R}c^2$ and show that the TOV GR solution does strange things as $\alpha \to 8/9$.
4. If $\alpha = 8/9$, then what is $\mathcal{R}$, in km, as a function of $\mathcal{M}/\mathcal{M}_\odot$?

5. If $\mathcal{M}$ is equal to $\mathcal{M}_\odot$ what is the density (in g cm^{-3}) at $\alpha = 8/9$. (Obviously we're not fooling anyone here. This all has to do with black holes.)

Exercise 1.8. In the paragraph following the expression of the virial theorem (1.18) we stated that an extra term $-3P_S V_S$ should appear on the right hand side if we had chosen to consider only that part of the spherical star interior to $r = r_S$ having a volume V_S and a surface pressure P_S at r_S.

1. Prove this for the case of hydrostatic equilibrium; that is, show that the correct expression is
$$2K + \Omega - 3P_S V_S = 0.$$
Hint: Integrate (1.22) by parts using the equation of hydrostatic equilibrium (1.5 or 1.16) and the mass equation, and remember to only go out to r_S in that integration and the one for Ω (1.7).
2. Show explicitly that this amended version works for the constant density sphere.

Exercise 1.9. Redo the analysis of §1.3.5 and compute the period, Π, of the sun assuming it has constant density. Take $\Gamma_1 = 5/3$. Note that this involves an integration.

Exercise 1.10. A short article by G.P. Collins in the February 2000 issue of *Scientific American* (p. 20) on the equivalence principle suggests a slightly off-the-wall, but easy, problem. Gravitational binding energies are negative but, by mc^2 arguments, so should the mass associated with this energy be negative. Thus, for example, the total mass of the sun should be less than the sum of its material parts when that negative mass is taken into account. Assuming the sun to be a constant density sphere, or anything else that is reasonable, by what fraction is the sun's mass decreased when gravitational binding energies are included?

Exercise 1.11. (This version of the stability problem is due to Cole Miller of the University of Maryland.) Take the second variation of the total energy $W_{\rm ad}$ (compare with §1.2) and derive a condition for stability. Assume that the equation of state is given by $P = c_1 \rho^{\Gamma_1}$, where c_1 is some constant.

1. Write $\delta^2 W_{\rm ad}$ in its general form, assuming this equation of state. At the end it will be most convenient to change variables from $\mathcal{M}_r$ to V, so you will also need to express factors like $(\delta r)^2$ in terms of V and δV, where V is the regular volume.
2. If the adiabatic index Γ_1 is a constant throughout the star, then it is possible to show that the volume perturbations are proportional to the volume; i.e., $\delta V = kV$, where k is a constant throughout the star. Use this to determine a simplified condition for stability.

1.11 References and Suggested Readings

The following format for references will be used throughout this text. In addition to listing the sources, we will occasionally make editorial comments leading the reader to where we believe especially good discussions of some material can be found. General references are usually listed first. These are then followed by those keyed to sections within a chapter. Appendix C provides a key to the journal abbreviations and the sequencing of volume and page numbers used here.

General References

Many of the quotes found at the beginning of the chapters are from

▷ *The Oxford Dictionary of Quotations*, 3rd ed. 1980 (Oxford: Oxford University Press)

▷ Metcalf, F. 1986, *The Penguin Dictionary of Modern Humorous Quotations* (London: Penguin Books Ltd.).

The monograph by

▷ Cox, J.P. 1968, *Principles of Stellar Structure*, in two volumes (New York: Gordon & Breach),

which was written with the aid of R.T. Giuli, is a classic monograph on stellar structure. You can sometimes find it in used bookstores, but even then its price is beyond the means of the average student. What you will *not* find in this work are modern discussions of topics such as evolution in close binary systems, supernova models, magnetic fields, rotation, etc. Don't let this discourage you. The care paid to detail and accuracy, and the clarity of style, are worth it. You will note, incidentally, that we have attempted to conform to Cox's nomenclature for various quantities but there is no true standard. You may have to do some translation if you consult other texts.

We must also guide you to the excellent text by

▷ Kippenhahn, R. & Weigert, A. 1990, *Stellar Structure and Evolution* (Berlin: Springer-Verlag).

The authors pioneered much of the work in stellar structure and evolution, and their text contains a wealth of detail regarding the results of stellar modeling. Although much of their philosophy and nomenclature differ from what you will find here, both texts supplement each other in many respects.

You can now purchase a paperback version (1983) of the text by

▷ Clayton, D.D. 1968, *Principles of Stellar Evolution and Nucleosynthesis* (New York: McGraw-Hill).

It, like Cox (1968), is a bit outdated, but the last four chapters on nuclear reactions and nucleosynthesis are still the clearest and most complete. There are also excellent sections on the calculation of opacities and other quantities discussed from a nice physical viewpoint.

We recommend the recent text

▹ Rose, W.K. 1998, *Advanced Stellar Astrophysics* (Cambridge: Cambridge University Press),

which covers several topics in more depth than we do here. The physics also tends to be at a higher level.

Other texts worthy of mention are

▹ Huang, R.Q., & Yu, K.N. 1998, *Stellar Astrophysics* (Singapore: Springer-Verlag)

▹ Böhm-Vitense, E. 1992, *Introduction to Stellar Astrophysics: Stellar Structure and Evolution* (Cambridge: Cambridge University Press)

which is the third volume in a three-volume series, and

▹ DeLoore, C.W. & Doom, C. 1992, *Structure and Evolution of Single and Binary Stars* (Hingham, Mass.: Kluwer).

This last text will prove especially useful for its treatment of binary systems, which is a topic we only touch upon in Chapter 2. Another general text is

▹ Collins, G.W. 1989, *The Fundamentals of Stellar Astrophysics* (New York: Freeman).

The text by

▹ Mihalas, D., & Binney, J. 1981, *Galactic Astronomy*, 2nd ed. (San Francisco: Freeman)

has a wealth of material on stars and other matters astronomical and astrophysical. We recommend it strongly as a general reference for all students. Yet another is the monograph by

▹ Jaschek, C., & Jaschek, M. 1987, *The Classification of Stars* (Cambridge: Cambridge University Press).

As the title implies, this work describes how and why stars are classified observationally. Most sciences start off with observation and classification so the importance of such work should not be underestimated.

The text (in two volumes)

▹ Shu, F.H. 1991, 1992, *The Physics of Astrophysics*, Vols. 1–2 (Mill Valley, CA: University Science Books)

offers an interesting alternative to gathering together many texts to fill in the physics you need for astrophysics. The two volumes are at the graduate level but Shu gives enough introductory material for an undergraduate to follow the presentation. Not all topics are covered but this work may fit many of your needs. The total cost, however, is not insubstantial.

▹ Allen, C.W. 1973, *Astrophysical Quantities*, 3rd ed. (London: Athlone)

is a popular compendium of astrophysical lore, tables, etc., in a single volume, although it is rapidly getting out of date. It should be on your shelf (if you can find an affordable used copy). A newer version of Allen (without Allen) is

▹ Cox, A.N. (editor) 1999, *Allen's Astrophysical Quantities* (New York: Springer-Verlag).

It too has a hefty price tag. Another reference to look into is

▹ Lang, K.R. 1991, *Astrophysical Data: Planets and Stars* (Berlin: Springer-Verlag).

Useful intermediate texts at the undergraduate level are hard to come by because many are written for the nonscientist. Among those few that we suggest are

▹ Shu, F.H. 1982, *The Physical Universe: An Introduction to Astronomy* (Mill Valley, CA: University Science Books)

and

▹ Carroll, B.W., & Ostlie, D.A. 1996, *An Introduction to Modern Astrophysics* (Reading: Addison-Wesley).

On a more elementary level are the many undergraduate first-year astronomy texts for the nonscience major. There are so many available we shall not go out on a limb and recommend one. However, just picking (almost) randomly from our bookshelf, we have

▹ Chaisson, E., & McMillan, S. 1999, *Astronomy Today*, 3rd ed. (New Jersey: Prentice-Hall, Inc.).

It (among other texts of its kind) has fancy acetate overlays and a CD-ROM containing a hyperlinked version of the text plus videos and animations that bring the discussion to life. A continuously updated website is also associated with the text that enables the student to access resources on the WWW. (We wish we could have done all this but, with the restricted market for advanced texts, you would not have been able to afford ours had our publisher gone along with such an idea!)

You might wish to check out

▹ Zel'dovich, Ya.B., & Raizer, Yu.P. 1966, *Physics of Shock Waves and High Temperature Hydrodynamic Phenomena*, Vols. 1–2 (New York & London: Academic Press)

from your library. It is by no means a text on astronomy but it contains a wealth of material, of all kinds, that bears on the subject. It is written in the Russian style, that is, clear, but not that easy.

§1.2: An Energy Principle

▹ Goldstein, H. 1981, *Classical Mechanics*, 2nd ed. (Reading: Addison-Wesley)

is a standard text on classical mechanics. Many of us were raised on it.

If you can find

▹ Chiu, H.-Y. 1968, *Stellar Physics*, Vol. 1 (Waltham, MA: Blaisdell)

we suggest you browse through its chapters. It is unfortunate that the second volume never appeared. The first volume covers topics you cannot find in other standard texts in stellar astrophysics. It is now out of print.

§1.3: The Virial Theorem and Its Applications

The short monograph by

▹ Collins, G.W., II 1978, *The Virial Theorem in Stellar Astrophysics* (Tucson: Pachart)

contains many applications and variations on the virial theorem plus detailed derivations. Our discussion of the theorem does not include important topics such as magnetic fields, rotation, and relativistic effects. You will find them in Collins. The references to Cox (1968) and Clayton (1968) are listed above.

§1.4: The Constant-Density Model

Take a good look at

▹ Stein, R.F. 1966, in *Stellar Evolution*, eds. Stein & Cameron, (New York: Plenum Press), pp. 3–82.

If you can find this symposium volume, Stein's article is worth the effort. In it he uses simple models to bring out important points in stellar structure and evolution.

▹ Clayton, D.D. 1986, AmJPhys, 54, 354,

titled *Solar Structure without Computers* goes at least one step further than Stein (1966) in constructing a model. It has lots of potential for homework problems!

▹ Grevesse, N., & Noels, A. 1993, in *Origin and Evolution of the Elements*, eds. Pratze, Vangioni-Flam, & Casse (Cambridge: Cambridge University Press), p. 15

is an excellent article to consult for abundances. (And see also the references in Chap. 2.)

§1.6: Stellar Dimensional Analysis

The *Observatory* often publishes useful short articles that deserve more exposure. Among these are often amusing commentaries on astronomical subjects and historical articles. The reference to

▹ Carson, T.R. 1986, Obs, 106, 71

may be found there.

The observational data for the mass–luminosity and mass–radius relations of Fig. 1.3 are from Allen (1973),

▹ Harris, D.L., III, Strand, K.Aa., & Worley, C.E. 1963, in *Basic Astronomical Data*, ed. K.Aa. Strand (Chicago: University of Chicago Press), p. 273

and

▹ Böhm, C. 1989, Ap&SS, 155, 241.

The second reference is one in a series of books which, though somewhat outdated, still contain much useful material.

▹ Kovalevsky, J. 1998, ARA&A, 36, 99

reviews what has been learned to date from the Hipparcos mission along with a brief description of the hardware. The material used for Fig 1.4 was taken from

▷ Martin, C., & Mignard, F. 1998, A&A , 330, 585

and

▷ Martin, C., Mignard, F., Hartkopf, W.I., & McAlister, H.A. 1998, A&AS, 133, 149.

See also

▷ Lebreton, Y. 2001, ARA&A, 38, 35

for a review of the impact of Hipparcos on our understanding of stars.

§1.8: The Hertzsprung–Russell Diagram

Figure 1.4 is from

▷ Iben, I. Jr. 1991, ApJS, 76, 55.

This paper contains a personal account of Iben's work and, as is usual in his papers, the reference list is exhaustive.

2 An Overview of Stellar Evolution

"And now for something completely different."
— Monty Python's Flying Circus (Oct. 1969–Dec. 1974)

"I never know how much of what I say is true."
— Bette Midler (1980)

"You'd look pretty simple from ten parsecs too."
— Attributed to Fred Hoyle in response to a question from someone who was puzzled why we do not seem to understand so simple a thing as a star. (c. 1955)

The structure and evolution of stars is the one part of modern astrophysics that can be described as, to a considerable extent, a solved problem. This means we can–

1. write down a set of four differential equations (see the previous chapter or §7.1) that describe gradients of conditions inside a star,
2. insert into these equations the necessary physics of nuclear reactions (Chap. 6), transport of energy by radiation, convection, and occasionally conduction (Chaps. 4 and 5), and the relationship among the thermodynamic variables T, ρ, and P, also called the equation of state (Chap. 3).
3. We can choose reasonable boundary conditions (e.g., see §4.3),
4. integrate the equations numerically (Chap. 7) to find out what stars should look like, and
5. compare model stars with real ones.

The happy result is that the calculated stars are very much like the observed ones in mass, brightness, size, temperature, surface composition, age, and the correlations between these.

This chapter is supposed to provide you with the vocabulary and a few other tools needed to work through the rest of the book and, meanwhile, to be able to start reading some of the research literature in stellar astronomy without having to look something up every other sentence. Appendix A, however, has a short glossary in case you need to and, of course, we hope you have read the basic material in Chapter 1. The approach is that of a teller of folk tales, beginning with the birth of the heroine, "once upon a time," and ending with "they exploded happily ever after." Keep in mind, however, that most of what is said can, nevertheless, be documented, calculated, and otherwise shown to be the honest story. Where the accompanying pictures are blurry or vital details are hidden in shadows, we will try to tell you so.

You will note that the main body of this chapter is almost entirely in narrative form. At most, you will see hardly more than fragments of an equation.

The idea is not to interrupt the narrative with distracting mathematics. We will, however, often suggest that you attempt an exercise at the end of this chapter that bears directly on some point we are making (and that's where we hide the equations).

The stellar formation process is actually the least well understood part of the lives of stars and we defer its discussion until the end. You might think that this would be a fatal flaw in all that follows. Curiously, it is not. The structure of a newborn star is simpler than that of any later stage and is well-explained by the "five-step process" described above. A standard analogy is a human one: you can do a good job of talking about babies and human life without knowing much about conception and embryology. The analogy extends to regions of current star formation sometimes being called stellar nurseries or even wombs. Even the difficulty in studying the two formation processes is somewhat similar. Baby formation is hidden in the uterus and even more private places; star formation is generally hidden by dust, at least to the observer of visible light. Indeed the advent of high-resolution infrared astronomy has begun to draw back some of the veils.

2.1 Young Stellar Objects (YSOs)

A protostar becomes a star when the energy released by thermonuclear fusion (hydrogen to helium) exceeds that released by contraction from the supply of gravitational potential energy. This is not something we can directly observe. Thus protostars and young stars are put into classes 0, 1, and 2 based on things we can see—ratio of infrared to visible light, amount of molecular gas around, how the gas is moving, and so forth. The class 0's are still contracting, and very few members are known, primarily because of the short time scales involved. The 1's and 2's are already living on nuclear energy and, typically, blowing material off their surfaces in bipolar or jet-like outflows.

These jets gradually clear away surrounding afterbirth, opening out from narrow beams to wide cones, until visible light can find its way from the stellar surface (photosphere) to us without being absorbed and re-emitted as infrared. A cartoon of this sequence in shown in Fig. 2.1.

Signatures of the YSO phase include the following:

1. We see variability in the visible light, because material ia still falling down onto the surface of the star from a residual disk, so that both the stellar surface and the disk have temperature irregularities that change in times from hours to days and longer.
2. Emission lines are observed in their spectra, from the disk, or the bipolar outflow, or both.
3. YSOs have more infrared luminosity than older stars of the same mass, because there is more dust around.

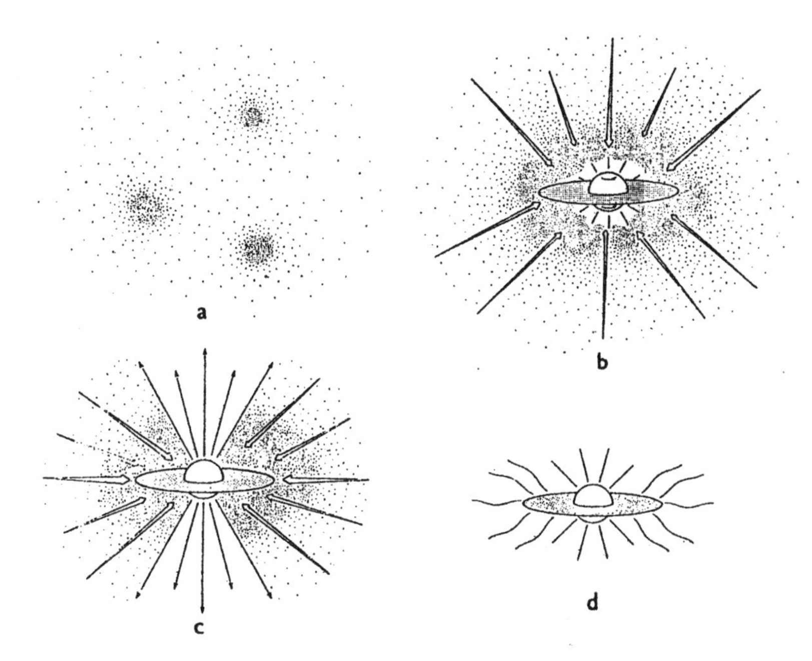

Fig. 2.1. This cartoon illustrates the four stages of star formation. (a) First protostar cores form within molecular clouds. Then, in (b), the protostar builds up from the inside out while the surrounding nebular disk rotates around it. (c) Bipolar flows break out along the rotation axis of the system. Finally, in (d), the surrounding nebular material is swept away, and the newly formed star, with disk, is revealed. From Shu et al. (1987). Reproduced with permisson, from the Annual Review of Astronomy and Astrophysics, Vol. 25, ©1987 by Annual Reviews.

4. A high level of what is called activity is seen, meaning flares, star spots, emission from a hot corona, and so forth, all of which are found at a low level in the sun and other stars. The reason seems to be two-fold: young stars are often rapid rotators (rotation periods from hours to days, versus a month for the sun) and, because they are cooler than they will be when settled onto the main sequence, they have surface convection that extends deeper. The combination results in a strong dipole magnetic field, which, in turn, drives the activity.
5. X-ray emission is seen from the hot corona. There is also radio emission, but it is too faint to see except from very nearby, very active stars.

YSOs were first recognized from the combination of variability, emission lines, location on the HR diagram, and location in space near clouds of gas and dust. The prototype low-mass YSO is T Tauri.[1] T Tauri stars (or T Taus, which rhymes with "cows") is a common name for the whole class.

The things you need to know about the protostar stage to carry on from here are–

1. The energy source is gravitational potential energy and the total lifetime therefore is short. The masses, luminosities, and radii of the YSOs are not terribly different from solar values, so using the Kelvin–Helmholtz and nuclear time scales of (1.32) and (1.90), the contraction life of a "typical" YSO is only about 0.1% of its potential nuclear life.
2. Protostars are convective throughout. Thus a new star is chemically homogeneous. This will change as it ages.

We shall discuss more details of the structure and evolution of protostars in Chapter 7, but, for now, Fig. 2.2 shows some evolutionary tracks on an HR diagram for a variety of masses. The evolution starts in the upper right (luminous but cool) and proceeds to the point where hydrogen is ignited on the main sequence.

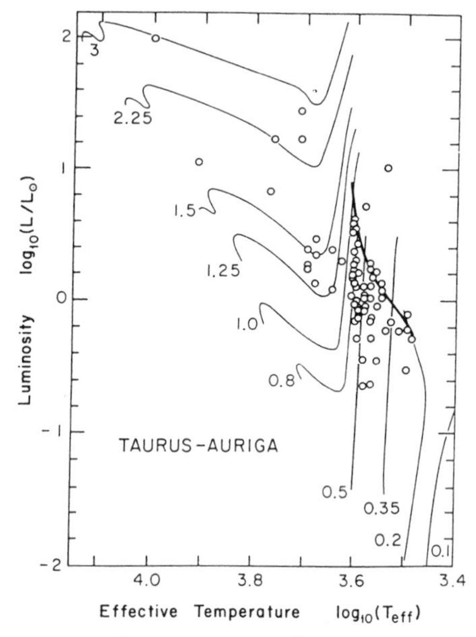

Fig. 2.2. Shown are pre-main sequence evolutionary tracks adopted by Stahler (1988) from various sources. Masses are in solar units. Also shown are the observed locations of a number of T Tauri stars. Reproduced with permission.

---

[1] For those of you unfamiliar with how and why stars are named, we suggest you look through some non-technical books such as Allen (1963) or Burnham (1978). Another source of named stars in the spirit of how stars are classified is Jaschek and Jaschek (1987).

2.2 The Zero-Age Main Sequence (ZAMS)

Most people pronounce this as a word, rhyming with "hams." The main sequence gets its name because it includes most of the stars in an honest sample. An "honest sample" means all the stars in a particular stellar cluster or in a given volume of space. The "naked-eye" stars are not an honest sample because they include not only the nearest few intrinsically faint stars, but lots of intrinsically bright stars that we can see out to large distances.

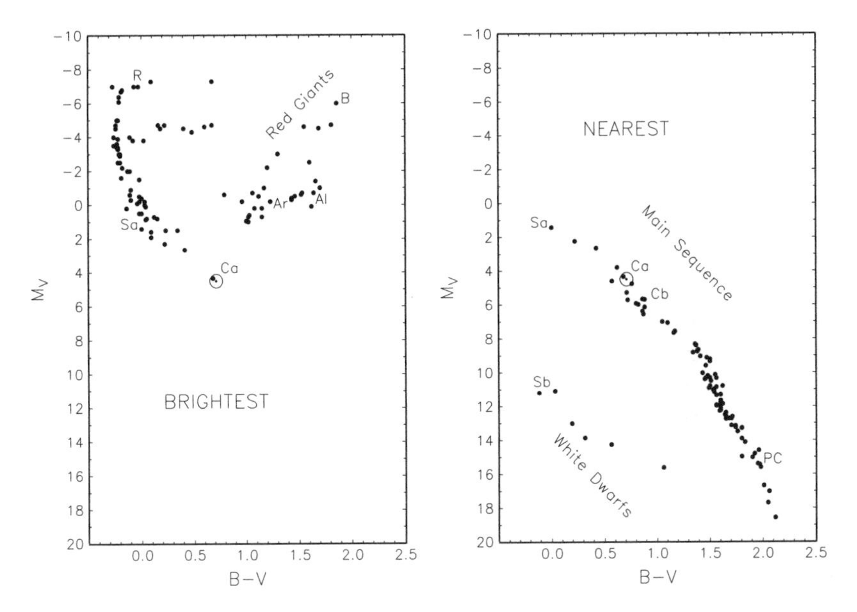

Fig. 2.3. Shown in the left panel is the HR diagram for the 101 brightest stars as seen from earth. Compare this to the right panel that shows what happens when only those stars (97 of them) within five parsecs are considered. The data are from Allen (1973) as supplemented from Lang (1991). The sun is indicated by the $\odot$, and the main sequence, white dwarfs, and red giants are labeled. The location of some well-known stars are also shown: Betelgeuse (B), Sirius A (Sa), the white dwarf Sirius B (Sb), α Cen A (Ca, very much like the sun), α Cen B (Cb), Proxima Cen (PC, the nearest known star), Rigel (R), Arcturus (Ac), and Aldebaran (Al).

To illustrate this point, the left-hand panel of Fig. 2.3 shows an observer's HR diagram for the brightest 101 stars as seen from our vantage point. The abscissa, B–V, is the difference in magnitudes of a star as measured using two of the Johnson–Morgan filters. "B" denotes blue whereas "V" means visual. Since blue is "hotter" than most wavelengths in the visible part of the spectrum, a large value of $B–V$ implies a cool object, once you remember that the magnitude scale is backward with respect to intensity. So, for example, a main sequence star with $B–V$ equal to zero has an effective temperature of $T_{\text{eff}} \approx 10,000$ K, whereas $B–V$ of unity implies a considerably cooler 4,700 K. The ordinate, M_V, is the absolute visual magnitude using a standard filter. With some effort, not to be described here, it can be converted to luminosity, although the conversion depends on the kind of star in question. (See Fig. 1.4, which shows a composite HR diagram.)

All the brightest star panel tells us is that the brightest stars are intrinsically luminous and most of them are distant. The *least* luminous of the bunch is the sun, indicated by the sun sign, and its near twin, Cen A. This is not an honest sample. On the other hand, the right-hand panel of Fig. 2.3 is more to the point. It shows what happens if you plunk down all the *nearest* stars on an HR diagram—at least those out to five parsecs. There are about 100 of them, not including objects so faint we cannot pick them out, but which are probably not true stars anyway. If you wish more nearby stars to play with, consult Lang (1991, p. 758ff.) for over 2,200 stars within 22 parsecs.

The majority of stars in the right-hand panel constitute the *main sequence*, and are so labeled. These stars are essentially unevolved even though they are converting hydrogen to helium. Were any of them formed very recently we would call them ZAMS stars. "Zero-age" in practice means that the star has changed so little in luminosity, radius, and T_{eff} since it first started hydrogen fusion that you cannot notice it. This might mean only a few thousand years for a massive star, 10^7 years for the sun, and 10^9 years or more for the least massive stars. (See §1.7 and especially Eq. 1.91 for t_{nuc}.) And, in a cluster of stars all formed at the same time, it is possible for the ones of 6 $\mathcal{M}_\odot$ to have long ago evolved to white dwarfs (as in Fig. 2.3), those of 4 $\mathcal{M}_\odot$ to be red giants (like those in the upper right of the left panel of Fig. 2.3), those of 2 $\mathcal{M}_\odot$ to be slightly off the ZAMS, those of 1 $\mathcal{M}_\odot$ still on it, and those of 0.4 $\mathcal{M}_\odot$ still not quite through the formation process. Note that the HR diagram says nothing directly about the mass.

To tidy up, note that you can still follow the main sequence on the brightest-stars panel. It continues on to hotter temperatures than the sun's, but then the stars plotted rise almost vertically. Part of this rise is due to evolution of those very bright (and massive, it turns out) stars.

2.2.1 Life on the Main Sequence

The single most important thing in the life of a star is its mass, with its initial mix of hydrogen, helium, and heavy elements a distant second. The mass determines luminosity, size, and surface temperature (as discussed at length in §1.6) and also which nuclear reactions will occur, how long they will last (§1.7), when and how material gets mixed through the star (convection), and how the star dies.

A ball of gas whose center never gets hot enough for nuclear fusion is, by *definition*, not a star. Thus you can say with great confidence that *all* stars spend most of their active (nuclear burning) lives on the main sequence, where[2] their energy source is the fusion of hydrogen (^{1}H) to helium (^{4}He) by one of two sequences of reactions, called the proton-proton chain and the CN or CNO cycle, bicycle, or tricycle. The last was originally intended to be funny but is an honest word for the three coupled sets of reactions among C, N, and O. The details appear in §6.4.

The proton-proton chain (pp-chain) is the main energy source in stars of less than about 1.5 $\mathcal{M}_\odot$ because it is easier for two protons to get close together then for a proton plus a carbon nucleus. Thus the pp-chain starts in gas that is too cool for the CNO cycle. But if a star is massive enough that the balance between gravity and pressure takes its central temperature into the CNO regime (above about 1.8×10^7 K), then CNO goes faster and produces most of the power. In either case, four hydrogen atoms are eventually converted to one helium atom (^{4}He), which is less massive by about 0.8%. The mass lost comes out as photons (mostly) and neutrinos. The neutrinos leave immediately and how much energy is lost in them has to calculated carefully for each relevant nuclear reaction. In the case of the sun, we observe the neutrinos as expected.[3]

Other differences among main sequence stars include–

1. The CNO cycle liberates energy in a much smaller region of the star than do the pp-chains (i.e., its rate depends on a steeper power of the temperature, and see §1.6 and Chap. 6) and so drives convection in the stellar center. The sun may have had a small convective core when it was young, but is now radiative there.
2. At about the same 1.5 $\mathcal{M}_\odot$ dividing line, there is also a difference in how energy is transported in the stellar envelope (loosely speaking, the outer layers, which may turn out to be quite extensive). Less massive stars have neutral hydrogen near their surfaces. Neutral hydrogen impedes the flow of ultraviolet photons, and convection transports most of the power. In more massive stars, the surface gas is hotter, the hydrogen largely ionized, and radiation carries the power. Notice that few stars (beyond the pre-main sequence stage) are convective throughout. Thus we do not see nuclear reaction products on their surfaces for most of their lives.

[2] You might think that this should say "*when* their energy source is ...." In practice, we often say "where" (having in mind a location on the HR diagram), and one of the subsidiary goals of a book like this one is acculturation—to enable you to sound like one of the tribe.

[3] Until quite recently, it looked as if we were seeing fewer neutrinos than expected by a factor of three. This has now been sorted out; it was a problem in weak interaction physics, rather than in astrophysics. Section 9.3 includes more of the story. See Bahcall (2001) and Seife (2002) for short reviews.

Stars with masses less than about 0.3 $\mathcal{M}_\odot$ are an exception and remain fully convective for all their lives.
3. The 1.5 $\mathcal{M}_\odot$ figure comes up yet again when main sequence stellar rotation is considered. Stars of mass greater than this tend to rotate comparatively rapidly compared to lower mass stars (see Fig. 9.9). There is definitely a good story lurking here!
4. Most of the surface opacity comes from the elements heavier than hydrogen and helium, namely the "metals." (See Fig. 1.2 and discussion for a typical "mix" of metals. And, yes, we do say "metals" in astronomer's primitive lingo—remember the tribes who are supposed to count one, two, many; we count hydrogen, helium, metals.) Thus the more metals there are the less deeply you can see into the star. The deeper you go, the hotter the gas, and so metal–poor stars look bluer than metal–rich stars of the same mass and age.[4] This can be conspicuous enough to show even in the integrated light of a whole cluster or galaxy of stars.
5. All of the correlations of $\mathcal{L}$, T, $\mathcal{R}$, and lifetime with $\mathcal{M}$ explored in §1.6 seem to hold up quite well.
6. The "Supplemental Material" section near the end of this chapter lists relevant properties of ZAMS models for your reading pleasure. You can reproduce some of these using the "ZAMS" code found on the CD-ROM.

Most main sequence stars change only very slowly (with exceptions due to mass loss from really massive stars), in both interior structure and external appearance. When four (ionized) hydrogen atoms fuse to one (ionized) helium atom, eight separate particles (including electrons of course) become only three. Thus, since $P = nkT$ (Eq. 1.35) and a fixed central pressure is needed to balance gravity, the stellar core must slowly contract and heat up. This makes the nuclear reactions go faster, and the star gradually brightens. You might think the surface temperature would increase too, but it does not, it goes down. Thus the sun, at formation 4.6 Gyr ago, was about 25% fainter but also somewhat bluer (and a better source for ultraviolet light for primitive biochemistry) than the present sun. Earth has somehow adjusted and kept its surface temperature nearly constant, despite the 20 K increase you might expect (and see Ex. 2.7).

2.2.2 Brown Dwarfs

A gas mass that does not get hot enough to fuse hydrogen all the way to helium is called a brown dwarf, meaning something between red and black. Really, of course, they are infrared dwarfs. For solar composition, the brown dwarf/main sequence cut is about 0.085 $\mathcal{M}_\odot$. It is about 0.1 $\mathcal{M}_\odot$ (around 75 times the mass of Jupiter) for a very metal poor star with less opacity to keep

[4] To show this, do Ex. 2.3 at the end of this chapter where, using homology arguments, $T_{\rm eff} \propto Z^{-0.35}$ for low-mass ZAMS stars; that is, the lower the metal content, the hotter (bluer) the star.

the light and heat in. While ordinary hydrogen fusion sets in close to 10^7 K, deuterium (^{2}H or ^{2}D) will burn at 10^6 K, via the second reaction in Table 6.1. Only one atom in a hundred thousand of hydrogen is deuterium, left over from the early universe, but this is enough to slow the contraction and death of brown dwarfs a good deal. Indeed, for stars less than a billion years old, the main sequence is almost continuous in appearance across the dividing line. A few young brown dwarfs even show some signs of "stellar" activity. But with time, they fade, while a true star of 0.1 $\mathcal{M}_\odot$ will live 10^{11}–10^{12} years.

The first certain brown dwarf, Gliese 229B, was discovered in 1995 and is the companion to the red dwarf star Gliese 229A. One of the keys to its identification, besides its being far fainter than its faint companion, was the presence of methane in its spectrum. Any "real" star is far too hot to allow methane to form in its atmosphere, although it is a common molecule in the atmospheres of gas planets in our Solar System. The effective temperature of Gliese 229B is about (only!) 900 K and it is a member of the newly minted spectral class T (see §4.7 for additional information). The year 1995 also saw reports of other brown dwarfs (not then confirmed to be such but some showed lines of lithium, an element consumed by nuclear reactions in true stars) and, by now, there are dozens in the known zoo. For reviews, see Basri (2000a,b), Gizis (2001) and, for low-mass stars and substellar objects, Chabrier and Baraffe (2000), and Burrows et al. (2001).

2.3 Leaving the Main Sequence

From now on, we will be making further distinctions by mass. Only stars of initial mass more than about 0.3 $\mathcal{M}_\odot$ will be coming with us to §2.4 and beyond. The smaller, fully convective, ones keep fusing and mixing until all the hydrogen is converted to helium after 10^{12} years. The universe is not old enough for this to have happened to any real star (except a member of a binary pair that gets stripped at some intermediate time), so we are telling you the result of a calculation here, but it is hard to escape.[5]

Stars of more than 0.3 $\mathcal{M}_\odot$ will eventually use up all the hydrogen fuel at their centers while much still remains in their outer envelopes. The star is, of course, still radiating (losing energy, a conserved quantity). The core cannot cool down to conserve the energy supply, or pressure would cease to balance gravity. Indeed this starts to happen, and so the core contracts, releasing gravitational potential energy to keep the star shining.

As a prelude to what is coming next, Fig. 2.4 summarizes what single, or essentially single, stars do in their ZAMS and later lives as a function of a series of "mass cuts" (a term you probably won't see in the literature, but we like it). Thus, stars of about 0.85 $\mathcal{M}_\odot$, or less, take a time of about the

---

[5] You might wish to use some of the computer codes on the CD-ROM to check on what we say here.

present age of the universe ($\tau_{\rm hubble}$) to evolve off the main sequence. Stars of initial mass of 1.5 $\mathcal{M}_\odot$, or less, use the pp-chains to burn hydrogen, whereas more massive stars use the CNO cycles to the same end. And so on. All will become clear, we hope, as this chapter continues. The masses at the mass cuts do depend on factors such as metallicity, which causes most mass cuts to increase as metals increase. Mass loss ($\dot{\mathcal{M}}$) has the same effect. If the star is in a close binary system, then the story can change dramatically. Note also that we have not included the energy releasing effects of gravitational contraction, which introduces a bunch of stuff at various stages that would require a very large, separate, diagram. In any case, consider this figure a sort of "crib sheet" to be consulted as we go along.

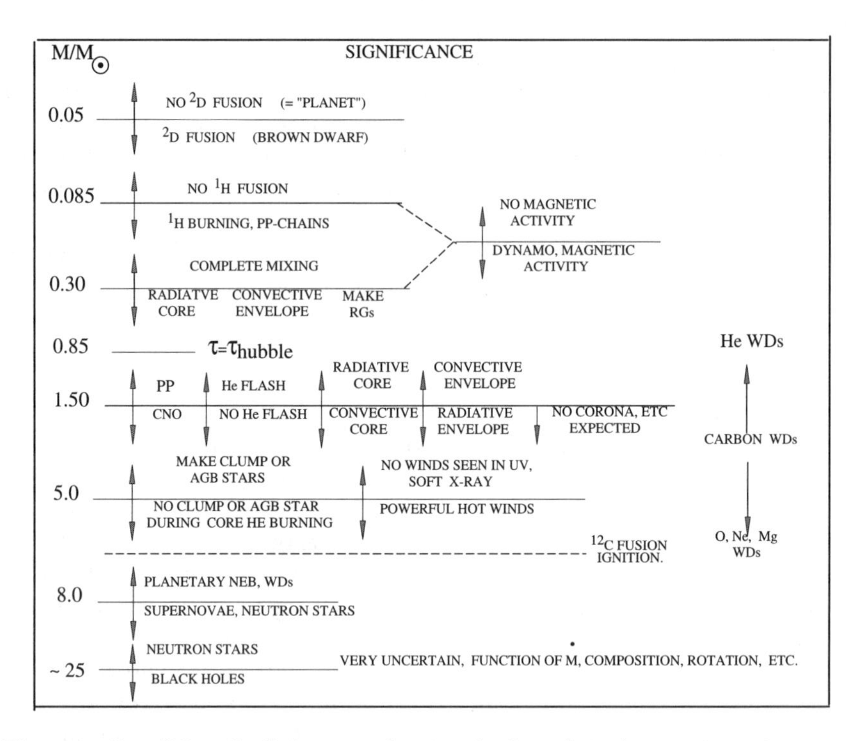

Fig. 2.4. Our "Mass Cut" diagram showing the fate of single stars in various mass classes. See text.

Some distinctions by mass again need to be made and we shall use the evolutionary tracks from Iben (1967) shown in Fig. 2.5 and the time scales listed in Table 2.1 as a guide. Stars of more than 1.5 $\mathcal{M}_\odot$ go immediately to the phase described in the next section, whereas those of more than about 5 $\mathcal{M}_\odot$ evolve without changing luminosity very much. It is not a coincidence that this 1.5 $\mathcal{M}_\odot$ cut is the same as the dividing line between pp-chain and CNO cycle energy generation on the ZAMS. Larger mass $\Longrightarrow$ higher temperatures $\Longrightarrow$ nuclear reactions occur more readily. We have never been able to decide whether agreement in mass cut between the nuclear issues and the mode of energy transport in the envelope is a coincidence or not.

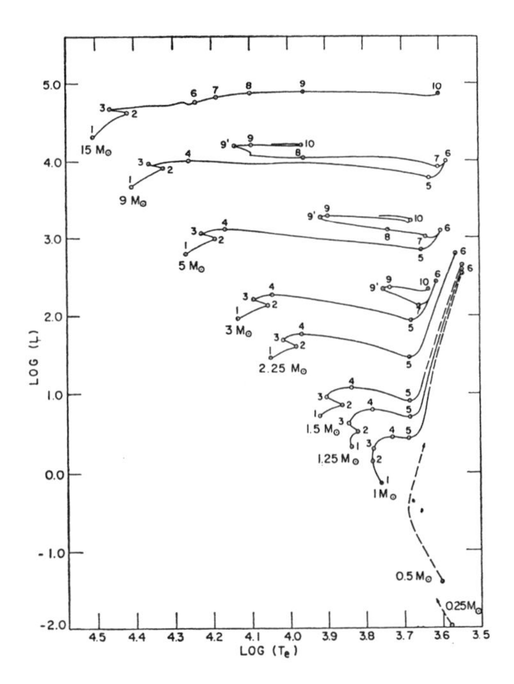

Fig. 2.5. Representative theoretical evolutionary tracks for stars of different masses. Reproduced, with permission, from Iben (1967), Annual Review of Astronomy and Astrophysics, Vol. 5, ©1967 by Annual Reviews. Luminosity is in units of $\mathcal{L}_\odot$. The stages 1, $\cdots$,5 are referred to in the text and Table 2.1.

Stars of less than 1.5 $\mathcal{M}_\odot$ require some heating of the hydrogen outside their inert helium cores to reach stage 4 in Fig. 2.5 and reacquire a nuclear energy source. Thus their cores contract rapidly. Their outer layers simultaneously expand, soaking up more energy. The result is change in structure on the thermal or Kelvin–Helmholtz time scale (see §1.3.2), so that we catch very few stars doing this. In a HR diagram for the members of a cluster of not more than 1,000 or so stars, therefore, there will be a gap, occupied by at most one or two stars. It is called the *Hertzsprung gap.* The "one in a thousand" factor is roughly the same as the ratio of thermal to nuclear time scales. (Compare Eqs. 1.32 and 1.90 for a star like the sun.) Globular clusters have so many stars (10^4–10^5) that there is no actual gap, just relative sparsity.

Table 2.1. Stellar Lifetimes Where (i)–(i+1) Is Interval (in yr)

$\mathcal{M}/\mathcal{M}_\odot$	(1)–(2)	(2)–(3)	(3)–(4)	(4)–(5)
9.00	2.14×10^7	6.05×10^5	9.11×10^4	1.48×10^5
5.00	6.55×10^7	2.17×10^6	1.37×10^6	7.53×10^5
3.00	2.21×10^8	1.04×10^7	1.03×10^7	4.51×10^6
2.25	4.80×10^8	1.65×10^7	3.70×10^7	1.31×10^7
1.50	1.55×10^9	8.10×10^7	3.49×10^8	1.05×10^8
1.25	2.80×10^9	1.82×10^8	1.05×10^9	1.46×10^8

2.3.1 Cluster HR Diagrams

As we are talking about HR diagrams for clusters, it is time we showed some examples. Much of what we know of stellar evolution derives from observation of these usually closely knit groups of stars. Since, as is usually assumed, the stars in a cluster form at very nearly the same time, their locations on the HR diagram represent a snapshot taken at the present moment of where the stars have gotten to since the cluster's formation, the primary determinant being the initial stellar mass on the ZAMS.

The first example is shown in Fig. 2.6, where the HR diagrams of two "young" *open* clusters are superimposed. The dots are for the stars in the Pleiades (the most prominent open cluster in the northern winter sky, also known as M45, C0355+239, the latter an International Astronomical Union number giving Right Ascension and Declination, etc.) and M67 (aka NGC 2682, C0847+120, etc.) gets the triangles. M67 is much further away from us than the Pleiades (some 720 versus 125 pc) so we have moved M67 upward so the main sequences of the two clusters coincide.[6]

There are problems with cluster diagrams such as these. For example, do all the stars shown (all 652 of them) really belong to the cluster, or are they stars that happen to be in the same field? Have the effects of interstellar absorption by dust and gas been properly taken into account? For example, if M67 is at 720 pc how has this affected the V magnitude (thus dimming the stars), and has B–V been altered because absorption affects different spectral bands in different ways (altering what we guess to be T_{eff})? Answering questions like these is beyond the scope of this text.

Figure 2.7 shows the HR diagram for the globular cluster M3 (NGC 5272, C1339+286), which is a popular object for amateur (and professional) astronomers. There are estimated to be some 3×10^5 stars in M3. The figure shows only a little over 10,000 because of the difficulty in obtaining UBV photometry in a such a crowded and faint stellar field. The radius of M3 is only about 25 pc, giving an average stellar density of some 20 stars pc^{-3},

[6] If the initial abundances of the clusters were the same this would be called "main sequence fitting," a technique used to find the relative distances of clusters. In this case the abundances are close enough for our purposes.

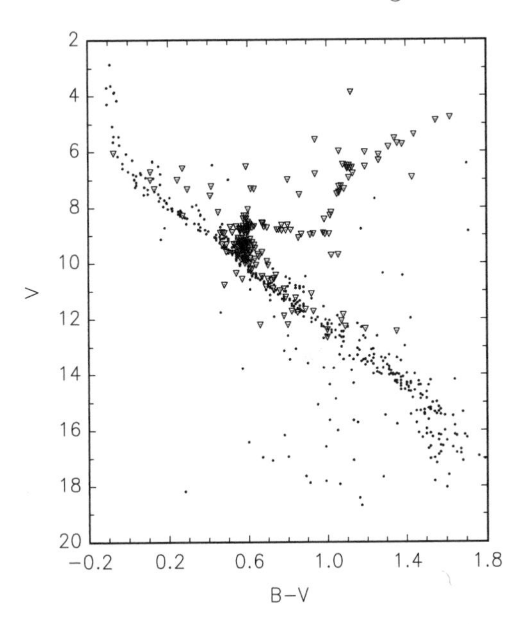

Fig. 2.6. Shown are the HR diagrams of the open clusters The Pleiades and M67. The data for the figure were downloaded from the WWW (see `ftp://cdsarc.u-strasbg.fr/cats/II/124A`) and is based on Mermilliod (1986) supplemented by Mermilliod and Bretschi (1997) for the Pleiades. The Strasbourg catalogs may be queried from `http://vizier.u-strasbg.fr/viz-bin/VizieR`.

which is more than 20 times as crowded as the solar neighborhood. The central regions of M3 are even denser. (For SciFi fans, Isaac Asimov's *Nightfall* tells it all.) The *turnoff point* (labeled "TOP" here but "TO" is frequently used) from the main sequence (MS) for M3 seems to be easy to find but to pin it down to a "point" is impossible. Thus, converting a color at the TOP to a $T_{\rm eff}$ for metal poor ZAMS stars, then to a luminosity, and finally to a lifetime on the ZAMS and an age for the cluster (see later), is an enterprise for optimists (but you should try Ex. 2.8, where you are to estimate the age of M67). This is an important point because stars in globular clusters (along with most "halo" stars permeating and surrounding our galaxy in a roughly spherical halo) are the oldest known surviving stars with large populations in our galaxy. Thus pinning down the age of globular clusters is tantamount to finding a lower limit to the age of the galaxy and, by hopeful extension, the ages of other galaxies in the universe.

Besides the turnoff point, there are other indicators that are used to determine ages of globular clusters, although we shall not go into any real detail at this point. These indicators include the combination of the location of the tip of the *red giant branch* (RGB in the figure) and the *asymptotic giant branch* (AGB), both of which represent advanced evolutionary stages. RGB stars are burning hydrogen in a shell surrounding a dormant helium core af-

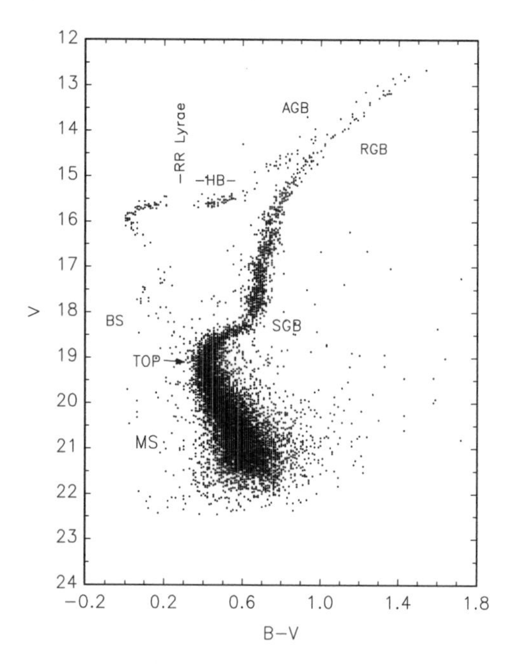

Fig. 2.7. The HR diagram of the globular cluster M3. The data for the figure were downloaded from the VizieR WWW site (see Fig. 2.6) and originally were reported in Buonanno et al. (1994).

ter having left the *subgiant branch* (SGB), whereas AGB stars are evolving rapidly while burning helium deep inside. The *horizontal branch* (HB) consists of stars evolving more leisurely while burning core helium. Its height above the ZAMS is an age indicator. The strange gap to the left of the HB is the home of the RR Lyrae variable stars: they are there but their color and luminosity vary periodically. They are standard candles that can be used to determine distances to a cluster. The figure also shows the location of the enigmatic *blue stragglers* (as "BS") that seem never to have left the ZAMS. See Ex. 2.5 for one guess at their history.

Cluster and Galactics Ages

So, how well are globular cluster ages known? At the time of the first edition of this text, in 1994, ages in the range 13–18 Gyr were the norm for clusters in our galaxy. The trouble, though, is that this range places the time of formation of globular clusters before the Big Bang if generally accepted (but not by all) values of the Hubble constant are used. (An excellent semi-popular review titled *Cosmology in the New Millenium* by Freedman & Turner may be found in the October 2003 issue of Sky & Telescope, p. 30. They give a value for the Hubble constant of $H_0 = 72^{+4}_{-3}$ km s^{-1} Mpc^{-1} corresponding to a time of 13.7 ± 0.2 Gyr since the Big Bang.) This embarrassing situation seems now to be resolved: the ages of the clusters are now in the range 10–13 Gyr, which now has them being formed at a reasonable time after the Big Bang.[7] Is this magic? No, a combination of factors have conspired to reduce the ages by a few Gyr. (For reviews see, e.g., Lebreton, 2000, and, in more popular form, Chaboyer, 2001, and the 3 Jan 2003 edition of *Science* starting on p. 59.)

Firstly, newer observations with larger telescopes using greatly improved instrumentation coupled with measurements by the astrometric satellite Hipparcos (see §1.6) put the globular clusters perhaps 10% further from us than previously thought. This means that they are more luminous than supposed. More luminous means the stars evolve faster and, hence, the clusters are younger. But stellar evolution studies must be consistent with these new ages. As is often the case in this business, a prod in the right direction gets results. A combination of realistic adjustments in microphysics (equations of state and opacities) and stellar atomosphere calculations (and this all includes better determination of stellar surface abundances, which, especially for helium, is not easy) now give consistent ages. If a third edition of this text ever appears, we hope we will not have to rewrite this story yet again.

Why do the HR diagrams for young and old clusters look the way they do? The following sections will give more details of the story, but imagine a cluster consisting of many stars all formed on the ZAMS at the same moment. As time goes on, the more massive stars will evolve off the ZAMS first, followed by successively less massive members. If we take snapshots of the cluster as time progresses and superimpose these on the same HR diagram, we get something that should look like Fig. 2.8 if the evolution is not carried on to the very later stages (or ignored for now to simplify the diagram). Notice how the stars have peeled off the main sequence, with each time frame (0.2–20 Gyr) having its own turnoff point. These are called *isochrones* and they do demonstrate, in effect, the evolution of HR diagrams for a sample cluster.

2.3.2 Mass Loss From Massive Stars

Before we truly leave the main sequence, we must mention the early evolution of very massive stars. Not many are made in the course of star formation, but their high luminosities have a profound effect on the formation of their nearby, less massive, siblings. The early course of evolution of massive stars is governed not only by the transmutation of hydrogen into helium, but also by how fast mass is driven from their surfaces by radiation pressure acting on strong spectral lines. This topic is worthy of a text by itself (as are many

[7] Note that not all globular clusters are old. Galactic collisions may induce intense star formation, including the formation of globulars, and this is going on even as we speak. See the article by S.E. Zepf and K.M. Ashman in SciAm, October, 2003, p. 46.

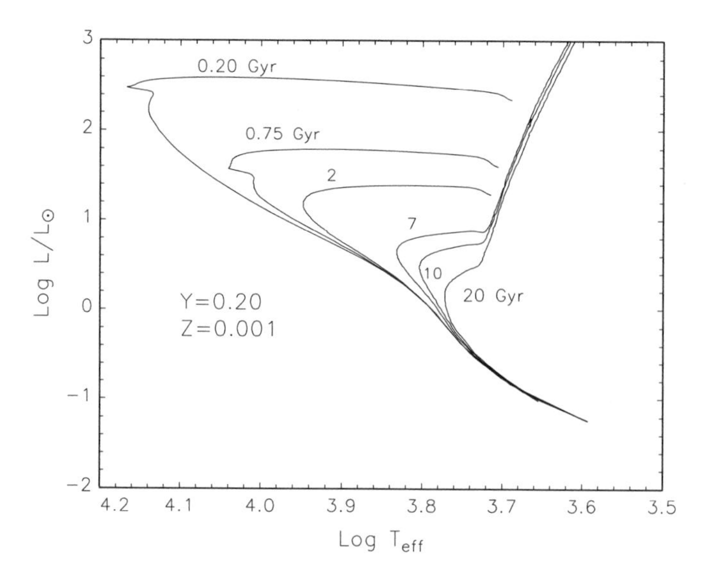

Fig. 2.8. Representative theoretical isochrones for Pop II clusters of the indicated ages. Adapted from the *Revised Yale Isochrones*, Green E.M., Demarque, P., and King, C.R., 1987, *The Revised Yale Isochrones and Luminosity Functions* (New Haven, CT).

other topics only mentioned here) and we recommend the text by Lamers and Cassinelli (1999), and, for a review, Kudritzki and Puls (2000).

The rate at which mass is driven off, $\dot{\mathcal{M}} = d\mathcal{M}/dt$ (pronounced "Mdot"), depends primarily on luminosity, the escape velocity at the stellar surface, and the metal abundance Z. (And, of course, all but the last of these depend on mass.) There is an excellent correlation between $\dot{\mathcal{M}} v_\infty \mathcal{R}^{1/2}$ and $\mathcal{L}$ for (roughly) fixed Z. Here v_∞ is proportional to the escape velocity at the stellar surface. This correlation is shown in Fig. 2.9, adapted from Lamers and Cassinelli (1999).

An application of Fig. 2.9, derived from further material in Lamers and Cassinelli (1999), is shown in Fig. 2.10. Although we have taken some liberties with their analysis to simplify matters, the figure gives the flavor of what happens (but don't use it for a robust meal). Also shown is the ZAMS for $Z = 0.02$ in mass–luminosity space; that is, choose a mass and where the dashed line intersects a line for luminosity (only a small number are shown), that is the luminosity for that mass. Note that the ZAMS result represents the newborn star with mass loss having been neglected. (You may wish to attempt Ex. 2.13 to see where Fig. 2.10 comes from.)

As an example of what this figure implies, suppose we consider an 80 $\mathcal{M}_\odot$ star on the ZAMS. Its luminosity is about 10^6 $\mathcal{L}_\odot$. At that luminosity and mass, the mass loss rate is a staggering $\dot{\mathcal{M}} \approx 2 \times 10^{-5}$ $\mathcal{M}_\odot$ yr^{-1}. From this we can compute a mass–loss timescale of $t_{\rm ML} = \mathcal{M}/\dot{\mathcal{M}}$ of about 4×10^6 years;

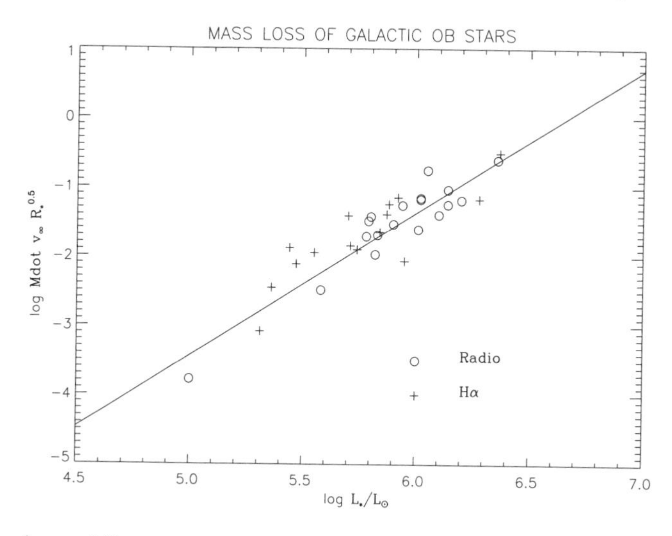

Fig. 2.9. $\dot{\mathcal{M}} v_\infty \mathcal{R}^{1/2}$ is shown plotted against $\mathcal{L}/\mathcal{L}_\odot$ for spectral class O and B stars in our galaxy from radio and Hα observations. Here $\dot{\mathcal{M}}$ is in $\mathcal{M}_\odot$ yr^{-1}, v_∞ in km s^{-1}, and $\mathcal{R}$ in solar units. See Ex. 2.13 for more on what this figure implies. We thank Henny Lamers and Joe Cassinelli for providing the PostScript file from which this figure was made. From Lamers and Cassinelli (1999) and reprinted with the permission of Cambridge University Press.

that is, an e-folding time of only 4×10^6 years with that $\dot{\mathcal{M}}$ would effectively evaporate almost half of the star. Compare this to our naive ZAMS lifetime estimate ($t_{\rm nuc}$) of Eq. (1.91), which yields 2×10^5 years. In other words, $t_{\rm ML}/t_{\rm nuc}$ is near 20. This would seem to imply that the effect of mass loss is minor. However, we must remember that any reduction in mass means a decrease in central temperature and, hence, for the CNO cycles, a more dramatic decrease in energy generation, and so on. The conclusion is that any self-respecting evolutionary calculation for massive stars must include mass loss.

There is good observational evidence for mass loss—and not only direct observation of winds—because in the stellar zoo we find the *Wolf–Rayet* stars, which are luminous stars characterized by strong emission lines that dominate the optical spectrum. They are near the main sequence but their spectra indicate a strong deficiency or absence of hydrogen, and there is also evidence that hydrogen and/or helium burning products have either been brought to their surfaces, or that mass loss has exposed deeper layers. Even the categories assigned to them ("WC," with lots of carbon in their spectra, or "WN" for nitrogen) tell the story. Because of vigorous mass loss, these massive stars are highly evolved but, in some sense, young.

An example of the (theoretical) effects of mass loss is shown in the HR diagram of Fig. 2.11 for a 40 $\mathcal{M}_\odot$ star on or very near the ZAMS subject

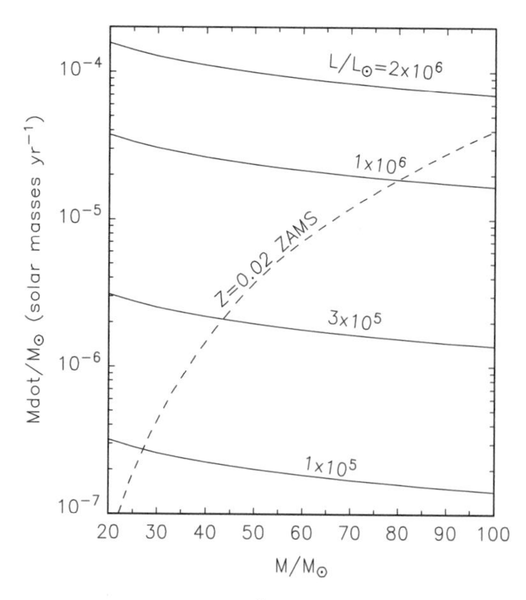

Fig. 2.10. Shown are mass loss rates, $\dot{\mathcal{M}}$, as a function of mass for massive stars, along with a ZAMS. See text for discussion.

to varying rates of loss. The rate is parameterized by the multiplier N as $\dot{\mathcal{M}} = N\mathcal{L}/c^2$ (cgs units). At the start on the ZAMS, N should be about 75 using the results of Fig. 2.10 (but N may change as mass is lost). Also shown (in the insert) is how mass changes with time. The last points shown correspond to the time the star can be said to have left the main sequence. In any case, the naive no–mass–loss case ($N = 0$) differs appreciably from what happens when mass loss is included.

2.4 Red Giants and Supergiants

For all stars that have made it this far, the hydrogen just outside the built-up helium core is soon hot enough for fusion to continue ($\mathcal{M} \gtrsim 1.5\mathcal{M}_\odot$) or resume ($\mathcal{M} \lesssim 1.5\mathcal{M}_\odot$). This "hydrogen shell burning" always occurs via the CNO cycle. We make a point of this because CNO hydrogen burning is the main (perhaps only) source of nitrogen in the universe, so without it you would not be able to eat a high protein diet. Energy production in the thin, hot shell also drives some convection. We do not expect actual mixing to the surface yet. Nevertheless, particularly among globular cluster members, stars not very far up the red giant branch (RGB, which nobody tries to pronounce as a word, and see, e.g., stage 5–6 in Fig. 2.5) often have surfaces somewhat enriched in nitrogen and/or carbon. The code phrase is "meridional circulation" (meaning gas flows that head north and south as well as up and

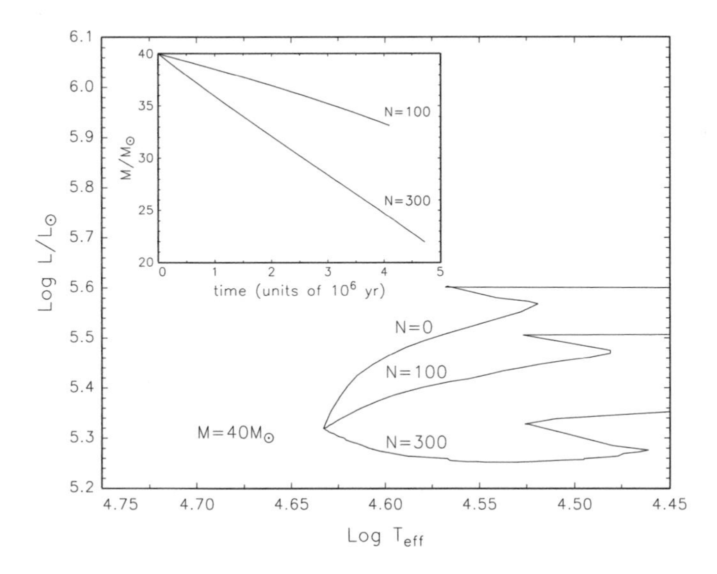

Fig. 2.11. This shows the effect of differing rates of mass loss on main sequence evolutionary tracks for a star with initial mass of 40 $\mathcal{M}_{\odot}$. The rate is parameterized by N where $N = 0$ means no mass lost. (See text and Lamers and Cassinelli, 1999, discussion for their Fig. 13.1.) The insert shows stellar mass as a function of time.

down). A one-dimensional stellar model cannot reveal whether mixing of this sort is expected. More complex models say that it probably is, though still not so much as we see perhaps.

Do we understand why stars become red giants, that is, why the envelope expands and the core contracts (and conversely a few stages downstream)? Well, everybody who has solved the differential equations for stars that have helium cores, or followed evolving stars as such cores develop, have found bright stars with extended envelopes. Thus red giants are implied by the underlying physics that we think we understand.

Indeed even in analytical form, the equations can be juggled to show that, if mean molecular weight drops sharply between a core and an envelope, the ratio of core to envelope density will be large. No one, however, has found a set of words that answers the question "Why do stars become red giants?" in a way that satisfies most of the community. So, rather than our attempting to do this, we refer you to Sugimoto and Fujimoto (2000), and the many difficult references therein.[8]

[8] If you try Ex. 2.22, you will get an idea that something must happen if the inert helium core gets too massive. That exercise discusses the "Chandrasekhar–Schönberg limit," which places limits on how much mass can be built up in an isothermal helium core before the core can no longer support the overlying envelope and must thus contract and heat up.

The red giant phase lasts, on average, about 10% as long as the main sequence phase, because there is a comparable amount of hydrogen fuel available (about 10% of the mass of the star) and the stars are ten times as bright. Indeed, the red giants, and supergiants, were recognized precisely because they are so much brighter than main sequence stars of the same surface temperature. This means they must be bigger as well, and, just as on the main sequence, there are correlations among mass, luminosity, size, and lifetimes that we understand. Larger mass means brighter, bigger, shorter lived, and hotter but less dense at the center. The hydrogen burning shell works its way out through the star, so that the mass of inert helium gradually increases. Both central density and central temperature (ρ_c and T_c) also increase. They are somehow in a race to reach §2.5. This "race" is summarized below (adapted from Kippenhahn and Weigert, 1990); that is, the destiny of the cores of most stars consists of nuclear burning, followed by fuel exhaustion, a phase of core contraction and heating, and then the ignition of built up fuel, and so on.

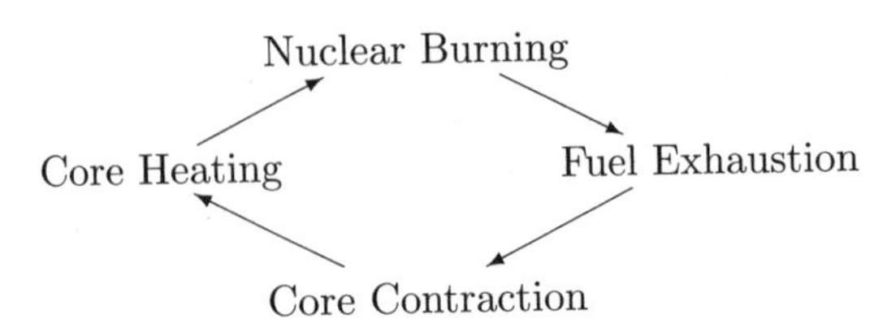

The sun will, of course, become a red giant in due course. Its surface will probably reach beyond the present orbit of Venus, but not as far as the earth (though some supergiants have radii of 1 AU and more). Other things being equal, the equilibrium surface temperature of the earth will increase in proportion as

$$\frac{T(\text{earth, now})}{T(\text{earth, future})} = \left[\frac{\mathcal{L}(\text{sun, now})}{\mathcal{L}(\text{sun, future})}\right]^{1/4}$$

and your first thought may be, oh, well, if it's only the 1/4 power, what difference does it make? But $10^{1/4} = 1.78$ and 1.78×290 K (the present average earth surface temperature) is 517 K, approaching the molten lead and sulfur regimes.

2.5 Helium Flash or Fizzle

What happens next depends on the outcome of the race between central temperature to increase from about 10^7 K to about 10^8 K and central density to increase from its main sequence value of $10^{2\pm1}$ to 10^6 g cm^{-3}. If density wins, the helium gas becomes degenerate (see §3.5), with pressure support provided by the electron momentum that is locked in by the uncertainty relation $\Delta x \, \Delta p = \hbar$. Such gas can contract no further and can heat no further. Some envelope may leave in a wind, but the core will be left as a helium white dwarf. Stars of initial mass less than about 0.4 $\mathcal{M}_\odot$ will meet this fate. (Remember that small mass goes with low T_c but high ρ_c.) The universe is not yet old enough for this to have happened to any star left to itself. We do, however, find helium white dwarfs as members of binary star pairs, where rapid mass transfer to the companion has stripped an initially more massive star down to a helium core of less than 0.4 $\mathcal{M}_\odot$. In globular clusters, the binary may be later disrupted by close encounters with other stars, so that we find single helium white dwarfs in some of the clusters as well.

The other extreme is that T_c reaches 10^8 K, while ρ_c is still considerably less than 10^6 g cm^{-3}. This happens in stars of more than about 1.5 $\mathcal{M}_\odot$ which, therefore, experience peaceful helium ignition and do not change their structure rapidly at this point in their lives, although the exact mass cut for this does depend on initial composition. Indeed, the more massive the star, the less it is shaken up by sequential nuclear fusion episodes (aside from explosions later on!), and those of more than about 3 $\mathcal{M}_\odot$ do not even get enormously brighter upon leaving the main sequence, as may be seen in Fig. 2.5, where the tip of the RGB at point 6 is not very much more luminous than point 1 on the ZAMS. Compare this to the situation for 1 $\mathcal{M}_\odot$.

Notice that our sun belongs to the intermediate type regime between 0.4 and 1.5 $\mathcal{M}_\odot$. (The mass of the sun is very precisely one, at least in these units, in case we haven't told you enough times before.) Such stars, which also include those now leaving the main sequence in globular clusters, ignite helium while the fuel is partly degenerate, that is, T_c reaches 10^8 K, the minimum temperature for barrier penetration to allow helium nuclei to cuddle up, when ρ_c is close to 10^6 g cm^{-3}. "Helium flash" is the phrase used to describe the resulting nuclear explosion.

Why an explosion? Well, in a gas where $P = nkT$, if you heat it up a bit, the pressure goes up, the gas expands and all is well. But, if P is a function only of density (or, nearly only) as it is in degenerate materials, ignition of a new nuclear fuel will heat the gas, and it will not expand. The increased temperature, however, makes the reaction go faster, which further heats the gas, which makes the reaction go faster, which $\cdots$. No, this does not go on forever. But it does continue until thermal pressure exceeds degenerate pressure, at which time the gas "notices" that it is wildly out of equilibrium and expands with vigor. Exactly how vigorous is this expansion is still a matter of some controversy. It certainly is not enough to disrupt the star because we see stars that have (peacefully?) survived the event. However, the usual evolutionary calculations sidestep possible hydrodynamical consequences by enforcing mechanical equilibrium so questions remain. We have also sidestepped the issue of exactly where the flash starts. In many calculations it does not take off at stellar center but a bit off-center. This is due to the emission of neutrinos, which are lost to the star and carry off energy, thus depressing the central temperature somewhat.

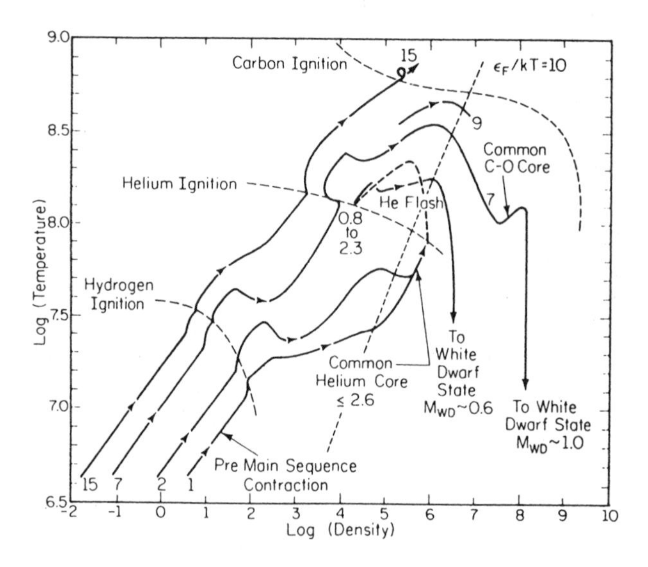

Fig. 2.12. Central density versus central temperature for evolving stellar models. Reproduced, with permission, from Iben (1985).

In the helium flash case, the central density of the star is reduced to about 10^3–10^4 g cm^{-3} and the core expansion is mirrored by envelope contraction (the inverse of "Why does a star become a red giant?"). This is reflected in a rapid decrease in stellar luminosity and increase of effective temperature and the star heads to the left in the HR diagram. By the way, other nuclear fuels may be ignited when they are degenerate, or partly so, and will also explode. This happens to hydrogen of the surfaces of white dwarfs (nova explosions), and to carbon and oxygen at the cores of white dwarfs that are driven above a critical mass limit.

A summary of what we have talked about appears in Fig. 2.12, which shows the evolution of central density and temperature for three stellar masses. Note that the helium flash takes place shortly after lower mass stars cross the dashed line labeled $\epsilon_F/kT = 10$. This is the "degeneracy boundary" to the right of which matter is degenerate (see §3.5.3 and Fig. 3.7 for the boundary at a different composition). The line "Helium Ignition" is self-explanatory, but it does dip to lower temperatures as density increases. (Higher densities mean the helium nuclei are closer together, although Chapter 6 has more to the story.) Note, in this figure, a star of 2 $\mathcal{M}_\odot$ does suffer the helium flash contrary to our mass cut of 1.5 $\mathcal{M}_\odot$. This is partially due to a higher metallicity than we are thinking of, but it also shows the fuzziness of some of the mass cuts, which, in part, depends on who is doing the calculating (and in what year).

An example of a helium flash calculation is shown in Fig. 2.13, where central density is plotted versus elapsed time for a 1 $\mathcal{M}_\odot$, $Z = 0.02$ model (which could just as well be the sun in a few billion years). Note the acceleration of core evolution as the RGB is climbed, then the flash, followed by core expansion. The label "1/2 & HB" refers to the time when helium is half-exhausted in the center of the core and the star is on the horizontal branch (as in Fig. 2.7, although stars in M3 are Pop II rather than Pop I). The density then increases again as helium is completely used up in the central core but the central temperature is not high enough to ignite the C–O mixture produced by the burning of helium. The AGB will be discussed shortly, but the ultimate fate of the star will likely be that shown in Fig. 2.12; that is, it will end up as a white dwarf of about 0.6 $\mathcal{M}_\odot$—meaning, as we will discuss later, almost half of the mass of the original star will have to be lost.

All stars that succeed in igniting helium, peacefully or explosively, are entitled to go on to §2.6.

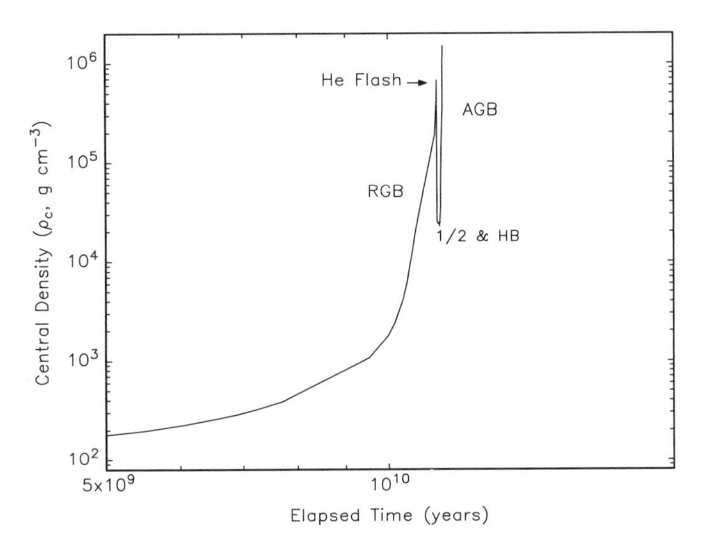

Fig. 2.13. Shown is the central density evolution of a 1 $\mathcal{M}_\odot$ model ($Z = 0.02$) from Charbonel et al. (1996). Time is measured from the ZAMS. The data for the figure were downloaded from the VizieR WWW site (see Fig. 2.6).

2.5.1 Helium Core Burning, Clumps, & Horizontal Branches

Because there is no bound nucleus containing eight nucleons, helium fusion can occur only when it is both hot enough for two helium nuclei to crawl through their mutual coulomb barrier to fuse together and dense enough for a third helium to come along in the 10^{-16} s that the unbound "di-alpha" (aka ^{8}Be) holds together (see §6.5 for the "triple-α" reaction, which we are describing). At only slightly higher temperatures, the product ^{12}C can capture another helium nucleus to make ^{16}O, so that the reactions

$$3\,^4\text{He} \Longrightarrow {}^{12}\text{C} \quad \text{and} \quad {}^{12}\text{C}(^4\text{He},\gamma)^{16}\text{O}$$

in effect compete with one another until all the helium is gone. The balance (which is very sensitive to the excited level structures of the two product nuclei) yields more carbon in stars of the smallest mass that fuse helium at all and more oxygen in the more massive stars. The nucleus ^{20}Ne has no appropriate excited level into which the product of $^{16}\text{O}(^4\text{He},\gamma)^{20}\text{Ne}$ can land, so helium fusion effectively stops with carbon and oxygen; that is, the reaction is very slow unless temperatures are very high.

Helium-burning stars have an additional energy source, because hydrogen burning (by the CNO cycle) continues in a thin shell around the helium core. The energy available from helium fusion is, however, considerably less than from hydrogen fusion and the stars are brighter than they were on the ZAMS, so that this phase has a lifetime of a few percent of the main sequence lifetime of the same star.

Stars that are not badly shaken up by helium ignition continue on their way through the red (super) giant part of the HR diagram, though now or later they may make loops back to bluer colors (higher surface temperatures), when a burning shell works its way out to a radius in the star where there is a composition discontinuity. Such a discontinuity will occur if a massive star, with a convective core, has that core shrink as it ages. The progenitor of SN1987A (the first supernova observed in 1987) had made such a loop and so exploded while blue rather than red.

Less massive stars that were shaken up by helium ignition find themselves with colors and effective temperatures characteristic of horizontal branch stars if their metal abundances are less than about 10% that of the sun. Higher metallicity, low mass, helium core burners look redder because opacity rises with Z (as discussed in full in §4.5). Thus they nestle up against the red giant branch in HR diagrams for old, open clusters, making a clump of dots because they are all about the same brightness. Hence the charming name "clump stars" or "red clump stars." In case we forget to tell you, this is also the reason that subdwarfs (main sequence stars of low metallicity) are bluer than ordinary main sequence stars. The color difference is fairly easy to observe, but models that take you between observed colors and calculated effective temperatures are less than perfect.

Helium Core Exhaustion

You have heard this story before. A time comes when the central fuel supply is exhausted and the star must again readjust its structure, because energy is flowing outward, but none is being liberated by nuclear reactions at the center. The adjustment is, however, less drastic than at the end of hydrogen core burning, because the thin shell of CNO cycle continues to supply some energy.

2.5.2 Double Shell Burning Phase or Asymptotic Giant Branch

Meanwhile, the center of the star continues to contract, getting hotter; hydrogen fusion continues in a thin shell between core and envelope; and, soon, the helium just outside the carbon–oxygen core is hot enough for helium fusion to continue there. The phase is, therefore, sometimes called "double shell burning" (as in Fig. 2.14). As in the case of the red giants of small and moderate mass, the contracting core and composition (molecular weight) discontinuity cause the envelope to expand and cool again. Thus the phase is also called AGB or "asymptotic giant branch," meaning that, in an HR diagram, either an evolutionary track or the points representing individual stars stretch into a sort of asymptote to the main RGB. If you use real observers' data to check this, you will discover that the two branches can be separated only if you have good, modern, photoelectric colors for the stars. Thus older texts (or ones whose authors have borrowed figures dating back to the era of photographic photometry) will show a single, broad red giant branch. An example of a well-defined separation of the two branches can be seen in Fig. 2.7 for the globular cluster M3.

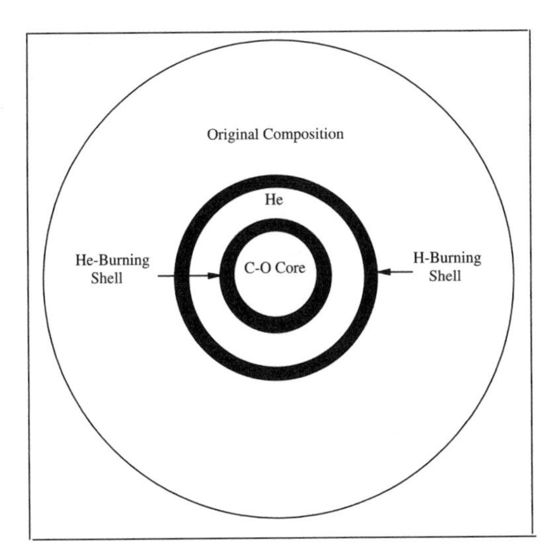

Fig. 2.14. Double shell burning. Not to scale (by quite a bit) because the outer envelope is huge, while shells and core are much smaller in real life.

Because the stars are now quite bright, this phase will last less than 1% of the main sequence lifetime. You can check the relative lengths of lifetimes by, for instance, counting all the AGB, horizontal branch, and red giant stars in a globular cluster and comparing them to the number of stars still on the main sequence. Yes, the numbers fit. Many or most stars with double shells are not only very bright but also unstable to pulsation. The combination results in a wind blowing off the surface. The wind speed is comparable with the escape velocity from the surface, 10–30 km s^{-1} for instance (this is not a coincidence), and the wind density is large enough that the star is likely to lose 10–50% of its mass before something else happens to stop the wind. Toward the end, the density gets even larger (up to 10^{-4} $\mathcal{M}_\odot$ yr^{-1} in extreme cases) and may be called a "superwind." This helps explain the seeming discrepancy between the 1 $\mathcal{M}_\odot$ initial mass of the star in Fig. 2.12 versus its final mass of about 0.6 $\mathcal{M}_\odot$ as a white dwarf. A recent review of winds from cool, but luminous, stars is due to Willson (2000).

The double shell phase is the first in which we definitely expect the surface of the star to show evidence of the nuclear reactions occurring inside. The reason is that each shell drives a convection zone just above itself—remember convection occurs if you (well, all right, the nuclear reactions, not you personally) dump too much energy into a small region and thus create a steep temperature gradient. As a result, the shell sources flash on and off (on the dynamical time scale) and chase each other back and forth in radius, with the outer one sometimes making contact with the convective envelope (which is, of course, caused by the large opacity of neutral hydrogen). Each zone in turn brings up processed material from below and leaves some where the next zone can pick it up and carry it on out. But the star is never fully convective, hence the continuing onion-like structure shown in Fig. 2.14.

What fusion products do you expect to see? Make a mental list before going on.

a. Helium, you said, from pp-chain and/or CNO cycles fusion. True, but helium atoms are so hard to excite that they introduce spectral lines only in a very hot gas. (We will find the extra helium later, though.)
b. Carbon, you said, from the triple–α reaction. True, quite often, and seen in the form of molecular bands due to C_2 and CN.
c. Nitrogen, which you might have forgotten about, because the CNO cycle, no matter what mix of CNO it starts with, leaves most of the catalyst as nitrogen (because $^{14}N(p,\gamma)^{15}O$ is the slowest capture reaction in the cycle). Indeed, one expects, and finds, that AGB stars have turned a good deal of their oxygen into nitrogen, and strong CN features are also common.
d. And, finally, one you would never have guessed. During the double shell burning phase, the relatively abundant atoms of iron are gradually converted to heavier elements, and those too are available for mixing. Thus AGB stars are distinguished by strong features due to barium, yttrium,

and other elements that you may not include in your favorites. Some even show absorption lines produced by atoms of technetium. Tc has no stable isotopes, only ones with lifetimes of a million years or less. Thus the subset of stars with technetium spectral features constitute direct evidence that stars live on nuclear reactions that are going on right now (by astronomical standards) before our very eyes (or at least spectrographs). (You should peruse the pioneering observations of Tc reported by Merrill, 1952.) You may ask where the "relatively abundant atoms of iron" come from, to say nothing of exotics like yttrium, etc., when, by now, we have only really gotten to carbon and oxygen. Please wait until §2.8.1 and §2.11.2 for the story (unless you can't wait and want to peek now).

The AGB phase, and even this section, eventually ends. But tracing out what happens next requires us to make another mass cut. The precise value depends on stellar metallicity and perhaps rotation and other factors. Calculations indicate it should fall somewhere between 6 and 10 $\mathcal{M}_\odot$, and this is confirmed by observations of star clusters of different ages, where stars of different masses have just left the main sequence. The mass cut is larger in close binaries, where mass transfer removes the envelope when the star tries to become a giant or AGB star.

2.6 Later Phases, Initial Masses $\leq$6–10 $\mathcal{M}_\odot$

Notice first that we talk about *initial mass*, because the wind has been removing envelope material for thousands to millions of years (mass dependent as, always, shorter times but more vigorous loss for more massive stars). We speak of post-AGB stars when enough material has been removed by the superwind, pulsations, and the last few flashes of the helium burning shell that we can see down to hotter (bluer) layers. The ejecta can harbor dust (so that much of the light is reprocessed into infrared before we see it) and OH molecular masers. The entities suffering all this are called OH/IR stars, and it is possible to earn a precarious living by studying them (as is true of every phase mentioned so far, and the ones to come).

As more and more envelope blows off, hotter and hotter layers are uncovered (and the escape velocity and wind speed increase as well). After perhaps 10^4 years, photons are leaving directly from a layer that is at a temperature of 50,000 K or hotter. These photons begin dissociating the molecules and ionizing the atoms in the ejecta. The expanding, ionized ejecta radiate a line spectrum characterized by ordinary hydrogen lines and forbidden lines of O, N, and C, and other species, mostly light and relatively abundant. They looked, through the telescopes of Herschel and other early astronomers, like the disks of Uranus and Neptune and so were called *planetary nebulae* (PN). They have nothing to do with planets (though when our sun does all this the stuff will envelop Earth, Uranus, and Neptune). They look greenish because of strong emission lines of oxygen, while Uranus and Neptune look greenish because of methane in their atmospheres.

After about 10,000 years, the residual core, called a *planetary nebula nucleus* (PPN), has cooled to the point of emitting few ionizing photons and the expanding ejecta have dispersed back into the general interstellar material. The PN is gone, and the PNN remains as a young white dwarf. Simultaneously, the central stellar temperature declines from 10^8 to 10^7 K as the nuclear reactions turn off. (A long review of the evolution to the PNN stage may be found in Iben, 1995.)

Stars of less than 6–10 $\mathcal{M}_\odot$ thus end their lives as carbon/oxygen white dwarfs, because their centers do not get hot enough for any reactions beyond helium fusion by the time their centers are dense enough to be degenerate (remember the temperature–density competition that has occurred before helium burning started).

The core is now officially a C–O white dwarf of 0.55–1.3 $\mathcal{M}_\odot$ (both as measured and as observed). Its only available energy source is the residual heat of the atomic nuclei, and it will cool and fade to about 10^{-5} $\mathcal{L}_\odot$ in about 10^{10} years. The smallest masses have the largest radii but have the smallest amount of heat stored and fade fastest.

There is a narrow mass range (not well defined) in which nuclear reactions can proceed one more stage, leaving a core consisting mostly of O, Ne, and Mg. We see evidence for all three sorts of white dwarf compositions (He, C/O, O/Ne/Mg) when material is torn off their surfaces by nova explosions in binary systems.

2.6.1 A Bit About White Dwarfs

This text has a whole chapter (Chap. 10) devoted to white dwarfs, but since they keep popping up in our discussion, we'll now give some more background on these stars.

Single white dwarfs have an average mass of 0.6 ± 0.1 $\mathcal{M}_\odot$ and radii not too different from the Earth's (about 10^{-2} $\mathcal{R}_\odot$). Doing the arithmetic yields an average density near 10^6 g cm^{-3}. As discussed in §3.5.2, radius decreases as mass increases—a bizarre consequence of their being supported by electron degeneracy pressure. They cool by transporting heat by conduction (as in a metal) through their interiors until, near the surface, photons must laboriously work their way out by diffusion. And, following the usual rules of thermodynamics, the hotter they are, the faster they cool.

Figure 2.15 shows a color–magnitude HR diagram for 782 well-observed white dwarfs. You can detect some tight correlations (like beads on a string) in the figure and this is due, in many instances, to using an empirical relation between color and magnitude: not quite fair, but it doesn't destroy the utility of what's shown. Most of the stars are "DA" white dwarfs, which have hydrogen Balmer lines in their spectra but no sign of helium or metals. There may be some undetected contamination but consider their visible atmospheres to

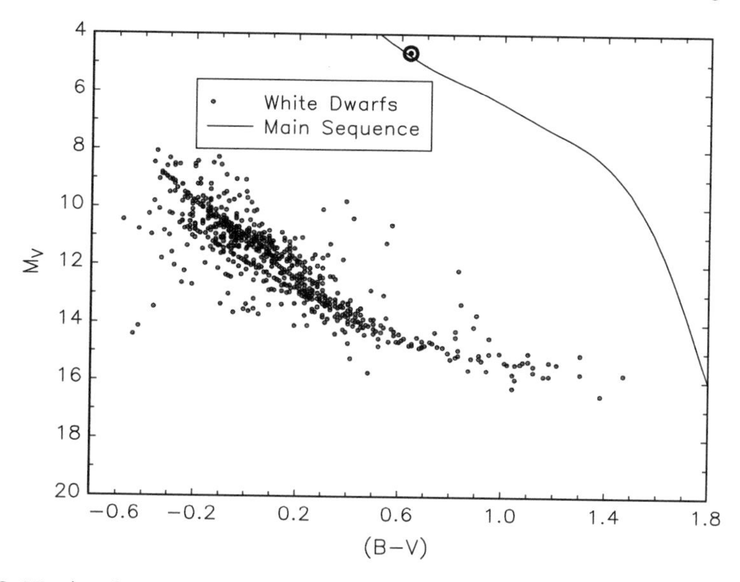

Fig. 2.15. A color–magnitude HR diagram for white dwarfs from data given by McCook and Sion (1999). A ZAMS with the sun is shown to help orient you. An effective temperature of 10,000 K is at B–$V \approx 0$.

be pure hydrogen. Pure helium atmosphere white dwarfs ("DB") are also present, but in smaller numbers. A still not completely understood phenomenon is the *complete* absence of DBs in the effective temperature range from 30,000 K to about 45,000 K. A similar, but cooler, gap in non-DA white dwarfs shows up between about 5,000 K and 6,000 K. It may well be that DAs can change (in their spectra) to DBs, and visa versa. As Fontaine et al. (2001) put it, "It is suspected that a complex interplay between mechanisms such as hydrogen and helium separation (through diffusion) and convective dilution is responsible for the fact that a white dwarf may show different 'faces' during its lifetime."

Following the maxim that hot things cool faster, we infer that cool things cool more slowly. Thus the cooler (larger B–V), and intrinsically dimmer, white dwarfs in Fig. 2.15 should take longer and longer to cool as time goes on. If the universe and our galaxy have a finite age, which appears to be the case, then the very oldest stars formed should have, by now, become the intrinsically dimmest white dwarfs. That is, we expect to see a cutoff in white dwarf luminosity below which there are no dimmer objects. This is indeed the case for white dwarfs near (on a galactic scale) the sun in the disk of our galaxy. Such a sampling is shown in Fig. 2.16 where the space density of white dwarfs drops off precipitously around $10^{-4.4}$ $\mathcal{L}_\odot$. The effective temperature at the drop-off point is near a very cool 4.600 K. If we then do evolutionary calculations from the ZAMS to the cool white dwarf stage, we can then determine the age of the least luminous white dwarfs in the sample; that is,

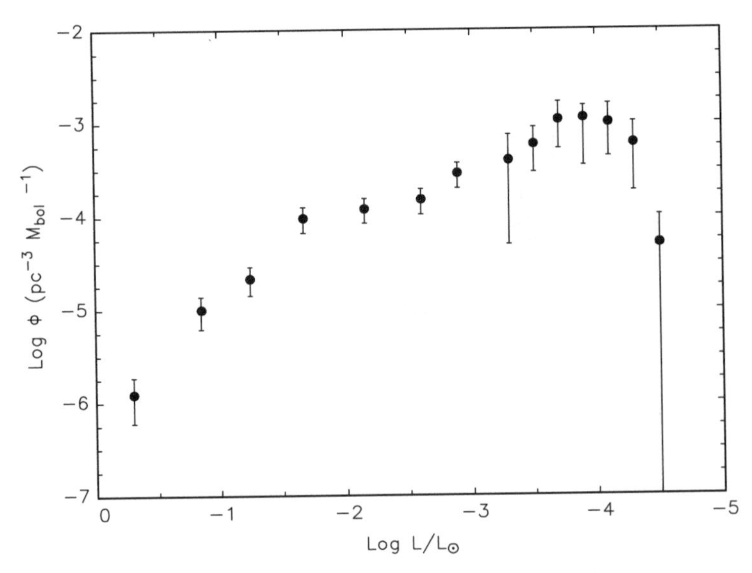

Fig. 2.16. Shown is the Liebert et al. (1988) and Leggett et al. (1998) luminosity function, ϕ, for white dwarfs in the local galactic disk. The units of ϕ are the number of white dwarfs per parsec cubed per unit interval in M_{bol} (as a surrogate for luminosity). The dropoff in luminosity is around $10^{-4.4}$ $\mathcal{L}/\mathcal{L}_{\odot}$, although this figure is fuzzy because of poor statistics at the cool end and possible choices in binning the data (and see Fontaine et al., 2001, for more on this important point). Note also the paucity of white dwarfs at high luminosities. This is to be expected because those stars cool so rapidly.

we can, in effect, measure the age of the local galactic disk. This has been done by several groups and the answer lies in the range 8.5–11 Gyr (see the review by Fontaine et al., 2001). The spread is a bit uncomfortable and results from uncertainties in the true mix of carbon and oxygen in white dwarf C/O cores, micro-physics, and all the usual suspects. Yet, it shows the potential of the method—and the oldest age in the spread does turn out to be less than the age of the universe! Of interest for the future are observations of very cool (perhaps in the 3,000–4,000 K $T_{\rm eff}$ range) galactic halo white dwarfs, which should be older than our neighbors in the disk.[9]

[9] As practically a note in proof, we suggest you check out Hansen et al. (2002, and see Hansen and Liebert, 2003) who report observations of white dwarfs in M4, which is a relatively close globular cluster. They find an age for the cluster of 12.7±0.7 Gyr using their cooling sequences. In this regard, do try Ex. 2.9.

2.7 Advanced Evolutionary Phases, Initial Masses Greater Than 6–10 $\mathcal{M}_\odot$

In the time it takes the sun to go from its T Tauri birth to its planetary nebula death, several thousand generations of massive stars can come and go, for they do everything more rapidly, spending only millions of years on the main sequence and thousands in later stages. They have, of course, more fuel to start out with, but they use it more profligately (see Eq. 1.7). This is not changed by their ability to fuse more elements on beyond the hydrogen and helium fuels available to the sun, for 90% of the available energy going from hydrogen to iron (the most tightly bound nucleus) comes out in the first, hydrogen-fusion reactions, and half the rest when helium burns to carbon and oxygen. In addition, the cores of massive stars, by the time they are fusing carbon and heavier elements, are so hot that they produce copious fluxes of neutrinos, made in several different processes, which, like the solar neutrinos, depart promptly, doing nothing to maintain the stellar photon luminosity.

Table 2.2. Advanced Nuclear Burning Phases of Massive stars

Dominant fuel	T_c	Duration	Important products
Carbon	5×10^8 K	10^3–10^4 yr	Ne, Na
Neon	8×10^8 K	10^2–10^3 yr	Mg, some O
Oxygen	1×10^9 K	< 1 yr	Si, some S, etc.
Silicon	3×10^9 K	days	^{56}Ni

Table 2.2 summarizes the stages of heavy element burning. These stages are only partially separated (that is, a bit of residual oxygen may still be around when the dominant process is silicon burning) and the central temperatures, T_c, required for ignition, do depend on density. The dominant products are not always what you would guess. Two ^{12}C nuclei, for instance, do not often make a ^{24}Mg because there is no level at the right energy, spin, and parity to make this reaction a probable one. Instead, the main reactions are (and see Table 6.3)

$$\begin{aligned} {}^{12}\mathrm{C} + {}^{12}\mathrm{C} &\longrightarrow {}^{20}\mathrm{Ne} + {}^{4}\mathrm{He} \\ &\longrightarrow {}^{23}\mathrm{Na} + \mathrm{p} \\ \text{followed by } \cdots\ & {}^{23}\mathrm{Na}(\mathrm{p},\alpha)^{20}\mathrm{Ne} \quad \text{and} \quad {}^{23}\mathrm{Na}(\mathrm{p},\gamma)^{24}\mathrm{Mg}\,. \end{aligned}$$

Notice that free protons, neutrons, and alpha particles will be floating around, inviting capture by the assorted heavier nuclei in the soup. This remains true through neon, oxygen, and silicon burning, so that the full range of stable nuclides from ^{12}C up to about zinc or gallium is produced. The group from Mn to Zn is called the "iron peak," because they are more abundant than those on either side, and Fe is both in the middle and most abundant.

Silicon burning is not achieved by two ^{28}Si meeting head-on and fusing to ^{56}Ni, with beta decays to ^{56}Fe (the stable one). The coulomb barrier caused by the mutual repulsion of the two charged ^{28}Si is so high that, by the time the stellar core is hot enough to raise their velocities to overcome it, high energy photons are busily photodisintegrating (analogous to ionization of atoms) some of the ^{28}Si back to alpha particles. As they are liberated, most of the alphas are quickly captured by remaining ^{28}Si nuclei, which build up to ^{32}S, ^{36}Ar, ^{40}Ca, ^{44}Ti, ^{48}Cr, ^{52}Fe, and ^{56}Ni, all of which are "alpha nuclei" containing, in effect, many alpha particles stuck together. Capture of free neutrons and protons yields smaller amounts of adjacent nuclides. This process was historically called "nuclear statistical equilibrium (NSE)" or the "e-process" as the nuclear version of chemical equilibrium (as discussed in §3.4).

In fact, however, there is no time for all the nuclei to come into the equilibrium abundances that would be set by beta decays. ^{56}Ni, for example, has a half-life of about seven days, and ^{44}Ti a half-life measured in years. And the whole silicon-burning phase lasts only a few days according to calculations. However, many reactions do take place rapidly enough that the term "quasi-equilibrium" has been used to describe at least some aspects of this burning stage.

At this stage, a color–magnitude (or HR) diagram is no longer useful in interpreting the star's life cycle. Everything beyond late helium fusion goes faster than the outer layers of the star can find out about interior events and respond to them. Indeed the star is nearly doomed. The various nuclear reactions of the previous sections are working outward though the star, leaving their ashes behind (still in fairly discrete layers), and the core of iron-peak elements is growing relentlessly toward the maximum mass that can be supported by the pressure of degenerate electrons (§3.5.2; about 1.2 $\mathcal{M}_\odot$ for heavy elements). Earlier cores of He, C+O, and so forth had also approached this mass limit, but always before a bit of additional contraction heated them until the next fuel ignited. But nuclei with mass numbers around A of 56 are the most tightly bound, and there is no more nuclear energy to be got out. The core must collapse.

As an illustration of where we have gotten to, Fig. 2.17 shows the layering of composition in the inner 8.5 $\mathcal{M}_\odot$ of a 25 $\mathcal{M}_\odot$ pre-supernova model of Arnett (1996). Following his description of the key letters at the top of the figure, we have–

A. The core is composed of iron-peak elements.

B. Silicon burning is taking place, adding to the iron core.

C. Oxygen is being burned, leading to Si and Ca.

D. This zone contains neon, which is being burned, Mg and O, but no carbon.

E. Neon and Mg are being produced in this carbon burning shell.

F. Nothing much goes on here but this zone contains C waiting to be burned.

G. Helium is being burned, producing C and O in this radiative zone.

H. This is an active convective zone, where He burning is going on, and which mixes burning products outward into the star.

I. This is what's left of the old helium core with no burning going on.

J. The letter **J** is not indicated but it is the rest of the star above a hydrogen burning shell.

The quantity Y_e is the number of electrons (presumed ionized) per nucleon in the mixture. (We shall use $\mu_e = 1/Y_e$ in Chap. 3.) For most of the star it is equal to 0.5, corresponding to helium or nuclei composed of multiple helium nuclei but, in the iron core, it decreases a small amount indicating,in effect, that electrons have been captured by protons to make neutrons.

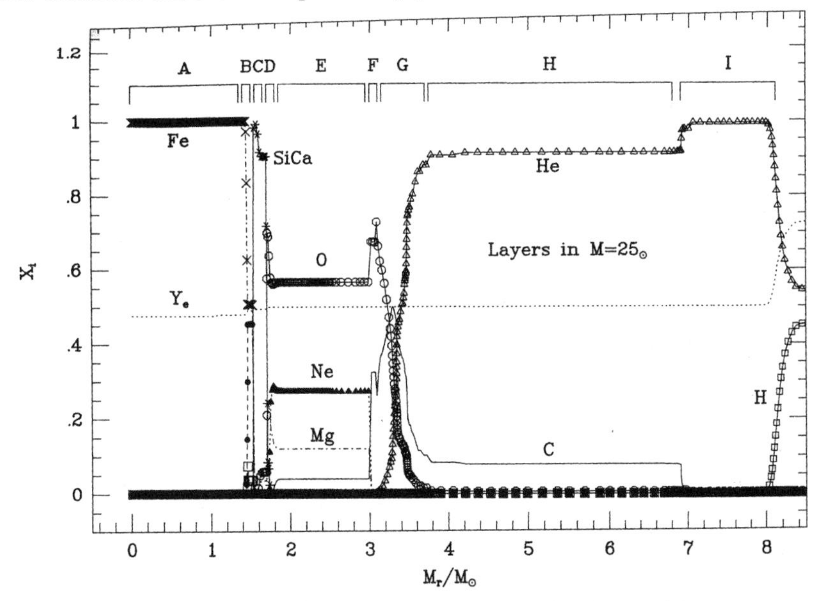

Fig. 2.17. The compositional layering in the inner core of a 25 $\mathcal{M}_\odot$ pre-supernova model versus interior mass. X_i is the mass fraction. See text for an explanation and further commentary. Reprinted with permission from Arnett (1996, his Fig. 10.8), ©1996 by Princeton Univerity Press.

2.8 Core Collapse and Nucleosynthesis

Two triggers can contribute to core collapse. First, photodisintegrations cool the gas, removing the support of thermal pressure. Second, the increasing density forces electrons into ever-higher momentum states—hence higher energy states—until some of them have kinetic energies exceeding the neutron-proton mass difference. Electron capture (inverse beta-decay) sets in, turning protons to neutrons and, with fewer electrons around, the degeneracy pressure also drops.

At this juncture we are just past the end point of the trajectory of central temperature, T_c, versus central density, ρ_c, shown in Fig. 2.18. This figure, from Arnett (1996), derives from evolutionary calculations of an initially pure helium 8 $\mathcal{M}_\odot$ model. It is not a completely realistic star, but the evolution of such models mimics very well what eventually happens in pre-supernova calculations. Various burning stages are indicated by "He", "C," etc. (Note that the temperatures for the burning stages are not quite what we quoted in Table 2.2. As we warned you, density does count.) At the last point (near $\rho_c \approx 6 \times 10^9$ g cm^{-3} and $T_c \approx 8 \times 10^9$ K) collapse is about to begin. As Arnett (1996) puts it, "The trajectories end at the point of hydrodynamic instability; the time-scales are now so short that these stars are not so much *objects* as *events!*" We shall defer until later what this implies when we go supernova, so to speak.

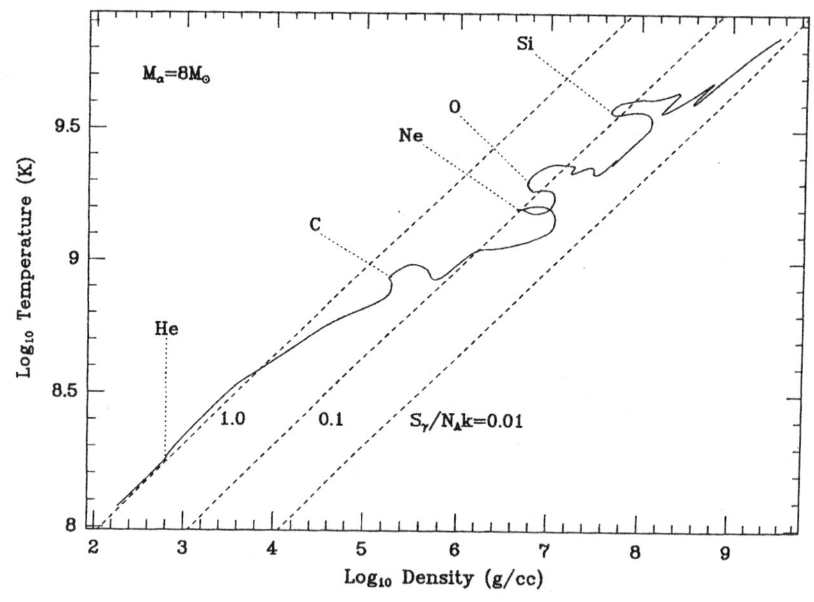

Fig. 2.18. The time trajectory of density and temperature at the center of an initially pure helium 8 $\mathcal{M}_\odot$ model. (The lines labeled $S_\gamma/N_A k$ are lines of constant radiation field entropy, which we will ignore for now.) Reprinted with permission from Arnett (1996, his Fig. 10.3), ©1996 by Princeton University Press.

Anyway, the core collapses suddenly and catastrophically. You can figure out what these words mean for yourself. The time scale will be something like the dynamic time scale, t_{dyn}, of (1.33). Evaluate it for a density of 10^9 g cm^{-3} to see that we are talking about seconds, at most. And the available energy will be the change in gravitational potential for the inner 1.2 $\mathcal{M}_\odot$ (or a bit more in practice) of the star contracting from a density of 10^9 g cm^{-3} to 10^{15} g cm^{-3}, which is the density of a normal nucleus or of the gigantic nucleus that is a whole star made of 10^{57} neutrons. Put in the numbers to show that this is going to be something like 10^{53} ergs. Compare this to the energy released by the sun over its whole main sequence life (and see Ex. 2.12).

The products are a core–collapse supernova and a neutron star, both of which you will meet again below. You will probably feel that, if there is something called a supernova, there must also be something called a not-so-super or plain old nova. There is, and it lives in §2.11.

Meanwhile, however, it is clear that the core collapse process just described is going to bring together lots of iron-peak elements, lots of neutrons, and lots of energy. This is just the set of conditions required for another sort of neutron capture, the *r-process* where "r" stands for "rapid." The definition of rapid in this context is that successive neutrons are captured before there is time for beta decays, until the next neutron wouldn't be bound at all. Then the process hangs up until one of the nuclear neutrons decays to a proton (and electron plus electron anti-neutrino), and the neutron capture continues. The immediate products are highly unstable nuclides. But, at leisure, after ejection from the star in a supernova explosion, they decay back to the most neutron-rich stable isotopes of heavy elements like ^{176}Yb (ytterbium) or ^{186}W (tungsten). The r-process is also the only source in our universe of thorium and uranium, because the reactions have to leap over polonium (Po, 84 protons), radon (Rn, 86 protons), and a bunch of other unstable elements to get there from bismuth (Bi, 83 protons), the last stable one.

Curiously, we have now made just about all the chemical elements that are found on earth or in the periodic table, though you probably have not been keeping count. A score sheet is presented in Table 2.3, which includes the few not previously mentioned.

Table 2.3. Sources of element production

Elements	Source
H and He	Left from Big Bang (including ^{2}H & ^{3}H); also a bit of ^{7}Li from early universe
Li, Be, B	Made by cosmic ray CNO fragmentation in interstellar medium
^{12}C, ^{16}O	Helium burning
^{13}C, ^{14}N, ^{15}N, ^{17}C, ^{19}C, F	CNO cycle burning and its extension to higher temperature
Ne to iron peak	Carbon, neon, oxygen, and silicon burning
$Z = 30$ and beyond	s-, r-, and p-processes, the latter two primarily in supernovae

2.8.1 Abundances and Nucleosynthesis

If we take an inventory of the abundances of the elements in the solar system we find what we show in Fig. 2.19 (and do read the caption). Hydrogen and helium are, by far, the most abundant (by at least two orders of magnitude) and, as indicated in Table 2.3, they are leftovers from the Big Bang. Their abundances may have been modified by stellar processes, but not in any significant way. The next three elements, Li, Be, and B, are CNO cosmic ray fragmentation products. Then follow elements produced by the CNO cycles, helium burning, and the peak around iron from carbon, oxygen, neon, and silicon burning. Heavy elements, with nuclear charge greater than about 30, are the responsibility of neutron (and some photon and proton) capture processes with the iron peak providing most of the "seed" nuclei that do the initial capturing (and see below). An excellent review of abundances (among many other matters), and the various means of establishing them, is given in Chapter 2 of Arnett (1996). Other reviews, with perhaps a slightly different slant are Trimble (1991, 1996, 1997).

A closer view of abundances is shown in Fig. 2.20 (and continued in Fig. 2.21), where individual odd ($\triangle$s) and even ($\bullet$ s) nuclide abundances are plotted versus nuclear mass number $1 \leq A \leq 90$. By "odd" and "even" we mean odd and even numbers of nucleons within a given nucleus; for example, ^{3}He is odd, whereas ^{4}He is even. Because of the large number of nuclides in the figure, we have shifted the odd nuclides down by two decades (N multiplied by 10^{-2}) for clarity, even though some of the impact of the information may be lost. (Odd nuclei tend to be less abundant than their even counterparts in any case.) For many nuclides, we indicate by what nuclear burning process they were made (see figure caption). Note, however, some nuclides may be produced by more than one process. For a more complete listing for each nuclide, see Arnett (1966, App. A).

The general impression given by Fig. 2.20 is a trend downward in abundance starting from ^{12}C (ignoring hydrogen and helium as cosmological remnants, and lithium, beryllium and boron, which may be involved in the pp-chains but usually do not survive that phase of burning). Some nuclides seem to be preferred, causing the alternating peaks at moderate mass number, such as ^{12}C, ^{16}O, ^{20}Ne, ^{24}Mg, ^{28}Si, ^{32}S, ^{36}Ar, $\cdots$, and ^{48}Ti. These are the α-particle nuclei composed of integral numbers of ^{4}He nuclei. The trend is interrupted, however, at the iron peak where, as pointed out earlier, it takes a lot of energy to remove or add on nucleons. The iron isotope ^{56}Fe is the most abundant by far. This may seem surprising since it is *not* an α-particle nucleus, but it is the β-decay product of ^{56}Ni, which is, and the latter is an important product of supernova explosions. The iron peak dribbles downward until we reach nuclei around zinc (30 protons) where the p-, r-, and s-processes begin to take over—which we will discuss more fully shortly.

This brings us to Fig. 2.21, which continues the story. Here we separate out those nuclei produced by the p-, r-, and s-processes. In some cases, if not

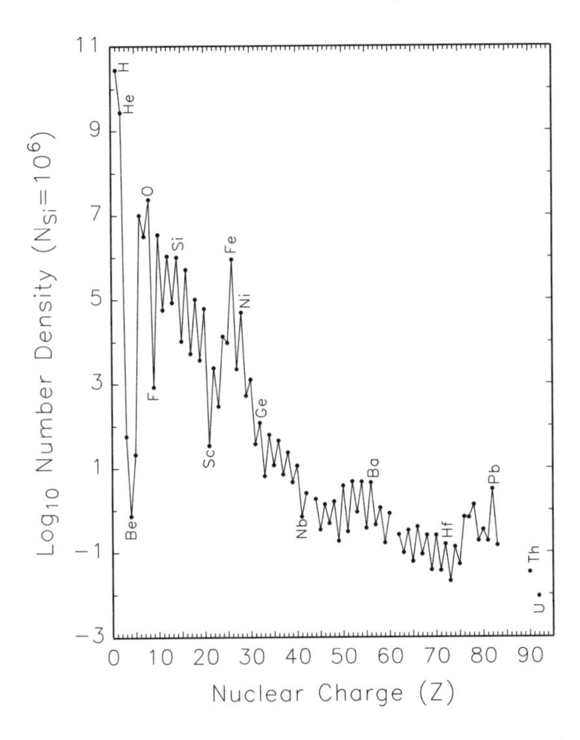

Fig. 2.19. "Solar" elemental abundances are plotted against nuclear charge, with some representative element names shown as a guide. These abundances, from Anders and Grevesse (1989), are derived from a combination of observed solar values, carbonaceous chondrite meteorites, and with some values folded in from ISM observations. The normalization used is that the number density of Si $= 10^6$. Note that some elements are missing (e.g., technetium, promethium. polonium, etc.) because all their isotopes are radioactive with half-lives short compared to the age of the solar system.

many, the addition of a proton in the p-process (i.e., proton capture) may in fact be the result of the high-energy photon field inducing the ejection of a neutron—that is, a (γ, n) reaction—from a different nuclide to give the same product. P-process nuclei tend not to be very abundant and this may be a reason why it is not that well-understood. Some peaks are obvious in the curves and these are associated with the neutron magic numbers 50, 82, and 126 (forming closed shells within the nucleus not unlike closed electron shells in the noble gases). Three nuclei usually assigned to the p-process seem to be anomalous—namely, ^{138}La, ^{176}Lu, and especially ^{180}Ta. The last may be underabundant because it is particularly susceptible to destruction by (γ, n). Note that (look carefully) ^{235}U is shown with an abundance much larger than you find at the present time (0.7% of all uranium). Anders and Grevesse (1989) have backdated some radioactive nuclei to 4.6 Gyr ago to

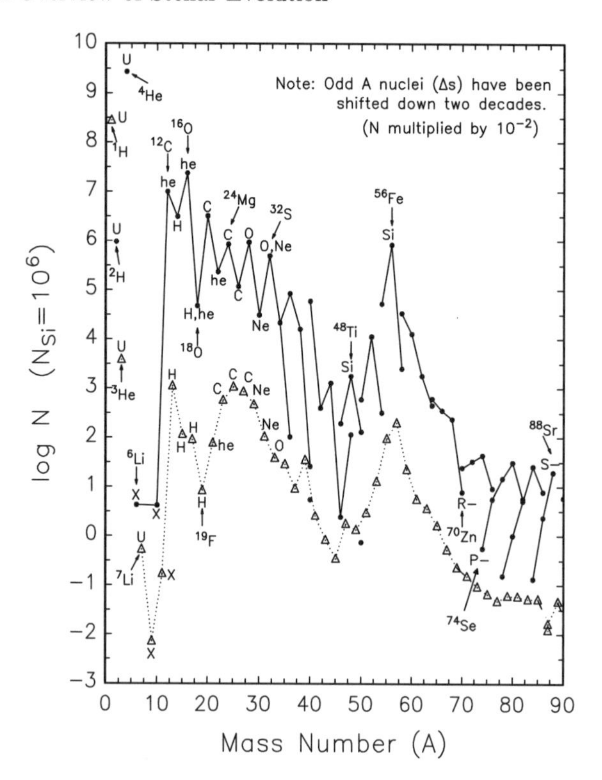

Fig. 2.20. Individual nuclide abundances for odd and even nuclides, taken primarily from Anders and Grevesse (1989). Triangles denote odd nuclides (which have been shifted down by two decades), whereas "dots" are for even. Some important nuclides are labeled. Sources (e.g., nuclear burning stages) are indicated for some nuclides: U means Big Bang; X from fragmentation of cosmic rays; H for hot (and hotter) hydrogen burning; he=helium burning; C, O, Ne, or Si=carbon, oxygen, neon, or silicon burning; and an occasional P-, S-, or R- for p-, s-, or r-process.

give a better idea of what nuclide abundances were available at the time of formation of the solar system. (^{235}U has a half-life of 7×10^8 yr, whereas ^{238}U's half-life is a long 4.5×10^9 yr. Both are important chronometers for dating the formation of the earth, moon, and some meteorites.)

Our understanding of the neutron capture s-process seems to be well in hand. Since it is "slow" (hence the "s"), relatively cool temperatures are indicated. The first outline of the process was given in the classic paper by Burbidge, Burbidge, Fowler and Hoyle (1957; hereafter, and forever, known as B^{2}FH). (A.G.W. Cameron also discussed the process at nearly the same time, but the original paper is a Chalk River report not readily available.) In its simplest form it is associated with helium shell burning and mixing by convection. The neutrons are produced in the reaction sequence

$$^{12}\mathrm{C}(\mathrm{p},\gamma)^{13}\mathrm{N}(\mathrm{e}^+,\nu)^{13}\mathrm{C}(\alpha,\mathrm{n})^{16}\mathrm{O}$$

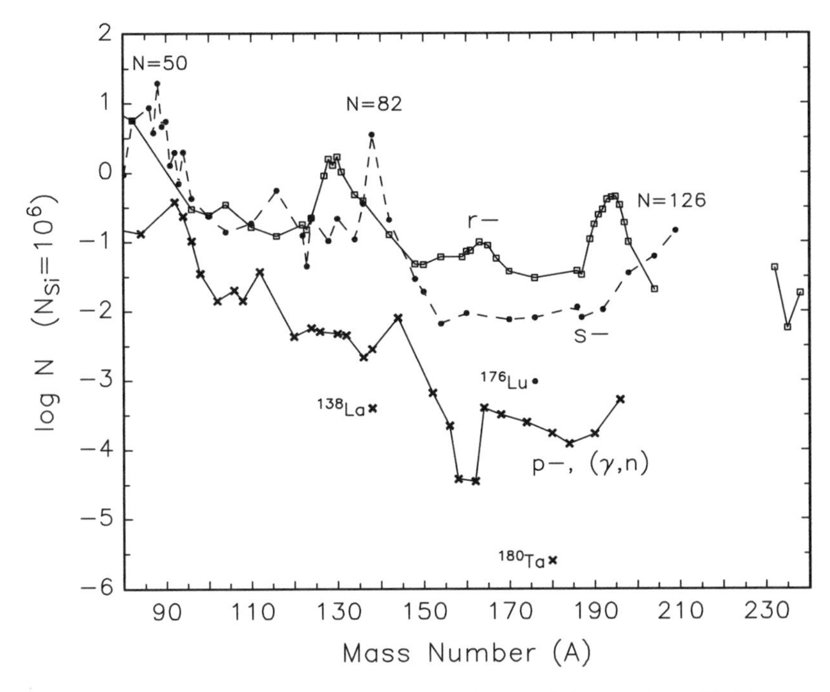

Fig. 2.21. This is a continuation of Fig. 2.16 but nuclides are identified by source mechanism, s- (•s), r- (□s), and p-process or (γ, n) (×s).

with $^{22}Ne(\alpha, n)^{25}Mg$ also making a contribution (among perhaps others). These neutrons are then captured by iron peak nuclei (mostly ^{56}Fe) to form heavier nuclei, which, in turn, may capture neutrons, etc. This chain may occasionally be interrupted by the β-decay of a short-lived radioactive nuclide. (Extensive reviews may be found in Käppeler et al., 1989, 1990; and Meyer, 1994). A sample path is shown in Fig. 2.22. We suppose that ^{78}Se has been produced by neutron capture (indicated by solid arrows pointing to the right) and is ready to capture its own neutron producing ^{79}Se. In the laboratory, ^{79}Se has a half-life against β-decay of nearly 10^5 years, which would seem to allow plenty of time for it to capture a neutron. In the stellar environment, with a temperature of around 3×10^8 K, however, the radiation field can cause an excited state in ^{79}Se to be populated, thus reducing the life time to less than a year. The dashed arrow labeled λ_β shows the path of the β-decay to ^{79}Br, which now competes with the neutron capture (λ_n) to ^{80}Se. And so it goes. Note that ^{82}Se cannot be reached because of a β-decay from ^{81}Se. It is "shielded" from the s-process and is classified as a r-process nucleus. ^{80}Kr, on the other hand, is shielded from the r-process and is a s-process nucleus (or perhaps p-process).

The r-process, as mentioned earlier, is associated with more violent, higher temperature, environments than the s-process (remember, the "r" stands for

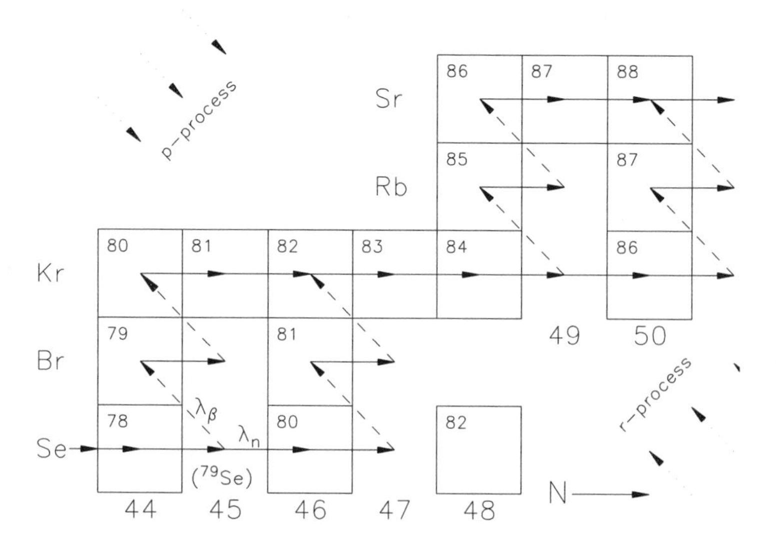

Fig. 2.22. Here is a portion of the Chart of the Nuclides in the selenium to strontium region showing sample s-process paths of neutron capture and β-decay. Neutron number is on the abscissa and element name on the ordinate. The boxes (with total nucleon number indicated) are for either stable nuclei or those with very long half-lives. Solid arrows are neutron capture paths, whereas dashed represent β-decays. The source directions for p-process (or $[\gamma,n]$) and r-process are also shown schematically. This is a standard figure and is modeled after Fig. 2 of Käppeler et al. (1989), and see Clayton (1968, Fig. 2–26).

"rapid"). All in all, it is also more complicated because it depends on the details of silicon burning, which, in turn, are dependent on temperature, density, and the time scale of evolution. The final abundances of the later phases of silicon burning, whether they are in near equilibrium or close to it, are further determined by the ratio of total neutrons in nuclei (or free neutrons) to the total number of neutrons plus protons. And, in an explosive situation, how fast expansion takes place is also critical. At some point, when the material cools sufficiently (but it is still hot!), charged particle reactions tend to shut off because particles are not swift enough to overcome Coulomb barriers. The reactions are said to "freeze out." The neutrons, as electrically neutral particles, do not have this problem, and they can busily go about sticking onto nuclei. And we need plenty of neutrons for the r-process. If a ^{56}Ni nucleus is a typical product of silicon burning, then to make one ^{238}U nucleus requires nearly 200 additional neutrons, nearly half of which must decay into protons to keep the charge right. A detailed calculation of the r-process is no easy matter because of the many reactions that are possible. In addition, the physics required is at the limits of what we know about nuclei. If we were to try to keep adding on neutrons to some seed nucleus, at some point a final capture would not be possible because the resulting nucleus would be unstable against neutron decay. Those nuclei that are unstable in this sense constitute the "neutron drip line" on the chart of the nuclides, and although we have a good idea of where that is (from experiment and theory), it would be better to do better.

2.9 Variable Stars: A Brief Overview

A few naked eye stars, including Mira (o Ceti) and Algol (β Persei) are, once you are looking for the effect, wildly variable in their brightness over times from a year or two (Mira) to a week or so (Algol). One suspects that their behavior must have been known to the ancients (as the names "wonderful" for Mira in Latin, and "the demon" for Algol in Arabic suggest). The first firm records of anyone's being aware of these recurrent and more-or-less periodic variables date, however, only from a few decades before the invention of the telescope. The Greeks and Chinese recorded a few "new" or "guest" stars that appeared where none had been seen before—and then faded away in weeks to years—from about the beginning of the Common Era,[10] and that was it.

It is now clear that, at some level, the light output of every star, including our sun, varies with time, over virtually every length of time possible from the dynamical time scale, $t_{\rm dyn}$ of §1.3.3 (about an hour for the sun) to the nuclear time scale, $t_{\rm nuc}$ of §1.7 (which, for the sun, is about 10 billion years as it gradually gets brighter). Both causes and manifestations are many and varied; a recent count identified more than 70 classes of variable stars, most of them named for a particular star (called the prototype), which belongs to the class, or was anyhow thought to belong when the class was established. Examples are T Tauri stars, Cepheids (for δ Cep), FK Comae stars, Miras, and U Geminorum stars.

Here we summarize some interesting sorts, but save those of more direct consequence for this text for later (in §9.10). These more "interesting sorts" include the pulsational variables (§2.10) and their violent relatives the novae and supernovae (§2.11).

2.9.1 Eclipsing and Ellipsoidal Variables

If a pair of stars orbit each other, you, the observer, may be located close enough to the orbit plane to have one pass in front of the other and block its light for a portion of each orbital period. Because the eclipse tells us that the system is nearly edge on, eclipsing binaries are among the sorts particularly useful in measuring stellar masses. Even if there is no eclipse, the gravitational

[10] We shall use the nondenominational designation Common Era (C.E.) instead of the more usual A.D. B.C. is then B.C.E. or Before Common Era.

field of one star may distort the shape of its companion into an ellipsoid, so that you see a larger star area when the stars are side-on to you than when they are end on. This will also result in periodic variability, though of a less useful sort.

2.9.2 Spotted, Rotating Stars

The sun is an example of this class. Its brightness varies both at its rotation period (about a month) and through the 11-year sunspot cycle. The variation is, however, only about 0.1% (and, curiously, the sun is brighter when it has more spots because the extra brightness of the bright vein-like facular regions of the photosphere more than makes up for the darker spots).

Larger fluctuations in brightness happen among younger, rapidly rotating stars of types G, K, and M, and among close binary pairs, where the rotation period is locked to the orbit period. The majority view is that rapid rotation plus the convective atmosphere of these cool-surfaced stars permits the operation of a dynamo, producing a magnetic field, which, in turn, drives spot formation and other kinds of stellar activity. Relatively strong variability of this sort is associated with emission of x-rays and radio waves from the solar or stellar corona and other indications of youth and activity.

2.9.3 T Tauri Stars, FU Orionis Stars (FUORs), and Luminous Blue Variables

These are stars that are very young (pre-main sequence) or very massive and bright, or both. They are probably both accreting material from a disk and blowing off material at their poles, and may be heavily spotted as well. The result is nonperiodic flaring and variability. Surrounding gas and dust frequently show up in images and spectra of these stars, and very occasionally it is possible to tell which bits are flowing in and which are being ejected, sometimes in jets. Rapid rotation and magnetic fields (which together collimate the jets) are also part of the picture.

2.9.4 Last Helium Flash and Formation of Atmospheric Dust

These two physically different causes of variability appear together because one is often precursor to the other. A star that has already left the AGB phase can experience one last flash of its helium burning shell. This puffs up the envelope so that the star quickly comes to resemble a red giant again and to brighten (because more of the light comes out in the visible part of the spectrum). Only three or four stars have been caught doing this in historic times. The prototype is FG Sge, which began galloping from blue to red and faint to bright across the HR diagram shortly after 1900. It and other members of the class also begin to display unusual elemental abundances in their spectra, as a result of the flash driving an outer convection zone. Extra carbon, s-process elements, and sometimes lithium appear in their atmospheres in a matter of years. A popular report of a closely related star (Sakurai's Object or V4334 Sgr) is due to Kerber and Asplund (2001), and see the special issue of Ap&SS, Vol. 279 (2002) for a series of articles. An evolutionary relationship between these (sometimes called) "born-again" stars is discussed in Lawlor and MacDonald (2003).

Highly evolved stars with carbon–rich atmospheres (including FG Sge stars) occasionally and unpredictably fade by many magnitudes in a few weeks and gradually recover over months. The missing light comes out as infrared, and the cause is sudden condensation of carbon dust in the cool stellar atmosphere, which then gets blown out again by radiation pressure. These are the R CrB variables (sometimes called inverse novae). About 40 are known, and two stars of the FG Sge type have recently displayed R CrB-type fading.

2.10 Pulsational Variables

These are the most useful variable stars because the length of time in which they brighten and fade again—their periods—are frequently correlated with their absolute brightnesses, so that they can be used to measure distances to star clusters anywhere in the Milky Way and to nearby galaxies. The period–luminosity relation (see §8.2.3) for one sort, the Classical Cepheid variables, is generally regarded as the first, fundamental rung on the ladder that leads to measuring distances to distant galaxies and so to measuring the speed of expansion of the universe, its age, and other things you might want to know. In the following, we shall set aside many technical details and refer to our Chapter 8. For texts and reviews we recommend Cox (1980), Unno et al. (1989), and Gautschy and Saio (1995, 1996).

Stellar pulsation can be purely in and out (radial modes) or include material slopping around in latitude and longitude as well (nonradial modes). The pulsation must be driven by some instability (often the repeated ionization and recombination of atoms of a common element, which acts like a faucet, letting radiation out and shutting it up). The restoring force that brings the gas back where it started from (only to overshoot the other direction, over and over again) can be gravity or pressure or (perhaps) magnetic fields. In this sense, the instability is *intrinsic* to the star and not due to external influences; hence, such stars are often referred to as intrinsic variables. A really complex gravity (g-) or pressure (p-) mode with both radial and nonradial motions is described by the number of nodes in the r, θ, and ϕ directions (like the eigenstates of the hydrogen atom in spherical coordinates), and stars sometimes show more than one at a time, up to dozens, or (in the case of the sun, for which we can observe the tiniest displacements) millions.

For purely radial pressure modes the underlying time scale is the dynamical time (t_{dyn}). The *fundamental mode*, which is a simple breathing in and out with the only node at the stellar center (you don't want the center moving off some place), is the only radial mode that shows a simple period–mean density relation (as in §1.3.5). But, if all the stars in a class have about the same mass and temperature, density depends only on radius, and so there is period–luminosity relation of the sort we see (and you might try Exs. 2.2 and 2.4). If you can pick out a fundamental mode and an overtone (more than one radial node), their ratio gives you one more handle on the star's properties. This can be used to learn masses (or, alternatively, to check that you have figured out what is really going on).

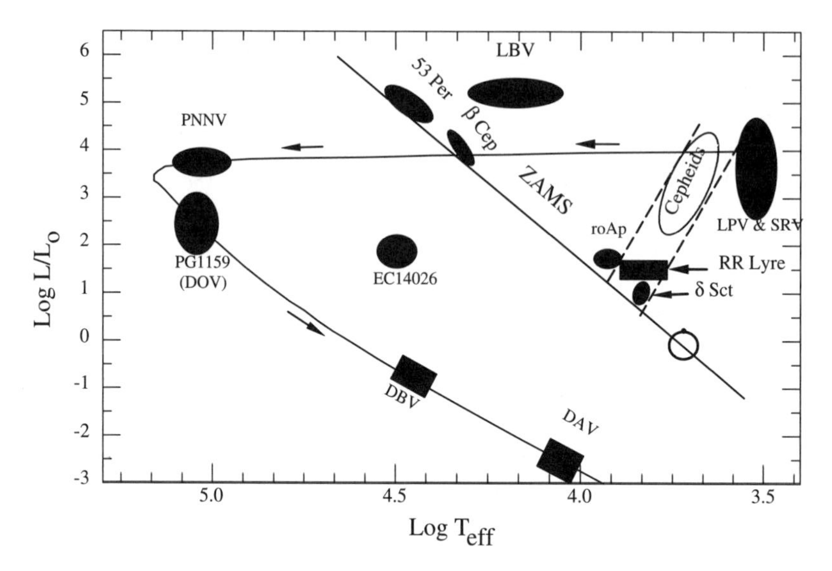

Fig. 2.23. A selection of intrinsically variable stars placed on an HR diagram. Were we to include all known classes (real or imagined), you wouldn't be able to read it. The sun, itself a variable, is indicated on the ZAMS. A schematic evolutionary track from AGB-PNN-WD stage helps place some variables. The "Cepheid Strip" is shown by the dashed lines demarcating the Cepheid variables.

An *instability strip* (driven by helium ionization) extends diagonally across the upper right of the HR diagram, and probably all in it are pulsational variables to some extent. (The name "Cepheid Strip" is often given to this strip.) A selection of variable stars, in and out of this strip, include (and see Fig 2.23)–

- *Classical Cepheids*, usually called just Cepheids (or Type I Cepheids), are young, metal-rich stars crossing the instability strip with spectral types F6–K2, with periods of days to months. The more luminous (and more massive) ones have the longer periods, which gives the period–luminosity relation. They appear to be purely radial pulsators. The cause of their pulsation was unraveled by the pioneering studies of Cox and Whitney (1958) and Zhevakin (1953).

- *W Vir* variables (or Type II Cepheids) have the same sort of behavior as the Classical Cepheids but are metal-poor, older, and lower-mass stars.
- *RR Lyrae* variables are horizontal branch stars in the instability strip with spectral types A2–F2 (see Fig. 2.7). They were formerly known as cluster-type variables or Cluster Cepheids because they are common in globular clusters. They have periods of a day down to a couple of hours and are useful in determining distances to globular clusters in our galaxy and in nearby galaxies. They are subdivided into subclasses ("Baily types"), depending on details of their light curves (i.e., their curve of magnitude versus time). Since Smith (1995) has written an excellent monograph on the whole subject, we defer to him (but you might try Ex. 2.11).
- *β Cepheid* and *γ Doradus* stars are main sequence stars (or close) of spectral types A and B. They are nonradial pulsators (and, arguably, radial) and multiple modes are common. Periods tend to be longer than the dynamic time scale, which suggests that the modes are gravity (g-) modes.[11]
- *δ Scuti* variables are spectral class A to early F (i.e., at the bluer end of spectral class F) stars on or near the main sequence (luminosity class V to III) in the instability strip. They have periods ranging from about 30 minutes to 8 hours and pulsate in radial and nonradial pressure modes, although gravity modes may be present. Amplitudes tend to be low. As with some other variable stars, the δ Sct variety have a well-organized network devoted to their observation and theory. In their case, the network is presently centered in Vienna, Austria. You may wish to access their WWW site at `http://www.deltascuti.net`.
- *SX Phe* stars are blue stragglers (stars brighter than the main sequence turnoff, perhaps arising from mass transfer or mergers in close binary systems) found in globular clusters and among old field stars, and are fundamental or first overtone pulsators.
- *ZZ Ceti* variable stars are white dwarfs with hydrogen atmospheres (DA white dwarfs) sitting in their own instability strip (see Fig. 2.23) on the white dwarf cooling curve. Only nonradial gravity modes are seen, often many in a single star. A corresponding strip exists for white dwarfs with helium atmospheres (the DB variables) and for carbon–oxygen–helium atmospheres (the PG1159 stars) and their close relatives the variable PNN stars. Chapter 10 will consider these in more detail partly because analysis of their often complex light curves lets us probe their interiors. The var-

[11] A recent review of some nonradial near-main sequence pulsators is due to M.–A. Dupret, part of whose Ph.D. dissertation appears in the *Bulletin of the Royal Society of Sciences of Liège*, Vol. 71, 249 (2002).

iable white dwarfs also have a network ("The Whole Earth Telescope," aka WET), which observes them (along with some other variables including δ Scts). Now coordinated from Ames, Iowa, its WWW site is `http://wet.iitap.iastate.edu`. As an example of a light curve of a variable white dwarf see Fig. 2.24 where a night's worth of data is shown for PG1159–035 (the prototype of its class). We shall see more of this star in Chapter 10 but, for now, let it be known that it pulsates over 100 ways at once, as is evidenced by the "beating" (the peculiar waves superimposed over the regular ups and downs).

- *Mira* variables, as mentioned previously, are luminous red supergiants belonging in the class of *Long Period Variables* (LPV in Fig. 2.23) with periods ranging from roughly 100 to 700 days. Radial modes seem to be the norm. Closely related to them are the *Semi-Regular Variables* (SRV), which act as their name implies. All stars of more than main sequence luminosity and redward of the Cepheid instability strip are to some extent variable. Usually the period is long (as a result of low density atmospheres) and somewhat irregular. RV Tauri stars are an extreme version, with low mass and large luminosity, so that a second pulse starts before the atmosphere has had time to fall down from the previous one. They display alternating large and small amplitudes in their light variability. Many R CrB stars are also pulsational variables of this sort.
- The *Rapidly Oscillating Ap* stars (roAp in Fig. 2.23) are characterized by low amplitude, short period photometric variations (typically around 10 minutes), strong magnetic fields, and enhanced surface abundances of exotic elements such as strontium and europium (among others less exotic). The observed light variations are modulated in amplitude by the rotation of the star and it is thought that the pulsations are carried around by an off-axis magnetic field as the star rotates. This is the "oblique rotator model," reviewed (and named) by Kurtz (1990). We have shifted their location in Fig. 2.23 for clarity; they should be close to the δ Scts.
- The sun is indicated in Fig. 2.23 (as a pressure mode variable with millions of nonradial modes with periods around five minutes—see Chap. 9). But it may be just our closest example. The sun-like star α Cen A (see Fig. 2.3) is also variable, as discovered by Bouchy and Carrier (2001).
- Finally, in this short list, we welcome the *EC14026* variable stars, which may be the newest class discovered (see Kilkenny et al., 1997). They are low amplitude subdwarf B (sdB) pulsators with periods of around 150 s. Their evolutionary state is uncertain but they are probably closely related to other sdB variables (as yet unnamed) having periods ten times as long (and which must be g-mode pulsators, as discussed in Green et al., 2003).

The variability of pulsational variables shows in their radial velocities as well as their light. In principle, one can integrate the velocity curves to get radius as a function of time and then, with a temperature from their colors or spectra, calculate the absolute luminosities. This is called the Baade–

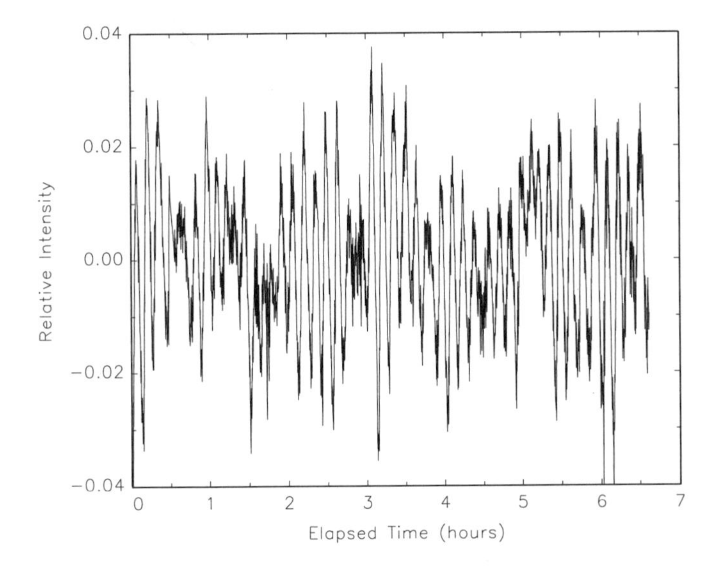

Fig. 2.24. The light curve of the variable white dwarf PG1159–035 shows periodic behavior with a period of around 500 s modulated by the the interference of many modes. The data is from a 1989 observing run by the Whole Earth Telescope consortium. More will be said about this star in Chapter 10 (and see Fig. 10.5).

Wesselink method, and the results (mercifully) more or less agree with the results of parallax measurements and other ways of getting the brightnesses and distances of the stars concerned (and try Ex. 2.14).

2.11 Explosive Variables

These are the stars that release a great deal of (nuclear or gravitational) energy in a hurry. They include the first variables ("new" or "guest" stars) to be recognized. The cataclysmic variables (see Warner, 1995; for an exhaustive review, Sparks et al., 1999, for a brief overview; and Sion, 1999, for the role of white dwarfs) are close star pairs with a white dwarf in orbit with a main sequence or red giant companion. The white dwarf accepts material from its companion. One sort of variability arises when the rate of acceptance or accretion and therefore the rate of release of gravitational potential energy changes. (Think of the erratic splashing of a waterfall if the river at the stop alternates between carrying a little water and lots.) When enough hydrogen–rich material has accumulated on the surface of the white dwarf, it fuses explosively. Remember what happened when we ignited degenerate helium in a helium shell flash. Degenerate hydrogen is even worse, and need not even be very hot when it is as dense as a white dwarf. Even more spectacular are supernovae, as evidenced by explosions that may temporarily outshine the galaxy in which they occur.

2.11.1 Novae

Nova and recurrent nova explosions are the names given to the outbursts fueled by degenerate hydrogen ignition. Actually, of course, the names were given to the phenomena long ago (when people still used Latin for scientific nomenclature), and understanding came later. Novae can recur, since more hydrogen can be accreted as long as the companion star exists, and "recurrent novae" simply means that more than one nuclear-fueled outburst has been

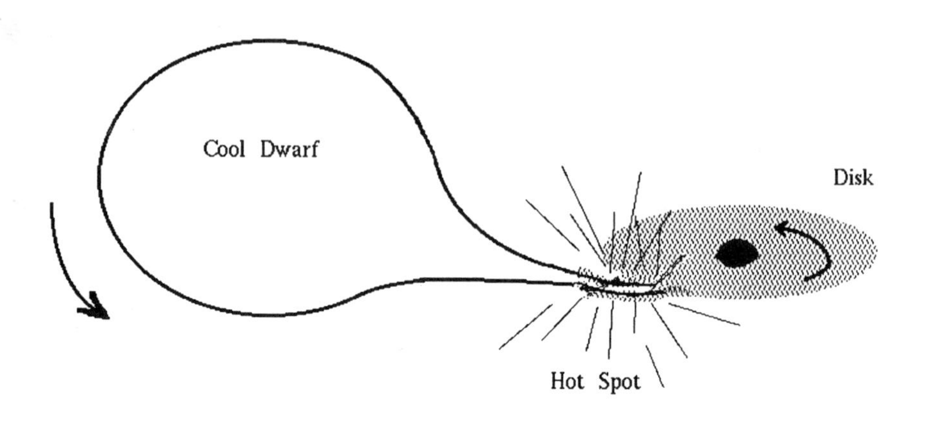

Fig. 2.25. This drawing illustrates mass accretion from a donor star to a cataclysmic variable with an accretion disk as an intermediary. The "hot spot" is where the incoming material meets the already formed disk.

A drawing of a cataclysmic system is shown in Fig. 2.25. Mass is gravitationally drawn off a donor star and forms an accretion disk around the cataclysmic variable (of one sort or another). The material gradually makes its way through the disk and is eventually deposited onto the variable. How a system gets itself into this predicament we reserve for §2.13.

Novae (often referred to as "classical novae") may yield a total of 10^{44}–10^{45} ergs upon eruption with some 10^{38} ergs being radiated in the optical. Mass lost during the explosion ranges from 10^{-5} to 10^{-4} solar masses. The odd thing about these objects is that the pre- and post-nova stars (after things have finally quieted down) appear to be identical. This suggests that the star has not suffered too much from the explosion and is waiting for enough mass to be accreted so that it can do its thing all over again. With estimated mass transfer rates of $\dot{\mathcal{M}} \sim 10^{-9}$–$10^{-8}$ $\mathcal{M}_\odot$ yr^{-1} and the mass losses quoted above, the time between explosions comes out to be (by simple division) thousands to millions of years. If we see one go off, it is unlikely that we will see it explode again in our lifetime.

A schematic nova light curve is shown in Fig. 2.26. There is a fast rise, lasting perhaps a day, followed by a decline in brightness that may be quite variable from nova to nova. In the "fast" novae the decline may take a couple of weeks to reduce the visual brightness by two magnitudes. "Slow" novae may take a few months to accomplish the same thing. Between the initial rise and eventual decline, there may be a plateau near the peak lasting an hour or so in fast novae but extending over days for slow ones. As indicated in the figure, the decline phase may be interrupted by oscillations or a pronounced trough. This complexity should not be too surprising because we are dealing with four objects at the same time: donor star, accretion steam, accretion disk, and white dwarf, all interacting with one another. For example, how the ejected material and radiation field from the explosion affect the donor star and/or accretion process is a formidable multi-dimensional problem.

The course of the explosion itself has been modeled with fair success. After 10^{-5}–10^{-4} $\mathcal{M}_\odot$ of hydrogen-rich donated material has been deposited on the high gravity surface of the white dwarf, gravitational compression has heated it up to $\sim 10^8$ K with densities 10^3–10^4 g cm^{-3}. This is a combination sufficient to initiate a runaway thermonuclear explosion using hydrogen as fuel. It has been established that CNO nuclei play a crucial role in controlling how the explosion proceeds. As in main sequence hydrogen burning, the CNO nuclei act as catalysts but, unlike quiescent burning, the time scales are so short that many intermediate nuclei produced do not have time to decay by positron emission until the explosion is well underway and the material is already being ejected. The energy released by decays at later times helps power the expansion of the ejecta.

It also appears that a successful fast nova requires that the material accreted from the secondary be overabundant in CNO nuclei as compared to the sun and the atmospheres of most other normal stars. What causes this overabundance in the outer layers of the secondary is not known but observations of the ejected matter confirm that C, N, and O are indeed overabundant along with other nuclei such as neon and magnesium (although the presence of the latter may reflect the composition of the underlying white dwarf).

Dwarf novae erupt repetitively (but not with a regular period) with intervals between outbursts of tens to hundreds of days. The duration of the outburst may vary but, for the majority of systems, there is a correlation between the duration of outburst and the interval of time before the next one takes place: the longer the duration, the longer the interval. At outburst

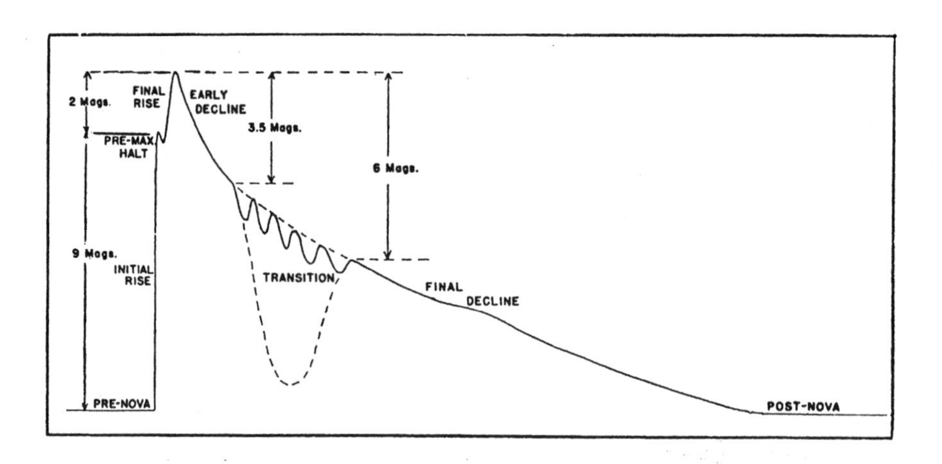

Fig. 2.26. Shown is a schematic light curve for classical novae. Reproduced, with permission, from McLaughlin (1960), ©1960, University of Chicago Press.

peak, the optical luminosity is typically around $\mathcal{L} \approx 10^{34}$ erg s^{-1} with a (usually) rapid rise to the peak and a slower decline. The total energy released for an outburst is estimated to be of the order 10^{38}–10^{39} ergs. With this combination of readily identifiable characteristics, it is not surprising that over 300 dwarf novae are now known.

Because the dwarf novae are binary systems we can view them from different aspects as the two stars orbit and perhaps eclipse each other. The following is a generic description of what information has been derived from observations of the binary light curves of many systems. (We shall note later some variations on the following themes.) There is clear evidence for the "hot spot" (as in Fig. 2.25) that reveals itself by a "hump" in the light curve. In most systems, the emergence of the hot spot also coincides with a noisy "flickering" in the light output, which most likely reflects the violence of the collision process. In strong support of the presence of a disk is the spectrum of the light emitted by the system: it is consistent with that expected from a thin but (often) optically thick bright disk and this light from the disk, in most cases, outshines both white dwarf and secondary stars. In rare cases Doppler-shifted atomic lines are observed that directly indicate rotation of material around the white dwarf.

What is not so clear, and is still controversial, is what causes the eruptions. Two perfectly reasonable models are prime contenders. If the mass–losing secondary star is subject to instabilities that cause variations in the amount of mass fed to the disk, then a higher than normal transfer rate will cause the disk to gain more energy and brighten and supplement the amount of material crashing down onto the white dwarf. A lull in the transfer rate, on the other hand, will result in a quiescent state. The rhythm of outbursts is then set by the secondary. But if the secondary is a well-behaved star, we are led to the competing model. As mass is steadily fed to the disk in this model, the disk grows in size and gradually brightens as it stores mass. Theoretical calculations have shown that this is a potentially unstable situation. If the accretion continues unabated, conditions in the disk may reach a point where the physics of ionization of hydrogen and helium cause what may best be described as a phase transition in the properties of the disk. The end result is a change in the mass–storage capabilities of the disk from one where additional mass may be easily accommodated and the disk is cool to one in which the disk rapidly heats up, glows more brightly, and dumps material down onto the white dwarf. Neither theory nor observations are yet up to discriminating between these models. The combination of observation and modeling of disk structures do lead to estimates of how much the total mass flow through the disk is modulated between outbursts and quiescent states. During quiescence the mass transfer rate estimate is $\dot{\mathcal{M}} \sim 5 \times 10^{-11}\ \mathcal{M}_\odot\ \mathrm{yr}^{-1}$, which is boosted to $\dot{\mathcal{M}} \sim 5 \times 10^{-9}\ \mathcal{M}_\odot\ \mathrm{yr}^{-1}$ during outburst. The factor of 100 between these numbers is roughly consistent with the difference of power output between the two states. An order of magnitude (or so) estimate for the mass of the disk is found by multiplying the quiescent mass transfer rate by a typical time between outbursts (say a month) and yields $\mathcal{M}_{\mathrm{disk}} \sim 10 \times 10^{-11}\ \mathcal{M}_\odot$. Compared to the mass of the stars involved, this is a remarkably small number considering what the disk is able to do.

The dwarf novae are not a completely homogeneous class of objects and there are well-recognized subclasses named after their prototypes. The "Z Camelopardalis" systems, for example, have normal outbursts but, occasionally, instead of returning to the usual quiescent state, the light output from the system stays roughly constant at a level intermediate between quiescence and outburst peak for a few days to months. (It is somewhat difficult to reconcile these "standstills" with a long-lasting disk instability.) "SU Ursae Majoris" dwarf novae sometimes undergo "super-outbursts" during which the light output far exceeds that of a normal outburst. Finally—and this is as far as we shall go—the "U Geminorum" variables are those that fit into neither of the above subclasses and which may be thought of as more the prototype dwarf novae. Sample visual light curves for the three major classes are shown in Fig. 2.27.

Note that we have not discussed variations on these systems in which the white dwarf has a measurably strong magnetic field, in the range 10–50 MG (and the field channels the accretion, so that there is additional variability at the rotation period resulting in objects called polars, DG Her stars, intermediate polars, and other things).

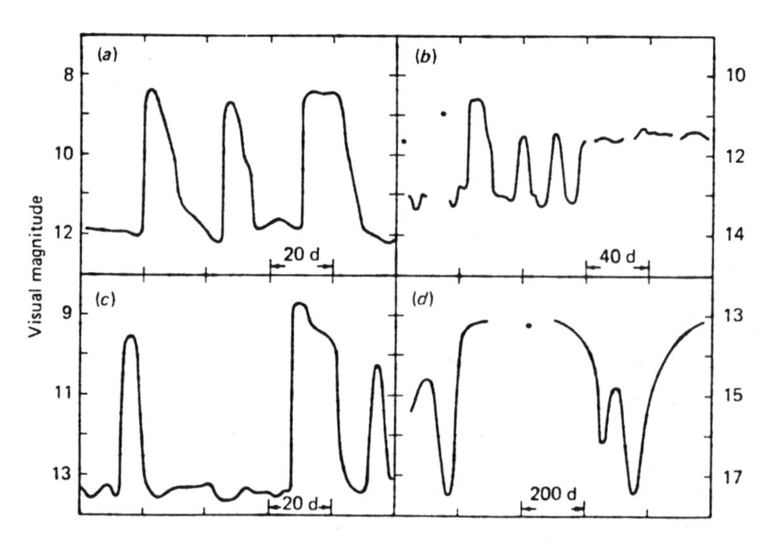

Fig. 2.27. Visual light curves for three classes of dwarf novae: (a) a U Gem DN, (b) Z Camelopardalis, (c) an SU UMa DN. (Panel d is for a weird beast.) From Wade and Ward (1985), ©1985 Cambridge University Press; reproduced with permission.

2.11.2 Supernovae

Supernovae (SN or, sometimes, SNe) are the most spectacular variables of all. At maximum light, they are as bright as a whole, smallish galaxy, and recognizing them for what they are was part of the total process between 1900 and 1925 C.E. that sorted out the approximate size of the Milky Way and demonstrated the existence of other galaxies. There is a sort of family resemblance among all supernovae—they get really bright in a matter of days and fade in months to years. Their spectra display very broad features (a combination and emission and absorption), indicating velocities of thousands of km s^{-1}. And they blow out a solar mass or more of material at these large velocities that can then be seen as a supernova remnant for thousands of years thereafter. A large galaxy experiences one to a few per century, though the Milky Way seems overdue for its next. Combing (mostly) Chinese, Japanese, and European records has established supernova events in 1006 C.E., 1054 (leaving the Crab Nebula), 1181, 1572 (seen by Tycho), 1604 (seen by Kepler), and 1685 (seen by Flamsteed, the first Astronomer Royal).[12]

A closer look at the spectra shows two basic supernova categories, called inevitably Type I and Type II. Type I spectra show no evidence of any hydrogen, though it is the most abundant element just about any place you

[12] We again refer you to Arnett's (1996) monograph for more information on supernovae. For a shorter review to start out with, see Burrows (2000). And, for a popular review of historical SN in our galaxy, see Stephenson & Green (2003).

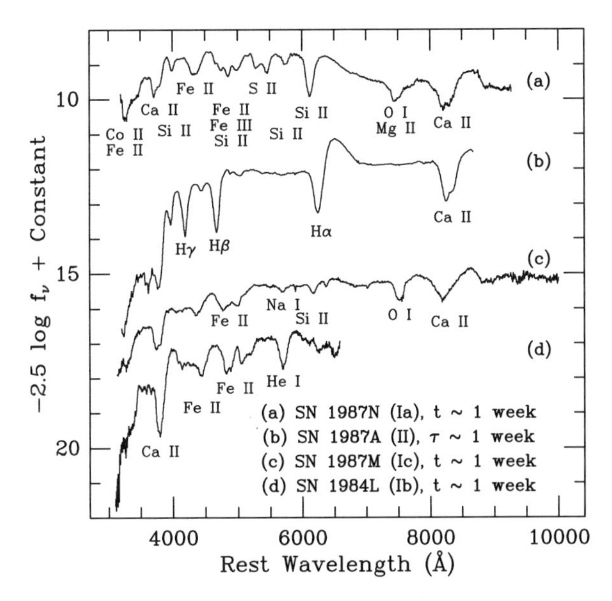

Fig. 2.28. Shown are the spectra of the four major types of supernovae captured about one week after maximum light in the B-band ("$t \sim 1$ week") or after core collapse ("$\tau \sim 1$ week"). The ordinate is essentially magnitudes in a spectral band (f_ν is flux) and the wavelengths of the abscissa are in the rest frame of the supernova. Reproduced with permission, from Filippenko (1997), Annual Review of Astronomy and Astrophysics, Vol. 35, ©1997 by Annual Reviews.

look in the universe, while Type II spectra have strong emission and absorption features due to hydrogen. This is illustrated in Fig. 2.28, where the spectra of four types are plotted against wavelength.[13] Three of the spectra, corresponding to Type I subtypes Ia, Ib, and Ic, show strong features due to ionized iron, calcium, etc., but not a sign of hydrogen. (If there is hydrogen between us and the SN we may see some, but this would be accidental.)

The Type II spectrum, on the other hand, has strong hydrogen absorption lines. In addition, Type II events always, or nearly always, occur in galaxies with recent, vigorous, star formation and in regions of that star formation (i.e., among Pop I stars), while Type I events can also occur in elliptical galaxies and galactic bulges and halos (i.e., among Pop II stars). Type II's expel more mass but at lower velocity, and there are also systematic but rather subtle differences between the two sorts of light curves. Type II's are considerably more likely to be picked up as radio and x-ray sources, usually at later times than the visible light peak.

[13] The University of Oklahoma hosts a WWW site devoted to archival spectra and light curves for supernovae. The access address is `http://tor.nhn.ou.edu/~suspect`.

The distinction between Type I and II supernova almost, but not exactly, corresponds to a very fundamental difference in what is going on in the two cases. Type II events (which are a commoner sort, though somewhat fainter and so harder to discover) are the products of the collapsing cores of massive stars (where we left you hanging at the end of §2.8). The basic energy source is the gravitational energy released in the collapse, often more than 10^{53} erg. Of this, most comes out in neutrinos, 1% or so in kinetic energy of the ejecta, and less than 0.1% in visible light and other electromagnetic radiation. Evidence for this mechanism includes the presence of the collapsed core (pulsar or rapidly rotating magnetized neutron star) at the center of the SN1054 remnant, the Crab Nebula, and the burst of neutrinos seen from SN1987A, which we will come back to shortly.[14]

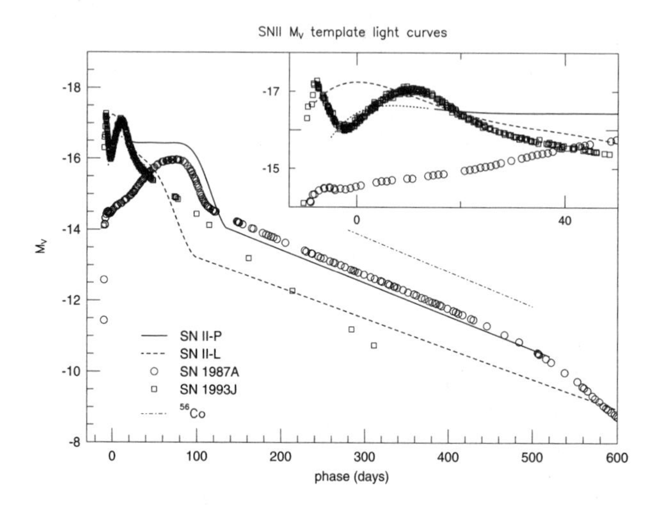

Fig. 2.29. A selection of light curves for Type II plateau and linear subtypes along with SN1987A and SN1993J. Also shown is "^{56}Co," whose radioactive decay helps power the light curve resulting in a decline rate of roughly one magnitude per 100 days (see text). The insert is on an expanded time scale. A Hubble constant of 75 km s^{-1} Mpc^{-1} is assumed. Reproduced with permission, from Wheeler and Benetti (1999), ©1999 by Springer–Verlag.

[14] Supernova are labeled by year of discovery and the order in which they were discovered. Thus SN1987A was not necessarily the most important event of the year, just the first. After 26 have been ordered, the labels are SN1998aa, ab, etc., to az, then SN1998ba to bz, and so forth. In recent years, discoveries have reached the e's and f's, and eventually a new system will be needed.

In an ordinary Type II event, there is a good deal of the original hydrogen-rich envelope left when the core collapses, which is heated and ionized by the outgoing blast wave, producing the hydrogen lines in the spectra. If a massive star has lost its hydrogen envelope (in a strong wind or by transfer to a companion star) before its core collapses, there will be no hydrogen lines. A composite set of Type II light curves is shown in Fig. 2.29. These include the "plateau" subtype, SN II-P, where the decline is held up for a while, and the "linear" subtype, SN II-L, which declines with essentially no hang-up. The best-studied event has been SN 1987A. Note that the figure has data for times well before maximum.

SN1987A

Rather than give a general discussion of Type II supernovae, we shall concentrate on SN1987A in the Large Magellanic Cloud (LMC), which was first observed visually and photographically on February 24, 1987. Despite the fact that this is not a typical SN II object—its light curve and spectrum are almost unique, and it is intrinsically dimmer than what is typical—the basic physical processes driving the explosion are most certainly those of other SN Type II. In addition, we have a wealth of information concerning this object because of its relative proximity. We shall rely on the review by Arnett et al. (1989, and see Arnett, 1996, §13.6, and Wheeler and Benetti, 1999) in what follows.

First of all, we know which star exploded. It was Sanduleak–69° 202, which was a B3 I blue supergiant with $\mathcal{L} \approx 1.1 \times 10^5\ \mathcal{L}_\odot$ and $T_{\text{eff}} \approx 16,500$ K. From various lines of evidence, it is estimated that the main sequence mass of Sanduleak–69° 202 was in the range 16 to 20 $\mathcal{M}_\odot$ and that during its pre-supernova evolutionary stages it lost perhaps a few solar masses of its hydrogen-rich envelope. Although the star was certainly a Pop I object, its original composition was metal-poor compared to objects of similar mass in our galaxy: low-metallicity stars are characteristic of stars in the neighborhood of SN1987A and for the LMC in general.

The whole story has not yet been unraveled, but SN1987A ended up as a blue star before it exploded. More usual SN II events are thought to involve red supergiants and this difference explains why SN1987A is peculiar. (However, comparatively low-luminosity supernovae such as SN1987A may be much more common than we think: we just have a harder time finding them than we do the "normal" brighter objects.)

Perhaps the two most important observations made of SN1987A are the detection of neutrinos *prior to* the optical discovery and the later detection of radioactive ^{56}Co. These are the two keys to our understanding of how the star exploded and both were anticipated by earlier theoretical work on the modeling of Type II events. To explain this we need to recall the thermonuclear burning stages of a star with a mass comparable to SN1987A. This is

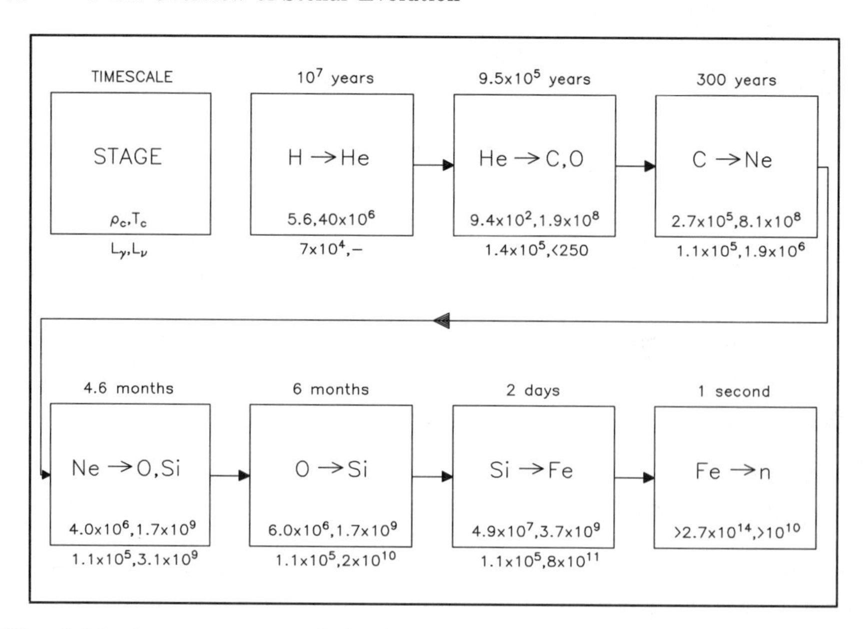

Fig. 2.30. A representation of the thermonuclear burning stages of a star similar to SN1987A. The first box is a key to the notations in the boxes following—each of which represents a stage. Central density is in the units of g cm^{-3}, T_c in K, and photon and neutrino luminosities are in the units of $\mathcal{L}_\odot$. The progression of arrows indicates the arrow of time. The figure is adapted from Table 1 of Arnett et al. (1989).

shown schematically in Fig. 2.30, where each box represents an active burning stage at the center of the star. Also indicated is the lifetime of each stage, the central density and temperature, the total stellar luminosity and, finally, the total power given off in the form of neutrinos. An "onion-skin" diagram for the last stage is shown in Fig. 2.31.

The amount of iron core that is formed by the burning of silicon is approximately 1 $\mathcal{M}_\odot$ and has a radius near that of a white dwarf. This core (or perhaps the innermost part of it), having no further source of nuclear energy production and losing energy from neutrinos, now collapses on a time scale of seconds or less. As it does so, the temperature rises rapidly until the radiation bath of high-energy photons (in the form of gamma rays) interacts with the iron peak nuclei and effectively boils off their constituent nucleons. What speeds the process along is that at the very high densities encountered during the collapse, electrons gain enough energy that they may be captured on nuclei, thus converting protons into neutrons plus neutrinos. These processes continue until the stellar plasma of the core is reduced to a sea of mostly neutrons at a density comparable to or exceeding that of nuclear matter (2.7×10^{14}g cm^{-3}) confined in a radius measured in only tens

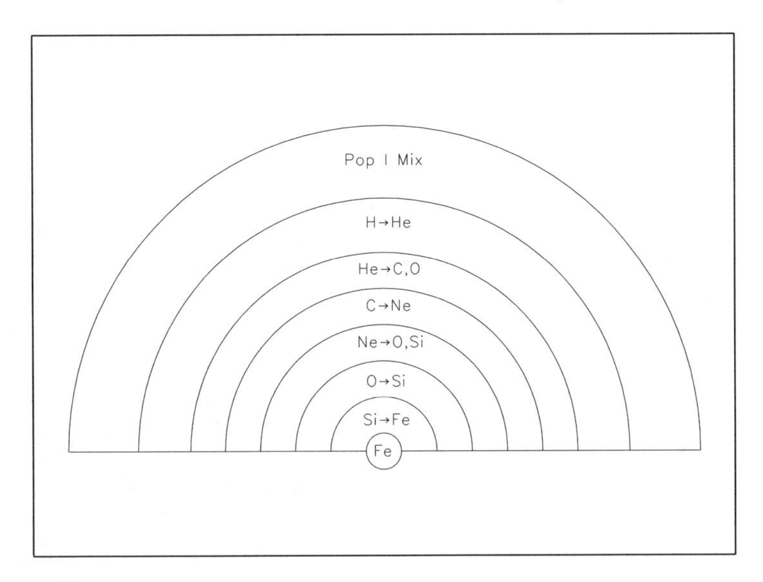

Fig. 2.31. An onion-skin diagram for the last stage of Type II presupernovae. The thickness of the layers is not to scale.

of kilometers. Further collapse is effectively halted by the very stiff equation of state of nuclear matter.

What has happened is that all the nuclear burning stages the core has experienced—from hydrogen-burning through to the production of iron peak material—have been unraveled. This means that all the energy produced in those stages must be repaid back along with the energy lost through the emission of neutrinos. To estimate what is owed we consider first the nuclear energetics. The binding energy per nucleon of nuclei in the iron peak is about 9 MeV/nucleon (see Chap. 6). If an average nucleus in this peak has 56 nucleons, it then requires about 8×10^{-4} ergs to reduce the nucleus to neutrons and protons and, for a unit solar mass containing about 2×10^{55} such nuclei, the total owed back for nuclear burning is about 2×10^{52} ergs. Finally (and as probably a lower limit), two days of neutrino emission at a rate of $8\times10^{11}\ \mathcal{L}_\odot$ (see Fig. 2.30) totals more than 10^{50} ergs. Where does all this energy come from? Actually, this is easy to answer because we have caused an object roughly the size of a white dwarf to collapse down its gravitational potential well to something the size of 10 km in radius. You may easily calculate that if 1 $\mathcal{M}_\odot$ is involved, the total release of gravitational energy is, and see §1.2,

$$|\Delta\Omega| \approx \frac{G(\mathcal{M}_\odot)^2}{10^6} \approx 3\times10^{53} \quad \text{erg}\,.$$

This is more than enough to repay debts. How this energy is used to blow up the star is still a matter of some controversy but what is involved is some way of abstracting a portion of the energy from out of the collapsed (or collapsing) core and depositing it further out in the star. This may be accomplished by having core material "bounce" as it reaches nuclear density (or beyond) and, as it bounces, collide with infalling material thus forming a shock that propagates outward lifting off most or all of the remainder of the star. The alternative is to produce enough high-energy neutrinos (by various processes ultimately relying on the energy released by collapse) so that some fraction of them might interact with overlying material to the extent that they effectively push off the outer layers. The calculations are very difficult and depend on many physical parameters (plus how the numerical work is done). Either way of doing things, however, and in the best of worlds, yields supernova energies in the proper range and, more to the point, predicts that high fluxes of neutrinos will pass out of the star before the event is seen optically, and that the violence of the event will cause nuclear processing of ejected material to iron peak nuclei—including large amounts of ^{56}Ni.

Both of these conditions are met for SN1987A. Neutrinos were detected about a quarter of a day before optical discovery, with energies within the proper range and over a time scale (5 to 10 s) that seems reasonable given the time scale estimated for their production. Furthermore, gamma-ray lines of ^{56}Co (half-life of 77 days) were detected well after the event (at about 160 days), consistent with the early production of ^{56}Ni. The significance of this is the decay sequence $^{56}\mathrm{Ni} \Rightarrow {}^{56}\mathrm{Co} \Rightarrow {}^{56}\mathrm{Fe}$, which not only makes ^{56}Fe (see Fig. 2.20) but also provides an energy source for the expanding ejecta. Modeling of the later light curve powered by these decays is a success story for supernova calculations (see Arnett 1996, §13.4).

One further question remains, and this is whether a compact neutron-rich or black hole remnant lurks within the exploding debris. All models point to a remnant neutron star but none has been observed as yet. In its simplest manifestation it would appear as a pulsar but this requires that the neutron star have a strong magnetic field and that it be rotating (as the least of the requirements).

SN Type I

In contrast to Type II events, Type I SN show no evidence for hydrogen in their spectra, strongly suggesting that the object that explodes has lost its hydrogen envelope (in a strong wind or by transfer to a companion star), leaving an "undressed" core. Two subsets of Type I events, called Types Ib and Ic (and characterized by anomalies in the lines produced by heavier elements), are thought to be powered by the collapse of such undressed cores, and they, like the Type II's, are confined to regions of recent star formation.

In contrast, classic Type I events, now called Type Ia, occur in a wide range of galaxies, locations, and stellar populations.[15] There is no evidence for

[15] A semi-popular discussion of Type Ia SN is by Maurer, S.M., & Howell, D.A. 2002, *Anatomy of a Supernova*, Sky&Tel, 104, 22. The graphics are excellent. Also see Branch, D., 2003, *Science*, 3 Jan 2003, p. 53.

formation of a neutron star or other condensed remnant. All observations are consistent with and even suggest that the energy source for these is explosive fusion of about one solar mass of carbon and oxygen to iron-peak elements, especially ^{56}Ni.

By now, you should be saying to yourself, "Aha! Explosive nuclear burning. The fuel must has been degenerate at ignition. Where might I find enough degenerate carbon and oxygen to do this?" And your better self will answer, "A white dwarf!" Of course a white dwarf of less than the limiting stable mass $\mathcal{M}_\infty$ (the Chandrasekhar limiting mass discussed in §3.5.2) will just sit and cool for the age of the universe.

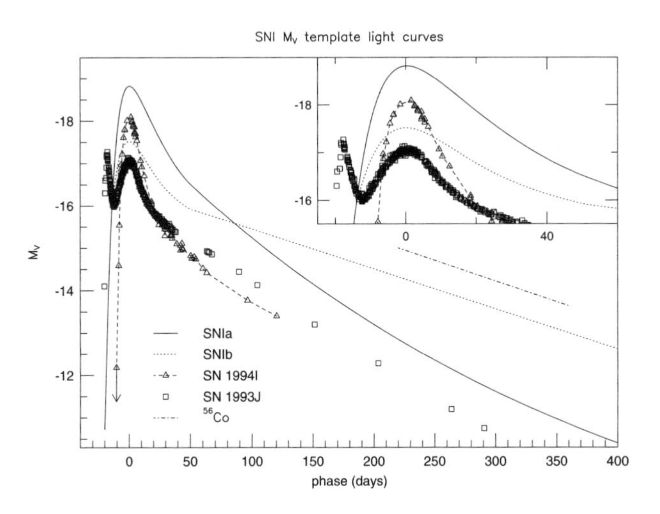

Fig. 2.32. A selection of light curves for Type I SN. The solid line is a composite of many Type Ia events, while SN1994I is Type Ic. Reproduced with permission, from Wheeler and Benetti (1999), ©1999 by Springer–Verlag.

Thus something must happen to increase the mass beyond the limiting stable mass. One possibility is the merger of a pair of white dwarfs, each of mass less than $\mathcal{M}_\infty$ but with the sum greater than $\mathcal{M}_\infty$. This is many people's "best buy" model. It accounts for the absence of hydrogen, the possibility of occurrence among even very old stars, and for the rather boring sameness of SN Ia light curves, as shown in the composite light curve of Fig. 2.32. SN Ia are not quite standard candles (all the same peak brightness), but they can be used as distance indicators out to redshifts of unity and beyond. (The peak blue magnitude of Type Ia SNs is estimated to be $-18.4 \pm 0.3 + 5 \log h$, where h is Hubble's constant in units of 100 km s^{-1} Mpc^{-1}.) "Boring," in reference to the light curve, means that they all decline in very much the same way with a slope of 0.012–0.015 mag day^{-1}. They are currently being used that way to provide hints of the distant past and long-range future of the expansion of the universe.

The only problem with the white dwarf pairs as SN Ia progenitors is that there don't seem to be very many (not any some would say) with both adequate mass and the ability to merge in the age of the universe. An alternative progenitor class is the recurrent novae, some of which, at least, have white dwarf masses very close to $\mathcal{M}_\infty$ and do not blow off in the explosion all the material they accreted to make it happen. Thus their masses are gradually increasing. It is not entirely clear why no hydrogen emission or absorption appears in the SN spectrum if this is what is going on, since the donor star (though not the white dwarf accretor) is made mostly of hydrogen.

If matter is accreted onto a white dwarf with mass close to $\mathcal{M}_\infty$, the star gets smaller. (Recall for white dwarfs that more massive means smaller, and less massive means bigger.) But this results in compressional heating. If the added heat cannot be transported away sufficiently rapidly, then potential thermonuclear fuels—such as the carbon and oxygen thought to make up most of a white dwarf—may be ignited. Because of the extremely high densities in the interior, any ignition of fuel initiates a runaway explosion and a supernova is born.

To demonstrate the possible energetics of such an explosion, consider the thermonuclear burning of pure carbon in the form of ^{12}C under these conditions. If the burning is not somehow controlled, then a sequence of reactions rapidly processes the carbon to elements in the mass range of iron with, yet again, ^{56}Ni as the most abundant. The energy released by the formation of ^{56}Ni (from 4+2/3 ^{12}C nuclei) is 8.25×10^{-5} erg. Were we to convert 1 $\mathcal{M}_\odot$ in this way, the total energy release would be about 10^{52} erg, with plenty to spare for a supernova.

Details of the explosion process have been explored by many investigators using state-of-the-art hydrodynamic computer codes but some uncertainties remain. (For a review of SN Type Ia models, see Hillebrandt and Niemeyer, 2000.) Two crucial parameters are the rate of mass accretion onto the white dwarf and the mass of the white dwarf. Different choices of combinations of these parameters lead to quite different events. It is possible, for example, completely to disrupt the white dwarf and leave no stellar remnant or to have a partial explosion that leaves behind a white dwarf of lesser mass. Other possibilities include the detonation of helium in a white dwarf that has not converted all of that element to carbon and oxygen. In any case, it is not thought that a neutron star—and, perhaps, pulsar—would be left behind as a remnant.

SN Remnants

The supernova remnants (SNRs) we see range in age from less than 20 years (SN1987A is just getting big enough to resolve) to 10^4 or more. Many are sources of radio waves and x-rays as well as visible light. Where there is a central pulsar, it continues to pour energy into the remnant until the gas is too dispersed to be seen. Pulsar-less remnants nevertheless remain bright for as long as the expanding ejecta are plowing into surrounding interstellar gas. They look brightest around the edges. The Crab Nebula, pulsar fed and left from the supernova seen by the Chinese, Arabs, and possibly Europeans in 1054 C.E., is the best known and most thoroughly studied supernova remnant. Recent x-ray images show energy from the pulsar being beamed out along the long axis of the prolate nebula, which is brightest at the center. Cas A (meaning the brightest radio source in Casseiopia) is the remnant of SN1685 (or thereabouts). Our view of its optical emission is partly blocked by dust, but the radio and x-ray images show that it is brightest around the edges, consistent with the absence of a detectable pulsar (though there is a central faint, point source, which could be a residual neutron star or black hole accreting some material from its surroundings). The gas in both of these SNRs includes lots of hydrogen, so presumably they would have been classified as Type II supernovae, though both Flamsteed and the medieval Chinese unaccountably neglected to photograph the spectrum for us.

Supernova remnants are important in the great scheme of things as heaters and stirrers of interstellar gas, probably as triggers to collapse gas clouds to initiate star formation, and probably as the accelerators of cosmic rays—particles, mostly protons, with kinetic energies greatly exceeding $m_{\mathrm{p}}c^2$ which pervade the Milky Way and other galaxies. Cosmic rays produce most of our lithium, beryllium, and boron, make ^{14}C in the upper atmosphere, and are a major source of mutations in terrestrial creatures.

2.12 White Dwarfs, Neutron Stars, and Black Holes

These are the three, and only three, ways we found in earlier sections that stars could end their lives. (Note that the title of this section is a permutation of the title of Shapiro and Teukolsky's 1983 text *Black Holes, White Dwarfs, and Neutron Stars*, which should be consulted.) Chapter 10 explains how to calculate what the inside of a white dwarf should be like, and neutron star structure is handled in much the same way. There are two catches, however.

First, neutron stars are compact enough that the equation in which pressure balances gravity must be rewritten with gravity described by general relativity, rather than its Newtonian approximation. This Tolman-Oppenheimer–Volkoff equation (so called because it was first written down by Lemaître but, in any case, you can play with it in Ex. 1.7) makes it clear that, deep down in the potential well, Einsteinian gravity is stronger than Newtonian by a factor of $\sqrt{(1 - 2\mathcal{M}G/rc^2)}$ in the denominator. This lowers the maximum mass that can be supported by degenerate pressure to about $0.7\ \mathcal{M}_\odot$, as Oppenheimer and Volkoff found in 1939.

Second, the equation of state, describing how pressure depends on density, temperature (and composition), must include the nuclear force as well as quantum mechanics and degeneracy pressure. At distances less than about 1 fm (10^{-15} m), the nuclear potential is repulsive and so helps balance gravity, raising the mass limit. This is good. Otherwise there would probably be no neutron stars; anything that made it past white dwarf density would continue down inside its own Schwarzschild horizon and become a black hole. The downside is that we have no closed, complete theory of the nuclear force to correspond to quantum electrodynamics for the electromagnetic force, and laboratory conditions cannot quite duplicate the enormous assemblages of neutrons needed to see just what the potential looks like. As a result, the literature includes many different equations of state for "dense nuclear matter" that imply different internal structures for neutron stars, different limiting masses, characteristic radii, break-up rotation velocities, and so forth. Indeed the interior of a neutron star need not even be primarily made of neutrons. Hyperons, muon condensates, and strange quark matter have all been suggested.

In contrast, the structure of a black hole is remarkably simple. It can be characterized by a mass (as measured by Kepler's third law from far away), an angular momentum (measured by how it drags space–time around up close), and electric charge (probably zero, since free electrons and protons can flow in). That's it. Of course calculating things (like the shortest period orbit that is possible for a test particle going around a black hole, or how synchrotron radiation will be modified if both the magnetic field and relativistic particles are being dragged around) is, at best, extremely difficult.

We can, however, summarize the expected and measured properties of white dwarfs, neutron stars, and stellar–mass black holes, as is done in Figs. 2.33 and 2.34. Notice that the three categories were initially defined in very different ways: white dwarfs by their location on an HR diagram, neutron stars in analogy as something supported by degenerate neutron pressure, and black holes (the idea predates general relativity by more than a century) by having an escape velocity larger than the speed of light. Given these definitions, white dwarfs must exist, neutron stars exist on the "walks like a duck, quacks like a duck" principle, and black holes exist on the Sherlock Holmes principle that, if you have eliminated the impossible (neutron stars of 10 $\mathcal{M}_{\odot}$ and so forth) whatever remains, no matter how improbable, must be the truth. Whether astrophysical black holes have all the internal properties implied by general relativity cannot be determined from outside (and all the students sent inside to do thesis research on this topic have so far failed to return and submit their theses) and for our purposes does not matter.

Property	White Dwarfs	Neutron Stars	Stellar Mass Black Holes
Definition	Position in HR diagram	Pressure from degenerate neutrons	v_{esc} greater than c
Prediction	———	Zwicky 1933-34	Michell 1784, Laplace 1798
Discovery			
wide binary	Bessell 1844, Clark 1862 WS Adams 1914 (Sir B)	———	———
single	van Maanen's star c. 1920	pulsar CP1919, Hewish, Bell 1967-68	tentative: 1999 MACHO events toward galactic bulge
interacting binaries	cataclysmic variables Struve, Kraft 1950s	interpretation of Sco X-1 Zeldovich et al. 1964+	orbit of Cyg X-1, Bolton, Murdin 1972
Masses	0.2-1.38 M_{sun} CBS orbits surface gravity, log g=8 gravitational redshifts 30-75 km s^{-1} probable correlation of M and interior compositions	0.8-1.44 M_{sun} CBS orbits gravitational redshifts	6-10 M_{sun} CBS orbits. BHXRB, duration of MACHO events
Sizes	radius ~ R_{earth}, $L=4\pi R^2\sigma T^4$ eclipse timing	R~10 km, X-ray L & T 10^{6-7} K, X-ray colors	some limits from rapid variability
Luminosity & energy sources	thermal (single WD)	thermal, ISM accretion (single, non-pulsar)	accretion disks 10^{6-7} K X-ray colors
	accretion (cataclysmic variables), nuclear (novae)	accretion (X-ray binaries), nuclear (X-ray bursters), magnetic extraction of rotational KE (pulsars),	accretion, perhaps Blandford-Znajek extraction of rotational KE, X-ray binaries
	10^{-4}-10^2 L_{sun} 10^4 L_{sun} in novae	0.001 of to greater than Eddington limit	very small (advection dominated accretion?) to greater than Eddington limit

Fig. 2.33. Our "crib sheet" for the properties of white dwarfs, neutron stars, and black holes. Codes used in this figure are: CBS = close binary systems, BHXRB = black hole x-ray binaries, ISM = interstellar medium, LMXRB = low mass x-ray binaries, HMXRB = high mass x-ray binaries, MACHO = MAssive Compact Halo Objects. Continued in Fig. 2.34.

2.13 Binary Stars

Half or more of all the dots of light you see in the sky actually represent two (or occasionally more) stars, and these gravitationally bound pairs, or binary stars, are nearly the only source of information we have about stellar masses, the most important thing in a star's life. In addition, some of the most spectacular of astronomical phenomena—novae, some supernovae, some gamma ray bursts—and many of the favorite objects of amateur astronomers (eclipsing variables, blue plus yellow pairs) are binaries. Why, then, are they left almost for last? Mostly because one needs to know about the evolution of single stars before one can put them together and follow the evolution of pairs.

2.13.1 Types of Binaries

Binary stars can tell us about their duplicity in many ways and are classified accordingly. Eclipsing binaries are those in which one star passes in front of

Property	White Dwarfs	Neutron Stars	Stellar Mass Black Holes
Rotation periods	minutes to century, most much slower than break-up; bimodal?	1.55 ms to 1500 s, some at breal-up	J/M^2=0.4 to 0.95 (c=G=1 units) of maximum allowed by general relativity
	v sin i, polarization variability	pulsar periods, $L_X(t)$ in XRBs	line profiles
	break-up is seconds to minutes	break-up=ms	J/M^2=1 maximum possible
Magnetic fields	hydrogen line splitting, circular polarization	channeling of gas accretion, P vs. dP/dt of pulsars	(attached to disks)
	10^4 to 10^9 G, bimodal	10^{8-9} G, LMXRB, recycled pulsars	
	binaries different, fields commoner, most 10^8 G	10^{11-13} G young pulsars, HMXRB	
	geometry: off-center dipole	geometry: inclined dipole	
Space motions	disk & halo populations	single=high velocity=runaways, binaries=low velocity	single=TBD, binaries=low velocity
Birthrates	1/year, by direct counts & cooling ages, #s of PNe and expansion ages	2-3/century: pulsar ages, SN rates	binaries:a few % of SN rate
Composition			
Core	He, CO, ONeMg (nova ejecta)	neutrons (assumed), superconductor, superfluid (confirmation from cooling and pulsar glitches)	no information & doesn't matter!
Surface	H, or He, or CO (often nearly pure; sometimes metal contaminants in H or He)	iron and neutron-rich isotopes	no surface

Fig. 2.34. Our WD, NS, & BH "crib sheet," continued.

the other from our point of view, wholly or partially blocking its light, so that we see periodic variability. Eclipsing systems are generally ones in which the stars are fairly close together, because this makes it more likely that we be close enough to the orbit plane to see the eclipse.

Spectroscopic binaries (called SB1 and SB2 or single- and double-lined spectroscopic binaries, depending on whether you see spectral features from one of the two stars or both) are systems in which the periodic change of stellar speed along our line of sight through the orbit is large enough for the Doppler shift to be detectable. Once that meant speeds projected along our line of sight of 10 km s^{-1} or more. The state of the art is now more like 1 km s^{-1} (and it is this capability that has enabled the detection of Jupiter-mass planets orbiting many dozens of nearby stars).

Spectroscopic binaries are of enormous value to astronomy, because the velocities of the two stars plus the orbital period, the information from the eclipse—which means the orbit is nearly edge on—and a bit of arithmetic permit calculating the masses of the two stars. The lengths of the eclipse tell us the radii of the stars; the spectral types contain information about the stellar temperatures, and with both $\mathcal{R}$ and $T_{\rm eff}$ you can immediately calculate the absolute luminosities of the two stars and compare all this information with evolutionary tracks of the masses you measured.

Pairs where you see two sets of spectral features corresponding to different types but no Doppler shifts are called *spectrum binaries* (and must have either very long orbital periods or orbits that we see nearly face on). *Visual binaries* are the ones where you can two dots of light moving around each other in the sky, and can also be analyzed to determine the masses of the stars, though you must be patient. Most visual orbit periods are years to decades to centuries, while spectrum binaries have periods of days to weeks.

If the two stars are too close to each other for you to resolve the two dots of light in the sky, the centroid of the dot may still move around, or you may see the light from a seemingly-single star wiggle back and forth across the sky as it orbits something too faint to resolve. These are the *astrometric binaries*. The first white dwarfs, Sirius B and Procyon B, were originally recognized from astrometric orbits of the bright stars Sirius and Procyon, otherwise known α CMa and α CMi. That the two dog stars both have white dwarf companions is thought to be a coincidence.

Still wider pairs of stars may still be gravitationally bound but have orbit periods too long and motions too slow to have been seen yet. But they will move together through space and are dignified by the name *common proper motions pairs*. Proper motion is not a term of social approbation but the standard name for the motion of a star (etc.) across the sky, measured in arcsec year^{-1}.

2.13.2 The Roche Geometry

The need for specific investigation of stellar evolution in binaries is made clear by the Algol (β Per) paradox. You know from earlier parts of this text that massive stars evolve faster than petite ones (as in $t_{\text{nuc}} \propto \mathcal{M}^{-2.9}$ of Eq. 1.91). Yet there is a large group of eclipsing binaries, of which Algol is the prototype, in which a clearly evolved giant (often of spectral class K) orbits a clearly less-evolved main sequence star (often of spectral type A or F). And the evolved giant star is the less massive of the two. This paradox has its resolution in the phenomenon of mass transfer in close binaries. That is, the initially more massive star has managed to off-load a good deal of its substance onto the initially less massive star. The donor, however, continues to evolve as if it still had its initial mass and so will finish first.

Just what happens is always discussed within the framework of what is called the Roche geometry (for the French mathematician; though Lagrange of Turin also comes into the story). Think of yourself as a test particle living somewhere near a binary star system and at rest in a coordinate system that rotates with its orbit period. The orbit of each star is assumed to be circular. That is, if the stars are going around their mutual center of mass in circles with period Π or angular velocity Ω, you too are going around the center of mass with period Π and angular velocity Ω. Now trace out the equipotential surfaces—the ones that you can walk (or fly or rocket) around without having to do any work, like the surface of the rotationally-distorted earth.

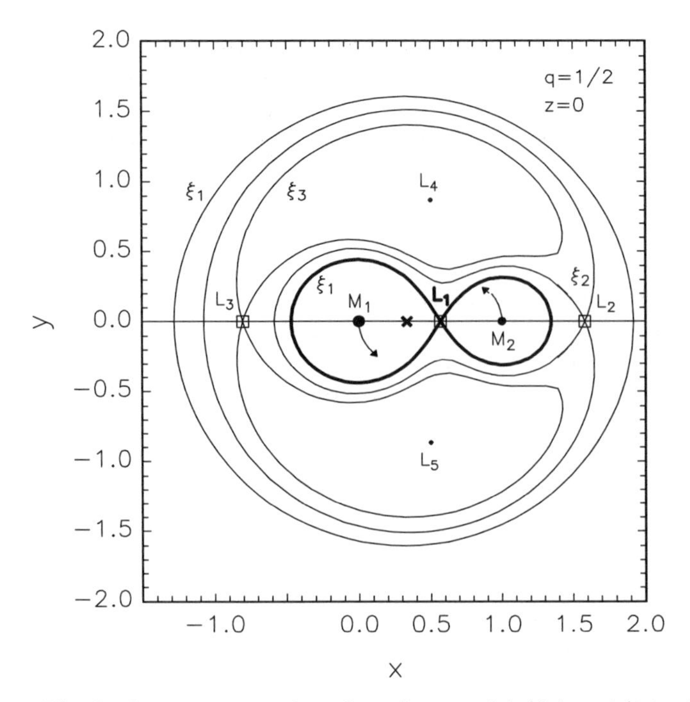

Fig. 2.35. The Roche equipotential surfaces for $q = \mathcal{M}_2/\mathcal{M}_1 = 1/2$ in the plane of rotation. Three surfaces (ξ_1, ξ_2, and ξ_3) are labeled. (To find out what the ξ_i represent, see the text and Ex. 2.19.) Boxes (□) enclose three of the Lagrangian points. The boldface × is the center–of–mass of the system. The coordinates x (the axis connecting the two stars) and y are in units of the distance between the stars.

What you will find is shown in cross-section in Fig. 2.35. (Ex. 2.19 explores some of the mechanics necessary to set up the equipotentials in this figure.) The two black dots are the stars, $\mathcal{M}_1$ the primary (more massive), and $\mathcal{M}_2$ the secondary with $\mathcal{M}_1 > \mathcal{M}_2$. (For simplicity it is assumed that the stars are just massive dots with spherically symmetric gravitational fields.) In this particular case $\mathcal{M}_1 = 2\mathcal{M}_2$, and the center of mass is at the point ×. We will adhere to the custom of theorists in binary astronomy of continuing to call $\mathcal{M}_1$ the primary no matter what happens to the stars subsequently. Working outward from the stars, first you find that the equipotentials are nearly spherical, with one sphere surrounding each star, just as for a single point mass. But soon you come to ξ_1, called the inner Lagrangian surface, which is made up of two *Roche lobes*, one surrounding each star. The point L_1 between them is the first Lagrangian point. Clearly if one star *fills its Roche lobe* and the other doesn't, material is free to flow from the star with the filled lobe to its companion. and if both lobes are filled, it would seem that gas could slop back and forth between the two stars.

The Roche geometry is responsible for another set of names for types of binaries. In detached systems, neither star fills its lobe. In semi-detached systems (like Algol) one does. And in contact systems both lobes are filled. The W UMa variables are contact systems of relatively low mass, and you see a bit of an eclipse almost no matter what direction you look at them from. Observed binary systems have separations (that is, major axes of their relative elliptical orbits) that range in size from the sum of the radii (contact systems) up to about 0.1 pc or 20,000 AU. Wider systems may well have formed but are very easily disrupted by any passing star or gas cloud.

Still further out from the stars comes the first equipotential that is open to the rest of space at the point L_2 (on the equipotential surface ξ_2). Gas that reaches this point is free to leave the system completely, as is even more true for gas that reaches L_3 (on ξ_3). The points L_4 and L_5 have no special significance for binary evolution but are analogous to the Trojan asteroid orbit locations for the Sun–Jupiter system and might be good places to put artificial satellites if you were living in the system. (In this context they are called the "Trojan points" and form equilateral triangles with the sun and Jupiter at two of the corners. Generally they are referred to as "triangle solutions.") A pair in which both stars over-fill their Roches lobes is a common envelope pair (CEB). Gas that leaves through L_2 and L_3 will have more than its fair share of angular momentum (because they are further from the center-of-mass than the main bodies of the stars). The residual stars will spiral together, perhaps rather rapidly. We know this must happen because there are pairs of evolved stars (white dwarf plus red dwarf for instance) whose present separation is smaller than the sum of their radii when they were both on the main sequence.

2.13.3 Formation and Early Evolution

Since we do not understand star formation very well this is obviously doubly so for binary star formation. It is probable, however, that two or more modes are in operation. One might, for instance, imagine two (or more) dense cores in a molecular cloud that happen to condense close enough together for the resulting star pair to be gravitationally bound. Such binaries should be wide pairs. Second, in a crowded environment with many protostars milling about, tidal capture may occur. The extra kinetic energy of the initially hyperbolic encounter has to go some place, via dissipation in the protostars, and extended disks make the process more efficient, while in turn perhaps disrupting the disks and making planet formation unlikely. This is OK, though you wouldn't want to try to live on a planet in a close binary system anyway. The orbit, like most three body processes, is likely to be unstable and send your planet careening out into distant space or crashing into one of the stars.

Third and last, a contracting core may fragment into two pieces, especially if it is rotating rapidly. Even the gentle rotation observed for many molecular clouds is enough that, with a contraction process that conserves angular momentum, many cores will be close to the break-up rotation speed (as in Ex. 2.20, and see Bodenheimer, 1995],for a discussion of angular momentum problems for young stars and disks). Fragmentation or fission of a single core is expected to produce close binary pairs, arguably with the two components of rather similar mass. Tidal capture, on the other hand, might "choose" stars at random out of the total ensemble and favor unequal pairs. In practice, we observe virtually all sorts of pairs that are physically possible. Notice that this excludes pairs of main sequence stars of such different mass that one has completed core hydrogen burning before the other even reaches the main sequence (mass ratios of less than 0.2 or thereabouts).

Ongoing loss of angular momentum is characteristic of most stars whether single or binary (including the sun). The usual mechanism, at least for stars with convective envelopes, is an outgoing stellar wind, magnetically locked to the rotation of the star close in, which then breaks loose and so carries off a ratio of total angular momentum to mass ($J/\mathcal{M}$) larger than the star (or binary) average. In close binary systems, the rotation and orbital periods are locked much of the time (like the moon to the earth), thus angular momentum loss affects the system as well as the individual stars, and members gradually spiral together. This must be how W UMa (contact) binaries form, since the systems we see are smaller than the stars were during protostellar collapse. Confirming evidence comes from looking at very young star clusters, which have lots of wide pairs but no W UMa stars. Older clusters (including globular clusters) do have W UMa pairs, and it looks like there is a steady supply of them forming from previously detached pairs. But they then merge after another billion years or so, only to be replaced by others coming through the spiraling-in process.

In a system that is initially detached, the two stars live their early lives much as they would in isolation, although they are likely to display more activity than average for their masses and ages both because of the reduced surface gravitational potential and because of their rotation being kept at the orbital period, faster than they might otherwise rotate. The extremity of this is the category called RS CVn stars, where both are slightly evolved (F, G, or K stars as a rule) with winds vigorous enough that the collision region is a source of both x-rays and radio emission, while the individual stars have vigorous chromospheres (detectable as emission features inside strong absorption lines of hydrogen, calcium, etc.).

2.13.4 The First Mass Transfer Phase and its Consequences

Sooner or later, $\mathcal{M}_1$ (the primary) is going to try to become a red giant, as all stars massive enough to have done anything at all in the age of the universe do. If its companion is sufficiently far away, neither star cares about the other, and you can go on to the next section. A "close" binary is, by definition, one in which the primary fills its Roche lobe at some evolutionary phase. Aficionados distinguish three cases: Case A where the lobe is filled during immediate post-main sequence evolution, Case B where filling occurs while the primary is a red giant, and Case C where Roche lobe overflow sets in even later, for instance when $\mathcal{M}_1$ becomes a AGB star. There are people who earn a precarious living by numerically integrating through all of the possible combinations. We will sort of average over them here. Much more detail is given in de Loore and Doom (1992).

When $\mathcal{M}_1$ first fills its Roche lobe, material begins spilling over onto $\mathcal{M}_2$ rapidly because (if you do the arithmetic—as you should in Ex. 2.21) the mass ratio coming closer to unity brings the two stars closer together so that the lobe is shrinking at the same time that the star is trying to expand. Now $\mathcal{M}_1$ can adjust its structure and dump on its thermal or Kelvin–Helmholtz time scale (see Eq. 1.32), but $\mathcal{M}_2$ can only adjust and accept on its thermal time scale, which is longer, since things all scale roughly as $\mathcal{M}^{-2}$ (combining Eqs. 1.32, 1.87, and 1.88). Thus its envelope puffs up until it, too, fills and overfills its Roche lobe, producing a contact or common envelope binary. The first person to discover this was a semi-mythical Berkeley graduate student named Benson, who had intended to couple two stellar evolution codes and follow both stars to the bitter end. Instead, by the time a tenth of a solar mass had passed through L_1, both lobes were full and the codes broke down. It is said that he submitted his thesis and left astronomy forever in about 1971.

But, leaving Benson to his fate, Fig. 2.36 shows the earlier stages of Case A evolution (from de Loore and Doom, 1992). Starting with $\mathcal{M}_1 = 9\ \mathcal{M}_\odot$ and $\mathcal{M}_2 = 5\ \mathcal{M}_\odot$, mass exchange leaves them at $3\mathcal{M}_\odot$ and $11\mathcal{M}_\odot$, respectively, after only 18 million years of evolution. At a later stage, the (former) secondary will do its thing also.

Several stars, including β Lyrae, have been proposed as examples of this rapid mass transfer, when $\mathcal{M}_2$ is so deeply buried in an accretion disk that all it can do is eclipse the other star. Come back in a little while (we mean after a few Kelvin–Helmholtz times, not 2005), however, and the ratio of masses will have been reversed, with part of the outer layers of $\mathcal{M}_1$ lost forever to the system and part settled in on top $\mathcal{M}_2$, and with $\mathcal{M}'_1 > \mathcal{M}'_2$ in the star's new guises. W Serpentis has been proposed as an example of a star at this evolutionary state (but don't bet money on it).

Now, further mass transfer will make the mass ratio more unequal, so that the system expands again and so does $\mathcal{M}'_1$'s lobe. Thus continued adjustment of its structure and further mass transfer occurs on the much longer nuclear time scale. The prototype of this phase is—had you almost forgotten about it?—Algol, with the evolved giant still gently overflowing its Roche lobe and transferring material to the now-more-massive main sequence star.

In due course, $\mathcal{M}'_1$, still evolving on the time scale set by its initial mass—at least for transfer in Cases B and C (sorry!)—completes its evolution as a white dwarf, neutron star, or black hole. The minimum main sequence masses to make a neutron star or black hole are probably somewhat larger

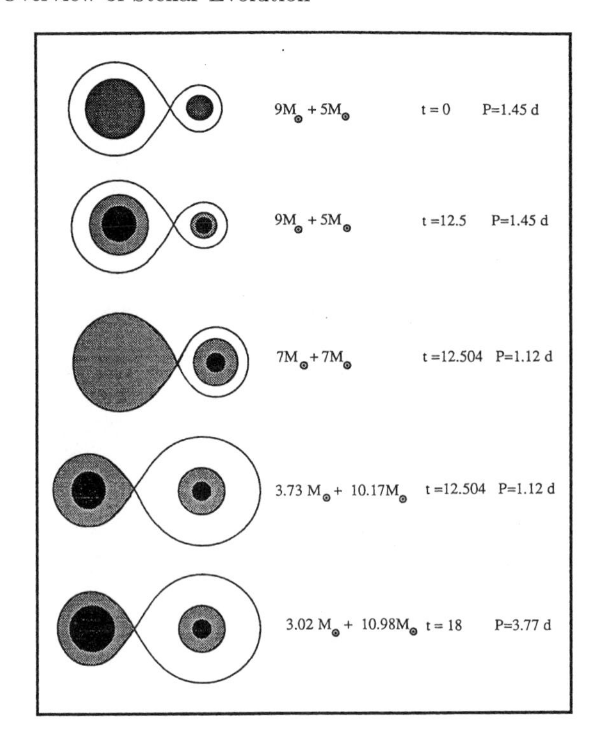

Fig. 2.36. An example of the earlier phases of Case A evolution. The two stars have initial masses of 9 and 5 $\mathcal{M}_\odot$ orbiting each other with a period of 1.45 days. Times (t) are given in millions of years. For the last four panels, the black circles represent helium-rich material, whereas gray is original hydrogen-rich. Note the somewhat dubious assumption here: no mass is supposed to have escaped from the system. Reprinted from de Loore and Doom (1992) with kind permisssion of the authors and Kluwer Academic Publishers (and see Kippenhahn and Weigert, 1967).

than for single stars, for instance 15 $\mathcal{M}_\odot$ versus 8–10 $\mathcal{M}_\odot$ for a neutron star (though this must also depend on initial composition and perhaps rotation and magnetic fields as well). The minimum single star mass needed to make a black hole is not known observationally, though numbers as small as 20 $\mathcal{M}_\odot$ have been proposed by theorists, but there is at least one x-ray binary with $\mathcal{M}_1$ now a neutron star, whose secondary is such that the initial mass of $\mathcal{M}_1$ must have been something like 50 $\mathcal{M}_\odot$.

You might suppose that the loss of mass in a planetary nebula or core collapse supernova would automatically unbind the system. This is, in general, not the case. Remember $\mathcal{M}_1$ is the less massive star by the time it dies, and to unbind a gravitational pair with sudden mass loss, you must remove at least half the total mass (which is left to you in Ex. 2.6). Thus white dwarfs nearly always remain attached to their secondaries, and neutron stars (and presumably black holes) will be liberated only if the supernova explosion is asymmetric and gives the core a reciprocal kick velocity (yes, that is the phrase normally used).

2.13.5 Systems With One Compact Component

Binaries at the start of this evolutionary stage can be rather difficult to recognize, particularly if $\mathcal{M}_1$ is a white dwarf, a good deal less luminous than its main sequence or giant $\mathcal{M}_2$. Consider Sirius. The white dwarf contributes roughly one photon in 10^4 of those reaching us, and although Bessell showed in 1844 that there must be a companion, it was not seen for nearly 20 years, and even now it still gets lost in the glare during the part of its 50-year elliptical orbit when it is closest to the main sequence star Sirius A. Incidentally, Sirius is a "close" binary in the sense that, if you trace the evolution backward, there was probably some mass transfer.

Nevertheless, binary nuclei in planetary nebulae are not uncommon (some are eclipsing binaries), and V471 Tau, in the Hyades, is the prototype of a class of temporarily non-interacting systems consisting of a white dwarf plus a spectral class M star. A Mickey Mouse handful of pulsars have spectral class B main sequence stars as companions, and there is a larger class of neutron star and black hole binaries in which the extended star is blowing off a vigorous wind, enough of which is captured by the neutron star or black hole to make a bright x-ray source. There must surely also be examples of non-accreting pairs in which a neutron star or black hole orbits an extended, bright $\mathcal{M}_2$, but they are not easy to distinguish from systems in which both stars are still on or near the main sequence and one is much more massive and brighter than the other.

2.13.6 The Second Phase of Mass Transfer

Binary stars are perhaps at their best when $\mathcal{M}_2$ in turn reaches the stage of expansion away from the main sequence. First an enhanced wind and then overflow of its Roche lobe provides a supply of hydrogen–rich gas for accretion onto the compact companion. And astronomy is at its most botanical in describing and classifying the cataclysmic variables. All are to some extent variable in light output, and, as discussed previously, display emission as well as absorption line features, and have at least some light coming from five locations—two stars, an accretion disk around the white dwarf, the stream of gas from L_1 down (in potential!) toward that disk, and a hot spot where the stream hits the disk. X-ray binaries, where the accretor is a neutron star or black hole have the same anatomical parts and also a corona of very hot gas above and below the accretion disk.

The x-ray binaries are classified along two axes. First, is the accretor a neutron star or black hole? The decision depends only on mass, and there seems to be a clean cut, with masses implied by the orbits' being either less than about 2 $\mathcal{M}_\odot$ (neutron stars) or more than 4 $\mathcal{M}_\odot$ (black holes). The former are more numerous by a factor of ten or so and are, in addition, sometimes characterized by variability at the rotation period (meaning that accretion is channeled by a magnetic field of 10^{11}–10^{12} G) and short bursts, lasting a few minutes, which represent degenerate ignition of helium on the neutron star surfaces. (The accreted hydrogen burns steadily, but the helium becomes degenerate before ignition and so explodes.) Some of the black hole sources display quasi-periodic oscillations, probably at periods characteristic of the last stable orbit around the black hole (before material plunges in freely as a result of general relativistic gravitational effects).

The second axis of classification is the mass of the donor star, resulting in the names "low-mass x-ray binary" (LMXRB) and "massive x-ray binary" (MXRB). Both can have either neutron star or black hole accretors. The LMXRB systems seem to be old, where the accretor is a neutron star, the magnetic field has decayed to 10^{10}–10^{9} G or less, so that accretion is not strongly channeled, and the evidence for both neutron star rotation and for the orbit period took some years to acquire. Curiously, this cut is also a fairly clean one, with donors of less than 1.5 $\mathcal{M}_\odot$ and more than about 8 $\mathcal{M}_\odot$ and not much in between. An exception is Her X-1 (meaning the brightest x-ray source in the direction of the constellation Hercules, previously known as the optical variable star HZ Her), where the donor is an A or F star. This system is a peculiarly complex and interesting one, having all the variability time scales you would expect plus another of about 35 days, which is probably precession of the accretion disk.

For reasons that now puzzle us, the XRBs with accretors of large mass were for many years called "black hole candidates" rather than BHXBs. This now seems rather silly, and perhaps arose out of some confusion about what is meant by a "black hole." To repeat, an astrophysicist's black hole is merely something with a size comparable with its Schwarzschild horizon. We make no promises about what, or who, is inside.

In due course, $\mathcal{M}_2$ will also complete its evolution, leaving another white dwarf, neutron star, or black hole. A core collapse and supernova on the part of $\mathcal{M}_2$ will now remove more than half the total mass of the system, which is more likely to be unbound at this stage than when $\mathcal{M}_1$ collapsed and exploded. Thus the products at this stage can include a runaway newly made pulsar, but also an old neutron star that has been spun back up to rapid rotation by accretion from its $\mathcal{M}_2$. This is thought to be the origin of at least some of the single, weak field, rapidly rotating, millisecond pulsars, of which the fastest has a rotation period of 1.55 ms.

2.13.7 Binaries With Two Compact Components

We can reasonably expect that some combination of initial star masses and separation will leave any of the combinations of compact stars that we think of by taking "one from column A and one from column B," where column A is

$\mathcal{M}_1$ = WD, or NS, or BH, and column B is $\mathcal{M}_2$ = WD, or NS, or BH. There is observational evidence for most of the sorts of pairs we would expect to see. Still missing from the inventory are WD + BH (not clear that any of these will form, as a large disparity in initial masses would seem to be required), NS+BH (but statistical considerations say that one of these should turn up by the time the current list of known binary pulsars has doubled or tripled), and BH+BH (but one is hard pressed to know how to look for these unless, by rare good chance, one acts as a gravitational lens for a background star while someone is watching).

White dwarf pairs often show spectroscopic or eclipse evidence for their duplicity, allowing for measurement of the component masses. Some of the systems with small total mass have separations comparable with the sum of the stellar radii. These *must* be products of common envelope binary evolution. The prototype is AM CVn. Such pairs will surely spiral together in time, since they are losing material (hence angular momentum). In addition, any orbiting binary radiates some gravitational radiation. For extended stars, this drains energy and angular momentum very slowly indeed. For compact pairs, it can be the dominant process, and will cause mergers of systems with periods of 12 hours or less in less than the present age of the universe.[16]

Pairs of white dwarfs with total mass in excess of 1.4 $\mathcal{M}_\odot$ (the Chandrasekhar limit) and orbit periods less than half a day should spiral together and, as noted earlier, continue to be many people's first choice as progenitors of Type Ia supernovae. As we go to press, there is known to exist somewhere between zero and one system with the requisite properties, several having been tentatively reported in the past and the system characteristics later corrected outside the target range.

The (rotation) periods of pulsars are a precise clock, whose periods show Doppler effects just as precise light wavelengths or frequencies do. This permits mass estimates for the pulsars and their companions, particularly where there is some additional information from precession of the perihelion or from an optical identification. White dwarf and neutron star companions are both found, and it is not always clear as we would like which of the two stars was initially the more massive.

Neutron star pairs (NSX2) and paired neutron stars and black holes will also eventually spiral together. The prototype is the first binary pulsar discovered (PSR 1913+16, where the two numbers refer to Right Ascension and Declination, respectively), with two neutron stars and an orbit period of about eight hours and a total mass of about 2.8 $\mathcal{M}_\odot$. A handful of other systems are known, though with the star masses, etc., less well measured. Thus the Milky Way (and presumably other large galaxies) should have a few such pairs merge every 10^8 years. The product would seem likely to include both some sort of very energetic explosion and a core that collapses

[16] You should work through §16.4 of Shapiro and Teukolsky (1983) to understand how binary orbits decay due to gravitational radiation.

to a black hole. Models exist in which observable manifestations include a spurt of r-process material (perhaps the source of the very heaviest nuclides, including ^{244}Pu, which existed when the solar system formed but has now decayed away) and an enormous burst of gamma rays, of 10^{53} erg or thereabouts, perhaps resembling a subset of observed gamma ray bursts (GRBs).[17] An efficient, all-sky detection system for such GRBs records a couple each day, coming from galaxies as far away as $z = 4$. NSX2 or NS+BH mergers are a candidate model; their main competitor is collapse of single, massive, rapidly rotating stars to rapidly rotating black holes. Arguably both happen and make different sorts of GRBs (the binary merger ones having such short durations that no x-ray, optical, or radio counterparts have yet been caught and no redshifts have yet been measured).

Finally, the merger of two stellar mass black holes in a binary system ought to make a burst of gravitational radiation describable as a chirp (that is, both the intensity and frequency of the radiation increase over a few moments to a few-second peak at close to a kHz). Detectors to look for these are being built several places, and at least one (an interferometer with a baseline of 300 meters, in Japan) had already reported some upper limits as this is being written.

2.14 Star Formation

Within the Milky Way at present, most star formation occurs within clouds of gas that are (a) molecular clouds (mostly H_2 but with CO as an important tracer), (b) cool (meaning 5–15 K), (c) dense with 10^3 or more H_2 cm^{-3}, which is thinner than thin air, but dense compared to the galactic gas average of about one hydrogen atom cm^{-3}, (d) largish (sizes of parsecs and masses up to 10^5 $\mathcal{M}_\odot$), and (e) primarily located in the spiral arms of the disk. Other sites of star formation must surely be important at other times and in other kinds of galaxies, but even less is known about them than the parochial sort.[18]

About 1% of the mass is invariably in dust, and this is more than enough to make the relevant clouds largely opaque to visible and ultraviolet light. Thus historically a major reason for our ignorance of star formation was that we couldn't really observe it. Constantly increasing sizes of collecting areas, improved angular and wavelength resolution, and better detectors for infrared and radio photons have largely ameliorated that situation. Emission

[17] The gamma ray bursters are beyond the scope of this text but you may wish to consult Schilling, G., 2001, *Science*, 294, 1816 for an introduction

[18] Check out the 4 January 2002 (Vol. 295) of *Science*, pages 63–91, where you will find a series of articles on various topics in star formation. For a conference proceedings devoted to this whole topic, see Holt and Mundy (1997). Going yet further into the formation of binary stars, see Tohline, J.E., 2002. ARA&A, 40, 349.

lines of CO and many other molecules (something like 130, with molecular weights up to 100 or so, are known, many familiar from earth, some distinctly odd) permit mapping out the gas clouds. An important discovery is that the clouds are never of uniform density. All have dense cores studded through them, some that already have protostars inside, and some that do not but, we suppose, eventually will. Infrared astronomy permits the detection and analysis of emission from the embedded stars themselves and the residual disks of dust and gas from which planets presumably form. A very short list of some of the more interesting, and perhaps surprising, interstellar and circumstellar molecules is given in Table 2.4 (material courtesy of Pat Thaddeus who periodically updates his list). Much more information about astronomical molecules, big and small, is available in the review by Ehrenfreund and Charnley (2000). For more on the interstellar medium (ISM) of our galaxy, see Ferrière (2001).

Table 2.4. Some Interstellar Molecules

Molecule	Other names
SiH_4	Silane
CH_4	Methane (marsh gas)
H_2CO	Formaldehyde (preservative, etc.)
NH_3	Ammonia
SiC	Carbide (whetstones, etc.)
H_2S	Hydrogen sulfide (rotten eggs)
CH_3CH_2OH	Ethanol, Ethyl alcohol (for cocktails)
CH_3OH	Methanol, Methyl alcohol (not for cocktails)
$CH_2{=}CH_2$	Ethylene (See Hale et al. 2003 for an unusual application.)
CH_2CHN	Vinyl cyanide

Another property of these giant molecular gas clouds is at least approximate balance between inward gravitational forces and outward pressure, the latter made up of contributions from microscopic gas kinetics of the molecules, turbulence, magnetic fields, and rotation (all of which are more or less observed). Thus the clouds typically last longer than their free-fall time, perhaps 10^8 years (the galactic rotation period) versus 10^5–10^6 years. What makes a given cloud to decide to start contracting and forming stars is sometimes posed as a question: "Is star formation triggered?" Possible triggers for contraction might include bumping up against another cloud, being zapped by an expanding supernova remnant or HII region (the expanding cloud of gas ionized and heated by a young, massive, hot star), or being swept up in the shock wave at the front of a spiral arm as the galaxy rotates. It is not obvious by looking but spiral arms are really sort of like standing waves, and their rotation speed differs from that of the galactic stars and gas. After years of careful observation and analysis, astronomers working on star formation have provided the answer to the question “Is star formation triggered?” The answer is, “Sometimes.”

Approaching the problem from the other side, we can look at the properties of a population of newly formed stars (a young cluster or association, for instance, or a ensemble average of a bunch of these). Important properties include (a) how much total mass goes into stars from a given mass molecular cloud?, (b) What is the distribution of stellar masses formed? This is called the *Initial Mass Function*, or IMF. (c) What fraction of the stars are in binary systems (or large hierarchies)? and, (d) what are the statistical properties of the binary ensemble (the distribution of separations, $\mathcal{M}_1$ and $\mathcal{M}_2$ or the ratio, and of eccentricities).

Most of these questions have at least approximate answers, though not all astronomers agree on precisely what they are or on the extent to which any of the properties varies from one star formation region to another. The IMF, for instance, looks rather like a power law,

$$\xi(\mathcal{M})\, d\mathcal{M} = \xi_0 \mathcal{M}^{-\alpha}\, d\mathcal{M}$$

where, for masses less than about 0.5 $\mathcal{M}_\odot$, α is typically around 1.35, and ξ_0 is a constant. The units of $\xi(\mathcal{M})$ are the number of stars per unit mass. (There is also the “birthrate function,” which describes the *rate* at which stars are formed in a given mass interval. See Ex. 2.17 for an example.) Notice that this diverges as mass becomes arbitrarily small. Another description is as a Gaussian, whose righthand side looks a lot like a power law form. This will not diverge, but the problem of finding the mass at which the peak occurs remains. A particular area of disagreement is how far toward small masses the IMF continues to rise and, therefore, how much matter is more or less hidden in very small stars or brown dwarfs whose lifetimes are longer than the age of the universe. And is this the same everywhere? (Almost certainly no, to the last part of the question. Some young clusters, including one near the center of the Milky Way and in the Large and Small Magellanic Clouds, appear to be making only rather massive stars, or at least more than their fair share.)

As for the binaries, we noted earlier that half or more of all stars (perhaps up to 90% in some places) are binaries and that the full range of possible separations occurs, with perhaps some preference for the middle of the range and periods of ten years or so (not very easy to study). The orbits are not all circular (except for contact systems) and not all extremely eccentric, but the distribution is not very well known. Older star populations have more circular orbits, but this is the result of gradual dynamical evolution, not of different initial conditions. The distribution of binary mass ratios is probably not the same everywhere or the same for systems of all possible separations. Some studies have found, for instance, that the binary members of some star clusters act like pairs of stars that were selected at random from the IMF to live together. Other studies, especially ones of short period systems, find an excess of pairs where the stars have roughly the same mass. Complete data to assess any of these distributions are very difficult to acquire.

Finally, the process is generally (again in the Milky Way context) not very efficient. Where giant molecular clouds have enough gas for 10^5 suns, young clusters and associations consist of dozens, to hundreds, or at most a few thousand stars, indicating an average efficiency of not more than 1%. Indeed if the efficiency were very high, probably all the gas would have been turned into stars long ago, as indeed apparently happened in elliptical galaxies. Part of the underlying reason for the inefficiency is that the first few massive stars that form will, via their winds and HII regions and eventual supernova explosions, dissipate the gas that has not already also gathered into stars very quickly.

The statement is often made by astronomers (including us) that there is no theory of star formation. Roughly what this means is that you can observe many of the details of molecular clouds (mass, rotation, internal distributions of density, temperature, turbulence, and magnetic fields) and tell a theorist about these details, but she or he will not, in turn, be able to tell you how efficient the star formation will be, what the resulting IMF and binary population characteristics will be, and so forth. "What makes the IMF?" in particular is one of those questions that astronomers have asked, and found a great many different answers to, over the years.

An early answer was that the stars should all have approximately the Jeans mass, the minimum that can hold itself together by gravity at a given gas density and temperature.

Another answer was that single clouds normally fragmented into little bits smaller than stars, which later collided and stuck. This sort of statistical process will indeed give you a power-law distribution of masses. Somewhat later came the answer that the mass of a star was set by the end of accretion from the surroundings and the onset of (probably collimated) mass loss. The transition is thought to happen when the core of the star gets hot enough (about 10^6 K) for deuterium to fuse, setting up a convection zone and permitting the generation of a dynamo magnetic field.

The whole story probably includes at least part of all these ideas, and also part of the idea that when gas is turbulent there will be bits that are tossed around and bits that are relatively quiescent backwaters, where cores might form and condense to make stars. Why, you ask, can't you just put it all on a computer and let it run? Surely there is no unknown fundamental physics in the process of star formation. True enough, but this concerns an enormous number of particles—perhaps 10^{62} molecules in a largish cloud. Clearly you cannot follow them individually through their gravitational and collisional interactions. Suppose instead you think of the cloud as being made up of many small fluid elements (small compared to a star, but large compared to an atom). You will still discover, if you want to resolve entities that will be a few AU across when the process is complete, that the largest cloud you can handle is perhaps 100 $\mathcal{M}_\odot$, and a good many Moore's doubling times in computing power will have to pass before there will be adequate dynamic range to simulate a whole giant molecular cloud divided into bits of a Jupiter mass or thereabouts. The problem is somewhat akin to weather forecasting, which is often done on a grid of 100-km squares, because more smaller ones would overwhelm number-crunching capabilities. Unfortunately, whether a given storm will pass 50 miles off the coast or right though your beachfront house is then impossible to predict.

Astronomers working on calculations of star formation currently receive many fewer complaints about this problem than do weather forecasters.

2.15 Supplemental Material

To give you a better idea of some properties of ZAMS models, Tables 2.5 and 2.6 list representative models from various sources. Along with the model mass and composition, each model is keyed by a model number (the first column) to help bridge across the tables. The fourth and fifth columns of Table 2.5 list the model luminosity and effective temperature and the sixth column gives the model radius in units of 10^{10} cm. (We shall occasionally use the subscript notation S_n to denote the value of a quantity S in units of 10^n.) The references in the last column are as follows:

(1) Models with this reference number were made by the authors using the computer code `ZAMS.FOR` that can be found on the CD-ROM on the end-cover of this text. It uses simple physics and analytic fits to opacities and energy generation rates. These models are perfectly fine for pedagogy.

(2) These models are from
 ▹ VandenBerg, D.A., Hartwick, F.D.A., Dawson, P., & Alexander, D.R. 1983, ApJ, 266, 747.

 As with the models of reference (3), they contain much more sophisticated physics than do our models.

(3) These very low mass models are from the "MM EOS" sequence of
 ▹ Dorman, B., Nelson, L.A., & Chau, W.Y. 1989, ApJ, 342, 1003

 and see
 ▹ Burrows, A., Hubbard, W.B., & Lunine, J.I. 1989, ApJ, 345, 939.

 A good review of the consequences of uncertainties in constructing models for low-mass stars may be found in
 ▹ Renzini, A., & Pecci, F.F. 1988, ARA&A, 26, 199.

The central temperature (in units of 10^6 K) is $T_{c,6}$, and ρ_c and P_c are, respectively, the central density and pressure in cgs units. These are listed in Table 2.6. Finally, the last two columns in that table list q_c and q_{env}. The quantity q_c is the fractional mass of a possible convective core in a model (see Chap. 5). For example, in a model of $\mathcal{M} = 60\ \mathcal{M}_\odot$ the inner 73% of the mass is convective starting from model center. The corresponding quantity q_{env} is

Table 2.5. Zero-Age Main Sequence Models

No.	$\mathcal{M}/\mathcal{M}_\odot$	(X, Y)	$\log \mathcal{L}/\mathcal{L}_\odot$	$\log T_{\rm eff}$	$\mathcal{R}_{10}$	Ref.
1	60	(0.74, 0.24)	5.701	4.683	70.96	(1)
2	40	(0.74, 0.24)	5.345	4.642	56.89	(1)
3	30	(0.74, 0.24)	5.066	4.606	48.53	(1)
4	20	(0.74, 0.24)	4.631	4.547	38.73	(1)
5	15	(0.74, 0.24)	4.292	4.498	32.89	(1)
6	10	(0.74, 0.24)	3.772	4.419	25.94	(1)
7	7	(0.74, 0.24)	3.275	4.341	20.99	(1)
8	5	(0.74, 0.24)	2.773	4.259	17.18	(1)
9	3	(0.74, 0.24)	1.951	4.118	12.76	(1)
10	2	(0.74, 0.24)	1.262	3.992	10.30	(1)
11	1.75	(0.74, 0.24)	1.031	3.948	9.695	(1)
12	1.50	(0.74, 0.24)	0.759	3.892	9.151	(1)
13	1.30	(0.74, 0.24)	0.496	3.834	8.827	(1)
14	1.20	(0.74, 0.24)	0.340	3.800	8.648	(1)
15	1.10	(0.74, 0.24)	0.160	3.771	8.032	(1)
16	1.00	(0.74, 0.24)	-0.042	3.752	6.931	(1)
17	0.90	(0.74, 0.24)	-0.262	3.732	5.902	(1)
18	0.75	(0.73, 0.25)	-0.728	3.659	4.834	(2)
19	0.60	(0.73, 0.25)	-1.172	3.594	3.908	(2)
20	0.50	(0.70, 0.28)	-1.419	3.553	3.553	(3)
21	0.40	(0.70, 0.28)	-1.723	3.542	2.640	(3)
22	0.30	(0.70, 0.28)	-1.957	3.538	2.054	(3)
23	0.20	(0.70, 0.28)	-2.238	3.533	1.519	(3)
24	0.10	(0.70, 0.28)	-3.023	3.475	0.805	(3)
25	0.08	(0.70, 0.28)	-3.803	3.327	0.650	(3)

the fractional mass contained in a fully or partially convective envelope. For our purpose here, if $q_{\rm env}$ is not zero, then it is the fractional mass measured from the model surface inward to a level where convection ceases. Thus, for example, the outer 0.35% of the mass of the model numbered 16 (1 $\mathcal{M}_\odot$) is entirely or partially convective. This is a ZAMS model of the sun and, it turns out, it is completely convective from just under the photosphere inward to that mass level. This corresponds, however, to the outer 17% of the radius. A listing of "neg." for $q_{\rm env}$ means that a negligible fraction of the envelope is convective (say, less than 10^{-8} in mass) but a "1" means the model is fully convective. A "0" in that column means that there is no convection. Finally, a "–" implies that the information was not available to us. Now for what may be learned from the models.

For the higher mass stars, and keeping composition fixed, radius is seen to increase with mass as expected from the homology relation (1.87) where $\mathcal{R} \propto \mathcal{M}^{0.75}$. Since strict hydrostatic equilibrium holds for these models, Equation (1.66) plus (1.87) implies that $P \propto \mathcal{M}^{-1}$. If this pressure is taken as the

Table 2.6. ZAMS Models (continued)

No.	$\mathcal{M}/\mathcal{M}_\odot$	$T_{c,6}$	ρ_c	$\log P_c$	q_c	q_{env}
1	60	39.28	1.93	16.22	0.73	0
2	40	37.59	2.49	16.26	0.64	0
3	30	36.28	3.05	16.29	0.56	0
4	20	34.27	4.21	16.37	0.46	0
5	15	32.75	5.48	16.44	0.40	0
6	10	30.48	8.33	16.57	0.33	0
7	7	28.41	12.6	16.71	0.27	0
8	5	26.43	19.0	16.84	0.23	0
9	3	23.47	35.8	17.06	0.18	0
10	2	21.09	47.0	17.21	0.13	neg.
11	1.75	20.22	66.5	17.25	0.11	neg.
12	1.50	19.05	76.7	17.28	0.07	neg.
13	1.30	17.66	84.1	17.28	0.03	neg.
14	1.20	16.67	85.7	17.26	0.01	10^{-7}
15	1.10	15.57	84.9	17.22	0	5×10^{-5}
16	1.00	14.42	82.2	17.17	0	0.0035
17	0.90	13.29	78.5	17.11	0	0.020
18	0.75	10.74	81.5	–	0	–
19	0.60	9.31	79.1	–	0	–
20	0.50	9.04	100	17.10	0	–
21	0.40	8.15	104	17.04	0	–
22	0.30	7.59	107	17.05	*	1
23	0.20	6.53	180	17.24	*	1
24	0.10	4.51	545	17.68	*	1
25	0.08	3.30	775	17.83	*	1

central pressure, then P_c should decrease with mass. It does, although not as fast as homology would imply. The relation of density to mass and radius of (1.62) combined with (1.87) yields $\rho \propto \mathcal{M}^{-5/4}$, and this general behavior is shown in Table 2.6 where ρ_c decreases with mass. We already know that luminosity increases with mass (from 1.88), and it is an easy matter to show that T_{eff} and T_c do so also. In summary, ZAMS stars of high mass get bigger, brighter, and less dense as mass increases.

2.16 Exercises

Exercise 2.1. By some combination of means, a binary system has been observed and the following parameters determined for it:

- The system has zero eccentricity; i.e., the orbits are circular.
- The mass of the primary (the brighter star) is $\mathcal{M}_1 = 5\,\mathcal{M}_\odot$.
- The inclination of the system is $i = 30°$.

- The period is $\Pi = 31.86$ days.
- The maximum velocity of the primary along the line of sight to us is $V_r = 10.17$ km s^{-1}.

We assume that both primary and secondary stars were formed at the same time on the ZAMS and that further evolution has been such that neither mass nor angular momentum has been lost from either star or the system since that time. This means that the orbital parameters have not changed since the system was formed. The following questions require that you read up on Kepler's laws as applied to binary systems: see, for example,

▷ Mihalas, D., & Binney, J. 1981, *Galactic Astronomy*, 2nd ed. (San Francisco: Freeman), pp. 79–86.

1. What is the numerical value of the semimajor axis, a, of the system? Compare this figure to the distance of the planet Mercury from the sun.
2. What is the mass, $\mathcal{M}_2$, of the secondary?
3. After what period of time following ZAMS formation will the primary expand to fill its Roche lobe as a result of normal evolution? You will need radius versus time information to answer this. This information can be found by reading $\mathcal{L}$ and T_{eff} from Fig. 2.5 and then computing. For a larger version of Fig. 2.5, see Fig. 6-16 of Clayton (1968). You may also "cheat" and use the original source for these figures:

   ▷ Iben, I. Jr. 1966, ApJ, 143, 483.

   The Roche lobe radius traditionally used is the radius of a sphere of volume equal to that of the Roche lobe. There are several versions: the one given below is due to

   ▷ Eggleton, P.P. 1983, ApJ, 268, 368.

   If $\mathcal{M}_1$ is the mass of the primary, $q = \mathcal{M}_2/\mathcal{M}_1 \leq 1$, and a the semimajor axis of the system, then the equivalent radius of the Roche lobe of the primary is

$$\frac{\mathcal{R}_{\text{RL},1}}{a} \approx \frac{0.49}{0.6 + q^{2/3} \ln\left(1 + 1/q^{1/3}\right)}. \tag{2.1}$$

Exercise 2.2. A Classical Cepheid variable with a period of 10 days is seen in a distant galaxy. Its observed color and apparent visual magnitudes are, respectively, $(B–V)_0 = 0.7$ and $m_V = 14$. If we assume there is no dust or gas between us and the star, estimate the distance to the galaxy using material from, say, Chapter 10 of Allen (1973, 3rd ed.) or Cox (1999). You should also check how well the following period–luminosity–color (PLC) relationship works given all the above information. From

▷ Iben, I. Jr. & Tuggle, R.S. 1972, ApJ, 173, 175,

we have

$$\log\left(\frac{\mathcal{L}}{\mathcal{L}_\odot}\right) = -17.1 + 1.49 \log \Pi + 5.15 \log T_{\text{eff}} \tag{2.2}$$

where Π is the period in days. A newer, but more complicated, version of this may be found in

▹ Iben, I. Jr. 2000, in *Variable Stars as Essential Astrophysical Tools*, ed. C. İbanoğlu (Dordrecht: Kluwer), p. 437.

Exercise 2.3. As an exercise of your skills in homology or dimensional analysis try the following:

1. Verify the following homology relations for the lower main sequence:

$$\begin{aligned}
\mathcal{L} &\propto Z^{0.35} X^{1.55} T_{\rm eff}^{4.12} \\
\mathcal{R} &\propto Z^{0.15} X^{0.68} \mathcal{M}^{1/13} \\
\mathcal{L} &\propto Z^{-1.1} X^{-5.0} \mathcal{M}^{5.46} \\
T_{\rm eff} &\propto Z^{-0.35} X^{-1.6} \mathcal{M}^{1.33} .
\end{aligned} \tag{2.3}$$

2. Now do the same thing for the upper main sequence where electron scattering and the CNO cycles are important. Still assume diffusive radiative transfer and the ideal gas equation of state. Since the rate for the CNO is proportional to the abundance of CNO nuclei times that of protons (see Chap. 6), take $\varepsilon_{\rm CNO} \propto XZ\rho T^{15}$. The HR diagram (Fig. 2.37) shows lower ZAMS models from
   ▹ Mengel, J.G., Sweigart, A.V., Demarque, P., & Gross, P.G. 1979, ApJS, 40, 733
   for metalicities $Z = 0.04$ and 10^{-4} ($Y = 0.3$). See how well your homology results compare with the figure. (Answer: "So, so, but not a disaster.")

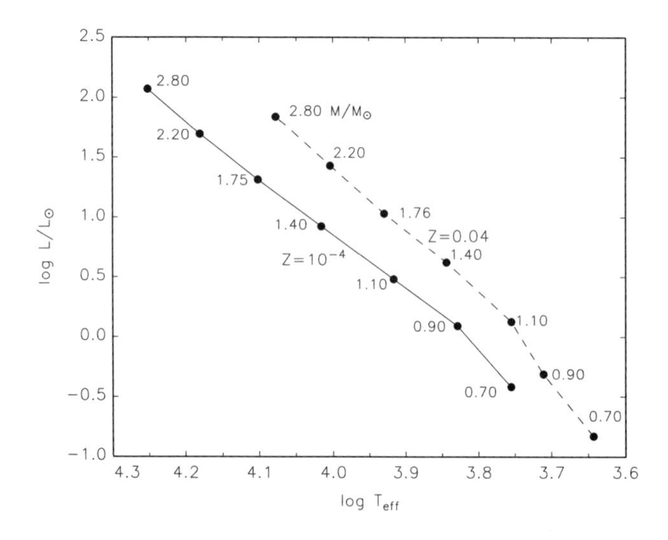

Fig. 2.37. The results of Mengel et al. (1979) for lower main sequence models are plotted on an HR diagram.

Exercise 2.4. This problem will explore a key property of Classical Cepheids pertaining to the distance scale and, in particular, you will find the distance to the galaxy M81, as follows. In the simplest scenario, there is a correlation between the period, Π, of variability of a Classical Cepheid Variable (CCV) and its luminosity, $\mathcal{L}$. This is the period–luminosity (Π–$\mathcal{L}$) relationship and it has the form

$$\Pi \propto \mathcal{L}^\alpha$$

where α is some constant.

1. To estimate α, use the period–mean density relation $\Pi \propto \langle\rho\rangle^{-1/2}$ and phrase this in terms of powers of $\mathcal{M}$ and $\mathcal{R}$. To eliminate $\mathcal{M}$ in what you found, note that there is a relation between $\mathcal{M}$ and $\mathcal{L}$ (derived from theoretical evolutionary tracks) for CCVs of the form $\mathcal{L} \propto \mathcal{M}^{7/2}$. Now use the blackbody relation between $\mathcal{L}$, $\mathcal{R}$ and T_{eff} to phrase Π as a power law in terms of only $\mathcal{L}$ and T_{eff}. Finally, there is a rough relation between T_{eff} and $\mathcal{L}$ on the Cepheid Strip of the form $\mathcal{L} \propto T_{\text{eff}}^{-15}$. Use this to eliminate T_{eff} to get $\Pi \propto \mathcal{L}^\alpha$ and thus derive the value of α.
2. To see whether your result for α is reasonable, consider Fig. 2.38, which shows the observed apparent visual magnitudes, m_V, versus $\log \Pi$ for CCVs in the galaxy M81. (The data have been massaged slightly to make this problem more tractable.) Fit this data to a straight line of the form $m_V = -a \log \Pi + b$, where a and b are constants and Π, as in the graph, is in the units of days. What are a and b? Convert your expression for m_V to a Π–$\mathcal{L}$ relation of the form given earlier; that is, you are to convert this to $\Pi \propto \mathcal{L}^\beta$. To do this you will need the following: $\log \mathcal{L}/\mathcal{L}_\odot = \left[M_{\text{bol}}(\odot) - M_{\text{bol}}\right]/2.5$, where $M_{\text{bol}} = M_V + B.C.$ is the absolute bolometric magnitude. [For the sun $M_{\text{bol}}(\odot) = 4.75$.] Take the bolometric correction $B.C.$ to be a constant for the data set. Neglect extinction and reddening in your analysis. Finally, you will need (here and elsewhere) $m_V - M_V = 5 \log d - 5$, where d is the distance to the star in pc. Find the value of β and compare this to your result for α.
3. Now to find the distance to M81. We use the LMC as a guide. The LMC is at a distance of 50.1 kpc and contains CCVs that are assumed to be identical in general properties to those in M81; that is, a CCV with a given $\log \Pi$ has the same M_V in either galaxy. In the LMC a CCV with a period of 10 days has an apparent magnitude $m_V = 14.4$. Neglecting extinction, etc., deduce the distance to M81. Give that distance in the units of Mpc.

Exercise 2.5. We now have the tools to investigate a curious class of stars called "blue stragglers" that continues to baffle astronomers. One model for these stars is that they mix up their insides somehow so that their composition is always homogeneous. Construct a family of homologous stars in which the mean molecular weight μ is kept as an independent variable (i.e., $\mathcal{L} \propto \mu^{\beta_\mathcal{L}} \mathcal{M}^{\alpha_\mathcal{L}}$, etc.). Assume CNO burning ($\varepsilon \propto \rho T^{15}$) and Kramers opacity ($\kappa \propto$

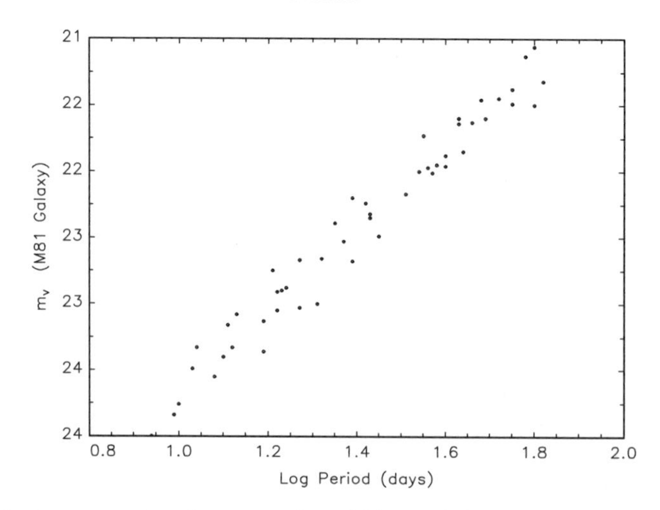

Fig. 2.38. Apparent visual magnitudes of Classical Cepheids in the galaxy M81.

$\rho T^{-3.5}$). For simplicity, neglect the weak composition dependence of ε and κ, and assume a nondegenerate ideal gas equation of state.

1. How does the main sequence $\mathcal{L}$ versus $T_{\rm eff}$ vary with μ? Explain what is happening in physical terms.
2. What is the functional form of an evolutionary track for a homogeneous star in the HR diagram? Draw an HR diagram showing the evolutionary track in relation to the main sequence.
3. How does the luminosity of such a homogeneously evolving star change with μ? Such a star burns about 10 times as much fuel as a normally evolving star before depleting hydrogen. Qualitatively, will the homogeneous star live 10 times as long as a normal star? Why or why not?
4. Draw a schematic HR diagram of a moderately old cluster as it would look if some small fraction of all stars underwent homogeneous evolution. Qualitatively, how would this diagram be modified if the "homogeneous" stars actually retained a small outer portion of their envelopes as unmixed hydrogen?
5. How much energy is required to mix the interior of such a star from gravitational considerations? Is such mixing then feasible? Can you suggest some ways in which intermediate-mass main sequence stars could mix themselves up?

You may wish to check out detailed computations of such mixed models. See, for example,

▷ Saio, H., & Wheeler, J.C. 1980, ApJ, 242, 1176.

Exercise 2.6. In §2.13.4 it was remarked that, in the simplest circumstances, "… to unbind a gravitational pair with sudden mass loss, you must remove at least half of the total mass." Thus consider a binary pair of masses $\mathcal{M}_1$ and $\mathcal{M}_2$ in circular orbits about their common center of gravity. The masses are separated by the distance a. One of them, say, $\mathcal{M}_1$, rapidly loses mass in a spherically symmetric way (in, for example, an ideal supernova explosion). The mass is lost to the system entirely. If the stars' initial velocities are v_1 and v_2, then the velocity of $\mathcal{M}_{\rm rem}$—as what's left of $\mathcal{M}_1$ as a remnant—will still be v_1 after the mass loss. (In the terms of §2.13.4, there will be no "reciprocal kick velocity.") Velocities and masses of the initial system are related by $\mathcal{M}_1 v_1 = \mathcal{M}_2 v_2$ (as simple dynamics will show). The aim here to show that the following is necessary to unbind the system:

$$\frac{\mathcal{M}_{\rm rem}}{\mathcal{M}_1 + \mathcal{M}_2} \leq \frac{1}{(1 + \mathcal{M}_2/\mathcal{M}_1)(2 + \mathcal{M}_2/\mathcal{M}_1)} < \frac{1}{2}.$$

In the following you may wish to consult, for example,

▷ Carroll, B.W., & Ostlie, D.A. 1996, *An Introduction to Modern Astrophysics* (Reading: Addison-Wesley), §17.5.

1. Convince yourself that the initial total energy, $W_{\rm init}$, of the system is

$$W_{\rm init} = U + \Omega = \tfrac{1}{2}\mathcal{M}_1 v_1^2 + \tfrac{1}{2}\mathcal{M}_2 v_2^2 - G\mathcal{M}_1\mathcal{M}_2/a\,.$$

   (Note the use of W, U, and Ω. The virial theorem is lurking here in one manifestation.) The final total energy of the system, $W_{\rm fin}$, is just the above with $\mathcal{M}_1$ replaced by $\mathcal{M}_{\rm rem}$.
2. For the final system to be unbound, $W_{\rm fin}$ must be greater than zero. You now have enough information to prove the result desired, after just a tad of algebra.

Exercise 2.7. It was stated that the sun's luminosity 4.6 Gyr ago was some 25% less than it is at present. The first reaction to this is that the earth's surface and atmosphere should have been cooler at that remote time. So let's put in numbers, examining the present epoch as a start. First calculate the *solar constant*, which is the flux of radiation incident upon a unit surface perpendicular to that beam of radiation at the top of the earth's atmosphere. (All you need is the sun's present luminosity and the distance to the sun; i.e., 1 AU.) To find out how much radiation is actually absorbed by the atmosphere, assume that the earth's *albedo* is 31%; that is, 31% of the solar flux is reflected back into space. Then, without further hints, assume that the earth is in thermal balance and re-radiates power into space as a blackbody. In that case, what is the effective temperature of the earth? (The answer for the present epoch will come out to about 254 K, which turns out to be very near the mean temperature of atmosphere for the earth we have grown to know and love.) Now do the same for the earth at 4.6 Gyr ago assuming nothing else has changed (including the composition of the atmosphere—which is a nonsensical assumption) except for $\mathcal{L}_\odot$. If the present mean temperature of the earth's surface is around 290 K (63° F), what, naively, might it have been

4.6 Gyr ago? (We really have no idea of the real answer to this but microscopic life began at a surprisingly early time in earth's history—perhaps as early as 3.6 Gyr ago, although there is now some controversy about the fossils.) For those of you interested in the earth's atmosphere, we warmly recommend the intermediate level text *Atmosphere, Weather and Climate* by R.G. Barry & R.J. Chorley, 6th (or earlier) ed., 1992 (London & New York: Methuen).

Exercise 2.8. It is obvious that of the two open clusters shown in Fig. 2.6, the Pleiades is younger than M67 because the turnoff point is at a cooler $T_{\rm eff}$ in M67.

▷ Lang, K.R. 1991, *Astrophysical Data: Planets and Stars* (New York: Springer–Verlag), Table 15.2

gives the age of the Pleiades as 0.08 Gyr and 4 Gyr for M67. Let's see how close we can come to the figure for M67. This is a tricky business, as you will see, but you should come within 50% or so (assuming the quoted figure is really correct to begin with!). First note that $(B–V)$ in the figure is as observed and is not corrected for interstellar absorption. Since light from different spectral bands is absorbed differently, we must make a correction to get to the colors as emitted from the stars before absorption take place. If $(B–V)_0$ is the true color index, then the correction is given by the *color excess*

$$E_{B-V} = (B-V) - (B-V)_0$$

where Lang gives $E_{B-V} = 0.08$ for M67. (This makes stars hotter than would be the case if you used the uncorrected color.) Thus look in the literature, and the tables in Lang are fine, to translate $(B–V)$ at the turnoff point in M67 to luminosity at the turnoff point—and remember that stars at that point are luminosity class V main sequence stars. (Note that this can be a bit messy. It's not easy to guess where the turnoff point might be in some HR diagrams. This is certain;y true for M67.) Next use the following arcane formula quoted by

▷ Iben, I. Jr., & Renzini, A. 1984, PhysRep, 40

between luminosity at the turnoff point and cluster age:

$$\log\left[\frac{\mathcal{L}_{\rm TO}}{\mathcal{L}_\odot}\right] \approx \left[0.019\,(\log Z)^2 + 0.065\log Z + 0.41Y - 1.179\right]\log t_9 + \\ + \; 1.246 - 0.028\,(\log Z)^2 - 0.272\log Z - 1.073Y \qquad (2.4)$$

where t_9 is the cluster age in units of 10^9 years and $\mathcal{L}_{\rm TO}/\mathcal{L}_\odot$ is at the turnoff point. This expression adequately reflects the results of evolutionary calculations for $-4 \le \log Z \le -1.4$, $0.2 \le Y \le 0.3$, and $0.2 \le t_9 \le 25$, which are ranges of general interest. If M67 is about 4 Gyr old, then a composition close to solar seems a good guess (perhaps); that is, try something like $Z = 0.02$ and $Y = 0.3$. What do you find for the age of M67?

Exercise 2.9. You ought to be able to estimate the age of our local galactic disk and other parts of our galaxy from the dropoff point of white dwarf luminosity functions, as discussed in §2.8.1. You will need the following kind of information: The time it takes to get to the PNN stage starting from the ZAMS has been estimated (from evolution studies) by

▹ Iben, I. Jr. & Laughlin, G. 1989, ApJ, 341, 312

to be

$$\log t_{\text{to PNN}} = 9.921 - 3.6648 \log\left(\frac{\mathcal{M}}{\mathcal{M}_\odot}\right) + \\ + 1.9697 \left[\log\left(\frac{\mathcal{M}}{\mathcal{M}_\odot}\right)\right]^2 - 0.9369 \left[\log\left(\frac{\mathcal{M}}{\mathcal{M}_\odot}\right)\right]^3 \quad (2.5)$$

for ZAMS masses $0.6 \leq \mathcal{M}/\mathcal{M}_\odot \leq 10$. Times are in units of years. (Note these are original ZAMS masses, and assume the sample is large enough that the WDs at dropoff started out at the maximum mass for stars ending up as WDs. Why make this last assumption?!) Once the star is a hot and luminous PNN, it has the mass it will have in its subsequent career as a white dwarf.

▹ Iben, I. Jr., & Tutukov, A.V. 1984, ApJ, 282, 615

estimate it takes the time t_{WD} (in years) for the WD to cool to some given luminosity with (and see Chap. 10)

$$t_{\text{WD}} = 8.8 \times 10^6 \left(\frac{A}{12}\right)^{-1} \left(\frac{\mathcal{M}}{\mathcal{M}_\odot}\right)^{5/7} \left(\frac{\mu}{2}\right)^{-2/7} \left(\frac{\mathcal{L}}{\mathcal{L}_\odot}\right)^{-5/7} \text{yr}\,. \quad (2.6)$$

Here A is the nuclear mass number (say for carbon) and μ is the mean molecular weight (both discussed in §1.4.1). Assume a standard WD mass. Having done this part of the problem, do the same for the halo using a T_{eff} of, say, 4,000 K while assuming a standard size and mass for the WD.

Exercise 2.10. In Chapter 1 (§1.7) we estimated t_{nuc}, the hydrogen ZAMS lifetime, by calculating the energy released by converting 10% of the available hydrogen to helium and then dividing by luminosity. We can do the same for a pure helium ZAMS or, more to the point, estimate the lifetime on the HB where helium is being converted to C/O. A typical HB star has a luminosity of 50 $\mathcal{L}_\odot$ and helium core mass of 0.5 $\mathcal{M}_\odot$. If we assume, for simplicity, that oxygen is the final product, then a total of 14.3 MeV is released when combining four ^{4}He nuclei (16 amu) to make one ^{16}O nucleus. With this information, find the lifetime of a star on the HB using the same 10% efficiency used on the hydrogen ZAMS. You might wish to experiment and use

$$\log\left(\frac{\mathcal{L}}{\mathcal{L}_\odot}\right) \approx 0.261 + 3.04 \frac{\mathcal{M}_{\text{core}}}{\mathcal{M}_\odot} \quad (2.7)$$

from

▹ Iben, I. Jr., & Renzini, A. 1984, PhysRep, 40,

where $\mathcal{M}_{\text{core}}$ is the helium core mass. This at least gives an idea of how the luminosity on the HB varies with core mass. (Some abundance information has been deleted by us in Eq. 2.7.) You might compare your results to an older estimate by

▷ Iben, I. Jr. 1974, ARA&A, 12, 215,
which is

$$\log t_{\rm HB} \approx 7.74 - 2.2 \left[\left(\frac{\mathcal{M}_{\rm core}}{\mathcal{M}_\odot} \right) - 0.5 \right] \ {\rm yr}\,.$$

Exercise 2.11. RR Lyrae variable stars are on the HB, have observed periods of roughly 2–24 hr in the "fundamental" mode of pulsation, and effective temperatures $T_{\rm eff} \approx 7,000 \pm 500$ K.

▷ Iben, I. Jr. 1971, PASP, 83, 697
has come up with the following expression (again a fit to calculations) that relates almost all interesting properties of these variables:

$$\begin{aligned} \log \Pi \approx\ & -0.340 + 0.825(\log \mathcal{L} - 1.7) - \\ & -3.34(\log T_{\rm eff} - 3.85) - 0.63(\log \mathcal{M} + 0.19) \end{aligned} \tag{2.8}$$

where $\mathcal{L}$ and $\mathcal{M}$ are in solar units, and the period Π is in days. With this information (and perhaps hints from Ex. 2.10), find a range of typical masses for RR Lyraes.

Exercise 2.12. In §2.8 we hinted that you should estimate the gravitational potential energy released in the collapse of a 1.2 $\mathcal{M}_\odot$ core from an initial density of 10^9 g cm^{-3} to a final 10^{15} g cm^{-3}. Do so.

Exercise 2.13. Figure 2.10 showed mass loss rates for massive and luminous stars. The numbers for this figure were derived from material in Lamers and Cassinelli (1999) in their §2.7. The key equation is their Eq. 2.38, which reads

$$\log \left(\dot{\mathcal{M}} v_\infty \mathcal{R}^{1/2} \right) = -1.37 + 2.07 \log \left(\mathcal{L}/10^6 \right) \tag{2.9}$$

where (in this equation alone) $\mathcal{M}$, $\mathcal{R}$, and $\mathcal{L}$ are in solar units, $\dot{\mathcal{M}}$ is in $\mathcal{M}_\odot$ yr^{-1}, and v_∞ (the terminal velocity of the wind far from the star) is in km s^{-1}. This is a semi-empirical formula based on observations of spectral class O and B stars in our galaxy, where $\log \left(\dot{\mathcal{M}} v_\infty \mathcal{R}^{1/2} \right)$ is well-fit by a straight line versus $\log \mathcal{L}/\mathcal{L}_\odot$ (see their Fig. 2.19). The terminal velocity v_∞, is found to be roughly proportional to the escape velocity, $v_{\rm esc}$, at the stellar surface, where

$$v_{\rm esc} = \sqrt{2(1-\Gamma_{\rm e}) G\mathcal{M}/\mathcal{R}}\,. \tag{2.10}$$

The curious term $(1 - \Gamma_{\rm e})$ arises from the levitating effect of the radiation field due to radiation pressure at the photosphere. It effectively lowers the escape velocity. For now—but see later—we shall set it to unity (i.e., set $\Gamma_{\rm e}$ to zero). For stars with $T_{\rm eff} \gtrsim 21{,}000$ K, $v_\infty \approx v_{\rm esc}/2.6$. (The factor 2.6 is less for cooler stars.)

1. With this information, check to see if we did our arithmetic correctly in producing Fig. 2.10.

2. Now back to Γ_e. This is given as

$$\Gamma_e = \frac{\kappa_p \mathcal{L}}{4\pi c G \mathcal{M}} \tag{2.11}$$

where $\kappa_p \approx 0.3$ cm^2 g^{-1} in the winds of hot stars. Look ahead to §4.3 where we discuss the "Eddington limit," and specifically to our (4.49), where you will find a numerical version of (2.11). Now redo you calculation of $\dot{\mathcal{M}}$ and see what are the effects of reducing the escape velocity because of radiation pressure.

Exercise 2.14. We briefly mentioned the Baade–Wesselink method for determining distances, luminosities, etc., of variable stars by examining how they pulsate. For the two original references see

▹ Baade, W. 1926, AstNachr, 228, 359

▹ Wesselink, A.J. 1946, BAN, 10, 91.

The technique goes back quite a ways. What you will do here is to look at a piece of the method. (Only a piece because a lot is really involved.) Imagine that you have observed the radial velocity of a spectral line in a RR Lyrae star; that is, you have determined the velocity of material, away or toward you in the line of sight, on the surface of the star as it pulsates. (Note that you may get somewhat different velocities were you to observe another line. This is one of the tricky points in the method.) Suppose the velocity curve you have observed is the one shown below (which is decidedly a fake, but for reasons to become obvious, it will make things simple). The velocity is plotted in Fig. 2.39 against phase, ϕ, meaning that you started observing at a zero time, said to be at $\phi = 0$, and then observed for one complete period of pulsation, Π, and call that $\phi = 1$. Thus ϕ measures time in units of Π. (Of course the curve goes on and on and we naively assume it repeats itself each Π.) It has been determined by other means that T_{eff} for the RR Lyrae is 7000 K (typical) and the luminosity is $\mathcal{L}/\mathcal{L}_\odot = 54$ (also typical), where both represent some average over time as the star pulsates. The pulsation period is $\Pi = 0.5$ days (also typical). What you are about to do is determine how the radius varies over one period. The radius of the star, $\mathcal{R}(t)$, is just the integral of the radial velocity, $v(\phi)$, making sure you get your units correct. It so happens that, thanks to us, your velocity curve is well-fit by the function

$$v(\phi) = -30 \cos(2\pi\phi) + 10 \sin(4\pi\phi) \quad \text{km s}^{-1}. \tag{2.12}$$

(Remember while doing this that a *negative* velocity means that the stellar surface is moving *toward* you.) Integrate this from $\phi = 0$ to some ϕ in order, in effect, to get $\mathcal{R}(t) - \mathcal{R}(0)$. If you assume that $\phi = 0$ represents the average radius (taken to be the radius the star would have were it not pulsating) derivable from the T_{eff} and $\mathcal{L}$ given above, then plot $\mathcal{R}(t)$ (in solar units) versus time. To see if you are on the right track, we find that the total excursion in radius from smallest to largest is $\Delta\mathcal{R} \approx 0.6\ \mathcal{R}_\odot$ (which is also typical for many RR Lyraes).

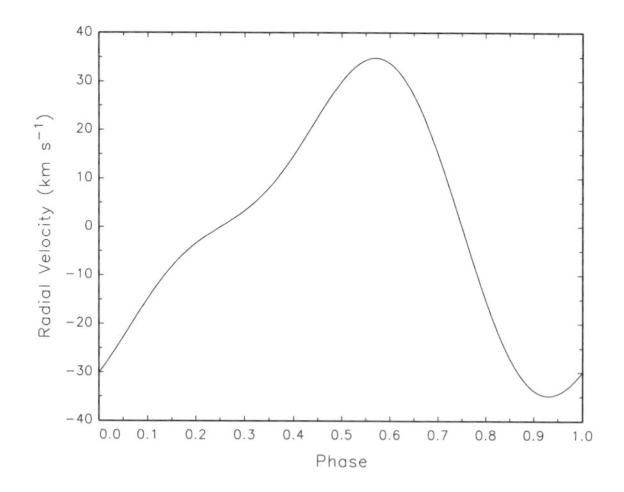

Fig. 2.39. Here is the velocity curve for your RR Lyrae variable. Note that velocity is in km s^{-1}.

Exercise 2.15. Material eventually accreted onto a white dwarf or neutron star first forms a luminous accretion disk surrounding the accretor.

▷ Shapiro, S.L., & Teukolsky, S.A. 1983, *Black Holes, White Dwarfs, and Neutron Stars* (New York: John Wiley & Sons)

give the following expressions for the luminosity of the disk (good to within an order of magnitude or so):

$$\mathcal{L}_{\rm disk}({\rm WD}) \sim \frac{1}{2}\frac{G\dot{\mathcal{M}}\mathcal{M}}{\mathcal{R}} \sim 10^{34}\,\dot{\mathcal{M}}_{-9} \quad {\rm erg\ s^{-1}} \tag{2.13}$$

for white dwarfs, and

$$\mathcal{L}_{\rm disk}({\rm NS}) \sim \frac{1}{2}\frac{G\dot{\mathcal{M}}\mathcal{M}}{\mathcal{R}} \sim 10^{37}\,\dot{\mathcal{M}}_{-9} \quad {\rm erg\ s^{-1}} \tag{2.14}$$

for neutron stars. Here $\dot{\mathcal{M}}_{-9}$ is the accretion rate in the units of $10^{-9}\mathcal{M}_\odot$ yr^{-1}. If you assume the material comes effectively from infinity and most of the luminosity comes from very near the accretor, take a stab at deriving these for a typical white dwarf and neutron star. (The first terms on the right-hand sides of the expressions are a giveaway.) A rate of $10^{-9}\mathcal{M}_\odot$ per year is in the right ballpark for these systems meaning that solar luminosities (or much higher) are easily achievable.

Exercise 2.16. Invent an eclipsing binary system of your choosing (but don't attempt a system where the two stars are in contact, or nearly so—too difficult!). So, choose your two stars, big, small, whatever, luminous, dim, etc., separation and eccentricity of orbit, inclination, etc., and plot the radial velocity and light curves over one cycle of the orbit. (Do have them eclipse, however.) To do this in real detail is difficult (problems of limb-darkening, etc.) so make it simple. A nice elementary discussion of binary systems may be found in

▷ Shu, F.H. 1982, *The Physical Universe: An Introduction to Astronomy* (Mill Valley, CA: University Science Books).

Exercise 2.17. This exercise is essentially Problem 1.6 of Shapiro and Teukolsky (1983, and we suggest you check their §1.3 for the necessary background on the whole question of how stellar statistics vary with time for the galactic disk): The

▷ Salpeter, E.E. 1955, ApJ, 121, 161

"birth rate function" is given by

$$\psi_\mathrm{s}\, d\left(\frac{\mathcal{M}}{\mathcal{M}_\odot}\right) = 2\times 10^{-12}\left(\frac{\mathcal{M}}{\mathcal{M}_\odot}\right)^{-2.35} d\left(\frac{\mathcal{M}}{\mathcal{M}_\odot}\right)\ \mathrm{pc}^{-3}\,\mathrm{yr}^{-1} \tag{2.15}$$

and it gives the *rate* at which stars are "birthed" in the galactic disk in a sample pc^{-3} and is supposed to be roughly valid for $0.4 \le \mathcal{M}/\mathcal{M}_\odot \le 10$. We will use this to estimate how much iron-peak material has been thrown off into the disk of our galaxy by hypothetical supernovae in the mass range $4 \le (\mathcal{M}/\mathcal{M}_\odot) \le 8$ where we assume that 1.4 $\mathcal{M}_\odot$ of iron-peak material is expelled from the star. (Note that the above mass range does *not* agree with what we have said about masses for pre-supernovae—and this is the main point to the problem.) We also assume that the rate of production of stars of a given mass has not changed during galactic history (which is somewhat unlikely). Take the age of the galactic disk as $\mathrm{T}_\mathrm{gal} = 10$ Gyr.

1. After a time T_gal, what is the concentration of iron-peak material in the units of $\mathcal{M}_\odot\ \mathrm{pc}^{-3}$?
2. Show that the main sequence lifetimes of the stars in question make little difference in the answer you obtain.
3. The Oort limit is the estimated total amount of matter in the solar neighborhood necessary to explain the motions of nearby stars. The corresponding density is 0.14 $\mathcal{M}_\odot\ \mathrm{pc}^{-3}$. Some of this must be composed of iron-peak material. If all of that material were produced from supernovae in the mass range used above, then what would be the mass fraction of iron in the disk?
4. Compare your result to the iron mass fraction given in our Fig. 1.2, assuming solar system abundances are representative of the disk. What do you conclude from this exercise?

Exercise 2.18. When a supernova explodes, a shock wave travels through the star heating up and compressing the stellar material. What we shall now explore is a simple model which gives a rough idea of how violent this process may be. We will use the *Hugoniot–Rankine* (HR) relations that tell us how hot and dense the material is after a shock has passed. For a first-rate reference see the first chapter of

▷ Zel'dovich, Ya.B., & Raizer, Yu.P. 1966, *Physics of Shock Waves and High Temperature Hydrodynamic Phenomena*, in two volumes, eds. W.D. Hayes and R.F. Probstein (New York: Academic Press).

The HR relations are

$$\rho(D - v) = \rho_0 D \quad \text{mass conservation} \tag{2.16}$$

$$P + \rho(D - v)^2 = P_0 + \rho_0 D^2 \quad \text{momentum conservation} \tag{2.17}$$

$$E + \frac{P}{\rho} + \frac{(D - v)^2}{2} = E_0 + \frac{P_0}{\rho_0} + \frac{D^2}{2} \quad \text{energy conservation .} \tag{2.18}$$

Here a zero subscript on pressure (P), internal energy per gram (E), and density (ρ) refers to the conditions of the stellar material before the shock hits (and this pre-shock material is assumed to be at rest), whereas no subscript refers to the post-shock conditions. The velocity v is that of the post-shock material and D is the velocity of the shock wave itself. (By the way, see if you can derive these equations.) The idea here is to specify the initial density ρ_0, temperature T_0, and composition of the pre-shock material, and then choose a value of D. All else should follow. We shall assume initial conditions $\rho_0 = 0.01$ g cm^{-3}, $T_0 = 10^5$ K, and a composition specified by a mean molecular weight of $\mu = 1/2$. This places us somewhere in the envelope of the supernova. We further assume that the material doesn't change its ionization state so that μ always remains a constant. For an equation of state we will take some combination (see below) of ideal gas and radiation. To tackle the problem as we will pose it, it is best to eliminate v in the HR relations so as to end up with only two equations. Then plug in directly the equation of state. The resulting equations will then have only ρ and T as unknowns with D as a parameter. The idea will be to choose D and then *numerically* solve for density and temperature. We suggest some kind of Newton–Raphson scheme for solving this problem. Chapter 7 of the text goes through some of this (and see the references there).

1. First assume that the equation of state is solely due to an ideal gas. Choose several values of D through the range 10^7–3×10^9 cm s^{-1} and find the corresponding post-shock densities and temperatures. Plot your results.
2. You will note that the post-shock temperatures continue to rise as D increases but, after a while, the density doesn't change. Show analytically that in the limit
$$\frac{D^2}{2} \gg \frac{N_{\rm A} k T_0}{\mu}$$
the density levels out to $\rho/\rho_0 = 4$ and the temperature is given by $T = 3\mu D^2/(16kN_{\rm A})$.
3. Now do the same thing but include radiation pressure and energy along with the ideal gas. Plot your results.

4. Here you find the same kind of behavior as in the ideal gas case; the density levels off but the temperature keeps increasing as D increases. However, the ratio $\rho/\rho_0 = 7$ is reached for this situation. Show this analytically. Note also that the temperature in this case ends up being less than for the ideal gas shock. Figure out why this happens and then show, in the limit of very large D, that $T^4 = 18\rho_0 D^2/(7a)$ where a is the radiation constant.

Exercise 2.19. Putting together the Roche geometry shown in Fig. 2.35 is not easy because the computations can be very tricky. This exercise will ask you to set up the problem (which involves a review of elementary mechanics), and then, at the end, request from you only a single number (but you may go on as far as you wish). The now-classic reference is

▹ Kitamura, M. 1970, Ap&SS, 7, 272,

which includes results for the rotation plane plus equipotentials off that plane (plus much more).

1. The starting point is to write down the potential, gravitational plus centrifugal, in the rotation plane (x–y plane, with $z = 0$) at some arbitrary point (x, y). For the moment, x and y are measured from the center-of-mass (CM) of the system. If Ω is the angular rotation velocity (with circular orbits), r_1 the distance from $\mathcal{M}_1$ to (x, y), and r_2 the corresponding distance from $\mathcal{M}_2$, then show that the total potential, $\Psi(x, y)$, in the CM system is

$$\Psi(x,y) = \frac{G\mathcal{M}_1}{r_1} + \frac{G\mathcal{M}_2}{r_2} + \frac{1}{2}\Omega^2\left(x^2+y^2\right) . \tag{2.19}$$

(Depending on how you define gravitational and centrifugal forces as derived from potentials, you may derive the negative of our Ψ. It doesn't matter, but keep track of your signs. Note also that an arbitrary constant may be added to your result.)

2. Now shift your reference point from the CM to the location of $\mathcal{M}_1$ and let x and y be measured from there. Show that the result is

$$\Psi(x,y) = \frac{G\mathcal{M}_1}{r_1} + \frac{G\mathcal{M}_2}{r_2} + \frac{1}{2}\Omega^2\left[\left(x - R\frac{\mathcal{M}_2}{\mathcal{M}_1+\mathcal{M}_2}\right)^2 + y^2\right] \tag{2.20}$$

where $r_1^2 = x^2 + y^2$, $r_2^2 = (R-x)^2 + y^2$, with R being the distance from $\mathcal{M}_1$ to $\mathcal{M}_2$. (You will have to recall the relation between masses and distances between masses in the CM to get this.)

3. Now to find out what the ξ_i are in Fig. 2.35. The aim is to get Ψ in a dimensionless form. Define $\xi(x, y)$ by

$$\xi(x,y) = \frac{R}{G\mathcal{M}_1}\Psi(x,y) - \frac{\mathcal{M}_2^2}{2\mathcal{M}_1\left(\mathcal{M}_1+\mathcal{M}_2\right)} . \tag{2.21}$$

Note that this only scales $\Psi(x,y)$ and adds a constant. Thus if $\Psi(x,y)$ is a constant, as an equipotential, then $\xi(x,y)$ is also constant and is the same equipotential but in different guise.

4. The final step is to introduce $q = \mathcal{M}_2/\mathcal{M}_1 \le 1$ (with $\mathcal{M}_1$ as the primary), and let x, y, r_1, and r_2 be given in units of R. Show that the final result is
$$\xi(x,y) = \frac{1}{r_1} + q\left(\frac{1}{r_2} - x\right) + \frac{1}{2}(1+q)\left(x^2+y^2\right) . \qquad (2.22)$$
(You will have to show that $\Omega^2 = G\left(\mathcal{M}_1 + \mathcal{M}_2\right)/R^3$ to get this. It follows from balancing centrifugal and gravitational forces in the CM.) In principle, this is all you need to find the curves of equipotentials such as the samples shown in Fig. 2.35. You may wish to try but, not being that cruel, we won't ask you to.
5. However, given that the equipotential, ξ_1, for the Roche lobe is $\xi_1 = 2.875845$ for $q = 1/2$, find the location of the first Lagrangian point L_1. (Check Fig. 2.35 to see if your answer is reasonable.)

Exercise 2.20. Suppose you wanted to make a star of one solar mass out of a very large sphere of interstellar matter of mass 1 $\mathcal{M}_\odot$ with a uniform density of 1 hydrogen atom cm^{-3}. First off, what is the radius of the sphere? If we want to make the star in the disk of our galaxy at the location of our sun (about 9 kpc from galactic center)—and the sun orbits the galactic center at a speed of about 230 km s^{-1}— what is the rotation period and angular velocity (in radians s^{-1}) at that location? By the way, a short listing of the properties of the Milky Way may be found in

▹ Trimble, V. 1999, in *Allen's Astrophysical Quantities, 4th ed.*, ed. A.N. Cox (New York: Springer–Verlag), p. 569.

What is the initial angular momentum of the cloud (assuming it participates with the galactic rotation)? If the cloud now condenses uniformly down to solar size, conserving angular momentum and not losing any by any means, what is the final rotation rate (in, say, Hz)? Estimate the breakup velocity at the sun's surface and the corresponding rotation rate. You should now realize why angular momentum has to be gotten rid of in star formation.

Exercise 2.21. It was stated that if the more massive star ($\mathcal{M}_1$) in a binary system loses mass to the secondary ($\mathcal{M}_2$), then the separation between them decreases. This is true if angular momentum isn't lost from the system (i.e., all the mass lost by $\mathcal{M}_1$ goes to $\mathcal{M}_2$). Using conservation of momentum, show that the separation does decrease. (You can use some of the results from Ex. 2.19.)

Exercise 2.22. (We thank Cole Miller of UMD for reminding us about this simple analysis and follow his version.) A very early, and now classic, paper by

▹ Schönberg, M., & Chandrasekhar, S. 1942, ApJ, 96, 61

discusses what happens if an isothermal core of helium tries to support a hydrogen-rich envelope but runs into trouble if the core gets too massive. (It is assumed that a hydrogen burning shell overlies the core.) The trouble point is the *Chandrasekhar–Schönberg limit* (in mass, and you may often see the names reversed). The following analysis is crude but is intended to give the idea of how this works. So, assume that you have a constant density star of total mass $\mathcal{M}$ and radius $\mathcal{R}$. These are kept fixed throughout the problem. The mass of the helium core, of mass $\mathcal{M}_{\text{core}}$, and its radius, $\mathcal{R}_{\text{core}}$, are, however, allowed to vary. The idea will be to examine what is the pressure at the inner surface at $\mathcal{R}_{\text{core}}$ and compare it to the pressure exterior to that surface where the composition has changed (and, hence, so has the mean molecular weight). To make things simple, assume that the ideal gas law holds everywhere. (You may wish to consult

▷ Stein, R.F. 1966, in *Stellar Evolution* (New York: Plenum Press), eds. Stein & Cameron, pp. 3–82

for a slightly different version of what follows.)

1. Use the virial theorem in the form

$$2K + \Omega = 3P_S V \tag{2.23}$$

as discussed briefly in §1.3 and Ex. 1.8 (where you are to derive it). P_S is the surface pressure at the inner surface at $\mathcal{R}_{\text{core}}$. Remembering that we have an isothermal, constant density, core made up of an ideal gas, find P_S as a function of $\mathcal{M}_{\text{core}}$, $\mathcal{R}_{\text{core}}$, μ_{core} (the mean molecular weight in the core), and the temperature T of the core.

2. This P_S is the pressure that is supposed to support the overlying layers. We want to maximize it to give the hardest push on the envelope. Show that the maximum, with respect to $\mathcal{R}_{\text{core}}$, is given by

$$P_S(\text{max}) = \frac{1}{4\pi}\frac{3}{4}\left(\frac{15}{4}\right)^3\left(\frac{N_{\text{A}}kT}{\mu_{\text{core}}}\right)^4\frac{1}{G^3\mathcal{M}_{\text{core}}^2}.$$

3. Now use the result (Eq. 1.42) for the central pressure of the constant density sphere to find the temperature T of the core; i.e., express T in terms of $\mathcal{M}$, $\mathcal{R}$, and μ_{core}. (Of course, for such a sphere, $\mathcal{M}_{\text{core}}/\mathcal{R}_{\text{core}}^3 = \mathcal{M}/\mathcal{R}^3$ so you have some leeway in how to express T. Hint: see Eq. 1.56)
4. Since we really don't want T to be discontinuous across $\mathcal{R}_{\text{core}}$ (otherwise heat would flow like crazy across $\mathcal{R}_{\text{core}}$), express $P_S(\text{max})$ in terms of $\mathcal{M}$, $\mathcal{M}_{\text{core}}$, μ_{core}, and μ_{env}, where the last is the mean molecular weight in the envelope.
5. In order that the core be able to support the envelope, we must have

$$P_S(\text{max}) \geq P_S(\text{env}) \tag{2.24}$$

where $P_S(\text{env})$ is the pressure just exterior to $\mathcal{R}_{\text{core}}$ (otherwise the envelope would push in the core). Now, if you will allow for a little inconsistency, since T and density are constant in the core the implication is that pressure is everywhere constant and, so that hydrostatic equilibrium is to be maintained at $\mathcal{R}_{\rm core}$, take $P_S(\text{env})$ to be the central pressure of the sphere. (For a different, but similar tack, see Stein 1966.) From this (somewhat dubious assumption) show that (2.24) is equivalent to

$$\frac{\mathcal{M}_{\rm core}}{\mathcal{M}} \leq \left(\frac{15}{4}\right)^3 \left(\frac{1}{2}\right)^5 \left(\frac{\mu_{\rm env}}{\mu_{\rm core}}\right)^2 . \tag{2.25}$$

Put in numbers for (2.25) using (1.55) assuming complete ionization, and compare your result to what was said in §1.7 about lifetimes on the main sequence. (The result isn't bad at all.) Note that an important element of what's going on here is that the envelope has the advantage to begin with. As we have emphasized several times, the mean molecular weight in the helium core is greater than in the hydrogen-rich envelope because there are fewer free particles per gram in the core. Thus the envelope pressure has more free particles to use.

2.17 References and Suggested Readings

§2.1: Young Stellar Objects

Figure 2.1 is from

▹ Shu, F.H., Adams, F.C., & Lizano, S. 1987, ARA&A, 25, 23

entitled "Star Formation in Molecular Clouds," which tells you what it's about. We recommend it highly. The evolutionary tracks and location of T Tauri stars of Fig. 2.2 are from the review article by

▹ Stahler, S.W. 1988, PASP, 100, 1474.

It's an excellent "easy" introduction to the subject.

Amateur astronomers are familiar with

▹ Allen, R.H. 1963, *Star Names* (New York: Dover)

and

▹ Burnham, R. Jr. 1978, *Burnham's Celestial Handbook*, in three volumes (New York: Dover)

but professionals also consult them—especially if teaching an undergraduate course. Of a different sort is

▹ Jaschek, C., & Jaschek, M. 1987, *Classification of Stars* (Cambridge University Press: Cambridge),

which is a compendium of, as titled, stellar classification. An excellent monograph to consult for spectral types, etc., although it is getting to be a bit dated.

§2.2: The Zero Age Main Sequence

The most widely consulted shorter compendia of astronomical tables, etc., are

▷ Allen, C.W. 1973, *Astrophysical Quantities* 3d ed. (London: Athlone)

▷ Cox, A.N. 1999, Editor of *Allen's Astrophysical Quantities* (New York: Springer–Verlag)

which replaces Allen(1973), and

▷ Lang, K.R. 1991, *Astrophysical Data: Planets and Stars* (Berlin: Springer–Verlag).

Their prices, however, are unfortunate.

Brief discussions of the "solar neutrino problem" include

▷ Bahcall, J.N. 2001, Nature, 412, 29

and

▷ Seife, C. 2002, Science, 296, 632.

As more are discovered, brown dwarfs will take up more and more space in textbooks such as this. For now, see

▷ Basri, G. 2000a, SciAm, 282, 77 (April 2000)

▷ Basri, G. 2000b, ARA&A, 38, 485

and

▷ Gizis, J.E. 2001, Science, 294, 801.

▷ Chabrier, G., & Baraffe, I. 2000, ARA&A, 38, 337

and

▷ Burrows, A., Hubbard, W.B., & Lunine, J.I. 2001, RevModPhys, 73, 719

discuss the general problem of substellar objects.

§2.3: Leaving the Main Sequence

Figure 2.5 is from

▷ Iben, I. Jr. 1967, ARA&A, 5, 571

which is now a classic and one of the first of many review articles by Icko Iben.

▷ Iben, I. Jr. 1991, ApJS, 76, 55

contains a personal account of Iben's work and, as is usual in his papers, the reference list is exhaustive. Figure 2.12 is from

▷ Iben, I. Jr. 1985, QJRAS, 26, 1.

Most of the data used for Fig. 2.6 is from

▷ Mermilliod, J.-C. 1986, A&AS, 24, 159

supplemented by

▷ Mermilliod, J.-C. & Bratschi, P. 1997, A&A, 320, 74.

The source for the HR diagram of M3 is

▷ Buonanno, R., Corsi, C.E., Buzzoni, A., Cacciari, C., Ferraro, F.R., & Fusi Pecci, F. 1994, A&A, 290, 68.

Much effort has made in recent years to determine globular cluster ages. Two convergent views are to be found in

▷ Chaboyer, B.C. 2001, *Rip Van Twinkle*, SciAm, 284, #5, 44 (May 2001)

and

▹ Lebreton, Y. 2000, ARA&A, 38, 35.

A very useful reference, though now somewhat outdated, that describes how cosmological distances and time scales are derived (usually virtually all the tools of astronomy) is

▹ Rowan-Robinson, M. 1985, *The Cosmological Distance Ladder* (New York: Freeman).

And, just to go back in time, see

▹ Larson, R.B., & Bromm, V. 2001, *The First Stars in the Universe*, SciAm, 285, #6, 64 (Dec 2001)

The isochrones of Fig. 2.8 were derived from data in

▹ Green, E.M., Demarque, P., & King, C.R. 1987, *The Revised Yale Isochrones and Luminosity Functions*, Yale University Observatory report (New Haven, CT).

You may also make up your own isochrones using the model results of, for example,

▹ Mengel, J.G., Sweigart, A.V., Demarque, P., & Gross, P.G. 1979, ApJS, 40, 733.

We are fortunate that

▹ Lamers, H.J.G.L.M, & Cassinelli, J.P. 1999, *Introduction to Stellar Winds* (Cambridge: Cambridge University Press)

was published so we could refer to it so often. For more on winds from hot stars see

▹ Kudritzki, R.-P., & Puls, J. 2000, ARA&A, 38, 613.

§2.4: Red Giants and Supergiants

With such a basic question as "Why do stars becomes giants?" it is surprising that people are still asking it, as in

▹ Sugimoto,D., & Fujimoto, M.Y. 2000, ApJ, 538, 857.

Our little diagram illustrating the cycle of nuclear burning, exhaustion, followed by contraction and heating, is adapted from the discussion in

▹ Kippenhahn, R., & Weigert, A. 1990, *Stellar Structure & Evolution* (Berlin: Springer–Verlag).

§2.5: Helium Flash or Fizzle

It is often instructive to follow ρ_c versus T_c through time and we have chosen

▹ Iben, I. Jr. 1991, ApJS, 76, 55

for Fig. 2.12 to illustrate this. The evolution of ρ_c for a solar model is based on data from

▹ Charbonel, C., Meynet, G., Maeder, & Schaerer, D. 1996, A&AS, 115, 339.

The discovery of Technetium in the atmospheres of some cooler stars has to be one of the most important in stellar astronomy; see

▹ Merrill, P.W. 1952, Science, 115, 484.

▷ Willson, L.A. 2000, ARA&A, 38, 573
offers an extensive review of winds from red supergiants.

§2.6: Later Phases, Initial Masses $\mathcal{M} \leq$6–10 $\mathcal{M}_\odot$

Every few years George McCook & Ed Sion update their very useful white dwarf catalog. The latest is

▷ McCook, G.P., & Sion, E.M. 1999, ApJS, 121, 1

and it is the basis for our Fig. 2.15. For white dwarf luminosity functions and ages, see

▷ Fontaine, G., Brassard, P., & Bergeron, P. 2001, PASP, 113, 409

▷ Liebert, J., Dahn, C.C., & Monet, D.G. 1988, ApJ, 332, 891

and

▷ Leggett, S.K, Ruiz, M.T., & Bergeron, P. 1998, ApJ, 497, 294.

The Fontaine et al. reference discusses the prospects of using white dwarfs to determine the age of our galaxy (and others). Our Chapter 10 will dwell on this further. Observations of the globular cluster M4 are reported in

▷ Richer, H.B. et al. 2002, ApJ, 574, L151

▷ Hansen, B.M.S. et al. 2002, ApJ, 574, L155.

An excellent review of the cool white dwarfs is

▷ Hansen, B.M., & Liebert, J. 2003, ARA&A, 41, 465.

For a review of evolution to the PNNS, see

▷ Iben, I. Jr. 1995, PhysRep, 250, 1.

§2.7: Advanced Phases, Initial Masses $\mathcal{M}$ >6–10 $\mathcal{M}_\odot$

We are yet again pleased that one of our colleagues has written a text that we can reference—instead of our having to do all the work. In this case it is

▷ Arnett, D. 1996, *Supernovae and Nucleosynthesis* (Princeton: Princeton University Press).

We shall refer to it often because it covers supernovae and nucleosynthesis in clear and full detail.

§2.8: Core Collapse and Nucleosynthesis

Besides Arnett (1996), the student should, at some time in her life, read the classic

▷ Burbidge, E.M., Burbidge, G.R., Fowler, W.A., & Hoyle, F. 1957 (B^2FH), RevModPhys, 29, 547.

The science is important, but it also shows how papers should be written.

We have used several references in our discussion of elemental and nuclear abundances, and the r- and s-processes. These are

▷ Anders, E., & Grevesse, N. 1989, GeoCosmo, 53, 197

▷ Trimble, V. 1997, Origins of Life, 27, 3

▷ Trimble, V. 1991, A&ARev, 3, 1

▷ Trimble, V. 1996, in *Cosmic Abundances*, eds. Holt & Sonneborn, ASP Conf. Ser., 99, 3

▷ Käppeler, F., Beer, H., & Wisshak, K. 1989, RepProgPhys, 52, 945

▷ Käppeler, F., Gallino, R., Busso, M., Picchio, G., & Raiteri, C.M. 1990, ApJ, 354, 630

and

▷ Meyer, B.S. 1994, ARA&A, 32, 153.

The following should also be at your fingertips when we discuss nuclear physics and nucleosynthesis:

▷ Clayton, D.D. 1968, *Principles of Stellar Evolution and Nucleosynthesis* (New York: McGraw-Hill).

§2.9: Variable Stars

Sakurai's object is a fascinating example of evolution in action. For an easy introduction, see

▷ Kerber, F., & Asplund, M. 2001, *The Star Too Tough To Die*, Sky&Tel (Nov.), p. 48.

Evolutionary calculations for these kinds of stars are given in

▷ Lawlor, T.M., & MacDonald, J. 2003, ApJ, 583, 913.

§2.10: Pulsational Variables

The two major textbook references to intrinsic variable stars are

▷ Cox, J.P. 1980, *Theory of Stellar Pulsation* (Princeton: Princeton University Press)

and

▷ Unno, W., Osaki, Y., Ando, H., Saio, H., & Shibahashi, H. 1989, *Nonradial Oscillations of Stars*, 2d ed. (Tokyo: University of Tokyo Press).

Recent reviews include

▷ Gautschy, A., & Saio, H. 1995, ARA&A, 33, 75

▷ Gautschy, A., & Saio, H. 1996, ARA&A, 34, 551

and

▷ Brown, T.M., & Gilliland, R.L. 1994, ARA&A, 32, 37.

▷ Cox, J.P., & Whitney, C.A. 1958, ApJ, 127, 561

and

▷ Zhevakin, S.A. 1953, RusAJ, 30, 161

essentially solved the problem of what makes Classical Cepheids pulsate, which also gave the key to the pulsations of other variables in the Cepheid Strip.

▷ Smith, H.A. 1995, *RR Lyrae Stars* (Cambridge: Cambridge University Press)

has reviewed the properties of RR Lyrae variables in some detail. The oblique rotator model for roAp variables was first discussed in detail by

▷ Kurtz, D.W. 1990, ARA&Ap, 28, 607

and thanks are due

- ▷ Bouchy, F., & Carrier, F. 2001, A&A, 374, 5

and

- ▷ Kilkenny, D., Koen, C., O'Donoghue, D., & Stobie, R.S. 1997, MNRAS, 285, 640

for discovering pulsations in α Cen A and EC14026, respectively. Even newer, but unnamed pulsators are discussed in

- ▷ Green, E.M., et al. 2003, ApJ, 583, L31.

§2.11: Explosive Variables

The text by

- ▷ Warner, B. 1995, *Cataclysmic Variable Stars*, (Cambridge: Cambridge University Press)

is required reading for those interested in cataclysmic varaibles. Other references include

- ▷ Sparks, W.M., Starrfield, S.G., Sion, E.M., Shore, S.N., Chanmugam, G., & Webbink, R.F. 1999. in *Allen's Astrophysical Quantities, 4th ed.*, ed. A.N. Cox (New York: Springer–Verlag), p. 429
- ▷ McLaughlin, D.B. 1960, in *Stellar Atmospheres*, ed. J.L. Greenstein (Chicago: University of Chicago Press), p. 585

and

- ▷ Wade, R.A., & Ward, M.J. 1985, in *Interacting Binary Stars*, eds. J.E. Pringle and R.A. Wade (Cambridge: Cambridge University Press).

For more about white dwarfs in these systems, see

- ▷ Sion, E.M. 1999, PASP, 111, 532.

Figure 2.28 (sample SN spectra) is from

- ▷ Filippenko, A.V. 1996, ARA&A, 35, 312

and see

- ▷ Branch, D., Nomoto, K., & Filippenko, A.V. 1991, ComAp, 15, 221

while the SN Type I & II light curves of Figs. 2.29 & 2.32 appeared in

- ▷ Wheeler, J.C., & Benetti, S. 1999. in *Allen's Astrophysical Quantities, 4th ed.*, ed. A.N. Cox (New York: Springer–Verlag), p. 453 (Fig, 18.2), p. 455 (Fig. 18.4), ©AIP Press, Springer–Verlag.

An introductory review of SN may be found in

- ▷ Burrows, A. 2000, Nature, 403, 727 (Feb. 17, 2000)

and, for a review of historical galactic SN, see

- ▷ Stephenson, F.R., & Green, D.A. 2003 (May), Sky&Tel, p. 40.

Note that they do not list SN1685. Our discussion of SN1987A follows that of

- ▷ Arnett, W.D., Bahcall, J.N., Kirshner, R.P., & Woosley, S.E. 1989, ARA&A, 27, 629

with additional material from Arnett (1996). More up-to-date material on the explosion of massive stars may be found in

- ▷ Woosley, S.E., Heger, A., & Weaver, T.A. 2002, RevModPhys, 74, 1015.

A thorough discussion of Type Ia models is to be found in

▷ Hillebrandt, W., & Niemeyer, J.C. 2000, ARA&A, 38, 191.

§2.12: White Dwarfs, Neutron Stars and Black Holes

The following reference is a bit dated but it remains the classic text:

▷ Shapiro, S.L., & Teukolsky, S.A. 1983, *Black Holes, White Dwarfs, and Neutron Stars* (New York: John Wiley & Sons).

§2.13: Binary Stars

This is a huge subject that we have condensed down to a smidgen. For an excellent introduction (both observationally and theoretically), see

▷ de Loore, C.W.H. & Doom, C. 1992, *Structure and Evolution of Single and Binary stars* (Dordrecht: Kluwer Academic Publishers).

We also took Fig. 2.36 from their work (Fig. 16.1), and see

▷ Kippenhahn, R., & Weigert, A. 1967, ZeAp, 65, 221.

Calculating Roche surfaces is not for the weak of heart. See

▷ Kitamura, M. 1970, Ap&SS, 7, 272

for the best examples.

§2.14: Star Formation

For an overview of the subject see

▷ Holt, S.S. & Mundy, L.G. 1997, eds. *Star Formation Near and Far*, AIP Conf. Ser. 393.

Angular momentum problems (always a problem because they are always multi-dimensional) are discussed in

▷ Bodenheimer, P. 1995, ARA&A, 33, 199.

▷ Ehrenfreund, P., & Charnley, S.B. 2000, ARA&A, 38, 427

take us on an interesting voyage through a space teeming with molecules. The ISM of our galaxy is reviewed by

▷ Ferrière, K.M. 2001, RevModPhys, 73,1031.

▷ Hale, J.R. et al. (2003), *Questioning the Delphic Oracle*, SciAm, 289, 67 (Aug. 2003)

suggest that the oracles at Delphi in classical Greece unwittingly used ethylene emitted from vents in the earth to pass into a trance-like state to make their ambiguous pronouncements. Nothing to do with astronomy, but interesting nevertheless.

3 Equations of State

"The worth of a State, in the long run,
is the worth of the individuals composing it."
— *John Stuart Mill (1806–1873)*

"What is Matter?—Never mind.
What is Mind?—no matter."
— *from Punch (1855)*

The equations of state appropriate to the interiors of most stars are simple in one major respect: they may be derived using the assumption that the radiation, gas, fluid, or even solid, is in a state of *local thermodynamic equilibrium*, or LTE. By this we mean that at nearly any position in the star complete thermodynamic equilibrium is as very nearly true as we could wish. It is only near the stellar surface or in highly dynamic events, such as in supernovae, where this assumption may no longer be valid.

The reasons that LTE works so well are straightforward: particle–particle and photon–particle mean free paths are short and collision rates are rapid compared to other stellar length or time scales. (A major exception to this rule involves nuclear reactions, which are usually slow.) Thus two widely separated regions in the star are effectively isolated from one another as far as the thermodynamics are concerned and, for any one region, the Boltzmann populations of ion energy levels are consistent with the local electron kinetic temperature.[1] Note, however, that different regions cannot be *completely* isolated from one another in a real star because, otherwise, energy could not flow between them. Chapter 4 will go into this further.

One typical scale length in a star is the *pressure scale height*, λ_P, given by

$$\lambda_P = -\left(\frac{d\ln P}{dr}\right)^{-1} = \frac{P}{g\rho} \tag{3.1}$$

where the equation of hydrostatic equilibrium (1.6) has been used to eliminate dP/dr. The constant–density star discussed in the first chapter easily yields an estimate for this quantity of

$$\lambda_P\left(\rho = \text{constant}\right) = \frac{\mathcal{R}^2}{2r}\left[1 - \left(\frac{r}{\mathcal{R}}\right)^2\right]$$

using the run of pressure given by (1.41). The central value of λ_P is infinite but through most of the constant–density model it is of order $\mathcal{R}$. Near the

[1] For further discussions of the conditions for LTE see Cox (1968, Chap. 7) and Mihalas (1978, Chap. 5).

surface it decreases rapidly to zero. We compare these lengths to photon mean free paths, $\lambda_{\rm phot}$, which we construct from the opacity by

$$\lambda_{\rm phot} = (\kappa\rho)^{-1} \ \ {\rm cm} \, . \tag{3.2}$$

This quantity is a measure of how far a photon travels before it is either absorbed or scattered into a new direction (see Chap. 4). Note that opacity has the units of $\rm cm^2 \ g^{-1}$.

For Thomson electron scattering, which is the smallest opacity in most stellar interiors, later work will show that $\kappa \approx 1 \ \rm cm^2 \ g^{-1}$. If we consider the sun to be a typical star and set $\mathcal{R} = \mathcal{R}_\odot$ and $\rho = \langle \rho_\odot \rangle \approx 1 \ \rm g \ cm^{-3}$ in the above, we then find $\lambda_{\rm phot}$ is at most a centimeter and $\lambda_P \sim 10^{11}$ cm through the bulk of the interior. Thus $\lambda_{\rm phot}$ is smaller than λ_P by many orders of magnitude. We could also have compared $\lambda_{\rm phot}$ with a temperature scale height and found the same sort of thing because, for the sun, the temperature decreases by only 10^{-4} K cm^{-1} on average from center to surface.

Another simple calculation yields an estimate of how much of a star is *not* in LTE. If the photon mean free path is still of order 1 cm, then the relative radius at which the pressure scale height is equal to the photon mean free path is $(r/\mathcal{R}) \approx 1 - 10^{-11}$ using the constant–density model. This means, as a crude estimate, that it is within only the last one part in 10^{11} of the radius that the assumption of LTE fails. In realistic models, the assumption of LTE breaks down within the region of the stellar photosphere, which is the only part of a star we can see.

In the following sections we shall quote some results from statistical mechanics, which will eventually be used to derive equations of state for stellar material consisting of gases (including photons) in thermodynamic equilibrium. Because several excellent texts on statistical mechanics are available for reference, many results will be stated without proof. One particular text we recommend is Landau and Lifshitz (1958, or later editions) for its clean style and inclusion of many fundamental physical (and astrophysical) applications. Additional material may be found in Cox (1968), Kippenhahn and Weigert (1990), and Rose (1998, §3.2).

3.1 Distribution Functions

The "distribution function" for a species of particle measures the number density of that species in the combined six–dimensional space of coordinates plus momenta. If that function is known for a particular gas composed of a combination of species, then all other thermodynamic variables may be derived given the temperature, density, and composition. For the next few sections we shall assume that the gas, including electrons and photons, is a perfect (sometimes called ideal) gas in that particles comprising the gas interact so weakly that they may be regarded as noninteracting as far as their thermodynamics is concerned. They may, however, still exchange energy and other conserved properties. Before writing down the distribution function for a perfect gas we first introduce what may be an unfamiliar thermodynamic quantity.

The variables of thermodynamic consequence we have encountered thus far are P, T, ρ (or $V_\rho = 1/\rho$), S, E, Q, and various number densities, n_i (see §1.4.1). The latter have been, and will be, given in the units of number cm^{-3}. We now introduce N_i, which is the (specific) number density of an ith species in the units of number per gram of material with $N_i = n_i/\rho$. It is the Lagrangian version of n_i and it will prove useful because it remains constant even if volume changes.

Another very useful thermodynamic quantity is the *chemical potential*, μ_i, defined by[2]

$$\mu_i = \left(\frac{\partial E}{\partial N_i}\right)_{S,V} \tag{3.3}$$

as associated with an ith species in the material (and is not to be confused with μ_{I}, the ion molecular weight). If there are "chemical" reactions in the stellar mixture involving some subset of species (ions, electrons, photons, molecules, etc.) whose concentrations could, in principle, change by dN_i as a result of those reactions, then thermodynamic (and chemical) equilibrium requires that

$$\sum_i \mu_i \, dN_i = 0 \tag{3.4}$$

which we state without proof. Changing N_i by dN_i in a real mixture usually means that other components in the mixture must change by an amount related to dN_i so that not all the dN_i are independent.

As an example, consider the ionization–recombination reaction

$$\mathrm{H}^+ + \mathrm{e}^- \Longleftrightarrow \mathrm{H}^0 + \gamma \tag{3.5}$$

where H^0 is neutral hydrogen—assumed to have only one bound state in the following discussion—H^+ is the hydrogen ion (a proton), and e^- is an electron. We shall neglect the photon that appears on the righthand side of (3.5) in the following because, as we shall show, its chemical potential is zero and will not enter into the application of (3.4). The double-headed arrow is to remind us that the reaction proceeds equally rapidly in both directions in thermodynamic equilibrium. Now write (3.5) in the algebraic form

$$1\,\mathrm{H}^+ + 1\,\mathrm{e}^- - 1\,\mathrm{H}^0 = 0$$

where the coefficients count how many individual constituents are destroyed or created in a single reaction. A more general form for this equation is

---

[2] A simple example indicating why μ_i is a "potential" is given as Ex. 3.6.

$$\sum_i \nu_i \,\mathrm{C}_i = 0\,. \tag{3.6}$$

The C_i represent H^+, H^0, and e^- in the example and the ν_i, or *stoichiometric coefficients*, are the numerical coefficients. Obviously the concentrations, N_i, are constrained in the same way as the C_i. Thus if N_1 changes by some arbitrary amount dN_1, then the ith concentration changes according to

$$\frac{dN_i}{\nu_i} = \frac{dN_1}{\nu_1}\,.$$

Equation (3.4) then becomes

$$\sum_i \mu_i \frac{dN_1}{\nu_1}\nu_i = \frac{dN_1}{\nu_1}\sum_i \mu_i \nu_i = 0$$

or, since dN_1 is arbitrary,

$$\sum_i \mu_i \nu_i = 0\,. \tag{3.7}$$

This is the equation for *chemical equilibrium*, which must be part of thermodynamic equilibrium when reactions are taking place.[3]

As another simple, and useful, example consider a classical blackbody cavity filled with radiation in thermodynamic equilibrium with the walls of the cavity. Equilibrium is maintained by the interaction of the photons with material comprising the walls but the number of photons, N_γ, fluctuates about some mean value; that is, photon number is not strictly conserved. Therefore dN_γ need not be zero. Nevertheless, reactions in the cavity must satisfy a symbolic relation of the form $\sum \mu_i \, dN_i + \mu_\gamma \, dN_\gamma = 0$ with $dN_i = 0$. The last two statements can only be reconciled if

$$\mu_\gamma = 0 \quad \text{for photons.} \tag{3.8}$$

It is for this reason that photons were not included in the ionization and recombination reaction of (3.5): the vanishing of μ_γ makes its presence superfluous in the chemical equilibrium equation (3.7).

It is reasonable, and correct, to expect that given T, ρ, and a catalogue of what reactions are possible, we should be able to find all the N_i for a gas in thermodynamic equilibrium. In other words, information about N_i is contained in μ_i for the given T and ρ. In a real gas this connection is difficult to establish because it requires a detailed knowledge of how the particles in the system interact. For a perfect gas things are easier. Any text on statistical mechanics may be consulted for what follows.

---

[3] We exclude thermonuclear reactions from this discussion for the present because they may proceed very slowly and, usually, only in one direction during stellar nuclear burning.

The relation between the number density of some species of elementary nature (ions, photons, etc.) in coordinate–momentum space and its chemical potential in thermodynamic equilibrium is found from statistical mechanics to be

$$n(p) = \frac{1}{h^3} \sum_j \frac{g_j}{\exp\left\{\left[-\mu + \mathcal{E}_j + \mathcal{E}(p)\right]/kT\right\} \pm 1} . \qquad (3.9)$$

We call $n(p)$ the distribution function for the species (although you will often see this referred to as the "occupation number"). The various quantities are as follows:

- μ is the chemical potential of the species.
- j refers to the possible energy states of the species (e.g., energy levels of an ion).
- $\mathcal{E}_j$ is the energy of state j referred to some reference energy level.
- g_j is the degeneracy of state j (i.e., the number of states having the same energy $\mathcal{E}_j$).
- $\mathcal{E}(p)$ is the kinetic energy as a function of momentum p.
- a "+" in the denominator is used for Fermi–Dirac particles (fermions of half-integer spin) and a "−" for Bose–Einstein particles (bosons of zero or whole integer spin).
- h is Planck's constant $h = 6.6260688 \times 10^{-27}$ erg s.
- $n(p)$ is in the units of number per (cm−unit momentum)^3 where the differential element in coordinate–momentum space is $d^3\mathbf{r}\, d^3\mathbf{p}$.

As we shall demonstrate in the following discussion, (3.9) will lead to all the familiar results from elementary thermodynamics.

To retrieve the physical space number density, n (cm^{-3}), for the species from (3.9) we need only integrate over all momentum space, which, from standard arguments, is assumed to be spherically symmetric; that is,[4]

$$n = \int_p n(p)\, 4\pi p^2\, dp \;\; \text{cm}^{-3} . \qquad (3.10)$$

The factor of 4π (steradians) comes from the two angular integrations over the surface of a unit sphere.

Because we shall want eventually to consider relativistic particles, the correct form of the kinetic energy, $\mathcal{E}$, for a particle of rest mass m is given by

$$\mathcal{E}(p) = \left(p^2c^2 + m^2c^4\right)^{1/2} - mc^2 \qquad (3.11)$$

[4] We explicitly assume here that the distribution of particles is angularly isotropic in momentum. This is really part of LTE but the assumption will have to be reexamined in Chapter 4 when we put back angular information and partially unravel the integral.

which reduces to $\mathcal{E}(p) = p^2/2m$ for $pc \ll mc^2$ in the nonrelativistic limit, and $\mathcal{E}(p) = pc$ for extremely relativistic particles or those with zero rest mass.

We shall also need an expression for the velocity which, from Hamilton's equations (one of the more elegant and important subjects in the physical sciences), is

$$v = \frac{\partial \mathcal{E}}{\partial p} . \tag{3.12}$$

(As a simple check on this definition of v, note that $v \to p/m$ for $pc \ll mc^2$ and $v \to c$ for the relativistic case, both of which are elementary results.) This is the velocity to use in the following kinetic theory expression for isotropic pressure (as in 1.20)

$$P = \tfrac{1}{3} \int_p n(p)\, pv\, 4\pi p^2\, dp . \tag{3.13}$$

Finally, the internal energy is simply

$$E = \int_p n(p)\, \mathcal{E}(p)\, 4\pi p^2\, dp . \tag{3.14}$$

That completes all that we shall need to construct practical equations of state in the following applications.

3.2 Blackbody Radiation

Photons are massless bosons of unit spin. Since they travel at c, they only have two states (two spin orientations or polarizations) for a given energy and thus the degeneracy factor in (3.9) is $g = 2$. From before, $\mu_\gamma = 0$ and $\mathcal{E} = pc$. Because there is only one energy level (no excited states), $\mathcal{E}_j$ may be taken as zero. Putting this together, we find that the photon number density is given by[5]

$$n_\gamma = \frac{8\pi}{h^3} \int_0^\infty \frac{p^2\, dp}{\exp(pc/kT) - 1} \quad \text{cm}^{-3} . \tag{3.15}$$

Let $x = pc/kT$ and use the integral

[5] It may seem contradictory to give one number for the photon density whereas we stated earlier that the photon concentration fluctuates about some mean value—thus giving $\mu_\gamma = 0$. But the point is that photons must interact with matter to equilibrate (not with each other unless you delve into quantum electrodynamics) and this is a statistical process. What you get in (3.15) is an *average*. Fluctuations about that average depend on the particulars of the matter interactions but, as long as there are many interactions, the effect of fluctuations is very small. Much the same can be said about even the ideal gas except there we deal with various conservation rules involving particles, not photons. See, for example, Landau and Lifshitz (1958, Chap. XII).

$$\int_0^\infty \frac{x^2\,dx}{e^x - 1} = 2\,\zeta(3) = 2(1.202\cdots)$$

where $\zeta(3)$ is a Riemann Zeta function, to find

$$n_\gamma = 2\pi\,\zeta(3)\left(\frac{2kT}{ch}\right)^3 \approx 20.28\,T^3 \quad \text{cm}^{-3}\,. \tag{3.16}$$

Find, in similar fashion, that the radiation pressure is given by

$$P_{\text{rad}} = \left(\frac{k^4}{c^3h^3}\frac{8\pi^5}{15}\right)\frac{T^4}{3} = \frac{aT^4}{3} \quad \text{dyne cm}^{-2} \tag{3.17}$$

and that the energy density is

$$E_{\text{rad}} = aT^4 = 3P_{\text{rad}} \quad \text{erg cm}^{-3} \tag{3.18}$$

where a is the radiation constant $a = 7.56577 \times 10^{-15}$ erg cm^{-3} K^{-4}. Thus we recover the usual results for blackbody radiation. The nice thing about LTE radiation is that all you have to know is the ambient temperature. Matter density, composition, etc., don't matter, so to speak.

Note that (3.18) is a γ–law equation of state $P=(\gamma-1)E$ (as in 1.24 after E in that equation is converted to energy per unit volume) with $\gamma = 4/3$. Thus a star whose equation of state is dominated by radiation is in danger of approaching the $\gamma = 4/3$ limit discussed in Chapter 1.

It will be convenient for later purposes to define the energy density per unit frequency (ν) or wavelength (λ) in the radiation field. These energy densities are usually designated by u (with an appropriate subscript). Recall that frequency is given by $\nu = \mathcal{E}/h = pc/h$ and wavelength by $\lambda = c/\nu$. If u_p is the energy density per unit momentum (that is, the integrand of 3.14 with $E_{\text{rad}} = \int_0^\infty u_p\,dp$) and u_ν and u_λ are the corresponding densities per unit frequency and wavelength, then you may easily show

$$u_\nu\,d\nu = \frac{8\pi h\nu^3}{c^3}\frac{1}{e^{h\nu/kT} - 1}\,d\nu \quad \text{erg cm}^{-3}\ \text{Hz}^{-1}\ \text{Hz} \tag{3.19}$$

and

$$u_\lambda\,d\lambda = \frac{8\pi hc}{\lambda^5}\frac{1}{e^{hc/\lambda kT} - 1}\,d\lambda \quad \text{erg cm}^{-3}\ \text{cm}^{-1}\ \text{cm}\,. \tag{3.20}$$

Associated quantities are the *frequency-dependent Planck function*

$$B_\nu(T) = \frac{c}{4\pi}u_\nu \quad \text{erg cm}^{-2} \tag{3.21}$$

and the *integrated Planck function*

$$B(T) = \int_0^\infty B_\nu(T)\,d\nu = \frac{ca}{4\pi}T^4 = \frac{\sigma}{\pi}T^4 \quad \text{erg cm}^{-2}\ \text{s}^{-1}\,. \tag{3.22}$$

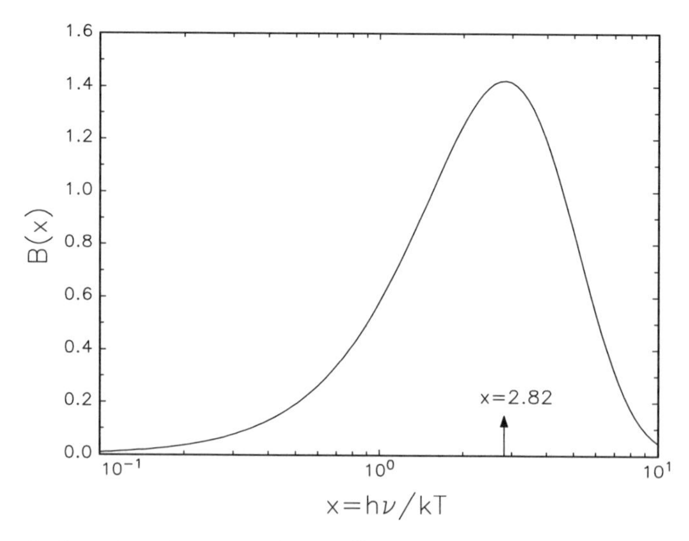

Fig. 3.1. A plot of the function $\mathrm{B}(x) = x^3/[\exp(x) - 1]$ corresponding to the vital part of either u_ν or the Planck function B_ν. The maximum is at $x = h\nu/kT = 2.821$.

The Stefan–Boltzmann constant $\sigma = 5.6704 \times 10^{-5}$ erg cm^{-2} K^{-4} s^{-1}. We shall make extensive use of these functions when we discuss radiative transfer in the next chapter.

To remind you of what u_ν or B_ν looks like, we plot the function $\mathrm{B}(x) = x^3/[\exp(x) - 1]$ (as part of 3.19) in Fig. 3.1 where $x = h\nu/kT$ and multiplicative constants have been ignored. The function is strongly peaked with a maximum at $x = 2.821\cdots$. For the center of the sun, with $T_\mathrm{c} \approx 10^7$ K, this peak corresponds to a photon energy of 2.4 keV. (For conversions to eV units see App. B.) Photons of these energies are capable of completely ionizing most of the lighter elements.

3.3 Ideal Monatomic Gas

As we shall soon show, the Boltzmann distribution for an ideal gas is characterized by $(\mu/kT) \ll -1$. We start off by asserting that this inequality holds for a sample of gas.

To make it simple, assume that the gas particles are nonrelativistic with $\mathcal{E} = p^2/2m$, $v = p/m$, and that they have only one energy state $\mathcal{E} = \mathcal{E}_0$. These could be, as examples, elementary particles, or a collection of one species of ion in a given state. If $(\mu/kT) \ll -1$, then the term ± 1 in the denominator of (3.9) may be neglected compared to the exponential and the gas becomes purely classical in character with no reference to quantum statistics. The expression for the number density is then

$$n = \frac{4\pi}{h^3} g \int_0^\infty p^2 e^{\mu/kT} e^{-\mathcal{E}_0/kT} e^{-p^2/2mkT} \, dp \, . \tag{3.23}$$

The integral is elementary and yields μ in terms of number density:

$$e^{\mu/kT} = \frac{nh^3}{g(2\pi mkT)^{3/2}} e^{\mathcal{E}_0/kT} . \tag{3.24}$$

Because we require $\exp(\mu/kT) \ll 1$ (since $\mu/kT \ll -1$), the righthand side of (3.24) must be small. Thus, $nT^{-3/2}$ cannot be too large. If this is not true, then other measures must be taken. For example, if μ/kT is negative but not terribly less than -1, it is possible to expand the original integrand for n (with the ± 1 statistics term retained) in a power series and then integrate. The additional terms obtained, assuming convergence of the series, represent Fermi–Dirac or Bose–Einstein corrections to the ideal gas. This is done for fermions in Chiu (1968, Chap. 3), and Chandrasekhar (1939, Chapt. X), for example. In any event, μ may be computed once n and T are given. We assume here that (3.24) is by far the largest contribution to any expansion leading to an expression for μ for given n and T.

It is easy to take logarithmic differentials of n that yield the following expressions, and you may easily verify from the literature that they are the distribution functions for a Maxwell–Boltzmann ideal gas:

$$\frac{dn(p)}{n} = \frac{4\pi}{(2\pi mkT)^{3/2}} e^{-p^2/2mkT} p^2 \, dp \tag{3.25}$$

and, in energy space,

$$\frac{dn(\mathcal{E})}{n} = \frac{2}{\pi^{1/2}} \frac{1}{(kT)^{3/2}} e^{\mathcal{E}/kT} \mathcal{E}^{1/2} \, d\mathcal{E} \, . \tag{3.26}$$

The relevant part of (3.26), $\mathrm{C}(x) = x^{1/2} \exp(-x)$, is shown in Fig. 3.2, where the maximum corresponds to $\mathcal{E} = kT/2$ (i.e., $x = 1/2$).

It is easy to show that the *average* kinetic energy of a particle in this distribution is just $3kT/2$, which gives rise to (3.29) below. (To get the average, multiply 3.26 by $\mathcal{E}$ and integrate from zero energy to infinity.) Thus the "important" particles, in a manner of speaking, of a Maxwell–Boltzmann gas are those with energies near kT. A major exception to this involves those partaking of fusion reactions. For the solar center kT is around 1 keV, which is small compared to nuclear energies measured in MeVs. It will turn out (in Chap. 6) that the important fusion reactants are those in the exponential right hand tail of Fig. 3.2, even though their population is small compared to those in the peak of the distribution.

A similar procedure involving the neglect of the ± 1 statistical factor equivalent to what was done for (3.23) yields the pressure

$$P = g \frac{4\pi}{h^3} \frac{\pi^{1/2}}{8m} (2mkT)^{5/2} e^{\mu/kT} e^{-\mathcal{E}_0/kT} \tag{3.27}$$

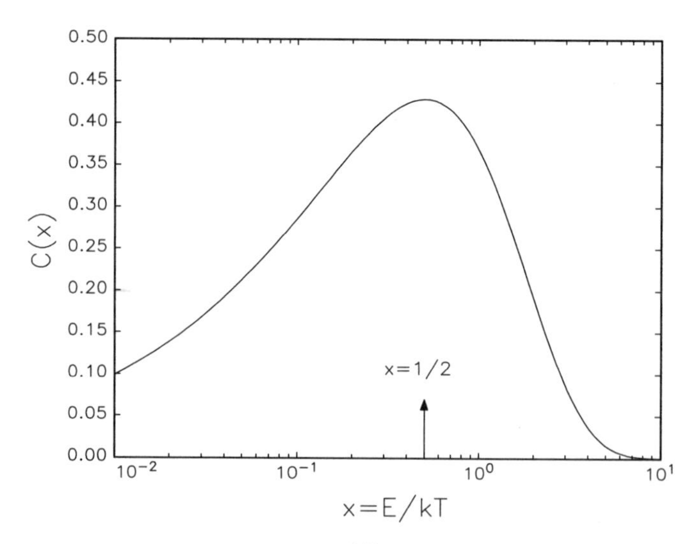

Fig. 3.2. A plot of the function $C(x) = x^{1/2} \exp(-x)$ corresponding to the exciting part of the Maxwell–Boltzmann distribution in energy space. The maximum is at $x = 1/2$ ($\mathcal{E} = kT/2$).

or, after substituting for $e^{\mu/kT}$ of (3.24),

$$P = nkT \quad \text{dyne cm}^{-2} \tag{3.28}$$

which comes as no surprise. This last result is true even if the particles are relativistic (as in Ex. 3.4). The internal energy is

$$E = \tfrac{3}{2} nkT \quad \text{erg cm}^{-3} \tag{3.29}$$

using the same procedures. (Note that if reactions are present that change the relative concentrations of particles, then E must contain information about the energetics of such reactions; see below.) These are all elementary results for the ideal gas so that, given n, T, and composition, then P, E, and μ immediately follow.

To tidy up, we return to a statement made at the beginning of this chapter; namely, that "the Boltzmann populations of ion energy levels are consistent with the local electron kinetic temperature" in LTE. We have implicitly assumed here that all species in a mixture have the same temperature, which, in some environments, is not warranted. For the stellar interior the assumption is fine. Thus consider an ion with two energy levels with $\mathcal{E}_1 > \mathcal{E}_2$. These levels are populated or depopulated by photon absorption or emission, for example. Because the photon chemical potential is zero, then $\mu_1 = \mu_2$. Dividing (3.24) for the two levels yields, after trivial algebra,

$$\frac{n_1}{n_2} = \frac{g_1}{g_2} e^{-(\mathcal{E}_1 - \mathcal{E}_2)/kT} \tag{3.30}$$

which is the Boltzmann population distribution and, if the statistical weights are not strange, means that levels become more sparsely populated as their energy increases.

3.4 The Saha Equation

In many situations the number densities of some species cannot be set a priori because "chemical" reactions are taking place. This is the problem referred to in §1.4 where mean molecular weights were computed. If the system is in thermodynamic equilibrium, however, then the chemical potentials of the reacting constituents depend on one another and this additional constraint is sufficient to determine the number densities.

As an example, consider the ionization–recombination reaction brought up earlier:[6]

$$\mathrm{H}^+ + \mathrm{e}^- \Longleftrightarrow \mathrm{H}^0 + \chi_{\mathrm{H}} \tag{3.31}$$

where $\chi_{\mathrm{H}} = 13.6$ eV is the ionization potential from the ground state of hydrogen (still assumed to have only one bound level). We assume that no other reactions are taking place that involve the above constituents and, in particular, that the gas is pure hydrogen. Reference to the photon in (3.31) has again been deleted because its chemical potential is zero and does not appear in the equilibrium condition (3.7), which will be invoked shortly.

To obtain the LTE number densities of the electrons and neutral and ionized versions of hydrogen, assume that all gases are ideal so that (3.24) applies. The reference energy levels for all species are established by taking the zero of energy as the just-ionized $\mathrm{H}^+ + \mathrm{e}^-$ state. (Other choices are possible of course.) Thus $\mathcal{E}_0$ for electrons and H^+ is zero, whereas for H^0 it is $-\chi_{\mathrm{H}} = -13.6$ eV lower on the energy scale. That is, we need 13.6 eV to convert H^0 to a free electron and a proton. The ground state of hydrogen has two near-degenerate states corresponding to spin-up or spin-down of the electron relative to the proton spin. For our purposes regard those states as having the same energy (but of course they do not, otherwise 21-cm HI radiation would not exist). Thus the degeneracy factor for H^0 is $g^0 = 2$. The situation for the free electron and H^+ is a bit more complicated because of the possible problem of double counting. If the spin axis of the proton is taken to be a fixed reference direction, then the free electron may have two spin directions relative to the free proton. Thus, $g^- = 2$ and $g^+ = 1$. The argument could be reversed without having any effect on the following results.

With μ^-, μ^+, and μ^0 denoting the chemical potentials of the components in (3.31), Equation (3.24) then yields

---

[6] See Ex. 3.1 for a more complicated problem.

$$n_e = \frac{2\left[2\pi m_e kT\right]^{3/2}}{h^3} e^{\mu^-/kT} \tag{3.32}$$

$$n^+ = \frac{\left[2\pi m_p kT\right]^{3/2}}{h^3} e^{\mu^+/kT} \tag{3.33}$$

$$n^0 = \frac{2[2\pi(m_e + m_p)kT]^{3/2}}{h^3} e^{\mu^0/kT} e^{\chi_H/kT} \tag{3.34}$$

where m_e and n_e denote, respectively, the electron mass and number density, m_p is the proton mass, and the neutral atom mass is set to $m_e + m_p$.

Now form the ratio $n^+ n_e / n^0$ and find

$$\frac{n^+ n_e}{n^0} = \frac{(2\pi kT)^{3/2}}{h^3} \left(\frac{m_e m_p}{m_e + m_p}\right)^{3/2} e^{(\mu^- + \mu^+ - \mu^0)/kT} e^{-\chi_H/kT}.$$

But $\mu^- + \mu^+ - \mu^0 = 0$ for equilibrium by application of (3.7), so that we obtain the *Saha equation* for the single-level pure hydrogen gas[7]

$$\frac{n^+ n_e}{n^0} = \left(\frac{2\pi m_e kT}{h^2}\right)^{3/2} e^{-\chi_H/kT} \tag{3.35}$$

where the *reduced mass approximation* $[m_e m/(m_e + m)] \approx m_e$ has been used. A numerical version of part of this equation is

$$\left(\frac{2\pi m_e kT}{h^2}\right)^{3/2} = 2.415 \times 10^{15}\, T^{3/2} \ \ \mathrm{cm}^{-3} \tag{3.36}$$

and note that

$$kT = 8.6173 \times 10^{-5}\, T \ \ \mathrm{eV} \tag{3.37}$$

where the eV units are handy for energies on the atomic scale.

To find the number densities, and not just ratios, further constraints must be placed on the system. A reasonable one is that of electrical neutrality, which requires that $n_e = n^+$ for a gas of pure hydrogen. Furthermore, nucleon number must be conserved so that $n^+ + n^0 = n$, where n is a constant if the density (ρ) is kept fixed.

We now define the degree of ionization (as in §1.4 and Eq. 1.47)

$$y = \frac{n^+}{n} = \frac{n_e}{n} \tag{3.38}$$

so that y is the fraction of all hydrogen that is ionized. The Saha equation (3.35) is then

[7] Clayton (1968, §1–2) extends this analysis to the case of multiple ionizations in many-electron atoms, which leads to a consideration of partition functions. We shall not need those functions but Clayton's discussion is worth looking into.

$$\frac{y^2}{1-y} = \frac{1}{n}\left(\frac{2\pi m_e kT}{h^2}\right)^{3/2} e^{-\chi_H/kT}. \tag{3.39}$$

For sufficiently high temperatures, with fixed density, we expect the radiation field or collisions effectively to ionize all the hydrogen. This is indeed the case because we see that as $T \to \infty$, then $y \to 1$. Similarly, low temperatures mean less intense radiation fields and recombination wins with $y \to 0$.

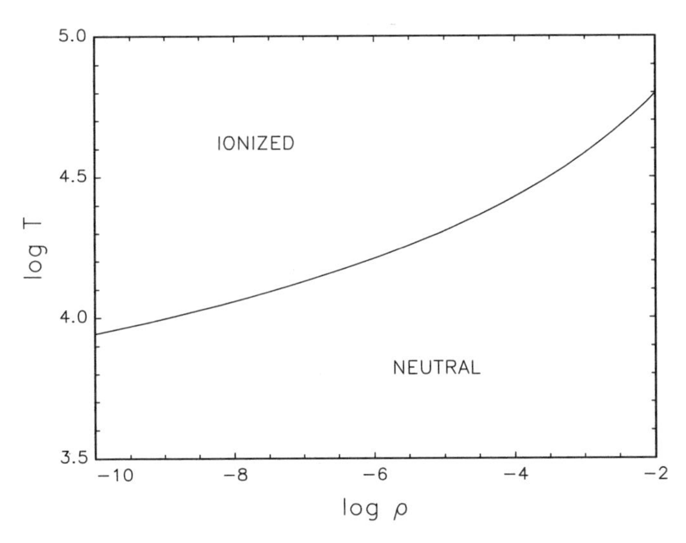

Fig. 3.3. The half-ionization curve for a mixture of pure hydrogen undergoing the recombination–ionization reaction $H^+ + e^- \Longleftrightarrow H^0 + \chi_H$ (ground state only).

For the pure hydrogen mixture $n = \rho N_A$ and (3.39) becomes

$$\frac{y^2}{1-y} = \frac{4.01 \times 10^{-9}}{\rho} T^{3/2} e^{-1.578\times10^5/T}. \tag{3.40}$$

The half-ionized ($y = 1/2$) path in the ρ-T plane for this mixture is then

$$\rho = 8.02 \times 10^{-9} T^{3/2} e^{-1.578\times10^5/T} \quad \text{g cm}^{-3} \tag{3.41}$$

and this is shown in Fig. 3.3 as a very shallow curve for a range of what are interesting densities.

The dominant factor in (3.40) and (3.41) is the exponential and this is what causes the half-ionization point to depend only weakly on density. For hydrogen ionization from the ground state, the characteristic temperature for ionization-recombination is around 10^4 K and you may readily check that the transition from $y = 0$ to $y = 1$ takes place very rapidly as the temperature scans across that value (or, more precisely, at the temperature corresponding to $y = 1/2$ at a particular density). This is shown in Fig. 3.4 for pure hydrogen at a density of 10^{-6} g cm^{-3}. A *rough* rule of thumb is that the transition temperature (where $y \approx 1/2$) is such that $\chi/kT \sim 10$ to within a factor of three or so depending on density. Thus, for example, the ionization potentials for removing the first and second electrons of helium are 24.6 eV and 54.4 eV, which correspond to transition temperatures of about 3×10^4 K and 6×10^4 K. (See Ex. 3.1.)

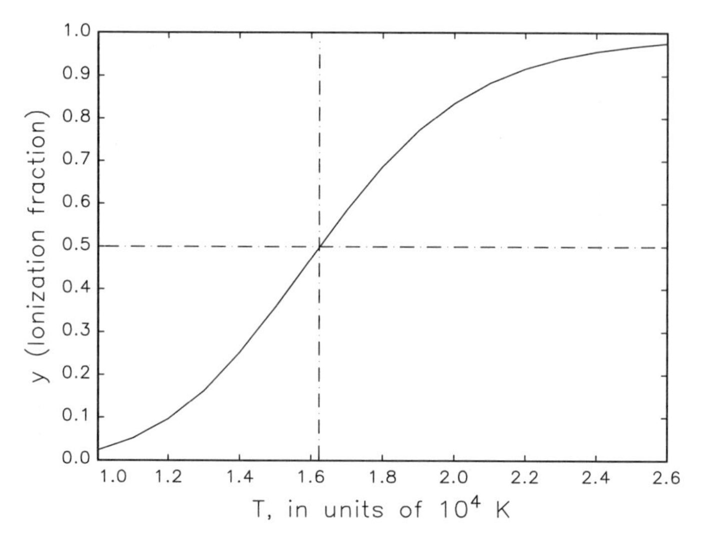

Fig. 3.4. Note how the ionization fraction y changes rapidly for pure hydrogen as temperature is varied through 1.62×10^4 K at which $y = 1/2$—as indicated by the dashed lines. The density is fixed at $\rho = 10^{-6}$ g cm^{-3}.

As we shall see, the presence of these zones of ionization have profound consequences for the structure of a star. You may wish to consider at this point a mixture of single-level hydrogen and helium (with two stages of ionization) and go through an analysis corresponding to the above to see how the various ions compete for electrons and to find out what the transition temperatures are for the three ionization stages involved. Even for this very practical, but simple, problem, you will find that a computer is essential for your sanity.

If the temperature and density of the hydrogen mixture are fixed, then (3.40) yields the ionization fraction y. The total hydrogen number density is clearly $n = \rho N_A$ and thus $n^+ = n_e = yn$ from (3.38). Chemical potentials, if required, follow from (3.32–3.34). The partial pressures and internal energies, which are additive, yield the total pressure

$$P = n(1 + y)kT \tag{3.42}$$

and total internal energy

$$E = \frac{3}{2}n(1+y)kT + y\,n\,\chi_{\rm H} \quad \text{erg cm}^{-3}. \tag{3.43}$$

The last term in E appears because we have to take account of the ionization energy. If we wish to ionize the gas ($y \to 1$) completely, then $(3nkT/2 + n\,\chi_{\rm H})$ erg cm^{-3} must be added to the system. Of this amount, $n\,\chi_{\rm H}$ strips off the electrons, and the remainder brings the system up to the common temperature T.

The real calculation of ionization equilibria is as difficult as that for real equations of state (and the two are intimately connected). In principle, all species, energy levels, and reactions must be considered. In addition, the effects of real interactions must be included (and these depend on composition, temperature, and density), which change the relations between concentration and chemical potential. For textbook examples see Cox (1968, §15.3), and Kippenhahn and Weigert (1990, Chap. 14), with the warning that, in practice, accurate analytic or semianalytic solutions are seldom possible: you are usually faced with computer-generated tables of pressure and the like and the task is to use them intelligently.

3.5 Fermi–Dirac Equations of State

The most commonly encountered Fermi–Dirac elementary particles of stellar astrophysics are electrons, protons, and neutrons; all have spin one-half. (Neutrinos also appear but in contexts not usually connected with equations of state.) The emphasis here will be on electrons, but (almost) all that follows may apply to the other fermions as well. The prime motivation for this discussion is that the equation of state in the inner regions of many highly evolved stars, including white dwarfs, is dominated by degenerate electrons and, to a great extent, this determines the structure of such stars.

The number density of Fermi–Dirac particles is given by (3.9) and (3.10) with the choice of $+1$ in (3.9) and an energy reference level of $\mathcal{E}_0 = mc^2$, where m is the mass of the fermion. (Other choices are indeed possible for $\mathcal{E}_0$. They lead to an additive constant in the definition of the chemical potential and you have to watch out for this in the literature.) For these spin $1/2$ particles, the statistical weight $g = 2$. Transcribing these statements then means that the number density is

$$n = \frac{8\pi}{h^3}\int_0^\infty \frac{p^2\,dp}{\exp\left\{\left[-\mu + mc^2 + \mathcal{E}(p)\right]/kT\right\} + 1} \tag{3.44}$$

where, in general, from (3.11) and (3.12),

$$\mathcal{E}(p) = mc^2\left[\sqrt{1 + \left(\frac{p}{mc}\right)^2} - 1\right] \tag{3.45}$$

and

$$v(p) = \frac{\partial \mathcal{E}}{\partial p} = \frac{p}{m}\left[1 + \left(\frac{p}{mc}\right)^2\right]^{-1/2} . \tag{3.46}$$

We now explore some consequences of the above.

3.5.1 The Completely Degenerate Gas

The "completely degenerate" part of the title of this subsection refers to the unrealistic assumption that the temperature of the gas is absolute zero.[8] In practice this does not happen but, under some circumstances, the gas effectively behaves as if it were at zero temperature and, for fermions in stars, these unusual circumstances are very important. So, in (3.44), note the peculiar behavior of the integrand as $T \to 0$. The exponential tends either to zero or infinity depending on, respectively, whether $-\mu + mc^2 + \mathcal{E}$ is <0 or >0. Therefore consider the interesting part of (3.9),

$$F(\mathcal{E}) = \frac{1}{\exp\left\{\left[\mathcal{E} - (\mu - mc^2)\right]/kT\right\} + 1} \tag{3.47}$$

where, as $T \to 0$, $F(\mathcal{E})$ approaches either zero or unity depending on whether $\mathcal{E}$ is greater or less than $\mu - mc^2$.

The critical kinetic energy at which $F(\mathcal{E})$ is discontinuous (for $T \to 0$) is called the "Fermi energy" and we denote it by $\mathcal{E}_F$; that is, where $\mathcal{E}_F = \mu - mc^2$. (But note that we have not yet described how μ is found.) The situation is depicted in Fig. 3.5 where, in the unit square corresponding to particle energies $0 \le \mathcal{E} \le \mathcal{E}_F$, $F(\mathcal{E})$ is unity. Fermions are contained only in that energy range and not at energies greater than $\mathcal{E}_F$ where the distribution function is zero. In this situation we refer to a "filled Fermi sea" of fermions because all the fermions present are swimming in that sea and nowhere else. (Ignore the dashed line for the moment. It shows what happens if the temperature is raised slightly above zero. See §3.5.3.)

The momentum corresponding to the Fermi energy is the Fermi momentum p_F. It is usually reduced to dimensionless form by setting $x = p/mc$ and defining $x_F = p_F/mc$. Then, from (3.45), we have

$$\mathcal{E}_F = mc^2\left[\left(1 + x_F^2\right)^{1/2} - 1\right] . \tag{3.48}$$

In this language, the chemical potential of the system is $\mu_F = \mathcal{E}_F + mc^2$ and it is the total energy, including rest mass energy, of the most energetic particle

[8] This has almost been achieved in the laboratory by the elegant experiments of DeMarco and Jin (1999), who, using atoms of ^{40}K at temperatures less than 300 nanoKelvin (!), have made a soup of fermions in their lowest energy states. Similar experiments by Anderson et al. (1995) have done the same for bosons by making a "Bose–Einstein Condensate" (BEC), a form of matter long thought possible but only now demonstrated actually to exist (and the work gained the two senior investigators a Nobel Prize). The two groups, not so incidentally, are in the same institute (JILA) at the University of Colorado.

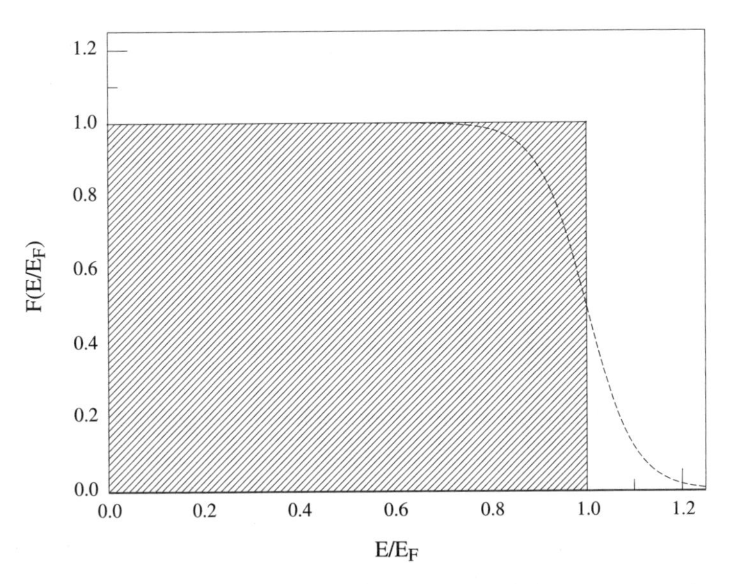

Fig. 3.5. The function $F(\mathcal{E}/\mathcal{E}_F)$ of (3.47) versus particle kinetic energy in units of $\mathcal{E}_F$ for zero temperature. Fermions are restricted to the shaded area of unit height and width and do not have energies greater than the Fermi energy $\mathcal{E}_F$. The dashed line shows how $F(\mathcal{E})$ is changed by raising the temperature slightly. (In this case $\mathcal{E}_F/kT = 20$.)

(or particles) in the system. If the spin is $1/2$ ($g = 2$), then all the rest of the particles are locked in pair-wise with spin-up and spin-down paired at each lower energy level by the Pauli exclusion principle.[9] The Fermi sea is then capped by the "Fermi surface" at $\mathcal{E}_F$.

The relation between particle number density and the Fermi energy, and thus μ_F, is found as follows. Because $F(\mathcal{E})$ is in the form of a unit step, (3.44) need only be integrated up to p_F. Hence

$$n = \frac{8\pi}{h^3}\int_0^{p_F} p^2\,dp = 8\pi\left(\frac{h}{mc}\right)^{-3}\int_0^{x_F} x^2\,dx = \frac{8\pi}{3}\left(\frac{h}{mc}\right)^{-3} x_F^3. \tag{3.49}$$

To deal with astrophysically interesting numbers we shall, from this point on, deal exclusively with electrons unless otherwise noted.

It is traditional, but admittedly confusing, to delete the F subscript on x_F so that (3.49) is written

$$n_\mathrm{e} = \frac{8\pi}{3}\left(\frac{h}{m_\mathrm{e}c}\right)^{-3} x^3 = 5.865\times 10^{29}\,x^3 \quad \mathrm{cm}^{-3} \tag{3.50}$$

[9] The most obvious application of the Pauli exclusion principle is for atoms. Were it not for this curious way nature works, electrons would all cascade down to the lowest energy level of atoms and we all would become *very* small entities indeed.

for electrons where $(h/m_e c)$ is the electron Compton wavelength equal to 2.426×10^{-10} cm. The transcription to other spin 1/2 fermions is accomplished merely by changing the mass in (3.50).

To convert this to density units we reintroduce the electron mean molecular weight, μ_e, of (1.48–1.49) with $n_e = \rho N_A / \mu_e$. Thus

$$\frac{\rho}{\mu_e} = B x^3 \tag{3.51}$$

with

$$B = \frac{8\pi}{3N_A} \left(\frac{h}{m_e c} \right)^{-3} = 9.739 \times 10^5 \quad \text{g cm}^{-3} \tag{3.52}$$

for electrons. This may be looked upon as a relation that yields x (i.e., x_F), and, hence, $\mathcal{E}_F$ and p_F, once ρ/μ_e is given.

Note that the demarcation between nonrelativistic and relativistic mechanics occurs when $p_F \approx m_e c$ or $x = x_F \approx 1$. The corresponding density is $\rho/\mu_e \approx 10^6$ g cm^{-3}, which, incidentally, is a typical central density for white dwarfs and is near the density at which the "helium flash" takes place (see §2.5). It remains to be shown, however, that temperatures in these contexts are sufficiently low to be effectively zero as far as electrons are concerned.

Looking ahead to neutron star matter, the numerical constant B in (3.51–3.52) is $B(\text{neutrons}) = 6.05 \times 10^{15}$ g cm^{-3} and μ_e in that expression is set to unity; that is, we must replace μ_e by the amu weight of the neutron (essentially unity). For typical densities in a neutron star (comparable to nuclear densities of $\rho \approx 2.7 \times 10^{14}$ g cm^{-3}), $x \approx 0.35$ and $\mathcal{E}_F \approx 57$ MeV. This implies that the neutrons are nonrelativistic because the neutron rest mass energy is 939.57 MeV.

The pressure of a completely degenerate electron gas is treated in the same way as that for the number density. It is the integral in (3.13) truncated at the Fermi momentum with $F(\mathcal{E})$ of (3.47) set to unity. A little work on (3.13) yields

$$P_e = \frac{8\pi}{3} \frac{m_e^4 c^5}{h^3} \int_0^{x_F} \frac{x^4 \, dx}{(1+x^2)^{1/2}} = A \, f(x) \tag{3.53}$$

where

$$A = \frac{\pi}{3} \left(\frac{h}{m_e c} \right)^{-3} m_e c^2 = 6.002 \times 10^{22} \text{ dyne cm}^{-2} \tag{3.54}$$

for electrons and

$$f(x) = x(2x^2 - 3)(1 + x^2)^{1/2} + 3 \sinh^{-1} x \, . \tag{3.55}$$

Similarly, the internal energy, from (3.14), is given by the integral

$$E_e = 8\pi \left(\frac{h}{m_e c} \right)^{-3} m_e c^2 \int_0^{x_F} x^2 \left[(1 + x^2)^{1/2} - 1 \right] dx = A g(x) \tag{3.56}$$

with

$$g(x) = 8x^3 \left[(1+x^2)^{1/2} - 1 \right] - f(x) \, . \tag{3.57}$$

The units for $E_{\rm e}$ are erg cm^{-3} as is that for A when dyne cm^{-2} is expressed in those units in (3.54).

It will often prove useful to have limiting forms for $f(x)$ and $g(x)$ that correspond to the limits of relativistic or nonrelativistic electrons. These are

$$f(x) \rightarrow \begin{cases} \frac{8}{5}x^5 - \frac{4}{7}x^7 + \cdots, & x \ll 1 \\ 2x^4 - 2x^2 + \cdots, & x \gg 1 \end{cases} \tag{3.58}$$

and

$$g(x) \rightarrow \begin{cases} \frac{12}{5}x^5 - \frac{3}{7}x^7 + \cdots, & x \ll 1 \\ 6x^4 - 8x^3 + \cdots, & x \gg 1 \, . \end{cases} \tag{3.59}$$

Note that $x \ll 1$ implies nonrelativistic particles, and $x \gg 1$ is the extreme relativistic limit. Also observe that

$$P_{\rm e} \propto E_{\rm e} \propto \begin{cases} (\rho/\mu_{\rm e})^{5/3}, & x \ll 1 \\ (\rho/\mu_{\rm e})^{4/3}, & x \gg 1 \end{cases} \tag{3.60}$$

and the limiting ratios of $E_{\rm e}$ to $P_{\rm e}$ are

$$\frac{E_{\rm e}}{P_{\rm e}} = \frac{g(x)}{f(x)} = \begin{cases} 3/2 \;\; (\gamma = 5/3), & x \ll 1 \\ 3 \;\;\;\; (\gamma = 4/3), & x \gg 1 \, . \end{cases} \tag{3.61}$$

The values for γ are included as a reminder that for a γ–law equation of state the completely degenerate nonrelativistic electron gas acts like a monatomic ideal gas whereas, in the extreme relativistic limit, it behaves like a photon gas.

3.5.2 Application to White Dwarfs

As a simple, but important, application of completely degenerate fermion statistics, consider zero temperature stars in hydrostatic equilibrium whose internal pressures are due solely to electron degenerate material and whose densities and composition are constant throughout.

The easiest way to look at this is to apply the virial theorem in the hydrostatic form $3(\gamma - 1)U = -\Omega$ from (1.25). Because the star is assumed to have constant density, $\Omega = -(3/5)(G\mathcal{M}^2/\mathcal{R})$. If $E_{\rm e}$ is the volumetric energy density (with no contribution from the zero temperature ions), then $U = VE_{\rm e}$ where V is the total stellar volume $V = (4\pi/3)\mathcal{R}^3$. In the nonrelativistic limit $E_{\rm e} = 12Ax^5/5$ from (3.56) and (3.59), x may be expressed in terms of $\rho/\mu_{\rm e}$ via (3.51) and ρ, in turn, may be eliminated in favor of $\mathcal{M}$ and $\mathcal{R}$ by $\rho = \mathcal{M}/(4\pi\mathcal{R}^3/3)$. If the entire virial theorem is also cast in a form containing only $\mathcal{M}$ and $\mathcal{R}$, and if the constants B and A of (3.52) and (3.54) are given in terms of fundamental constants, then a little algebra yields the nonrelativistic mass–radius relation

$$\mathcal{M} = \frac{1}{4}\left(\frac{3}{4\pi}\right)^4 \left(\frac{h^2 N_{\mathrm{A}}}{m_{\mathrm{e}} G}\right)^3 \frac{N_{\mathrm{A}}^2}{\mu_{\mathrm{e}}^5} \frac{1}{\mathcal{R}^3} \quad \text{for constant density.} \tag{3.62}$$

This relation has the remarkable property that *as mass increases, radius decreases* and is quite unlike the homology result for main sequence stars discussed in the first chapter. And this result is what we promised you several times in Chapter 2.

For electrons, this yields the numeric expression

$$\frac{\mathcal{M}}{\mathcal{M}_\odot} \approx 10^{-6} \left(\frac{\mathcal{R}}{\mathcal{R}_\odot}\right)^{-3} \left(\frac{2}{\mu_{\mathrm{e}}}\right)^5 . \tag{3.63}$$

We state, without proof for now, that the interiors of white dwarf stars are almost entirely supported by electron degeneracy pressure, and that they typically have masses around $0.6\,\mathcal{M}_\odot$. If the electrons are nonrelativistic, then (3.63) yields a typical radius of $\mathcal{R} \approx 0.01\,\mathcal{R}_\odot$ for $\mu_{\mathrm{e}} = 2$ (completely ionized ^{4}He, ^{12}C, ^{16}O, etc.). This radius is very close to that of the earth's with $\mathcal{R}_\oplus = 6.38\times10^8$ cm. An exact analysis involving integration of the hydrostatic equation using the nonrelativistic equation of state shows that (3.63) gives the correct result provided that the numerical coefficient is increased by (only!) a factor of two.

If μ_{e} in (3.62) is replaced by unity and the particle mass is taken to be that of the neutron, then the neutron star equivalent of (3.63) becomes

$$\frac{\mathcal{M}}{\mathcal{M}_\odot} \approx 5 \times 10^{-15} \left(\frac{\mathcal{R}}{\mathcal{R}_\odot}\right)^{-3} \quad \text{(neutron stars)} \tag{3.64}$$

in the nonrelativistic limit. For $\mathcal{M} = \mathcal{M}_\odot$, $\mathcal{R} \approx 11$ km, which is in the right ballpark. Note that general relativistic effects have been completely ignored, but this is the least of our sins because the nuclear force makes our noninteracting equation of state inaccurate.

You will have realized by now that the simple arguments outlined above for mass–radius relations contain a serious flaw. The nonrelativistic degenerate electron pressure depends solely on density and composition (through μ_{e}); that is, in numeric form and using (3.51), (3.53), and (3.58)

$$P_{\mathrm{e}} = 1.004 \times 10^{13} \left(\frac{\rho}{\mu_{\mathrm{e}}}\right)^{5/3} \quad \text{dyne cm}^{-2} \tag{3.65}$$

and, as may easily be verified, the corresponding extreme relativistic expression is

$$P_{\mathrm{e}} = 1.243 \times 10^{15} \left(\frac{\rho}{\mu_{\mathrm{e}}}\right)^{4/3} \quad \text{dyne cm}^{-2}. \tag{3.66}$$

Thus if ρ and μ_{e} are constant, then so is P_{e} by virtue of the equation of state. But a constant pressure is inconsistent with hydrostatic equilibrium and, in fact, (1.41) is the correct solution for the pressure through a constant–density star. Thus P_e is not a constant and neither is E_e as assumed above. The trouble is that we have overconstrained the problem by insisting on the constancy of ρ combined with the degenerate equation of state.

The correct way to construct equilibrium degenerate models is to use the general expression for the pressure given by (3.53) along with the relation between ρ/μ_e and dimensionless Fermi momentum of (3.51). This yields a pressure–density relation, which is then put into the equation of hydrostatic equilibrium. The resulting equation is then combined with the equation of mass conservation yielding a second-order differential equation that must be integrated numerically. We shall not go into the tedious details here because more than adequate discussions are given in Chandrasekhar (1939, Chap. 11) and Cox (1968, §25.1), and, in any case, such solutions are easy to come by using modern numerical techniques. (See, for example, Chap. 7.) Important results are summarized below.

In the limit of extreme relativistic degeneracy, where (3.66) is appropriate, you may easily convince yourself by using dimensional analysis that the total stellar mass depends only on μ_e and not on radius. An exact analysis yields

$$\frac{\mathcal{M}}{\mathcal{M}_\odot} = \frac{\mathcal{M}_\infty}{\mathcal{M}_\odot} = 1.456\left(\frac{2}{\mu_e}\right)^2 \tag{3.67}$$

where $\mathcal{M}_\infty$ is the *Chandrasekhar limiting mass.*[10] A virial analysis similar to that used to find (3.62), but done in the relativistic limit, yields a result differing from the above by only a change in the constant (a 1.75 instead of 1.456). We assume you will try to verify this and, if you do, you should also find that the full virial expression (1.25) implies d^2I/dt^2 becomes negative if the total mass exceeds $\mathcal{M}_\infty$. The interpretation is that electron degenerate objects (of fixed μ_e) cannot have masses exceeding the Chandrasekhar limit without collapsing the object. Increased densities and pressure cannot halt the collapse because the relativistic limit has already been reached. In the nonrelativistic limit, on the other hand, a new configuration may be reached by decreasing the radius as indicated by (3.63). Extreme relativistic equations of state, including that for photons, are too "soft" compared to the effects of self–gravity. (You can't make the particles exceed the speed of light to try to increase pressures!) This conclusion might have been anticipated because extreme relativistic effects imply $\gamma \to 4/3$.

[10] The exact value of this limiting mass depends on physics we have not included in our analysis. Hamada and Salpeter (1961), for example, consider the effects of electrostatic interactions and electron captures on various nuclei. For single white dwarfs with normal masses and compositions, these effects are not that significant. However, we can imagine massive objects formed by various means in binary systems where such effects could well give a stable maximum mass less than the Chandrasekhar limiting mass, as discussed earlier in §2.13.

The astrophysical significance of the Chandrasekhar limiting mass is just as we discussed in Chapter 2. If electron degenerate configurations are good representations of white dwarfs, and if those objects are the final end product of evolution for most stars, then the late stages of evolution are severely constrained. That is, if a star does not finally rid itself of enough mass to eventually leave a white dwarf with $\mathcal{M} \lesssim 1.46\mathcal{M}_\odot$ (assuming a reasonable value of μ_e near 2), then something catastrophic will happen at some time in its life. Since there are so many white dwarfs in the sky, a large fraction of stars either start off with sufficiently low masses, or they manage to rid themselves of the excess mass.

The regime intermediate between nonrelativistic and full relativistic degeneracy is intractable using simple means, and full-scale models must be calculated (and you may try this by using the code `WD.FOR` on the CD-ROM). The following useful and quite accurate mass–radius relation bridging the two regimes (fit to actual calculations) is based on one given by Eggleton (1982) for electrons:

$$\frac{\mathcal{R}}{\mathcal{R}_1} = 2.02 \left[1 - \left(\frac{\mathcal{M}}{\mathcal{M}_\infty}\right)^{4/3}\right]^{1/2} \left(\frac{\mathcal{M}}{\mathcal{M}_\infty}\right)^{-1/3} . \tag{3.68}$$

Here, $\mathcal{M}/\mathcal{M}_\infty$ is given by (3.67), and $\mathcal{R}_1$ is defined by

$$\frac{\mathcal{R}_1}{\mathcal{R}_\odot} = 5.585 \times 10^{-3} \left(\frac{2}{\mu_e}\right) . \tag{3.69}$$

This radius is a typical scale length for electron degenerate objects. The relativistic and nonrelativistic limits of (3.68) go to the correct values as $\mathcal{R} \to 0$ (relativistic) or $\mathcal{M}$ becomes small (nonrelativistic). It is shown plotted in Fig. 3.6.

We shall have more to say about white dwarfs in Chapter 10. One crucial item that has not been addressed here, and that pertains to these objects, is the effect of temperature on degeneracy. After all, if white dwarfs were really at zero temperature we wouldn't see them.

3.5.3 Effects of Temperature on Degeneracy

The crucial step in deriving some of the thermodynamics of the completely degenerate zero temperature fermion gas was the realization that the distribution function becomes a unit step function at a kinetic energy equal to $\mu - mc^2$. If the zero temperature condition is relaxed, the distribution function follows suit. Suppose the temperature is low—on some scale yet to determined—but not zero. Fermions deep in the Fermi sea, at energies much less than $\mathcal{E}_F$, need roughly an additional $\mathcal{E}_F$ energy units to move around in energy. That is, if the energy input to the system, as measured by kT, is much smaller than $\mathcal{E}_F$, then low–energy particles are excluded from promotion to already occupied upper energy levels by the Pauli exclusion principle.

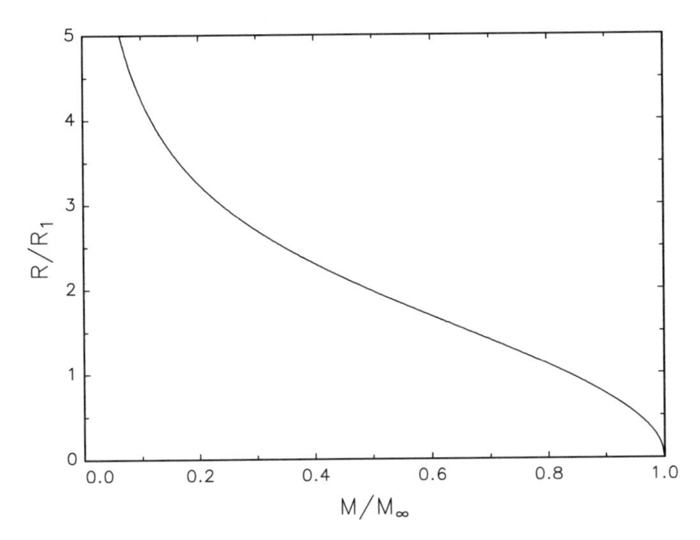

Fig. 3.6. The mass–radius relation for zero temperature white dwarfs with constant μ_e. (See Eqs. 3.68–3.69.)

Fermions near the top of the Fermi sea don't have that difficulty and they may find themselves elevated into states with energies greater than $\mathcal{E}_F$. Thus as temperature is raised from zero, the stepped end of the distribution function smooths out to higher energies. This is the effect shown in Fig. 3.5 by the dashed line. If temperatures rise high enough, we expect the effects of Fermi–Dirac statistics to be washed out completely and the gas should merge into a Maxwell–Boltzmann distribution. With this discussion as a guide, it should be apparent that a rough criterion for the transition from degeneracy to near- or nondegeneracy is $\mathcal{E}_F \approx kT$. The dashed line in Fig. 3.5 shows the effect of a rise in temperature corresponding to $\mathcal{E} = 20kT$. The effect on the distribution function is rather small, as would be expected, but the gas is no longer completely degenerate. A better description is that the gas is *partially degenerate.* As an example of the transition to nondegeneracy we apply the criterion $\mathcal{E} \approx kT$ to nonrelativistic electrons.

The Fermi energy of a nonrelativistic electron gas is $\mathcal{E}_F = mc^2 x_F^2/2$, which is easily obtained by expanding the radical in (3.48) for small x_F. The dimensionless Fermi momentum x_F is then converted to ρ/μ_e using (3.51). After this is applied to $\mathcal{E}_F \approx kT$, and numbers put in, the criterion becomes

$$\frac{\rho}{\mu_e} \approx 6.0 \times 10^{-9}\, T^{3/2} \quad \text{g cm}^{-3}. \tag{3.70}$$

If ρ/μ_e exceeds the value implied by the righthand side of (3.70) for a given temperature, then the gas is considered degenerate. Realize though that this is a rough statement: there is no clean demarcation line on the T–ρ/μ_e plane that distinguishes degenerate from nondegenerate electrons.

The extreme relativistic equivalent to (3.70) is

$$\frac{\rho}{\mu_e} \approx 4.6 \times 10^{-24} \, T^3 \quad \text{g cm}^{-3}. \tag{3.71}$$

The density near which special relativistic effects become important was estimated earlier as $\rho/\mu_e \approx 10^6$ g cm^{-3}. Equations (3.70) and (3.71) are illustrated in Fig. 3.7 where the transition near 10^6 g cm^{-3} has been smoothed. Note that the center of the present–day sun, as indicated in the figure, is nondegenerate but close enough to the transition line that good solar models include the effects of Fermi–Dirac statistics.

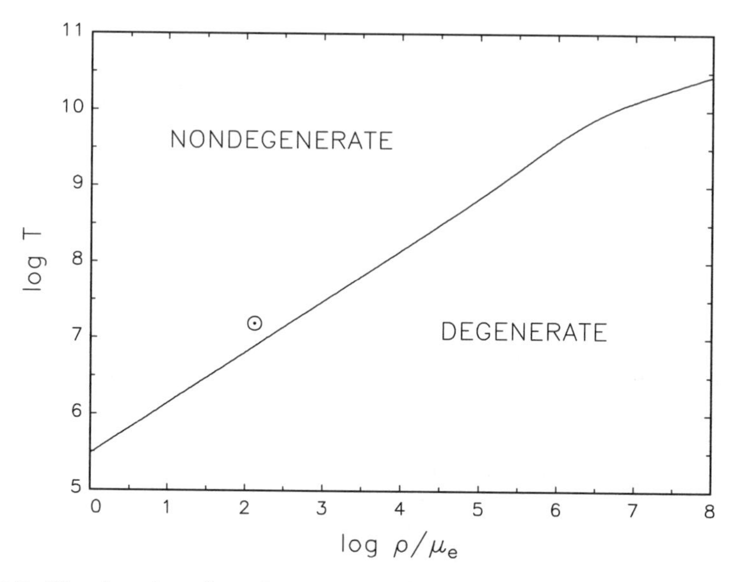

Fig. 3.7. The domains of nondegenerate and degenerate electrons in the T–ρ/μ_e plane. The location of the center of the present-day sun in these coordinates is indicated by the $\odot$ sign.

A better idea of how the transition from degeneracy to nondegeneracy takes place with respect to temperature and ρ/μ_e requires explicit evaluation of the Fermi–Dirac integrals. In general, this involves numeric integration, although there are some useful series expansions and we shall discuss one of these in a bit. The reader is referred to Cox (1968) and other references at the end of this chapter for a full discussion but the results are summarized in Fig. 3.8, which is derived from the numeric tabulations in App. A2 of Cox and his §24.4. Cloutman (1989) discusses some techniques for computing the Fermi–Dirac integrals and includes a **FORTRAN** program listing (see also Eggleton et al., 1973, and Antia, 1993).

Plotted versus ρ/μ_e in Fig. 3.8 is the ratio of electron pressure at nonzero temperature, $P_e(T, \rho/\mu_e)$, to the electron pressure for complete degeneracy

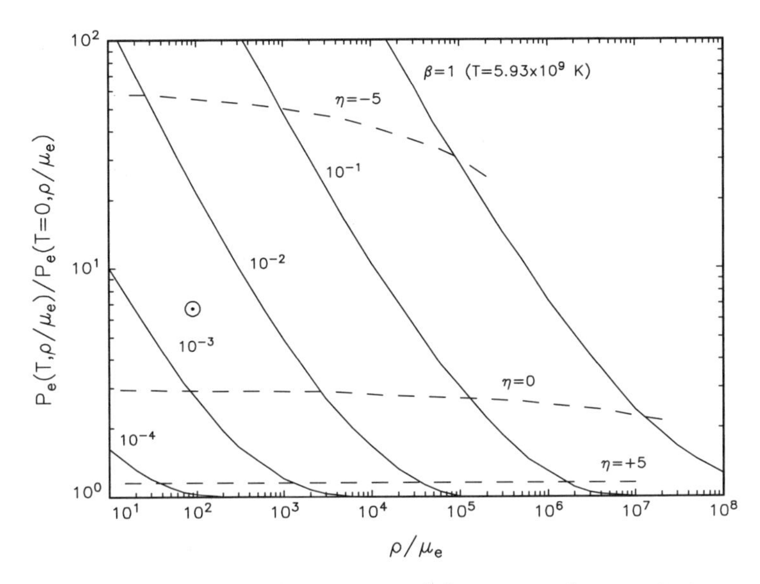

Fig. 3.8. The domains of nondegenerate and degenerate electrons in temperature and density as expressed by the ratio $P_e(T, \rho/\mu_e)/P_e(T=0, \rho/\mu_e)$. Temperatures are given in units of $\beta = kT/m_e c^2$, where $\beta = 1$ corresponds to 5.93×10^9 K. The dashed lines are lines of constant η, which is sometimes called the "degeneracy parameter" and is related to the chemical potential (see text). The position of the solar center is indicated by $\odot$.

at zero temperature, $P_e(T=0, \rho/\mu_e)$. Values near unity for this ratio imply strong degeneracy for $P_e(T, \rho/\mu_e)$, whereas large values mean that the gas is nondegenerate and, if large enough, the Maxwell–Boltzmann expression may be used. The solar center is indicated in the figure, and its position implies that degeneracy accounts for some 15% of the total pressure at that location.

Note that the effects of electron–positron pairs created by the radiation field are not included here. These become important if temperatures approach or exceed $kT \approx m_e c^2$ (i.e., $T \gtrsim 6 \times 10^9$ K). We shall discuss pair-created electrons briefly in Chapter 6, where they play a role in creating neutrinos.

A parameter called η is plotted as dashed lines on the figure and an η of five, for example, corresponds to the situation where the true pressure is only about 15% greater than if the gas were completely degenerate. Along the dashed line labeled "$\eta = 0$," a degenerate estimate for the pressure would be too low by about a factor of three. Transferring this line to the temperature versus density plane results in a plot that is very similar to that of Fig. 3.7. Finally, the parameter η, which is commonly used in the literature (but not by everyone), is related to the electron chemical potential defined here by $\eta = (\mu - m_e c^2)/kT$.

For strongly, but not completely, degenerate gases, there are useful expansions for number density, pressure, and internal energy that are often quoted in the literature. We shall not derive complete versions of those expansions here (see the references) but they all depend on the mild relaxation of the shape of the distribution function near $\mathcal{E}_F$. One of them is the following.

Following Landau and Lifshitz (1958, §57) we write any of the Fermi–Dirac integrals (for number density, etc.) in the kinetic energy-dependent form

$$I(\mu, T) = \int_0^\infty \frac{G(\mathcal{E})\, d\mathcal{E}}{\exp\left[\left(-\mu + mc^2 + \mathcal{E}\right)/kT\right] + 1} . \tag{3.72}$$

The integral I may be expressed as an asymptotic (but not necessarily convergent) series whose leading terms are

$$I(\mu, T) = \int_0^{\mu'} G(\mathcal{E})\, d\mathcal{E} + \frac{\pi^2}{6} \frac{\partial G}{\partial \mathcal{E}} (kT)^2 + \frac{7\pi^4}{360} \frac{\partial^3 G}{\partial \mathcal{E}^3} (kT)^4 + \cdots \tag{3.73}$$

where $\mu' = \mu - mc^2$ and all the partials are evaluated at μ'. It is assumed that μ'/kT is much larger than unity.

It is a simple, but tedious, exercise to transform the integrals for n, P, and E of, respectively, (3.10), (3.13), and (3.14), into their energy space counterparts and then to find $G(\mathcal{E})$. Another way, however, is to transform all of the elements in the expansion (3.73) into $x = p/mc$–space using (3.45); that is, $\mathcal{E} = mc^2\left[(1+x^2)^{1/2} - 1\right)]$. A big part of this was done when the expressions for the completely degenerate electron gas were written down in the equations for $n_{\rm e}$ (3.49), $P_{\rm e}$ (3.53), and $E_{\rm e}$ (3.56). Thus, for example, the leading term in the expansion of (3.73) for $n_{\rm e}$ is simply (neglecting constants)

$$n_{\rm e}\ (\text{first term}) \propto \int_0^{x_f} x^2\, dx .$$

Here x_f takes the place of $\mu' = \mu - mc^2$ and, since we have converted from energy to x-space, it should be obvious that the relation between x_f and μ' is

$$\mu' = \mu - m_{\rm e}c^2 = m_{\rm e}c^2 \left[\left(1 + x_f^2\right)^{1/2} - 1\right] . \tag{3.74}$$

This relation is given in the same spirit as was done for the completely degenerate case where the Fermi energy was related to the chemical potential by $\mathcal{E}_F = \mu - m_{\rm e}c^2$ and $\mathcal{E}_F$ was given in terms of x_F through (3.48). In that instance, x_F and, hence, μ were found by fixing the number density $n_{\rm e}$ and using (3.49). The same sort of thing can be done here except there is an additional complication because temperature also appears in the thermodynamics; that is, $n_{\rm e}$ must be a function of both x_f (or μ) and T. This all can be accomplished by performing the indicated operations in the expansion (3.73). Carrying out this enterprise is left to you as an exercise in elementary calculus, but the result, to second-order in temperature, is

$$n_{\rm e} = \frac{8\pi}{3} \left(\frac{h}{m_{\rm e}c}\right)^{-3} x_f^3 \left[1 + \pi^2 \frac{1 + 2x_f^2}{2x_f^4} \left(\frac{kT}{m_{\rm e}c^2}\right)^2 + \cdots\right] \text{cm}^{-3} . \tag{3.75}$$

This expansion is useful only if the second term in the brackets is small compared to unity. A useful rule of thumb is to be wary if it exceeds 0.1 to 0.2. In any case, given any two of n_e (or ρ/μ_e), T, or x_f (or μ), the third follows. Looked at another way (and we shall use this shortly), (3.75) may be used to find out how the chemical potential changes with respect to temperature for fixed n_e or ρ/μ_e. Note that as $T \to 0$, the number density approaches the completely degenerate expression (3.49) with $x_f \to x_F$, and $\mu' \to \mathcal{E}_F$.

The corresponding expansions truncated to second order in kT for pressure and internal energy are

$$P_e = Af(x_f)\left[1 + 4\pi^2 \frac{x_f(1+x_f^2)^{1/2}}{f(x_f)}\left(\frac{kT}{m_e c^2}\right)^2\right] \tag{3.76}$$

$$E_e = Ag(x_f)\left[1+4\pi^2 \frac{(1+3x_f^2)(1+x_f^2)^{1/2} - (1+2x_f^2)}{x_f\, g(x_f)}\left(\frac{kT}{m_e c^2}\right)^2\right] \tag{3.77}$$

where $f(x_f)$ and $g(x_f)$ are given, respectively, by (3.55) and (3.57). Note that P_e is in dyne cm^{-2} and E_e is the volumetric energy density in erg cm^{-3} (and *not* specific energy density in erg g^{-1}).

These equations will be used to find such things as specific heats and temperature exponents for the almost completely degenerate electron gas.

Note: As a matter of practicality, x_f is often computed as if the gas were completely degenerate. Thus if the correction term for temperature is very small, then x (or x_F) of (3.50) is used instead of x_f as a good approximation for direct calculation of n_e, P_e, and E_e in (3.75–3.77). This is what we shall usually do here.

3.6 "Almost Perfect" Equations of State

In real gases, interactions have to be taken into account that modify the "perfect" results given above. In addition, a stellar equation of state might consist of many components with radiation, Maxwell–Boltzmann, and degenerate gases competing in importance. This short section will not attempt to show how imperfections are treated in detail but will indicate where some are important in practical situations. The results of this discussion are summarized in Fig. 3.9 for a hypothetical gas composed of pure hydrogen.

In an almost-ideal gas, a measure of the interaction energy between ions is the Coulomb potential between two ions. If the ionic charge is Z, then the potential is Z^2e^2/a, where a is some typical separation between the ions. Coulomb effects are expected to become important when this energy is comparable to kT. Thus form the ratio

$$\Gamma_C \equiv \frac{Z^2e^2}{akT} \tag{3.78}$$

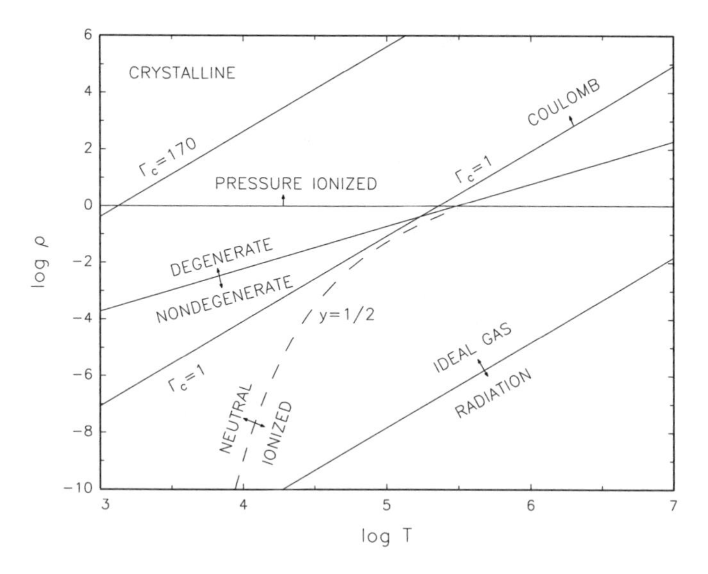

Fig. 3.9. A composite showing how the ρ–T plane is broken up into regions dominated by pressure ionization, degeneracy, radiation, ideal gas, crystallization, and ionization-recombination. The gas is assumed to be pure hydrogen.

where $\Gamma_C = 1$ is the rough demarcation between where Coulomb effects might be important or not, and a $\Gamma_C > 1$ implies they probably are important. The distance a is usually taken as the radius of a Wigner–Seitz sphere whereby $(4\pi a^3/3) = (1/n_\text{I})$ and n_I is the ion number density. If the gas consists of pure ionized hydrogen and $\Gamma_C = 1$, then (3.78) becomes

$$\rho = 8.49 \times 10^{-17} T^3 \quad \text{g cm}^{-3}. \tag{3.79}$$

If the density is greater than that implied by (3.79) for a given temperature, then you can be reasonably certain that a perfect gas is not as perfect as could be desired. This line is shown in Fig. 3.9. You may check, from the material given previously, that the centers of very low mass ZAMS stars are encroaching upon both this line and the one for degeneracy effects. Carefully done stellar models of these stars contain corrections for these effects.

If Γ_C becomes large enough, then Coulomb effects overwhelm those of thermal agitation and the gas settles down into a crystal. The best estimates as to how this takes place yield a Γ_C of around 170 for the transition. With this value of Γ_C in a hydrogen gas (which is kind of silly for a crystallizing composition but fine for talking purposes), (3.79) becomes

$$\rho = 4.2 \times 10^{-10} T^3 \quad \text{g cm}^{-3}. \tag{3.80}$$

This is not an academic issue because some portions of very cool white dwarfs are thought to turn crystalline, but with carbon and/or oxygen rather than hydrogen.

We have already discussed the Saha equation for pure hydrogen, and the density–temperature relation for half-ionization was given by (3.41). That relation is also shown in Fig. 3.9 as the dashed line. In deriving the Saha equation it was implicitly assumed that the energy levels of the hydrogen atom (had we included all of them) were known and that their energies were independent of conditions in the ambient environment. This cannot be true in general. If the gas is dense, then the electrostatic field of one atom should influence a neighboring atom and hence disturb atomic levels. In the extreme, we can imagine this continuing until electron clouds practically rub and electrons are ionized off the parent atoms. This is a crude description of *pressure ionization*. To estimate under what conditions this occurs, take the rubbing picture seriously and find at what density the Wigner–Seitz radius equals the radius of the first Bohr orbit of hydrogen (0.53×10^{-8} cm). A very easy calculation says that this takes place when

$$\rho \approx 1 \quad \text{g cm}^{-3}. \tag{3.81}$$

This density is shown in Fig. 3.9 as the line that terminates ordinary Saha ionization. Such densities are commonplace in stellar interiors and lead to the statement that the larger bulk of those interiors are ionized as far as the lighter elements are concerned independent of the effects of the radiation field.[11]

We finally ask under what conditions radiation pressure dominates over ideal gas pressure or the other way round. That is, where does $aT^4/3 = \rho N_A kT/\mu$? With the assumption of complete ionization in hydrogen this becomes

$$\rho = 1.5 \times 10^{-23}\, T^3 \quad \text{g cm}^{-3} \tag{3.82}$$

as shown in the figure. This ends the discussion of the major factors determining pressures and internal energies in simple environments.[12]

3.7 Adiabatic Exponents and Other Derivatives

For the most part, all we need in the way of thermodynamic variables to construct a simplified stellar model is the internal energy and pressure as a function of density, temperature, and composition (as was done in Chap. 1). To construct realistic models, and to evolve them in time, however, we need several thermodynamic derivatives. We shall assume, at first, that the detailed

---

[11] As a side comment, note that several lines in the figure cross at $T \approx 3 \times 10^5$ K and $\rho \approx 1$ g cm^{-3}. You can be assured that computing accurate equations of state in that region of the T–ρ plane is a nightmare.

[12] We have purposely ignored equations of state at ultrahigh densities such as are found in neutron stars and the collapsing cores of supernovae. This is a difficult subject itself worthy of a monograph. For further reading we suggest chapters 2 and 8 of Shapiro and Teukolsky (1983) and Bethe (1990, §§3-4).

composition, including concentrations of ions, etc., has been determined and that chemical reactions are not taking place. We also assume that you have some facility in transforming thermodynamic functions under reversible conditions and that you are familiar with their properties.

3.7.1 Keeping the Composition Fixed

If changes in temperature and density (or volume) do not cause corresponding changes in the relative concentrations of various species of atoms or ions in the stellar mixture, then the calculation of thermodynamic derivatives is not particularly difficult. We now examine this situation and ignore until later those complications arising from chemical reactions.

Specific Heats

The first derivatives encountered in elementary thermodynamics are specific heats. In general form these are defined by

$$c_\alpha = \left(\frac{dQ}{dT}\right)_\alpha \tag{3.83}$$

where α is kept fixed as T changes. In the following, Q will have the units of erg g^{-1} and thus the specific heats will have units of erg g^{-1} K^{-1}. The most useful variables for α for us are P, ρ, or the specific volume $V_\rho = 1/\rho$. (We shall also have occasion to use the ordinary volume, V.) From the first law for a reversible process (and see 1.11)

$$dQ = dE + P\,dV_\rho = dE + P\,d\left(\frac{1}{\rho}\right) = dE - \frac{P}{\rho^2}\,d\rho \tag{3.84}$$

so that

$$c_{V_\rho} = \left(\frac{dQ}{dT}\right)_\rho = \left(\frac{\partial E}{\partial T}\right)_\rho \quad \text{erg g}^{-1}\ \text{K}^{-1}. \tag{3.85}$$

For an ideal monatomic gas $E = 3N_\mathrm{A}kT/2\mu$ erg g^{-1} (from 3.29) so that $c_{V_\rho} = 3N_\mathrm{A}k/2\mu$ and $E = c_{V_\rho}T$. Note that the composition has not been mentioned here except in the mean molecular weight μ: it is kept fixed by assumption.

To find c_P, recall (from any of many thermodynamic texts) that c_P and c_{V_ρ} (or c_V) are related by

$$c_P - c_{V_\rho} = -T\left(\frac{\partial P}{\partial T}\right)^2_{(\rho \text{ or } V_\rho)}\left(\frac{\partial P}{\partial V_\rho}\right)^{-1}_T . \tag{3.86}$$

To cast this in a form that will prove more suitable for later purposes we reintroduce the power law expression for the equation of state given in Chapter 1 by (1.67):

$$P = P_0 \rho^{\chi_\rho} T^{\chi_T} \tag{3.87}$$

where P_0, χ_ρ, and χ_T are constants. This means that the last two are also defined by

$$\chi_T = \left(\frac{\partial \ln P}{\partial \ln T}\right)_{(\rho \text{ or } V_\rho)} = \frac{T}{P}\left(\frac{\partial P}{\partial T}\right)_{(\rho \text{ or } V_\rho)} \tag{3.88}$$

and

$$\chi_\rho = \left(\frac{\partial \ln P}{\partial \ln \rho}\right)_T = -\left(\frac{\partial \ln P}{\partial \ln V_\rho}\right)_T = \frac{\rho}{P}\left(\frac{\partial P}{\partial \rho}\right)_T = -\frac{1}{\rho P}\left(\frac{\partial P}{\partial V_\rho}\right)_T . \tag{3.89}$$

Thus

$$c_P - c_{V_\rho} = \frac{P}{\rho T}\frac{{\chi_T}^2}{\chi_\rho} \quad \text{erg g}^{-1}\text{ K}^{-1}. \tag{3.90}$$

For an ideal monatomic gas $\chi_\rho = \chi_T = 1$ and

$$c_P - c_{V_\rho} = \frac{N_{\rm A} k}{\mu} \quad \text{erg g}^{-1}\text{ K}^{-1} \quad \text{(ideal gas)}, \tag{3.91}$$

which gives the elementary result $c_P = 5N_{\rm A}k/2\mu$.

We also define γ (yes, another γ), the ratio of specific heats, to be

$$\gamma = \frac{c_P}{c_{V_\rho}} = 1 + \frac{P}{\rho T c_{V_\rho}}\frac{{\chi_T}^2}{\chi_\rho} \tag{3.92}$$

which will be discussed shortly. This γ need not be the γ of the γ–law equation of state, but sometimes it is—see later.

Adiabatic Exponents

The dimensionless adiabatic exponents, the "Γs," measure the thermodynamic response of the system to adiabatic changes and will be used extensively. (Two of them, Γ_1 and Γ_2, were already introduced in Chap. 1.) They are defined as follows:

$$\Gamma_1 = \left(\frac{\partial \ln P}{\partial \ln \rho}\right)_{\rm ad} = -\left(\frac{\partial \ln P}{\partial \ln V_\rho}\right)_{\rm ad} \tag{3.93}$$

$$\frac{\Gamma_2}{\Gamma_2 - 1} = \left(\frac{\partial \ln P}{\partial \ln T}\right)_{\rm ad} = \frac{1}{\nabla_{\rm ad}} \tag{3.94}$$

which also defines $\nabla_{\rm ad}$, and

$$\Gamma_3 - 1 = \left(\frac{\partial \ln T}{\partial \ln \rho}\right)_{\rm ad} = -\left(\frac{\partial \ln T}{\partial \ln V_\rho}\right)_{\rm ad} . \tag{3.95}$$

As in Chapter 1, the subscript "ad" means that the indicated partials are to be evaluated at constant entropy. (We shall not need it directly, but extensive use will be made of $\nabla_{\rm ad}$ in later chapters.) It will shortly become clear why the Γ_i appear in such curious combinations in the definitions, but first note that not all the Γ_i are independent. You may easily show that

$$\frac{\Gamma_3 - 1}{\Gamma_1} = \frac{\Gamma_2 - 1}{\Gamma_2} = \nabla_{\rm ad} \; . \tag{3.96}$$

Computation of the Γ_i is tedious and not particularly enlightening. Complete and clear derivations may be found in Cox (1968), but we suggest you try to derive the expressions that follow using the more compact methods given in Landau and Lifshitz (1958), for example. They start from fundamentals and then use powerful yet simple Jacobian transformations to derive what is needed. All you need watch out for is the distinction between V and V_ρ. When you get done, realize that there are many variations in the ways that the Γ_i may be expressed and the following may not always be the most efficient to use; that is, you may wish to rearrange things. The adiabatic exponents are

$$\Gamma_3 - 1 = \frac{P}{\rho T}\frac{\chi_T}{c_{V_\rho}} = \frac{1}{\rho}\left(\frac{\partial P}{\partial E}\right)_\rho \tag{3.97}$$

$$\Gamma_1 = \chi_T\left(\Gamma_3 - 1\right) + \chi_\rho = \frac{\chi_\rho}{1 - \chi_T \nabla_{\rm ad}} \tag{3.98}$$

$$\frac{\Gamma_2}{\Gamma_2 - 1} = \nabla_{\rm ad}^{-1} = c_P \frac{\rho T}{P}\frac{\chi_\rho}{\chi_T} = \frac{\chi_\rho}{\Gamma_3 - 1} + \chi_T . \tag{3.99}$$

The last exponent, γ, is given by

$$\gamma = \frac{c_P}{c_{V_\rho}} = \frac{\Gamma_1}{\chi_\rho} = 1 + \frac{\chi_T}{\chi_\rho}\left(\Gamma_3 - 1\right) = \frac{\Gamma_3 - 1}{\chi_\rho}\frac{1}{\nabla_{\rm ad}} \; . \tag{3.100}$$

Note that the righthand side result for Γ_3 implies that $P = (\Gamma_3 - 1)\rho E$ so that the γ in the γ–law equation of state of (1.24) is Γ_3 and, generally, not one of the other gammas. Lay the blame for any possible confusion here on the quirks of historical nomenclature.

Explicit values for all the exponents and specific heats, etc., for interesting gases follow below. Remember, however, that there are still no chemical reactions going on so that the relative concentrations of ions and electrons are fixed despite changes in temperature and density.

Mixtures of Ideal Gases and Radiation

For a monatomic ideal gas χ_ρ and χ_T are equal to unity and $\Gamma_1 = \Gamma_2 = \Gamma_3 = \gamma = 5/3$. A pure radiation "gas" has $\chi_\rho = 0$, $\chi_T = 4$, and $\Gamma_1 = \Gamma_2 = \Gamma_3 = 4/3$. Note that $\gamma = \Gamma_1/\chi_\rho \rightarrow \infty$ in this case.

If $\gamma = \Gamma_1 = \Gamma_2 = \Gamma_3$ of the same constant value, as can be satisfied by an ideal gas, then

$$P \propto \rho^{\gamma} \tag{3.101}$$
$$P \propto T^{\gamma/(\gamma-1)} \tag{3.102}$$
$$T \propto \rho^{(\gamma-1)} \tag{3.103}$$

along adiabats. This is the result usually quoted in elementary physics texts for adiabatic behavior: it is collectively true only if the exponents satisfy the above equality.

In modeling simple stars, it often turns out that an equation of state consisting of a mixture of ideal gas and radiation suffices:

$$P = \frac{\rho N_A kT}{\mu} + \frac{aT^4}{3} = P_g + P_{rad} \quad \text{dyne cm}^{-2} \tag{3.104}$$

and

$$E = \frac{3N_A kT}{2\mu} + \frac{aT^4}{\rho} \quad \text{erg g}^{-1}. \tag{3.105}$$

We can find the density and temperature exponents almost by inspection so that

$$\chi_\rho = \frac{P_g}{P} \equiv \beta \tag{3.106}$$

which also defines β, the ratio of gas (P_g) to total pressure, and

$$\chi_T = 4 - 3\beta\,. \tag{3.107}$$

(This β is not to be confused with $\beta = kT/m_e c^2$ introduced earlier.) Further analysis, using the general expressions given previously, yields

$$c_{V_\rho} = \frac{3N_A k}{2\mu}\left(\frac{8-7\beta}{\beta}\right) \quad \text{erg g}^{-1}\ \text{K}^{-1} \tag{3.108}$$
$$\Gamma_3 - 1 = \frac{2}{3}\left(\frac{4-3\beta}{8-7\beta}\right) \tag{3.109}$$
$$\Gamma_1 = \beta + (4-3\beta)(\Gamma_3 - 1) \tag{3.110}$$
$$\frac{\Gamma_2}{\Gamma_2 - 1} = \frac{32 - 24\beta - 3\beta^2}{2(4-3\beta)} \tag{3.111}$$

and, finally,

$$\gamma = \frac{\Gamma_1}{\beta}\,. \tag{3.112}$$

It is easy to confirm that all quantities go to their proper limits as $\beta \to 1$ (ideal gas) or $\beta \to 0$ (pure radiation) and that all quantities are intermediate between their pure gas and radiation values for intermediate β.

Mixtures of Degenerate and Ideal Gases

The first thing we shall find is the specific heat at constant volume for an almost completely degenerate electron gas. Recall our earlier discussion of the temperature corrections to such a gas where the number density, $n_{\rm e}$, was given as a function of T and x_f in (3.75). If the volume or density of the gas is fixed while temperature is varied, then $n_{\rm e}$ does not change but x_f must. Thus $(\partial n_{\rm e}/\partial T)_\rho = 0$. If this operation is performed on (3.75), then the righthand side of the resulting equation contains $(\partial x_f/\partial T)_\rho$, which may be solved to first order in T as

$$\left(\frac{\partial x_f}{\partial T}\right)_\rho = -\frac{\pi^2 k^2}{m_{\rm e}^2 c^4}\frac{1+2x_f^2}{3x_f^3}\,T\,. \tag{3.113}$$

When you derive this you will find that it is missing a denominator of the form $\left[1+\mathcal{O}\left(T^2\right)\right]$, where $\mathcal{O}\left(T^2\right)$ contains terms that are of order T^2. Those terms must be ignored because they are of the same order as other correction terms that would have appeared if the equation for $n_{\rm e}$ had been carried out to higher order in temperature. Thus (3.113) is correct to first order in T.

To find the specific heat we have to differentiate $E_{\rm e}$ of (3.77) with respect to T while keeping density fixed. This operation yields, through the chain rule, nasty terms such as $[dg(x_f)/dx_f]\,(\partial x_f/\partial T)_\rho$. When these are all straightened out (see Chandrasekhar 1939, Chap. 10, §6), we find

$$c_{V_\rho\,({\rm e})} = \frac{8\pi^3 m_{\rm e}^4 c^5}{3h^3 T\rho}\left(\frac{kT}{m_{\rm e}c^2}\right)^2 x_f\left(1+x_f^2\right)^{1/2} \tag{3.114}$$

for electrons or

$$c_{V_\rho\,({\rm e})} = \frac{1.35\times 10^5}{\rho}\,T\,x_f\left(1+x_f^2\right)^{1/2}\quad {\rm erg\ g^{-1}\ K^{-1}}. \tag{3.115}$$

Note the presence of ρ in the (3.115). It is required because this specific heat is a specific specific heat (from the units). As before, it is reasonable to replace x_f with x_F or x using (3.49–3.50) provided that temperature correction terms are small in all of $n_{\rm e}$, $P_{\rm e}$, and $E_{\rm e}$. In any case, note the important result that the electron specific heat for the nearly degenerate gas is proportional to temperature.

From here on, we have to make some reasonable physical assumptions about the nature of the stellar gas. Because of pressure ionization, we expect all or most of the nuclear species to be completely ionized so that all electrons are free to swim in the Fermi sea. Thus pressure and energy, as additive quantities, are determined by bare ions and the free electrons. Radiation should play no significant role because, if it did, the temperatures would be so high that electrons would no longer be nearly degenerate—which we assumed at the onset. (See Fig. 3.9.) Thus the total pressure consists of $P = P_{\rm e} + P_{\rm I}$, where "I" means "ions." Internal energies and specific heats are also additive. The reason we bring this up is because the rest of the thermodynamic derivatives are, for the most part, logarithmic (like the Γs) and we cannot simply add them together. It is best to give an example.

The temperature exponent of pressure, χ_T, is $(T/P)\,(\partial P/\partial T)_\rho$ from (3.88) where P is the *total* pressure. We cannot separate χ_T into components describing just the electrons or just the ions. We had the same problem when treating the gas and radiation mixture of the previous section but the calculations there were fairly straightforward. Here, however, the complexity of the electron gas equation of state makes things computationally more difficult. Nevertheless, we can compute all the derivatives fairly easily if we assume that temperatures are very low. If this is the case, then electron degeneracy pressure greatly exceeds that of the ions and $P_\mathrm{e} \gg P_\mathrm{I}$. The same is not true for the partials of pressure with respect to temperature. By following the same course of analysis as was outlined above for the specific heat, you should verify that $(\partial P_\mathrm{e}/\partial T)_\rho \propto T$. (See 3.76.) On the other hand, $(\partial P_\mathrm{I}/\partial T)_\rho = N_\mathrm{A} k\rho/\mu_\mathrm{I}$ where μ_I is the ion mean molecular weight. (The ions are still assumed to be ideal.) Thus for low enough temperatures the temperature derivative of electron pressure may be neglected compared to that of the ions. The net result is that for low temperatures

$$\chi_T \rightarrow \frac{N_\mathrm{A} k}{\mu_\mathrm{I}} \frac{\rho T}{P_\mathrm{e}} \tag{3.116}$$

and, as $T \rightarrow 0$, so does χ_T. The electrons have nothing to say in the matter.

The density exponent $\chi_\rho = (\rho/P)\,(\partial P/\partial \rho)_T$ of (3.89) is easier. The electron pressure dominates both terms for low temperatures so that

$$\chi_\rho \rightarrow \frac{\rho}{P_\mathrm{e}} \left(\frac{\partial P_\mathrm{e}}{\partial \rho}\right)_T \rightarrow \begin{cases} 5/3 & \text{nonrelativistic} \\ 4/3 & \text{relativistic.} \end{cases} \tag{3.117}$$

The limiting forms come directly from the pressure-density relations (3.60) for the degenerate gas.

The rest of the derivatives require that the specific heats be found. We already have $c_{V_\rho\,(\mathrm{e})}$ (from 3.114) and we know that the ion specific heat is $3N_\mathrm{A} k/2\mu_\mathrm{I}$ (from, e.g., 3.85) and it is a constant. Therefore, for sufficiently low temperatures

$$c_{V_\rho} \rightarrow c_{V_\rho\,(\mathrm{I})} = \frac{3N_\mathrm{A} k}{2\mu_\mathrm{I}} = \frac{1.247 \times 10^8}{\mu_\mathrm{I}} \quad \text{erg g}^{-1}\ \text{K}^{-1} \tag{3.118}$$

and the electrons do not matter. (But always check that the temperatures *are* "sufficiently low.") It may seem strange at first that the electrons, which may have a lot of total kinetic energy tied up in their Fermi sea, have a low specific heat. But most of that energy is locked in, so to speak, because of the exclusion principle and the vast majority of electrons have nowhere to go in energy space. Thus increasing or lowering the temperature of the electrons does little to change their total kinetic energy. The ideal gas ions do not have that constraint.

The combination of pressure dominance by electrons, low sensitivity of pressure to temperature (small χ_T), and low specific heats (only the ions matter), all add up to a potentially explosive situation when very reactant nuclear fuels are present, as in the helium flash.

Having found the above, it should be a simple matter for you to verify the following: $c_P = c_{V_\rho\,(\mathrm{I})}$, $\Gamma_3 - 1 = 2/3$, $\Gamma_1 = \chi_\rho$, and $\nabla_{\mathrm{ad}} = 2/3\chi_\rho$.

3.7.2 Allowing for Chemical Reactions

We now give an example of how the thermodynamic derivatives are found when chemical reactions are taking place. For simplicity, the ideal gas, one-state hydrogen atom will again be used, and radiation in the equation of state will be ignored. As usual, real calculations are very difficult and you are referred to Cox (1968, §9.18) for a fuller discussion. As you will see, even in the simple example given here, the analysis is made difficult because relative concentrations of particles vary as temperatures and densities change.

Because we assume that all changes in the system take place along paths in thermodynamic equilibrium, which implies chemical equilibrium, the Saha equation of (3.35) holds and

$$\frac{n^+ n_{\mathrm{e}}}{n^0} = \mathcal{B} T^{3/2}\, e^{-\chi_{\mathrm{H}}/kT} \tag{3.119}$$

where $\mathcal{B}$ is

$$\mathcal{B} = \left(\frac{2\pi m_{\mathrm{e}} k}{h^2}\right)^{3/2} = 2.415 \times 10^{15} \;\; \mathrm{cm}^{-3}\ \mathrm{K}^{-3/2} \tag{3.120}$$

and the other symbols are the same as those in §3.4. Define N (as in §3.1) so that $N\rho \equiv n = n^+ + n^0$. Thus N is the total ion plus neutral atom number density per unit mass and it is independent of density and will not change as the system is compressed or expanded. With the usual definition of $y = n^+/n = n_{\mathrm{e}}/n$, the pressure may be written

$$P = (n_{\mathrm{e}} + n^+ + n^0)kT = (1+y)N\rho kT \quad \mathrm{dyne\ cm}^{-2} \tag{3.121}$$

and the specific internal energy is (see 3.43)

$$E = (1+y)\frac{n}{\rho}\frac{3kT}{2} + y\frac{n}{\rho}\,\chi_{\mathrm{H}} \quad \mathrm{erg\ g}^{-1} \tag{3.122}$$

or

$$E = (1+y)N\frac{3kT}{2} + yN\,\chi_{\mathrm{H}} \quad \mathrm{erg\ g}^{-1} \tag{3.123}$$

where the energetics of the reaction are accounted for.

Having the pressure and internal energy now allows us to compute the thermodynamic derivatives. First note that the analysis leading to the determination of those derivatives in the previous discussion involved only taking partials with respect to either temperature or density with the other kept fixed: concentrations were never mentioned in that analysis. But this implies that partials with respect to concentrations (i.e., the N_i) were never needed. Thus the general expressions derived for the specific heats, the Γs, etc., are formally correct and all we need do is put in the correct pressures and internal energies that contain the information about chemical equilibrium. To carry this out in detail, however, still requires some effort. We start with easier quantities, χ_T and χ_ρ, and leave most of the rest of the work to you.

The ionization fraction y is given by a slightly rewritten version of (3.39):

$$\frac{y^2}{1-y} = \frac{\mathcal{B}}{N\rho} T^{3/2} \, e^{-\chi_{\mathrm{H}}/kT} . \tag{3.124}$$

We now have the three relations $P = P(\rho, T, y)$, $E = E(\rho, T, y)$, and the Saha equation. Take total differentials of the first two to find

$$dP = P \left[\frac{dT}{T} + \frac{d\rho}{\rho} + \frac{dy}{1+y} \right]$$

and

$$dE = \tfrac{3}{2} NkT(1+y) \left[\frac{dT}{T} + \frac{2}{3} \left(\frac{3}{2} + \frac{\chi_{\mathrm{H}}}{kT} \right) \frac{dy}{1+y} \right] .$$

Recall that N remains fixed because it is the number of hydrogen nuclei per gram and cannot change with temperature, density, or volume.

Also take the differential of the Saha equation (3.124) and divide the result by the Saha equation itself to find

$$\frac{dy}{1+y} = \mathcal{D}(y) \left[\left(\frac{3}{2} + \frac{\chi_{\mathrm{H}}}{kT} \right) \frac{dT}{T} - \frac{d\rho}{\rho} \right]$$

where

$$\mathcal{D}(y) = \frac{y(1-y)}{(2-y)(1+y)} . \tag{3.125}$$

Note that $\mathcal{D}(1) = \mathcal{D}(0) = 0$ and, for general $0 \le y \le 1$, $\mathcal{D}(y) \ge 0$. It reaches a maximum at the half-ionization point $y = 1/2$ where $\mathcal{D}(1/2) = 1/9$.

The lefthand side of the differentiated Saha equation appears explicitly in the expressions for dP and dE. Therefore, use that equation to eliminate any reference to dy in dP and dE and find, for dE,

$$dE = \frac{3}{2} NkT(1+y) \left\{ \left[1 + \mathcal{D}\frac{2}{3} \left(\frac{3}{2} + \frac{\chi_{\mathrm{H}}}{kT} \right)^2 \right] \frac{dT}{T} - \mathcal{D}\frac{2}{3} \left(\frac{3}{2} + \frac{\chi_{\mathrm{H}}}{kT} \right) \frac{d\rho}{\rho} \right\} .$$

From this find directly

$$c_{V_\rho} = \frac{3}{2}Nk(1+y)\left[1+\mathcal{D}(y)\frac{2}{3}\left(\frac{3}{2}+\frac{\chi_{\text{H}}}{kT}\right)^2\right] \quad \text{erg g}^{-1}\ \text{K}^{-1}. \tag{3.126}$$

Note that $Nk = N_{\text{A}}k/\mu_{\text{I}}$ and $Nky = N_{\text{A}}k/\mu_{\text{e}}$ from which may be found μ_{I} of (1.45) and μ_{e} of (1.48).

Treating the pressure differential in like fashion we find

$$\frac{dP}{P} = \left[1+\mathcal{D}\left(\frac{3}{2}+\frac{\chi_{\text{H}}}{kT}\right)\right]\frac{dT}{T} + (1-\mathcal{D})\frac{d\rho}{\rho}$$

so that

$$\chi_\rho = 1 - \mathcal{D}(y) \tag{3.127}$$

and

$$\chi_T = 1 + \mathcal{D}(y)\left(\frac{3}{2}+\frac{\chi_{\text{H}}}{kT}\right). \tag{3.128}$$

Because $\mathcal{D} \geq 0$, we have $\chi_\rho \leq 1$ and $\chi_T \geq 1$. The interpretation here is that if temperature rises, keeping density fixed, we get more free electrons liberated and the pressure rises more so than the rise due to temperature alone. Hence χ_T increases above its nominal value of unity without ionization or recombination; that is, χ_T must be greater than or equal to unity. If density increases, keeping temperature constant, then recombination decreases the number of free electrons per gram and thus χ_ρ can fall below unity.

The Γ_i may now be calculated using equations (3.97) through (3.99) in the forms that contain χ_ρ, χ_T, and c_{V_ρ}. After a bit of algebra the results are

$$\Gamma_3 - 1 = \frac{2 + 2\mathcal{D}(y)\ (3/2 + \chi_{\text{H}}/kT)}{3 + 2\mathcal{D}(y)\ (3/2 + \chi_{\text{H}}/kT)^2} \tag{3.129}$$

$$\frac{\Gamma_2}{\Gamma_2 - 1} = \frac{5 + 2\mathcal{D}(y)\ \{\chi_{\text{H}}/kT + (3/2 + \chi_{\text{H}}/kT)\,(5/2 + \chi_{\text{H}}/kT)\}}{2 + 2\mathcal{D}(y)\ (3/2 + \chi_{\text{H}}/kT)} \tag{3.130}$$

and Γ_1 follows from (3.96).

Note that as y approaches zero or unity (so that $\mathcal{D} \to 0$) all the Γ_i approach their ideal gas values of 5/3. This is as it should be. If the gas is completely neutral or totally ionized, then the equation of state is of its usual ideal gas form since y is not changing. It is the intermediate case that is interesting.

To compute the Γ_i the scheme is, choose ρ (or n) and T, find y from the Saha equation (3.124) (and $\mathcal{D}$ by means of 3.125), and then apply the above expressions. A typical result is shown in Fig. 3.10, where Γ_3 is plotted as a function of temperature for three densities. The half-ionization point, $y = 1/2$, is indicated. Note that if T is near the typical hydrogen ionization temperature of 10^4 K, Γ_3 drops rapidly from its value of 5/3 to much lower values. Even the dangerous 4/3 may be passed by in the process. A word to the wise: always watch out for temperatures near 10^4 K in a hydrogen-rich mixture.

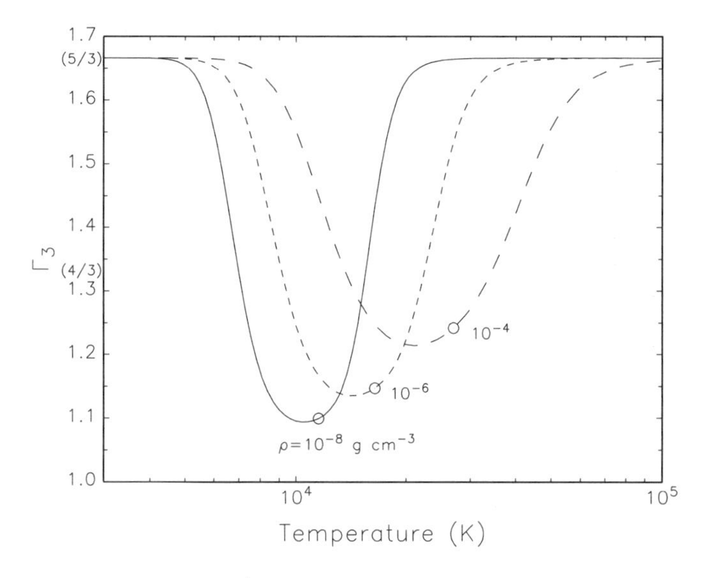

Fig. 3.10. The adiabatic exponent Γ_3 for an ionizing pure one–state hydrogen gas is plotted as a function of temperature. Results are shown for three densities. The ∘ indicates the half-ionization point. The fiducial points 5/3 and 4/3 are also shown.

The reason why Γ_3 (and the other Γs) behaves the way it does when ionization is taking place is quite simple. First suppose no ionization or recombination processes can operate in an almost completely neutral gas so that concentrations remain constant as the system is compressed adiabatically. In that case $\Gamma_3 = 5/3$, $T \sim \rho^{2/3}$, (as in 3.103) and the gas heats up. If, however, we allow ionization to take place, then compression may still heat up the gas, but ionization is much more sensitive to temperature changes than to changes in density. Hence, ionization is accelerated. But this takes energy and that energy is paid for at the expense of the thermal motion in the gas. Thus the temperature tends not to rise as rapidly as $\rho^{2/3}$ and Γ_3 is smaller than its value with no ionization.

As we shall see in chapters to come, all the Γ_i are important in some respect or another: Γ_3 says something about how the heat content of the gas responds to compression; Γ_1 is intimately tied up with dynamics (partially through the sound speed); the behavior of Γ_2 and ∇_{ad} may be a deciding factor in whether convection may take place. As an example, Fig. 3.11 shows the run of ∇_{ad} through a ZAMS model sun. The abscissa is $-\log(1-\mathcal{M}_r/\mathcal{M})$, the stellar center is at the left, and the surface is to the right. Such an axis emphasizes the outer layers of the model. Thus, a value of "9" on this axis corresponds to $1-\mathcal{M}_r/\mathcal{M} = 10^{-9}$ or a mass point that is within 10^{-9} of the total mass. The dips in ∇_{ad} signal ionization. The one at "5" takes place at a temperature of about 10^5 K and is the first ^{4}He ionization zone. The broad trough around 8–9 is at about 10^4 K and corresponds to a combination of

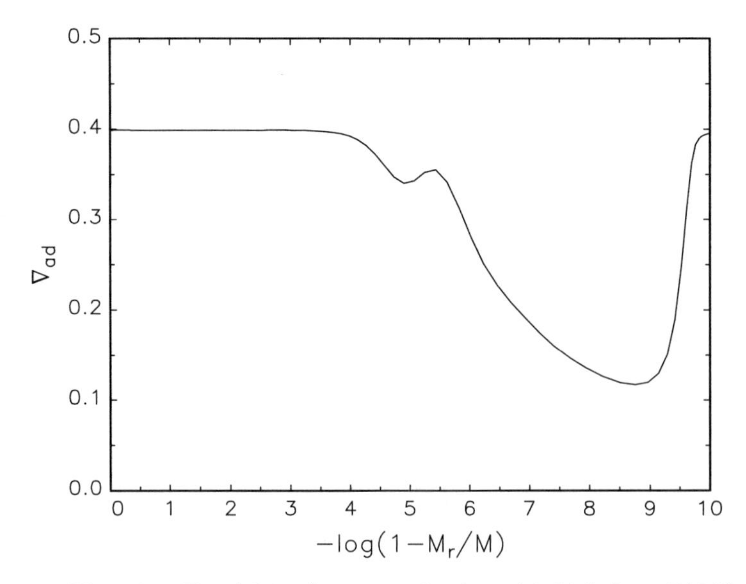

Fig. 3.11. Plotted is $\nabla_{\rm ad}$ (of 3.96) versus $-\log(1 - \mathcal{M}_r/\mathcal{M})$ for a ZAMS model sun. The region from roughly "3" on this scale to almost the surface is convective due, in part, to the depression of $\nabla_{\rm ad}$.

second ^{4}He and ^{1}H ionization. Suffice it to say that the whole region with depressed $\nabla_{\rm ad}$ is convective, for reasons that will be explained in in Chapter 5.

The effects of radiation or other ionizing species and energy levels are included in more complete analyses than what we have done here (see Cox, 1968). In addition, the effects of pressure ionization (among other things) have to be included in many situations. Even though the results we have obtained are useful for many calculations in stellar structure, you should be aware that real models are usually constructed using tabular equations of state with P, E, and, sometimes, various derivatives given as functions of temperature and density for a fixed nuclear composition. Very often these are included in tabulations of opacities—which we discuss in the next chapter.

3.8 Exercises

Exercise 3.1. We have already explored the Saha equation using a pure hydrogen gas as an example. Now consider the more complicated ^{4}He atom with its two electrons. Assume, as in the hydrogenic example, that the neutral atom and first ionized ion are in their respective ground states. The ionization potential to remove the first (second) electron is $\chi_1 = 24.587$ eV ($\chi_2 = 54.416$ eV). To agree on a common nomenclature, let $n_{\rm e}$, n_0, n_1, and n_2 be the number densities of, respectively, electrons, neutral atoms, and first and second ionized ions. The total number density of atoms plus ions of the pure helium gas is denoted by n. Furthermore, define z_e as the ratio n_e/n; and, in like manner, let z_i be n_i/n, where $i = 0, 1, 2$. The gas is assumed to be electrically neutral. For the following you will also need the degeneracy factors for the atoms and ions and these are to be found in Allen (1973), Lang (1991), or Cox (1999) as the data-type references given at the end of Chapter 1.

1. Following the hydrogenic case, construct the ratios $n_e n_1/n_0$ and $n_e n_2/n_1$. In doing so you must take care to establish the zero points of energy for the various constituents. One way to do this is to use mc^2 arguments. For example, the first ionization has $m_e c^2 + m_1 c^2 = m_0 c^2 + \chi_1$ in obvious notation. The reference energies $\mathcal{E}$ to be used in (3.24) for each constituent are then taken to be the mc^2s. This establishes the relative order of the $\mathcal{E}$s. The final form you obtain should not contain chemical potentials (and you must show why this is true).
2. Apply $n = n_0 + n_1 + n_2$, overall charge neutrality, and recast the above Saha equations so that only z_1 and z_2 appear as unknowns. The resulting two equations have temperature and n or, equivalently, $\rho = 4n/N_A$ as independent parameters.
3. Simultaneously solve the two Saha equations for z_1 and z_2 for temperatures in the range $1 \times 10^4 \leq T \leq 2 \times 10^5$ K with a fixed value of density from among the choices $\rho = 10^{-4}$, 10^{-6}, or 10^{-8} g cm^{-3}. Choose a dense grid in temperature because you will soon plot the results. (These cases will prove useful when discussing pure helium opacities in Chap. 4.) Once you have found z_1 and z_2, also find z_e and z_0 for the same range of temperature. Note that this is a numerical exercise and use of a computer is strongly advised.
4. Now plot all your z_i as a function of temperature for your chosen value of ρ. (Plot z_0, z_1, and z_2 on the same graph.) This is an essential step because it will make clear how the ionization responds to temperature changes.
5. Find the half-ionization points on your plot. The two temperatures you obtain (for fixed density) will correspond to the single half-ionization point for pure hydrogen.

Exercise 3.2. We earlier established that photon mean free paths were very short in a star except in the very outermost layers. This means that photons must follow a tortuous path to escape eventually from a star and must take a long time in doing so. To estimate this time, assume that a photon is created at the center of a star and thereafter undergoes a long series of random walk scatterings off electrons until it finally reaches the surface. The mean free path associated with each scattering is $\lambda_{\text{phot}} = (n_e \sigma_e)^{-1}$, where σ_e is the Thomson scattering cross section $\sigma_e \approx 0.7 \times 10^{-24}$ cm^2 (see §4.4.1). For simplicity, assume that the star has a constant density so that λ_{phot} is also constant. This is an order-of-magnitude problem, so don't worry too much about constants of order unity. (Real diffusion in stars is much more complicated and your estimate for the time will be an underestimate.)

1. Using one-dimensional random walk arguments, show that $L \approx \mathcal{R}^2/\lambda_{\text{phot}}$ is the *total* distance a photon must travel if it starts its scattering career at stellar center and eventually ends up at the surface at $\mathcal{R}$.
2. Since the photons travel at the speed of light, c, find the time, τ_{phot}, required for the photon to travel from the stellar center to the surface. (Assume that any scattering process takes place instantaneously.)
3. Give an estimate for τ_{phot}, in the units of years, for a star of mass $\mathcal{M}/\mathcal{M}_\odot$ and radius $\mathcal{R}/\mathcal{R}_\odot$.

Exercise 3.3. Neutron stars are assumed to be objects with $\mathcal{M} \sim \mathcal{M}_\odot$, $\mathcal{R} \sim 10$ km ($\langle\rho\rangle \sim 10^{14}$ g cm^{-3}) where internal temperatures (kT) are small compared to the Fermi energies of electrons, protons, and neutrons (which are assumed to be the only particles present). To demonstrate that the name neutron star is apt, consider the following. Assume that the stellar temperature is zero and that chemical equilibrium exists between electrons, protons, and neutrons. The reaction connecting them is

$$\mathrm{n} \Longleftrightarrow \mathrm{p} + \mathrm{e}^- + Q$$

where $Q = 0.782$ MeV and we are neglecting the electron anti-neutrino, which should appear on the right-side of the reaction. Further assume that the electrons are completely relativistic but that protons and neutrons are nonrelativistic.

1. Convince yourself that the "Saha" equation is

$$\mathcal{E}_\mathrm{n} + Q = \mathcal{E}_\mathrm{p} + \mathcal{E}_\mathrm{e}$$

   where the $\mathcal{E}$s are the Fermi energies of the respective particles. Do your "self-convincing" two ways: (a) argue from the chemical potential equation of the reaction; (b) make a physical argument based on the energetics of the reaction and the Pauli exclusion principle.
2. Now find the number densities of the particles as a function of density. Assume charge neutrality, so that $n_\mathrm{e} = n_\mathrm{p}$, and use the Saha equation to find n_e, n_p, and n_n for densities in the range $10^{13} \lesssim \rho \lesssim 2 \times 10^{14}$ g cm^{-3}. You may take the density as being $\rho = (n_\mathrm{p} + n_\mathrm{n})m$ where m is the mass of either proton or neutron.
3. Plot your number density results as a function of density and, if possible, compare to what you might find in the literature.

Exercise 3.4. Show for the ideal gas ($\mu/kT \ll -1$) that $P = nkT$ is a general result independent of whether the particles are relativistic, nonrelativistic, or anything in between. (Hint: integrate 3.13 by parts after inserting 3.12.)

Exercise 3.5. Verify (3.29) by computing the average kinetic energy of a Maxwell–Boltzmann distribution.

Exercise 3.6. To give an idea why the chemical potential is referred to as a "potential," consider the following, as discussed in Landau and Lifshitz (1958, §25). They state that a body subject to an external field is in equilibrium if the sum of the local chemical potential at every position in the body—here call it $\mu_{\text{local}}(\mathbf{r})$—and the potential of the external field , $\psi(\mathbf{r})$, is a constant; that is,

$$\mu_{\text{tot}} \equiv \mu_{\text{local}}(\mathbf{r}) + \psi(\mathbf{r}) = \text{constant}.$$

To make things simple, consider a one-dimensional situation where the external field is gravitational and the local gravity, g, is everywhere constant so that $\psi(\mathbf{r}) = -mgz$ where z is height and m is the mass of a particle in the body. Further assume that the particles compose an ideal gas.

1. Using the ideal gas results, show that

$$\left(\frac{\partial P}{\partial \mu_{\text{local}}}\right)_T = \frac{\rho}{m}.$$

2. Compute $d\mu_{\text{tot}}/dz$ and finally show that

$$\frac{dP}{dz} = -g\rho$$

which is the elementary result for the equation of hydrostatic equilibrium in a constant gravity field; that is, you have shown that the chemical potential is part of a potential! For a more complicated situation, see
▹ Aronson, E., & Hansen, C.J. 1972, ApJ, 177, 145,
who give an example of the "gravo–thermo catastrophe."

Exercise 3.7. This problem deals with corrections to Maxwell–Boltzmann thermodynamics due to the effects of weak electron degeneracy. Suppose μ/kT is still very much less than –1 as discussed in §3.3, but we wish to include some effects of Fermi–Dirac statistics; i.e., what are the effects due to the +1 in the distribution function (3.9)?

1. If the exponential term in (3.9) is still large then, we can use the expansion $1/(a+1) \approx (1-1/a)/a$ to first order in the large quantity a. If you assume, as an approximation, that μ/kT of (3.24) is still given by

$$e^{\mu/kT} = \frac{n_0 h^3}{2\left(2\pi m_e kT\right)^3/2} \equiv K$$

where n_0 is the electron number density in the pure Maxwell–Boltzmann limit, then show that the number density, n, for weak degeneracy is

$$n = n_0\left[1 - 2^{-3/2}K\right].$$

2. Similarly show that the new pressure is

$$P = n_0 kT \left[1 - 2^{-5/2} K\right].$$

Exercise 3.8. Section 3.6 discusses "imperfections" in equations of state that make life difficult for the stellar modeler. One of these imperfections arises from electrostatic interactions between ions. These cause modifications in the ideal gas equation of state. The severity of the modifications depends on density and temperature in the sense that low temperatures and/or high densities means you have to work harder. One method of attacking the problem is to use Debye–Hückel theory wherein it is assumed that (for, say, a one-component composition) that the average inter-ion spacing r_0 is large compared to the Debye length

$$r_D = \left(\frac{kT}{4\pi e^2 \rho \zeta N_{\rm A}}\right)^{1/2} .$$

Here $\zeta = Z(Z+1)/A$ where Z is the ion charge and A is its atomic mass (in rounded off amu's). This statement is equivalent to

$$n_Z << \left(\frac{kT}{Z^2 e^2}\right)^3$$

where n_Z is the ion number density. If this condition is satisfied then, we find the following expression for the pressure:

$$P = nkT \left[1 - \frac{e^3}{3} \frac{(\pi N_{\rm A} \rho)^{1/2}}{(kT)^{3/2}} \mu \zeta^{3/2}\right]$$

which becomes, after putting in numbers,

$$P = nkT \left[1 - 0.32 \frac{\rho^{1/2}}{T_6^{3/2}} \mu \zeta^{3/2}\right] .$$

Here n is the total number density (ions plus electrons), μ is the mean molecular weight, and T_6 is the temperature in units of 10^6 K. You may check these expressions by consulting Cox (1968, §15.5) or Clayton (1968, §2.3). In any case, write this as

$$P = nkT(1 - B)$$

where, for this analysis to work at all, B must be small compared to unity. If it gets moderately large (say 0.1 or larger), then electrostatic effects are considered to be significant. Now do the following.

1. Consult the literature (or the *Supplemental Material* section of Chap. 2) for properties of ZAMS models. Make believe these are composed of pure ionized hydrogen ($\mu = 1/2$) and compute B at model center for a selection of these models starting with $60\mathcal{M}_\odot$ and ending at $0.08\mathcal{M}_\odot$.

2. What do you conclude from this exercise? Where, in mass on the ZAMS, do you think electrostatic corrections begin to be important?

Exercise 3.9. Having already found the ionization fractions for pure helium in a previous exercise, let's go one step further—but "we" have done most of the work for you here. The FORTRAN program `GETEOS.F90` to be found on the CD-ROM on the endcover of this text was written by W. Dean Pesnell (an old colleague of ours) to compute the pressure and internal energy (among other things) for an ideal gas plus radiation. (This code is also part of `ZAMS.FOR` also found on the CD.) You input the hydrogen (one ionization state) mass fraction X, the helium (two ionization stages) mass fraction Y, the temperature T, and the specific volume V_ρ. The output from `GETEOS` consists of pressure P, internal energy E (in ergs g^{-1}), the electron pressure (PE), $(\partial P/\partial V_\rho)_T$ (PV), $(\partial P/\partial T)_{V_\rho}$ (PT), $(\partial E/\partial V_\rho)_T$ (EV), and $(\partial E/\partial T)_{V_\rho}$ (ET). The code is sparsely annotated but you should try to see what goes on. The variable GES is our $1/\mu_\mathrm{e}$ and it is iterated upon until all the Saha equations are satisfied. One way to unravel the code (in your mind, not when using the code) is to set XHE and XHE2 (the helium ionization potentials) to infinity, thus shutting off the ionization of that element. The metals consist partially of Mg, Si, and Fe, included as a single element, with potential XM. Set XM to infinity also. The rest of the metals are Na and Al, which are always assumed to be ionized. A driver code at the beginning is just an example and you will have to change it to get all the output quantities from EOS. Note that this is in FORTRAN 90. In using the code be aware that it doesn't always like X or Y (or $Z = 1 - X - Y$) to be zero. But you can set them to some very small number.

1. Use this code, with your version of the driver, to compute various pressures, etc., for interesting combinations of the input quantities.
2. Find the Γs for nearly pure hydrogen and compare to what was shown in Fig. 3.10. The output from `GETEOS` gives you all you need.
3. Do the same for nearly pure helium to show the effects of the two ionization stages. And we are sure your instructor can think of lots more things to keep you busy! **Note**: pressure ionization is not included in this code but, by the time the density reaches that level, the major constituents (H and He) are already ionized.

3.9 References and Suggested Readings

Introductory Remarks

The place to go for general information on stellar equations of state is

▷ Cox, J.P. 1968, *Principles of Stellar Structure*, in two volumes (New York: Gordon & Breach).

In particular, see his Chaps. 9–11, 15, and 24. We also recommend Part III of

▹ Kippenhahn, R., & Weigert, A. 1990, *Stellar Structure and Evolution* (Berlin: Springer–Verlag)

and §3.2 of

▹ Rose, W.K. 1998, *Advanced Stellar Astrophysics* (Cambridge: Cambridge University Press).

▹ Clayton, D.D. 1968, *Principles of Stellar Evolution and Nucleosynthesis* (New York: McGraw-Hill)

also contains useful material.

A favorite text of ours is

▹ Landau, L.D., & Lifshitz, E.M. 1958, *Statistical Physics* (London: Pergamon)

and its later editions. We recommend it for its clarity (but it is not easy) and wealth of practical applications. You will even find material about neutron stars in it.

A complete discussion of what conditions must be met to use the approximation of LTE sensibly may be found in

▹ Mihalas, D. 1978, *Stellar Atmospheres*, 2nd ed. (San Francisco: Freeman).

Anyone thinking seriously about studying stars should try to find a copy. The last we heard, it is out of print, but permission might be granted by the publisher to reproduce it (but check for royalty fees).

§3.3: Ideal Monatomic Gas

A complete monograph discussion of Fermi–Dirac equations of state for use in stars was first published by

▹ Chandrasekhar S. 1939, *An Introduction to the Study of Stellar Structure* (Chicago: University of Chicago Press).

It should be available in paperback Dover editions and is well worth buying at modest cost. We shall refer to this work quite often. Other versions may be found in §3.5 of

▹ Chiu, H.-Y. 1968, *Stellar Physics*, Vol. 1. (Waltham, MA: Blaisdell)

and Chapter 24 of

▹ Cox, J.P. 1968, *Principles of Stellar Structure*, in two volumes (New York: Gordon & Breach).

§3.4: The Saha Equation

Systematic application of the Saha equation to multicomponent mixtures is not easy. The bookkeeping required to keep track of all the energy levels is a daunting task, to say nothing of getting information on level parameters. See Chapter 15 of

▹ Cox, J.P. 1968, *Principles of Stellar Structure*, in two volumes (New York: Gordon & Breach)

and Chapter 14 of

▷ Kippenhahn, R., & Weigert, A. 1990, *Stellar Structure and Evolution* (Berlin: Springer–Verlag).

§3.5: Fermi–Dirac Equations of State

Chandrasekhar (1939) and Cox (1968) (see above) are standard references. The references to fermionic matter and Bose–Einstein condensates are

▷ Anderson, M.H., Ensher, J.R., Matthews, M.R., Wieman, C.E., & Cornell, E.A. 1995, Science, 269, 198.

▷ DeMarco, B., & Jin, D.S. 1999, Science, 285, 1703.

The reference to Peter Eggleton (1982) is given as a private communication in

▷ Truran, J.T., & Livio, M. 1986, ApJ, 308, 721,

who use it in some work concerning nova systems. We have extended Eggleton's mass–radius fit for white dwarfs to accommodate general μ_e.

▷ Cloutman, L.D. 1989, ApJS, 71, 677

is a good source of numerical techniques for computing the Fermi–Dirac integrals. See also

▷ Eggleton, P., Faulkner, J., & Flannery, B. 1973, A&A, 23, 325

for a thermodynamically self-consistent and efficient computation of the equation of state for arbitrarily degenerate and arbitrarily relativistic ionized gases.

▷ Antia, H.M. 1993, ApJS, 84, 101

gives rational expansions for the Fermi–Dirac integrals. Early work on realistic corrections to the perfect Fermi–Dirac gas includes

▷ Hamada, T., & Salpeter, E.E. 1961, ApJ, 134, 683.

§3.6: "Almost Perfect" Equations of State

Our Fig. 3.9 is our version of Fig. 1 of

▷ Fontaine, G., Graboske, H.C., & Van Horn, H.M. 1977, ApJS, 35, 293.

This paper has an excellent discussion of the problems that arise when ionization (including pressure effects) and electron degeneracy must be accounted for. Their results are in the form of tables.

We have not discussed nuclear equations of state. To get an idea of what may be involved see chapters 2, 8, and 9 of

▷ Shapiro, S.L., & Teukolsky, S.A. 1983, *Black Holes, White Dwarfs, and Neutron Stars* (New York: John Wiley & Sons)

and the review article by

▷ Bethe, H.A. 1990, RevModPhys, 62, 801.

4 Radiative and Conductive Heat Transfer

"In an intuitive picture of diffusion,
one usually conceives of a slow leakage
from a reservoir of large capacity by means
of a seeping action. These ideas apply in the
radiative diffusion limit as well."

— Dimitri Mihalas in Stellar Atmospheres (1978)

OK, this plus a little math and I suppose we're done.

In this chapter we discuss two ways by which heat can be transported through stars: diffusive radiative transfer by photons, and heat conduction. The third mode of transport, which is by means of convective mixing of hot and cool material, will be discussed in Chapter 5. For references on the theory and application of energy transfer in stars, we recommend the following excellent texts by Mihalas (1978) and Mihalas and Mihalas (1984). Cox (1968), Rybicki and Lightman (1979), and Rose (1998) also contain some very useful material. The discussion here will barely scratch the surface of this complex subject and will be directed toward the specific end of finding approximations suitable for the stellar interior.

4.1 Radiative Transfer

In discussing blackbody radiation and equations of state we assumed LTE as a very good approximation. We do know, however, that LTE implies complete isotropy of the radiation field and this, in turn, means that radiant energy cannot be transported through the material of the star. Anisotropy in the field is *required* for that to happen. On the other hand, it is easy to demonstrate that only a small degree of anisotropy is needed to drive photons through most of the stellar interior. Another way to phrase this is that even small gradients in temperature can do the job. For example, a crude estimate of the overall temperature gradient in the sun is given by the ratio $(T_c/\mathcal{R}_\odot) \approx 10^{-4}$ K cm^{-1}. Although convection might augment heat transport in parts of the star, that small gradient is usually sufficient. At the solar photosphere, however, we shall see that gradients are large and, in any case, the radiation field must eventually become very anisotropic since radiation only leaves the star at the surface while none enters.

What we shall examine is what near-isotropy in the radiation field implies for the stellar interior. In the end, we shall find that the diffusion equation discussed in the first chapter (§1.5) is more than adequate for most of our purposes. Consideration of the very surface will be deferred until later. For a start, assume that all photons have the same frequency (ν). Amends will be made shortly.

Central to a discussion of radiative transfer is the *specific intensity*, $I(\vartheta)$. It is so defined that the product $I(\vartheta)\,d\Omega$ is the radiative energy flux (in erg cm^{-2} s^{-1}) passing through a solid angle $d\Omega$ (in sr=steradians) around a colatitude angle ϑ (in spherical coordinates ϑ and φ) at some position r or, in plane parallel geometry, z. We delete, for now, reference to r, z, and φ in the intensity and we make the important assumption that the energy transfer does not depend on time. The picture is that of a thin cone of radiation starting from r and passing through $d\Omega$ as shown in Fig. 4.1.

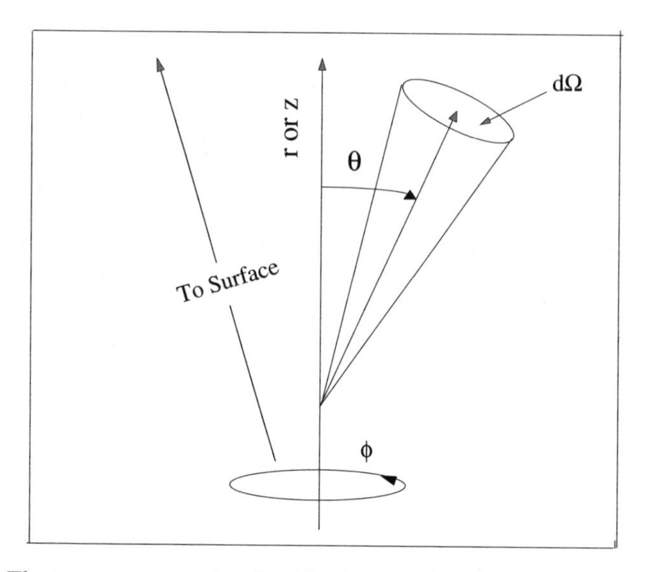

Fig. 4.1. The geometry associated with the specific intensity $I(\vartheta)$. The position coordinate may either be radius r in a spherically symmetric star or vertical distance z in a plane parallel "star." In the latter case, symmetry in the transverse x- and y-coordinates is assumed. The properties of the stellar medium are then independent of azimuthal angle φ for either choice of geometry.

The quantity $u(\vartheta)\,d\Omega$ is the corresponding energy density (in the units of erg cm^{-3}). This may be related to $I(\vartheta)$ by considering how much radiant energy is contained in a tube of unit cross section and length 1 sec $\times\, c$ along a thin cone in the direction ϑ. Thus

$$u(\vartheta)\,d\Omega = \frac{I(\vartheta)}{c}\,d\Omega\,. \tag{4.1}$$

The total energy density U is obtained by integrating (4.1) over 4π steradians with

$$U = \int_{4\pi} u(\vartheta)\,d\Omega = \frac{2\pi}{c}\int_{-1}^{1} I(\mu)\,d\mu \tag{4.2}$$

where azimuthal symmetry (for a round star or flat plane) in φ has been assumed and $\mu = \cos\vartheta$ with $-1 \leq \mu \leq 1$.

Because we are eventually interested in the net outflow (or inflow) of energy along z or r, define the *total flux*, $\mathcal{F}$, as follows. Place a unit area (1 cm^2) perpendicular to the z-direction (in planar geometry to simplify matters). The projection of $I(\vartheta)$ onto this area is then $I(\vartheta)\,\cos\vartheta \times 1\ \text{cm}^2$. If this last is integrated over $d\Omega$ and the result is divided by $1 \times \text{cm}^2$ we then obtain the total flux in the z-direction,

$$\mathcal{F} = \int_{4\pi} I(\vartheta)\,\cos\vartheta\,d\Omega = 2\pi \int_{-1}^{1} I(\mu)\mu\,d\mu \quad \text{erg s}^{-1}\ \text{cm}^{-2}. \tag{4.3}$$

Note that if I is a constant, then the total flux is zero because the same amount of radiation comes in as goes out. Hence I must vary with μ (or ϑ) for radiant energy to be transported; that is, $I(\vartheta)$ must be anisotropic.

What are the sources and sinks for $I(\vartheta)$? At any location z or r, $I(\vartheta)$ may be fed by radiation being scattered from other directions into ϑ or by direct emission from local atoms. We are not going to go into all the subtleties of what these different kinds of processes mean for radiative transfer, but will rather lump them together into the *mass emission coefficient*, $j(\vartheta)$ (in erg s^{-1} g^{-1}), constructed as follows.[1] If ds is a distance directed along $I(\vartheta)$ over which $I(\vartheta)$ is augmented by the amount $dI(\vartheta)$ due to scattering or emission, then $j(\vartheta)$ is defined by

$$dI(\text{put into } \vartheta) = j(\vartheta)\rho\,ds\,.$$

Photons can also be removed from the beam by absorption and scattering. These processes are accounted for by the *opacity* (or mass absorption coefficient), κ (in cm^2 g^{-1}), so that the amount removed from $I(\vartheta)$ is

$$dI(\text{taken out of } \vartheta) = -\kappa\rho I(\vartheta)\,ds$$

and it is proportional to $I(\vartheta)$ itself because it depends on the number of photons present locally. Note that if j is zero and κ and ρ are both constant, then

$$I \propto e^{-\kappa\rho s}$$

as simple attenuation of the beam. Recall that the product $(\kappa\rho)^{-1}$ was previously used to compute a typical mean free path in (3.2). We see now that it is an e–folding length for attenuation.

The net change in $I(\vartheta)$ per unit path length is then

$$\frac{1}{\rho}\frac{dI(\vartheta)}{ds} = j - \kappa I(\vartheta) \tag{4.4}$$

[1] Note that we do not discuss here many of the processes by which photons are emitted. See, for example, Ex. 4.9, where spontaneous versus stimulated emission is discussed in terms of the Einstein coefficients.

which is the *equation of transfer*. Note that in the form we give it, the transfer equation really only holds in planar geometry: a rigorous derivation in spherical geometry would have to contain curvature terms. Because we shall finally get the diffusion equation, such niceties are unnecessary here.

If we had LTE and complete isotropy and spatial uniformity of the radiation field, then $(dI/ds) = 0$ and $I = (j/\kappa)$ would be constant with no radiant energy transported. In that case the energy density of the radiation field is, from (4.2),

$$U = \frac{2\pi}{c}\int_{-1}^{1} I\, d\mu = \frac{4\pi}{c} I\,. \tag{4.5}$$

But in LTE, U follows from the blackbody result $U = aT^4$ (as in 3.18 with E_{rad} replaced by U) so that

$$I = \frac{j}{\kappa} = \frac{c}{4\pi} aT^4 = \frac{\sigma}{\pi} T^4 = B(T) \quad \text{(in LTE)} \tag{4.6}$$

where $B(T)$ is the integrated Planck function, introduced in Chapter 3 as (3.22). The frequency-dependent Planck function is

$$B_\nu(T) = \frac{2h\nu^3}{c^2} \frac{1}{\left(e^{h\nu/kT} - 1\right)} \quad \text{erg cm}^{-2} \tag{4.7}$$

as may be deduced from (3.19) and (3.21).

If, as we suppose, the radiation field is nearly isotropic through most of the star, then the intensity should closely resemble $B(T)$. The question is, by how much? We have to work a bit harder to answer this.

At this point we introduce frequency, ν, into all expressions and realize that quantities such as I, j, and κ must all depend on ν so that we can talk about photons of a given frequency being added to or subtracted from a beam in direction ϑ, etc. In addition, we introduce the *source function* (which, for us, will be merely a computational device),

$$S_\nu = j_\nu / \kappa_\nu$$

and the *optical depth*, τ_ν, with

$$d\tau_\nu = -\kappa_\nu \rho\, dz \ \text{(or } dr\text{)}\,. \tag{4.8}$$

The integrated version of (4.8) is

$$\tau_\nu(z) = \tau_{\nu,0} - \int_{z_0}^{z} \kappa_\nu \rho\, dz \tag{4.9}$$

where z_0 is some spatial reference level and $\tau_{\nu,0}$ is the optical depth evaluated at that level. If z_0 corresponds to the "true surface" of the star where density and pressure presumably go to zero, then $\tau_{\nu,0}$ is taken to be zero. We shall make this choice and thus $\tau_\nu(z)$ measures depth from the surface in curious, but dimensionless, units.

If we now recall that ds is measured along the direction of $I(\vartheta)$, then $dz = \cos\vartheta\, ds = \mu\, ds$ and, putting all together, the equation of transfer becomes

$$\mu \frac{dI_\nu(\tau,\mu)}{d\tau_\nu} = I_\nu(\tau,\mu) - S_\nu(\tau,\mu) \,. \tag{4.10}$$

The solution of the transfer equation is easy to pose but difficult to carry out in practice. As a first step, note that (4.10) admits of the integrating factor $\exp(-\tau/\mu)$ where here (and often elsewhere) the subscript ν and the arguments ϑ or μ will be deleted for visual clarity. Thus multiply through by that factor, recognize a perfect differential, and find

$$\frac{d}{d\tau}\left[e^{-\tau/\mu}\, I\right] = -e^{-\tau/\mu}\, \frac{S}{\mu} \,.$$

If we formally integrate from some reference level τ_0 to a general level τ, then the solution is

$$I(\tau,\mu) = e^{-(\tau_0-\tau)/\mu}\, I(\tau_0,\mu) + \int_\tau^{\tau_0} e^{-(t-\tau)/\mu}\, \frac{S(t)}{\mu}\, dt \tag{4.11}$$

where t is a dummy integration variable.

Depending on the range of μ (or ϑ), different values for τ_0 are chosen in seeking solutions for $I(\tau,\mu)$. For forward-directed radiation (heading out toward the surface) with $\mu \ge 0$ ($0 \le \vartheta \le \pi/2$), choose τ_0 to be very large and positive so that the reference level lies deep (at least with respect to optical depth) within the star. Thus with $\tau_0 \to \infty$,

$$I(\tau,\mu \ge 0) = \int_\tau^{\infty} e^{-(t-\tau)/\mu}\, \frac{S(t)}{\mu}\, dt \,. \tag{4.12}$$

If $\mu < 0$, signifying inwardly directed radiation, use $\tau_0 = 0$ so that

$$I(\tau,\mu < 0) = \int_\tau^{0} e^{-(t-\tau)/\mu}\, \frac{S(t)}{\mu}\, dt. \tag{4.13}$$

In the last expression, advantage has been taken of the fact that the level $\tau_0 = 0$ has been chosen to be the true surface of the star, where it is required that $I(0,\mu < 0) = 0$; that is, there is no incoming radiation at the surface. Note that this would be inapplicable to a star in a close binary where its companion might bathe the stellar surface with radiation. A similar situation holds for stellar winds where the wind itself may radiate profusely.

If the deep interior is to be nearly in LTE, we expect the source function $S = j/\kappa$ to be almost independent of angle and, from (4.6), to be near its Planckian value $B(T)$ at depth τ (assumed to be appropriately large and still leaving off reference to ν). If this is so, then it seems reasonable to expand $S(t)$ in a Taylor series in τ which, to first order, is

$$S(t) = B(\tau) + (t - \tau)\left(\frac{\partial B}{\partial \tau}\right)_\tau \tag{4.14}$$

where $B(\tau)$ stands for $B[T(\tau)]$. (A more exacting discussion of this, and the approach to the diffusion equation, may be found in Mihalas, 1978, §2–5.)

Inserting (4.14) into (4.12–4.13) yields

$$I(\tau, \mu \geq 0) = B(\tau) + \mu\left(\frac{\partial B}{\partial \tau}\right)_\tau \tag{4.15}$$

and

$$I(\tau, \mu < 0) = B(\tau)\left[1 - e^{\tau/\mu}\right] + \mu\left(\frac{\partial B}{\partial \tau}\right)_\tau \left[e^{\tau/\mu}\left(\frac{\tau}{\mu} - 1\right) + 1\right]. \tag{4.16}$$

Since $\mu < 0$ in (4.16), we may neglect the exponential $e^{\tau/\mu}$ for τ large and find that (4.15) *is valid for all* μ. You may easily verify that higher-order terms in the Taylor series expansion for $S(t)$ in (4.14) lead to additional terms in $I(\tau, \mu)$ that go as $|\partial^n B/\partial \tau^n| \sim B/\tau^n$ (and see Exs. 4.1–4.3 at the end of this chapter). Thus, roughly speaking, convergence is rapid if τ is greater than unity.

This looks promising. Since we expect temperature to increase inward, as does τ, then $\partial B/\partial \tau > 0$. Thus, because of the presence of the factor μ in (4.15) or (4.16), the outwardly directed intensity (with $\mu \geq 0$) is enhanced over its Planckian value, $B(\tau)$, whereas the intensity directed inward (with $\mu \leq 0$) is reduced. The net result is a flow of radiation outward when the two intensities are integrated over their respective angles and the two results are added. To compute the flux of this radiation, use (4.3) and find

$$\mathcal{F}(\tau) = 2\pi \int_{-1}^{1} \left[B(\tau) + \mu \frac{\partial B(\tau)}{\partial \tau}\right] \mu \, d\mu = \frac{4\pi}{3}\frac{\partial B(\tau)}{\partial \tau} \tag{4.17}$$

in the units of erg cm^{-2} s^{-1}. Since the integrated Planck function is proportional to T^4, the amount of energy flux carried by radiation depends only on how rapidly temperature varies with optical depth.

If (4.17) represents the total flux (implying that it contains all the frequency-dependent fluxes integrated over frequency), then the *luminosity* for a spherical star at radius $r(\tau)$ is

$$\mathcal{L}(r) = \mathcal{L}_r = 4\pi r^2 \mathcal{F}(r) \quad \text{erg s}^{-1}. \tag{4.18}$$

A measure of the anisotropy in the intensity is the comparison of $I(\tau)$ to $\partial B(\tau)/\partial \tau$ of (4.15). So, calculate $\partial B(\tau)/\partial \tau$ from the flux by means of

(4.17), and use $(\sigma/\pi)T^4(\tau)$ as an estimate for $I(\tau)$ from (4.6). As an example, consider the sun. Using global values for everything in sight, a typical flux may be found from $\mathcal{L}_\odot \sim 4\pi\mathcal{R}_\odot^2\mathcal{F}_\odot \sim 4 \times 10^{33}$ erg s^{-1} or, $\mathcal{F}_\odot \sim 7 \times 10^{10}$ erg cm^{-2} s^{-1}. Thus $\partial B(\tau)/\partial\tau_\odot \sim 2 \times 10^{10}$ erg cm^{-2} s^{-1}. A typical solar temperature is $T_\odot \sim 10^7$ K, so that $I_\odot = B_\odot \sim 2 \times 10^{23}$ erg cm^{-2} s^{-1}. The measure of anisotropy is then $\left[\partial B(\tau)/\partial\tau\right]_\odot / I_\odot \sim 10^{-13}$. (And see Ex. 4.2.) We have, of course, used estimates for various numbers here, but the final result is quite representative of the true situation in the deep interior.

The truncation to first order of the expansion for $I(\tau, \mu)$ is reasonable provided that $\tau \gtrsim 1$. We shall find that an optical depth of unity lies very close below the physical surface in most stars. Hence the approximations used here will be valid for just about all of a star. The thin region above $\tau \approx 1$ we call the *atmosphere* and it is the region where radiation is processed so that we ultimately see it. Except for some simple calculations to be considered shortly in §4.3, the atmosphere will be left to the specialists in that important subject.

4.2 The Diffusion Equation

To derive the diffusion approximation properly, we return to the expression for the flux but include the frequency dependence:

$$\mathcal{F}_\nu = \frac{4\pi}{3}\frac{\partial B_\nu}{\partial \tau_\nu} \tag{4.19}$$

which may be rewritten using the definition of $d\tau_\nu$ as

$$\mathcal{F}_\nu = -\frac{4\pi}{3}\frac{1}{\kappa_\nu \rho}\frac{\partial B_\nu}{\partial r} . \tag{4.20}$$

The derivative of B_ν is cast into more a convenient form by using the chain rule so that

$$\frac{\partial B_\nu}{\partial r} = \frac{\partial B_\nu}{\partial T}\frac{dT}{dr}$$

where, if desired, $\partial B_\nu/\partial T$ may be found using (4.7). The flux is then

$$\mathcal{F}_\nu = -\frac{4\pi}{3}\frac{1}{\rho}\frac{dT}{dr}\frac{1}{\kappa_\nu}\frac{\partial B_\nu}{\partial T} . \tag{4.21}$$

To obtain the total flux integrate over frequency and define the *Rosseland mean opacity*, κ, by

$$\frac{1}{\kappa} = \left[\int_0^\infty \frac{1}{\kappa_\nu}\frac{\partial B_\nu}{\partial T}\, d\nu\right]\left[\int_0^\infty \frac{\partial B_\nu}{\partial T}\, d\nu\right]^{-1} \tag{4.22}$$

so that the total flux is

$$\mathcal{F} = -\frac{4\pi}{3}\frac{1}{\kappa\rho}\frac{dT}{dr}\int_0^\infty \frac{\partial B_\nu}{\partial T}\, d\nu \ .$$

The last integral is eliminated by observing that

$$\int_0^\infty \frac{\partial B_\nu}{\partial T}\, d\nu = \frac{\partial}{\partial T}\int_0^\infty B_\nu(T)\, d\nu = \frac{\partial B}{\partial T} = \frac{ac}{\pi}T^3$$

where (4.6) and $\sigma = ac/4$ have been used. Thus, finally,

$$\mathcal{F}(r) = -\frac{4ac}{3}\frac{1}{\kappa\rho}T^3\frac{dT}{dr} = -\frac{c}{3\kappa\rho}\frac{d(aT^4)}{dr} \ . \tag{4.23}$$

This version of $\mathcal{F}$ is in the Fick's law form introduced in the first chapter (§1.5), where the diffusion coefficient $\mathcal{D}$ is now identified as $\mathcal{D} = c/(3\kappa\rho)$. The factor of 1/3 that appears is usual in diffusion theory and the remainder represents a velocity (c) times a mean free path $\lambda = 1/(\kappa\rho)$. The derivative term in (4.23) implies that the "driving" is caused by spatial gradients in the energy density (aT^4) of the radiation field.

The total luminosity in the diffusion approximation to radiative transfer is simply $\mathcal{L} = 4\pi r^2\mathcal{F}$ or

$$\mathcal{L}(r) = \mathcal{L}_r = -\frac{16\pi acr^2}{3\kappa\rho}T^3\frac{dT}{dr} = -\frac{4\pi acr^2}{3\kappa\rho}\frac{dT^4}{dr} \tag{4.24}$$

which is what was stated in (1.60). There are several other ways of expressing $\mathcal{L}_r$ which will prove useful for future work. Among these are the following.

The Lagrangian form of (4.24) is obtained by using the mass equation (1.2) to convert the radial derivative to one of mass:

$$\mathcal{L}_r = -\frac{\left(4\pi r^2\right)^2 ac}{3\kappa}\frac{dT^4}{d\mathcal{M}_r} \tag{4.25}$$

and this was used in the dimensional arguments of §1.6. Absorbing the factor of a and recognizing $P_{\text{rad}} = (1/3)aT^4$ also yields

$$\mathcal{L}_r = -\frac{\left(4\pi r^2\right)^2 c}{\kappa}\frac{dP_{\text{rad}}}{d\mathcal{M}_r}. \tag{4.26}$$

For still another version, introduce the equation of hydrostatic equilibrium (1.6) (in a slightly disguised form and note the presence of the pressure scale height of Eq. 3.1 in the middle term) so that

$$-\frac{d\ln P}{d\ln r} = \frac{r}{\lambda_P} = \frac{G\mathcal{M}_r\rho}{rP} \ . \tag{4.27}$$

Then divide both sides by $(d\ln T/d\ln r)$ to find

$$\frac{(d \ln P/d \ln r)}{(d \ln T/d \ln r)} = \frac{d \ln P}{d \ln T} = -\frac{G\mathcal{M}_r \rho}{rP} \frac{1}{(d \ln T/d \ln r)} .$$

The reciprocal of the derivative in the middle of this expression is used to define a new quantity, ∇, called "del" with

$$\nabla \equiv \frac{d \ln T}{d \ln P} = -\frac{r^2 P}{G\mathcal{M}_r \rho} \frac{1}{T} \frac{dT}{dr} . \tag{4.28}$$

Sometimes a subscript "act" is appended to ∇ to denote "actual." The implication is that ∇ represents the actual run, or logarithmic slope, of local temperature versus pressure in the star. If ∇ is known by some means or another, then a simple rearrangement of (4.24) yields

$$\mathcal{L}_r = \frac{16\pi acG}{3} \frac{T^4}{P\kappa} \mathcal{M}_r \nabla \tag{4.29}$$

as yet another way to express the relation between luminosity and a gradient. All these variations on luminosity will be used at some point or another.

4.2.1 A Brief Diversion into "∇s"

Besides the ∇ defined above, it is useful to define another logarithmic quantity, $\nabla_{\rm rad}$, called "delrad," as follows. Suppose $\mathcal{L}_r(\text{total})$ is the luminosity corresponding to an energy flux transported by *any* means and not necessarily just by radiation. Then define $\nabla_{\rm rad}$ by turning (4.29) around so that

$$\nabla_{\rm rad} \equiv \left(\frac{d \ln T}{d \ln P}\right)_{\rm rad} \equiv \frac{3}{16\pi acG} \frac{P\kappa}{T^4} \frac{\mathcal{L}_r(\text{total})}{\mathcal{M}_r} = \frac{3r^2}{4acG} \frac{P\kappa}{T^4} \frac{\mathcal{F}_{\rm tot}}{\mathcal{M}_r} . \tag{4.30}$$

(The flux $\mathcal{F}_{\rm tot}$ is the total flux $\mathcal{F}_{\rm tot} = 4\pi r^2/\mathcal{L}_r(\text{total})$, and will be ued later.) Thus $\nabla_{\rm rad}$ is the local logarithmic slope of temperature versus pressure that *would* be required *if* all the given luminosity were to be carried by radiation. This quantity will prove useful for future work, although, at the moment, it may seem to be superfluous baggage. But, for example, suppose you were given the run of density, temperature, opacity, and energy generation rate in a star and the luminosity and ∇ and $\nabla_{\rm rad}$ as functions of radius. But you don't know how the energy is transported. It could well be that the luminosity, $\mathcal{L}$, at a given radius consists of a part from diffusive transfer, $\mathcal{L}_{\rm rad}$, plus a contribution from other sources such as convection, $\mathcal{L}_{\rm conv}$, with $\mathcal{L} = \mathcal{L}_{\rm rad} + \mathcal{L}_{\rm conv}$. The luminosity of (4.29) is obviously $\mathcal{L}_{\rm rad}$ because it is that which is generated by Fick's law with, in the present nomenclature, a gradient term ∇; that is, ∇ is the actual driving gradient in the star and thus $\mathcal{L}_{\rm rad}$ follows from it. However, $\nabla_{\rm rad}$ derives from the total $\mathcal{L}$. Thus if $\nabla = \nabla_{\rm rad}$ then all the luminosity must be radiative, $\mathcal{L} = \mathcal{L}_{\rm rad}$, and $\mathcal{L}_{\rm conv} = 0$. If, on the other hand, $\nabla_{\rm rad} > \nabla$, then $\mathcal{L} > \mathcal{L}_{\rm rad}$, $\mathcal{L}_{\rm conv}$ is not zero and radiation does not transport all of the energy.

The preceding analysis will turn out to be not all that abstract. Recall that yet another "del" was defined in (3.94) and (3.96) of the previous chapter as $\nabla_{\rm ad} = (\partial \ln T/\partial \ln P)_{\rm ad}$. It has the same T–P structure as ∇ but it is a thermodynamic derivative. All three "dels" consist of logarithmic derivatives of temperature with respect to pressure except they are computed under different circumstances. When convection is discussed in the next chapter, these three derivatives will serve to establish one description of how energy is transported in the stellar interior.

4.3 A Simple Atmosphere

Later on we shall have to ask what boundary conditions should be applied to the stellar surface to make satisfactory models. In Chapter 1 we used "zero boundary conditions" as a first go. It will turn out that these are (barely) satisfactory for many stars but are completely inadequate for others. The purpose of this section is to take a small step forward and derive boundary conditions from a simple model atmosphere that are a great improvement over just setting everything in sight to zero at the surface. Note, however, that our efforts are not a real substitute for accurate stellar atmospheres and what we find should not be applied to all stars.

Recall from (4.18–4.20) that the relation between frequency-dependent radiative flux and the Planck function is

$$\mathcal{F}_\nu = -\frac{4\pi}{3}\frac{1}{\kappa_\nu \rho}\frac{\partial B_\nu}{\partial r} = \frac{\mathcal{L}_\nu}{4\pi r^2} \tag{4.31}$$

at large optical depths. This also defines the frequency-dependent luminosity, $\mathcal{L}_\nu$. At the same level of approximation it is easy to show that the frequency-dependent radiation pressure is given by

$$P_{{\rm rad},\nu} = \frac{4\pi}{3c} B_\nu \ . \tag{4.32}$$

This is consistent with the statement that $P_{{\rm rad},\nu} = U_\nu/3$ (with U_ν being the radiation energy density) because $U_\nu = 4\pi B_\nu/c$ in LTE (see Eqs. 3.19–3.21 and 4.5–4.6).

Putting this together gives

$$\frac{c}{\rho}\frac{\partial P_{{\rm rad},\nu}}{\partial r} = -\frac{\kappa_\nu \mathcal{L}_\nu}{4\pi r^2} \ . \tag{4.33}$$

Integrate this over frequency so that

$$\frac{dP_{\rm rad}}{dr} = -\frac{\kappa \rho \mathcal{L}}{4\pi r^2 c} \tag{4.34}$$

where κ is defined as

$$\kappa = \frac{1}{\mathcal{L}} \int_0^\infty \kappa_\nu \mathcal{L}_\nu \, d\nu \tag{4.35}$$

and it is *not* the Rosseland mean opacity defined previously by (4.22). (Later we shall use this opacity but make believe it is Rosseland as one of a series of approximations.)

Because at this juncture we are only interested in the radiation properties of the stellar material very near the photosphere, we simplify (4.34) by replacing r on the right-hand side with $\mathcal{R}$, which is defined as the radius at which we find the photosphere. We also remind you that the effective temperature, T_{eff}, is the temperature that satisfies the relation

$$\mathcal{L} = 4\pi\sigma\mathcal{R}^2 T_{\text{eff}}^4 \tag{4.36}$$

(and see §1.8). Thus T_{eff} is the temperature the photosphere (at $\mathcal{R}$) would have if that surface radiated as a black body. Note that of the three quantities in (4.36) only $\mathcal{L}$ is directly observable; $\mathcal{R}$ and T_{eff} may both turn out to be convenient fictions. This is because the term "visible surface" is really a spectrum-dependent statement (photons of one frequency may emerge to final visibility from different depths compared to other photons) and, in any case, no star really emits radiation into space as a pure blackbody, as we shall demonstrate later with real spectra.

One final point before we go on. Recall from our earlier discussion that we had defined the "true surface" as that level in the star where there was no incoming radiation. This was set at optical depth $\tau = 0$ (see Eqs. 4.8–4.9, and discussion). What we will have to determine is how that level is related to the level at $\mathcal{R}$ or how the photosphere differs, if at all, from the true surface.

With the above in mind, now integrate (4.34) with $r = \mathcal{R}$ from $\tau = 0$ to some arbitrary depth τ and find

$$P_{\text{rad}} = -\int_{\text{true surface}}^{\text{arbitrary point}} \frac{\mathcal{L}}{4\pi\mathcal{R}^2 c} \kappa\rho \, dr = \int_0^\tau \frac{\mathcal{L}}{4\pi\mathcal{R}^2 c} \, d\tau \tag{4.37}$$

or

$$P_{\text{rad}}(\tau) = \frac{\mathcal{L}}{4\pi\mathcal{R}^2 c}\,\tau + P_{\text{rad}}(\tau = 0) = \frac{\sigma T_{\text{eff}}^4}{c}\,\tau + P_{\text{rad}}(\tau = 0)\,. \tag{4.38}$$

We now have to determine what the radiation pressure at the true surface is. A general expression for radiation pressure may be constructed by considering the momentum transferred by radiation across an imaginary surface at some position. If we realize that $I(\theta)/c$ is that flux (energy flux/c), then

$$P_{\text{rad}} = \frac{2\pi}{c} \int_0^\pi I(\theta) \cos^2\theta \, \sin\theta \, d\theta \tag{4.39}$$

by arguments similar to those used in deriving the total flux (see 4.3). The additional factor of $\cos\theta$ comes about because we require a projection of the momentum to the radial direction. To make further progress, $I(\theta)$ must be specified. There are several strategies possible here but the most straightforward is to invoke a version of the Eddington approximation. (For a much fuller discussion of its virtues and faults, see Mihalas, 1978.) A primary consequence of this approximation is that the radiation pressure is given everywhere, *except at* $\tau = 0$, by its LTE value $P_{\rm rad} = aT^4/3$. This is the same result as would be obtained were $I(\theta)$ isotropic with $I = (\sigma/\pi)T^4$ (from 4.6), as should be apparent if you put this in (4.39). (Show this in Ex. 4.13.)

Assume, therefore, that $I(\theta)$ is isotropic everywhere except at the true surface. At $\tau = 0$ we compromise and let $I(\theta)$ be isotropic for all outgoing angles but set it to zero for $\pi \geq \theta > \pi/2$. Thus no radiation enters the true surface from the outside. Equation (4.39) then yields

$$P_{\rm rad}(\tau{=}0) = \frac{2\pi}{3c}\, I(\tau{=}0)\ . \tag{4.40}$$

We now find $I(\tau = 0)$ by computing the flux at zero optical depth and assuming, as a further minor approximation, that the position of the true surface is at $\mathcal{R}$ (and remember that a relatively large change in optical depth need not mean a correspondingly large change in radius). Using expression (4.3) for the flux, we have

$$\mathcal{L} = 4\pi\mathcal{R}^2\, 2\pi \int_0^{\pi/2} I(\tau{=}0)\cos\theta\, \sin\theta\, d\theta = 4\pi\mathcal{R}^2 \pi I(\tau{=}0)\ . \tag{4.41}$$

Use this to eliminate $I(\tau{=}0)$ in (4.40) and find

$$P_{\rm rad}(\tau{=}0) = \frac{2}{3c}\frac{\mathcal{L}}{4\pi\mathcal{R}^2} = \frac{2}{3c}\sigma T_{\rm eff}^4\ . \tag{4.42}$$

The complete expression for the radiation pressure at depth is then

$$P_{\rm rad}(\tau) = \frac{1}{3}aT^4(\tau) = \frac{\sigma}{c}(\tau + 2/3)\, T_{\rm eff}^4\ . \tag{4.43}$$

From this also obtain the run of temperature in the very outermost layers,

$$T^4(\tau) = \frac{1}{2}T_{\rm eff}^4\left(1 + \frac{3}{2}\tau\right) \tag{4.44}$$

after recalling that $a = 4\sigma/c$. Thus in these approximations the photosphere lies at the optical depth $\tau_{\rm p} = 2/3$, where $T(\tau_{\rm p}) = T_{\rm eff}$ (and "p" stands for "photosphere"). Note also that the temperature is nonzero even at the surface, where it has the value $2^{-1/4}T_{\rm eff}$, and not zero as assumed for zero boundary conditions.[2]

[2] Exercise 4.10 uses the Eddington result of (4.44) to examine a criterion for convection. It's a cute result. Try it.

Before we continue, it is worth pointing out that an optical depth at the photosphere of about unity ($2/3 \approx 1$ is close enough) is to be expected. If we see the photosphere, then it must be at a physical depth, Δr, of about one photon mean free path (the distance an average photon travels before something happens to it) $\lambda_{\rm phot} = (\kappa\rho)^{-1}$ (from 3.2 and the discussion leading to 4.4). But, by the definition of optical depth, $\Delta\tau \approx \kappa\rho\Delta r$; that is, the photosphere should be at $\tau_{\rm p} \approx 1$.

To find the run of total pressure in the outer layers requires solving the hydrostatic equilibrium equation

$$\frac{dP}{dr} = -g\rho \, . \tag{4.45}$$

If mass and radius are regarded as fixed in the local gravity, then g is a constant, with $g_s = G\mathcal{M}/\mathcal{R}^2$, and the hydrostatic equation can immediately be integrated from the true surface down to some optical depth to yield

$$P(\tau) = g_s \int_0^\tau \frac{d\tau}{\kappa} \, . \tag{4.46}$$

What we want is the pressure at the photosphere, which is now known to lie at $\tau = 2/3$ (or nearby, depending on how the previous analysis is done in detail). To again make matters simple, consider the case where opacity is constant (as a version of the "grey" atmosphere) and equal to its value at the photosphere. Denote that opacity by $\kappa_{\rm p}$. Equation (4.46) can then be integrated and becomes

$$P(\tau_{\rm p}) = \frac{2}{3}\frac{g_s}{\kappa_{\rm p}} + P(\tau{=}0) \, . \tag{4.47}$$

If the material gas contributes little or nothing to the total pressure at the true surface (as seems reasonable because nothing should act there to reverse the flow of radiation outward), then setting $P(\tau{=}0) = P_{\rm rad}(\tau{=}0)$ yields

$$P(\tau_{\rm p}) = \frac{2}{3}\frac{g_s}{\kappa_{\rm p}} \left(1 + \frac{\kappa_{\rm p}\mathcal{L}}{4\pi c G\mathcal{M}}\right) \tag{4.48}$$

after a little algebra and the use of (4.42).

For most stars the last factor in parentheses is small with

$$\frac{\kappa_{\rm p}\mathcal{L}}{4\pi c G\mathcal{M}} = 7.8 \times 10^{-5}\, \kappa_{\rm p} \left(\frac{\mathcal{L}}{\mathcal{L}_\odot}\right) \left(\frac{\mathcal{M}}{\mathcal{M}_\odot}\right)^{-1} \tag{4.49}$$

and it can almost always be ignored. For some very massive and luminous stars, however, it cannot ignored as the following argument shows.

Near the true surface where radiation pressure dominates, the hydrostatic equation is as given above but with dP/dr replaced by $dP_{\rm rad}/dr$. If the luminosity is very high and the radiation field very intense, we can imagine that the force due to radiation pressure might overwhelm the local gravitational force. This situation can be written as

$$-\frac{dP_{\rm rad}}{dr} > g_s \rho \, . \tag{4.50}$$

If we further make the (actually contradictory) assumption that radiative diffusion is still responsible for energy transport, then a slightly rewritten form of the transport equation (from 4.26) is

$$\mathcal{L} = -\frac{4\pi \mathcal{R}^2 c}{\kappa_{\rm p} \rho} \frac{dP_{\rm rad}}{dr} \, . \tag{4.51}$$

Eliminating the pressure gradient between the two equations then yields an estimate of how large the luminosity must be so that radiative forces exceed gravitational forces. That limiting luminosity, called the *Eddington critical luminosity* or *Eddington limit*, is

$$\mathcal{L}_{\rm Edd} = \frac{4\pi c G \mathcal{M}}{\kappa_{\rm p}} \tag{4.52}$$

and this overall combination is exactly that which appears as the second term of (4.48). If that term exceeds unity then the Eddington limit has been exceeded. It should be obvious that this subject is intimately connected with mass loss (and for more on the implications, see §2.3.2). As a practical matter, the opacity usually used in (4.52) is electron scattering because high luminosities usually imply high temperatures. With a hydrogen mass fraction of $X = 0.7$ and $\kappa_{\rm e} = 0.34 \text{ cm}^2 \text{ g}^{-1}$ used for the photospheric opacity, the Eddington limit is

$$\left(\frac{\mathcal{L}_{\rm Edd}}{\mathcal{L}_\odot} \right) \approx 3.5 \times 10^4 \left(\frac{\mathcal{M}}{\mathcal{M}_\odot} \right) . \tag{4.53}$$

It is to be understood that if the luminosity approaches 10%, or so, of this number, then a simple static stellar atmosphere will not adequately describe what is going on; the dynamics of momentum and energy transfer between the radiation field and matter must be done correctly and this is *very* difficult.

If the Eddington term is neglected, then the photospheric pressure is given by

$$P(\tau_{\rm p}) \approx \frac{2}{3} \frac{g_s}{\kappa_{\rm p}} \, . \tag{4.54}$$

This may now be used to find the density at the photosphere. If the gas is assumed to be composed only of the sum of ideal gas plus radiation, we set (4.54) equal to that sum and find

$$\frac{1}{3} a T_{\rm eff}^4 + \frac{N_{\rm A} k}{\mu} \rho_{\rm p} T_{\rm eff} = \frac{2}{3} \frac{g_s}{\kappa_0 \rho_{\rm p}^n T_{\rm eff}^{-s}} \tag{4.55}$$

where the power law expression $\kappa_0 \rho_{\rm p}^n T_{\rm eff}^{-s}$ has replaced the opacity. If n and s are known and $T_{\rm eff}$ is fixed, then $\rho_{\rm p}$ may be found using some iterative method. This implies, incidentally, that one must have some idea of photospheric conditions beforehand in order that the *kind* of opacity and its exponents be known. In the simple case, where it is assumed that radiation pressure is unimportant,

$$\rho_{\rm p}^{n+1} \approx \frac{2}{3} \frac{g_s}{\kappa_0} \frac{\mu}{N_{\rm A} k} T_{\rm eff}^{s-1} . \tag{4.56}$$

We know typical ranges for $T_{\rm eff}$ and gravity so we can easily find out what kinds of numbers are associated with photospheric densities. For example, if the gravity and $T_{\rm eff}$ are chosen as solar ($g_s \approx 2.7 \times 10^4$ cm s^{-2} and $T_{\rm eff} \approx 5780$ K), and the opacity is pure electron scattering ($n = s = 0$), then (4.56) yields $\rho_{\rm p} \sim 10^{-7}$ g cm^{-3}. Using the same conditions but with the more realistic $\rm H^-$ opacity (see 4.65) gives $\sim 10^{-6}$ g cm^{-3}, which is essentially the same number at our level of approximation. In any case, photospheric densities are far smaller than those deeper down.

Note in all the above that it has been assumed that convection plays no role in heat transport between the true and photospheric surfaces. This is consistent with our notion of a radiating, static, visible surface, and we shall continue to think of the photosphere in those terms. However, even in the sun the effects of underlying convection may easily be seen in the form of cells, granulation, etc., so that if the photospheric regions are to be modeled correctly, much care must be taken (and do not forget magnetic fields, and so on). We shall not go to such extremes, but we will find that convection can extend right up to the base of the photosphere.

When making stellar models in practice, things can get complicated. What is done is to construct a "grid" of realistic stellar atmospheres where each model atmosphere in the grid is labeled by, for example, a different combination of effective temperature and surface gravity. If, during the course of some sort of iterative procedure used in making a complete stellar model, a set of boundary conditions is required at the photosphere, then interpolation is done in the grid to yield these boundary conditions for a given effective temperature and gravity. A description of one strategy for such an interpolation is given in the classic paper of Kippenhahn, Weigert, and Hofmeister (1967, §IV).

4.4 Radiative Opacity Sources

The calculation of realistic stellar opacities is easily among the most difficult problems facing the stellar astrophysicist. At the present time, the most commonly used opacities for stellar mixtures are those generated at the Los Alamos National Laboratory (LANL), at the Lawrence Livermore National Laboratory (LLNL), and the "Opacity Project" (OP) group for both astronomers and nonastronomers. (The original need for opacities at LANL was, needless to say, prompted not by astrophysical considerations but rather by those of fission and fusion bomb work. How, for example, does the atmosphere respond to a blast of radiation?) Opacities are available in tabular form and include many stellar mixtures with opacities computed over wide ranges of density and temperature. The references at the end of this chapter include published sources and it would be a worthwhile exercise for you to plot up some opacities and get a feel for how they behave, as we shall do shortly.

The following discussion is by no means complete and will give only sketches (if even that) of what goes into the calculation of opacities. A physically clear, and not terribly difficult, description of the ingredients of the calculations may be found in Clayton (1968, Chap. 3). Cox (1968, Chap. 16) also contains some very useful material. The aim is to construct a total Rosseland mean opacity, $\kappa_{\rm rad}$, which is the sum of contributions from the following sources. We shall start with the simplest, which is electron scattering.

4.4.1 Electron Scattering

Equation (4.4) gave a prescription for calculating how much intensity is removed from a beam when an opacity source is present. In the instance where the opacity is independent of frequency, a simple relation may be found between the opacity and the cross section of the process responsible for beam attenuation. Before we proceed, recall that a cross section is a microscopic measure of how a particular reaction takes place, whereas the opacity is a macroscopic quantity that tells us how a large collection of such reactions modifies the flow of radiation. This distinction sometimes escapes the student's attention.

A cross section for a process may be defined quite generally as in this example of low-energy electron scattering. If a beam of photons of a given flux—now defined as the number of photons per cm^2 per second—is incident upon a collection of stationary electron targets, then the rate at which a given event (a photon scattered out of the beam) takes place per target is related to the cross section, σ, by

$$\sigma = \frac{\text{number of events per unit time per target}}{\text{incident flux of photons}} \ \ \text{cm}^2 . \tag{4.57}$$

As we shall soon indicate, the cross section for low-energy electron scattering is independent of energy, and the transfer equation that describes how a beam is attenuated is (4.4) with j set to zero. Thus if $n_{\rm e}$ is the number density of free target electrons, then the product $I\sigma n_{\rm e}\, ds$ is the number of scatterings in cm^{-2} s^{-1} erg over the path length ds (from the definition of σ) and this is to be equated to $I\kappa\rho\, ds$ of (4.4). The desired relation between κ and σ is then

$$\kappa = \frac{\sigma n_{\rm e}}{\rho} \quad {\rm cm}^2\ {\rm g}^{-1} \tag{4.58}$$

where $1/\sigma n_{\rm e}$ may be identified with a mean free path (as in 3.2).

For electron or photon thermal energies well below the rest mass energy of the electron ($kT \ll m_{\rm e}c^2$ or $T \ll 5.93 \times 10^9$ K), ordinary frequency-independent Thomson scattering describes the process very well, and the cross section for that is

$$\sigma_{\rm e} = \frac{8\pi}{3}\left(\frac{e^2}{m_{\rm e}c^2}\right)^2 = 0.6652 \times 10^{-24} \quad {\rm cm}^2 \tag{4.59}$$

where $\left(e^2/m_{\rm e}c^2\right)$ is the classical electron radius. Because, as it will turn out, electron scattering is most important when stellar material is almost completely ionized, it is customary to compute $n_{\rm e}$ according to the prescription of (1.48) and (1.53) if the composition is not unusual. Thus take $n_{\rm e} = \rho N_{\rm A}(1+X)/2$ where X is the hydrogen mass fraction. Folding this in with (4.58–4.59) we obtain the electron scattering opacity

$$\kappa_{\rm e} = 0.2(1+X) \quad {\rm cm}^2\ {\rm g}^{-1}. \tag{4.60}$$

If heavy elements are very abundant or ionization is not complete, then $n_{\rm e}$ must be calculated in a more general way using the ionization fractions, etc., of (1.48). Note also that in a mixture consisting mostly of hydrogen, this opacity decreases rapidly from the value implied by (4.60) at temperatures less than the hydrogen ionization temperature of 10^4 K: there are just too few free electrons left. The corresponding temperature for a gas consisting mostly of helium is around 5×10^4 K. (And see Figs. 4.2 and 4.3.)

As remarked upon in §1.5, this opacity depends neither on temperature nor density if ionization is complete and hence its temperature and density exponents s and n in $\kappa = \kappa_0 \rho^n T^{-s}$ (of 1.62) are $s = n = 0$.

Besides having to worry about exotic mixtures of elements and partial ionization, the electron scattering opacity presented above must be modified for high temperatures (relativistic effects with $kT \gtrsim m_{\rm e}c^2$) and for the effects of electron degeneracy at high densities where electrons may be inhibited from scattering into already occupied energy states.

4.4.2 Free–Free Absorption

As is well known from elementary physics, a free electron cannot absorb a photon because conservation of energy and momentum cannot both be satisfied during the process. If, however, a charged ion is in the vicinity of the electron, then electromagnetic coupling between the ion and the electron can serve as a bridge to transfer momentum and energy making the absorption possible. It should be apparent that this absorption process is the inverse of normal bremsstrahlung wherein an electron passing by and interacting with an ion emits a photon.

A complete derivation will not be presented here (which would deal with the quantum mechanics of the absorption) but a rough estimate of the opacity may be found classically. We first compute the emission rate for bremsstrahlung, and then turn the problem around.

Imagine an electron of charge e moving nonrelativistically at velocity v past a stationary ion of charge $Z_c e$. As the electron goes past, it is accelerated and radiates power according to the Larmor result

$$P(t) = \frac{2}{3}\frac{e^2}{c^3}\, a^2(t)$$

where $a(t)$ is the time-dependent acceleration. If we naively assume that the electron trajectory is roughly a straight line, then it is easy to show (as an E&M problem in Landau and Lifshitz 1971, §73, or Jackson, 1999, Prob. 14.7) that the time-integrated power, or energy, radiated is

$$E_s = \frac{Z_c^2 e^6 \pi}{3c^3 m_e^2}\,\frac{1}{v s^3}$$

where s is the impact parameter for the trajectory; that is, the distance of closest approach were the trajectory to remain straight.

The maximum energy radiated during the scattering will peak in angular frequency around $\omega \approx v/s$. Thus if E_ω is the energy emitted per unit frequency, then E_ω must be simply related to E_s, which is the energy emitted per unit impact parameter. If $2\pi s\, ds$ is the area of an annular target that intercepts a uniform velocity beam of electrons, then

$$E_\omega\, d\omega = -E_s\, 2\pi s\, ds = \frac{2Z_c^2 e^6}{3c^3 m_e^2}\,\frac{\pi^2}{v^2}\, d\omega$$

where ω has been set to v/s and the minus sign comes about because $ds > 0$ implies $d\omega < 0$.

To get a rate of emission per unit frequency, assume that the electron distribution is Maxwell–Boltzmann so that (3.25) applies and

$$n_e(v)\, dv = 4\pi n_e \left(\frac{m_e}{2\pi kT}\right)^{3/2} e^{-m_e v^2/2kT}\, v^2\, dv$$

after the transformation $p = m_e v$ is used in (3.25). The product $n_e(v)v$ is the flux of electrons per unit velocity so that $E_\omega n_e(v) v\, dv$ integrated over all permissible v is the desired rate per target ion per unit frequency. All that remains is to multiply by the ion number density, n_I, and to identify the result as being part of the mass emission coefficient j of (4.4). The total power emitted per unit frequency and volume is then

$$4\pi j_\omega \rho = n_I \int_v E_\omega n_e(v) v\, dv$$

where j_ω, assumed isotropic, has been integrated over 4π steradians.

The lower limit on the integral should correspond to the minimum velocity required to produce a photon of energy $\hbar\omega$, namely, $(1/2)m_e v_{\rm min}^2 = \hbar\omega$, with $\hbar = h/2\pi$. Even though we have assumed that the electrons are nonrelativistic, the upper limit is taken as infinity. (Unless temperatures are very high, the exponential in the Maxwell–Boltzmann distribution will serve as an effective cutoff.) The integral is elementary and yields

$$4\pi j_\omega \rho \, d\omega = \frac{2\pi}{3} \frac{Z_c^2 e^6}{m_e c^3} \left(\frac{2\pi}{m_e kT} \right)^{1/2} n_e n_{\rm I} \, e^{-\hbar\omega/kT} \, d\omega \, .$$

Finally, integrate over ω and find

$$4\pi j \rho = \frac{2\pi}{3} \frac{Z_c^2 e^6}{m_e c^3 \hbar} \left(\frac{2\pi kT}{m_e} \right)^{1/2} n_e n_{\rm I} \approx 10^{-27} \, Z_c^2 n_{\rm I} n_e T^{1/2} \ \ {\rm erg\ cm^{-3}\ s^{-1}}.$$

This result is very nearly correct; the numerical coefficient should be 1.4×10^{-27} and an additional quantum mechanical "gaunt factor" ($g_{\rm f}$), which is of order unity, should appear (as in, for example, Spitzer, 1962, §5.6)

To get the absorption coefficient, we assume that the radiation field is in LTE with $j/\kappa = S = B(T)$ and that κ is due only to free–free absorption. Thus, $\kappa = j/B(T) = \pi j/\sigma T^4$, or, putting in the numbers, the free–free opacity is

$$\kappa_{\rm ff} \approx 4 \times 10^{-24} \frac{Z_c^2 n_e n_{\rm I} T^{-3.5}}{\rho} \propto \rho T^{-3.5} \quad {\rm cm^2\ g^{-1}} \tag{4.61}$$

where the last proportionality arises from eliminating the number densities, both of which are proportional to density.

The functional relation of $\kappa_{\rm ff}$ to ρ and T of the above is basically correct. The numerical coefficient is too high by a factor of ten. To use this opacity as a Rosseland mean, we must really perform the integration indicated in (4.22) and put in the relevant atomic physics. All this is done when constructing opacity tables, and we defer to them. There is, however, a fair approximation to the free–free opacity, which does prove useful in working with simplified stellar models (and is only a factor of ten less than 4.61 if you put in the numbers); that is,

$$\kappa_{\rm ff} \approx 10^{23} \frac{\rho}{\mu_e} \frac{Z_c^2}{\mu_{\rm I}} T^{-3.5} \quad {\rm cm^2\ g^{-1}} \tag{4.62}$$

where Z_c is an average nuclear charge and μ_e and $\mu_{\rm I}$ are the mean atomic weights used previously on several occasions. Note that this opacity requires the presence of free electrons: if none are present, then $\kappa_{\rm ff}$ should be zero. This is effectively taken care of by μ_e, where, if all ions are neutral, then $\mu_e \to \infty$ from the definition of μ_e in (1.48–1.49, with $y_i = 0$). For a mixture composed of hydrogen and some helium (and traces of metals), we expect the free–free opacity to be negligible below temperatures of around 10^4 K (or perhaps a little higher if densities are relatively high: see the half-ionization curve for hydrogen of Fig. 3.10).

The main features of (4.62) are correct and the general form is that of a Kramers' opacity, which was used in Chapter 1 (§1.5). Recall from there that the opacity was written in power law form $\kappa = \kappa_0 \rho^n T^{-s}$ in Eq. (1.62) and please note the *sign* in the temperature dependence. Thus the free–free opacity may be characterized by $n = 1$ and $s = 3.5$.

The strongest dependence in $\kappa_{\rm ff}$ is that of temperature. In our quick and dirty derivation, this comes about because j is a weak function of temperature ($j \sim T^{1/2}$) whereas $B(T) \sim T^4$. Another closely related approach is to construct directly the cross section for the free–free process. The contribution to this quantity from electrons in the velocity band dv is $\sigma \propto n_{\rm e}(v)\, dv/v\nu^3$ where ν is the frequency of the absorbed photon. (Several factors varying relatively slowly with frequency or velocity have been neglected here.) An average for this cross section over velocity introduces a temperature dependence going as $T^{-1/2}$ (from integrating $n_{\rm e}[v]\, dv/v$). The Rosseland mean integral of (4.22) weights most heavily those photons with frequencies near $\nu \approx 4kT/h$ (as you may verify in Ex. 4.11). Thus ν^{-3} in the cross section gives a dependence of T^{-3} and this is folded in with the velocity average contribution to yield a factor of $T^{-3.5}$. The opacity is proportional to the cross section and, hence, $s = 3.5$.

4.4.3 Bound–Free and Bound–Bound Absorption

Bound–free absorption is absorption of a photon by a bound electron where the photon energy is sufficient to remove the electron from the atom or ion altogether. To do a proper job of opacity calculation, the atomic physics of all the atoms and ions in the mixture must be handled with great care. However, it may be shown that the frequency dependence of the opacity κ_ν is again $1/\nu^3$ and that the total bound–free opacity is again of Kramers' form. A rough-and-ready estimate, permissible for simple stellar calculations, has been given by Schwarzschild (1958), who gives (with some factors of order unity deleted)

$$\kappa_{\rm bf} \approx 4 \times 10^{25} Z(1+X)\rho T^{-3.5} \quad {\rm cm}^2\ {\rm g}^{-1} \tag{4.63}$$

where X and Z are, respectively, the hydrogen and metal mass fractions discussed in §1.4. This expression should not be applied if temperatures are much below $T \approx 10^4$ K because, as only part of the story, most photons are not energetic enough to ionize the electrons.

Bound–bound opacity is associated with photon-induced transitions between bound levels in atoms or ions. The calculation is quite complex because it involves detailed description of absorption line profiles under a wide variety of conditions of line broadening, etc. The form of the opacity is of a Kramers' type and could be included in with the preceding expressions. Since, however, it is usually of magnitude less than $\kappa_{\rm ff}$ or $\kappa_{\rm bf}$, we shall not give any estimates here.

Schwarzschild also gives an expression for the free–free opacity, which is quite useful although, it makes some assumptions about the composition:

$$\kappa_{\rm ff} \approx 4 \times 10^{22}(X+Y)(1+X)\rho T^{-3.5} \quad \text{cm}^2 \text{ g}^{-1} \tag{4.64}$$

with the usual warning not to use this for something serious.

4.4.4 H^- Opacity and Others

Among the more important sources of opacity in cooler stars is that resulting from free–free and bound–free transitions in the negative hydrogen ion, H^- ("H-minus"). It is, for example, the most important opacity source for the solar atmosphere. Because of the large polarizability of the neutral hydrogen atom, it is possible to attach an extra electron to it with an ionization potential of 0.75 eV. But this implies that the resulting negative ion is very fragile and is readily ionized if temperatures exceed a few thousand degrees ($kT \approx 0.75$ eV). Making the ion is not an easy task either because it requires both neutral hydrogen and free electrons. This means that some electrons must be made available from any existing ionized hydrogen (helium will be neutral for $T \lesssim 10^4$ K) or from outer shell electrons contributed from abundant metals such as Na, K, Ca, or Al. In this respect, the H^- opacity is sensitive not only to temperature but also to metal abundance. If temperatures are less than about 2,500 K, or if metal donors have very low abundances, then insufficient numbers of free electrons are available to make H^- and the opacity becomes very small.

An estimate of the opacity contributed by H^- can be obtained by using existing tabulations (to be discussed shortly, and see Fig. 4.4). The following power-law fit, eyeballed by us, gives reasonable results (within a factor of ten) for temperatures in the range $3,000 \lesssim T \lesssim 6,000$ K, densities $10^{-10} \lesssim \rho \lesssim 10^{-5}$ gm cm^{-3}, a hydrogen mass fraction of around $X \approx 0.7$ (corresponding to main sequence atmospheric hydrogen abundances), and a metal mass fraction $0.001 \lesssim Z \lesssim 0.03$, assuming a solar mix of individual metals:

$$\kappa_{\rm H^-} \approx 2.5 \times 10^{-31} \left(\frac{Z}{0.02}\right) \rho^{1/2} T^9 \quad \text{cm}^2 \text{ g}^{-1}. \tag{4.65}$$

This expression should only be used for estimates when tabulated opacities are not available. On the other hand, it does give the flavor of how this opacity operates and it will prove useful when we examine some properties of cool stars. Note that its power law exponents are $n = 1/2$ and $s = -9$. Unlike Kramers', it increases strongly with temperature until about 10^4 K, above which Kramers' and electron scattering take over (and, any case, most or all of the H^- is gone by this temperature). This will be apparent when curves of realistic opacities are presented later.

For very cool stars with effective temperatures of less than about 3000 K, opacity sources due to the presence of molecules or small grains become important. Because of the proliferation of complex molecules in cool stars and the difficulty in modeling their abundances and opacities, there is still a good deal of uncertainty about how the atmospheres of cool stars really work. This situation is likely to be with us for several more years.

This ends our discussion of opacities derived from atomic processes. In the interiors of dense objects, however, there are other processes that control the flow of energy.

4.5 Heat Transfer by Conduction

We have already stated that the structural support of the deep interior of a white dwarf or of some red supergiants is due to the presence of degenerate electrons. Not only do these electrons prevent the interior from collapsing, they also are the major means by which energy is transported outward (or, in some instances, inward). The mechanism is by means of electron heat conduction down a temperature gradient—as in a metal—and it is at this point that we must do a little solid-state physics (and that's why dealing with stellar interiors is so much fun: you get to do almost everything).

A good approximation to heat transfer in metals is, again, Fick's law of diffusion:

$$\mathcal{F}_{\text{cond}} = -\mathcal{D}_{\text{e}} \frac{dT}{dr} \, . \tag{4.66}$$

Here, $\mathcal{D}_{\text{e}}$ is a diffusion coefficient with "e" standing for electron. It is convenient to recast (4.66) into a form identical to that used in diffusive radiative transfer (i.e., Eq. 4.23 or 4.24) by defining a "conductive opacity," κ_{cond}, with

$$\kappa_{\text{cond}} = \frac{4acT^3}{3\mathcal{D}_{\text{e}}\rho} \, . \tag{4.67}$$

The conductive flux is then

$$\mathcal{F}_{\text{cond}} = -\frac{4ac}{3\kappa_{\text{cond}}\rho} T^3 \frac{dT}{dr} \tag{4.68}$$

so that κ_{cond} looks like a radiative opacity.

Assuming, for the moment, that we already know how to compute κ_{cond}, how do we combine this opacity with atomic opacities, since, if we have a temperature gradient, photons should also flow? The total energy flux, from radiation and conduction electrons combined, is additive. Thus, calling the radiative component $\mathcal{F}_{\text{rad}}$, the total is $\mathcal{F}_{\text{tot}} = \mathcal{F}_{\text{rad}} + \mathcal{F}_{\text{cond}}$ if convection is ignored. By inspection (see 4.24), the opacities are additive as in a parallel resistive circuit or

$$\frac{1}{\kappa_{\text{tot}}} = \frac{1}{\kappa_{\text{rad}}} + \frac{1}{\kappa_{\text{cond}}} \tag{4.69}$$

with

$$\mathcal{F}_{\text{tot}} = -\frac{4ac}{3\kappa_{\text{tot}}\rho} T^3 \frac{dT}{dr} \ . \tag{4.70}$$

Radiative opacities are added together in simple sums as in a series circuit. Note that whichever opacity in (4.69) is the *smaller* of κ_{rad} or κ_{cond}, it is also the more important in determining the total opacity and hence the heat flow (again, as in a current flowing through a parallel circuit). In normal stellar material κ_{cond} is large (conduction is negligible) compared to radiative opacities and, in those situations, only the latter need be considered. The opposite is usually true in dense degenerate material.

The diffusion coefficient, $\mathcal{D}_{\text{e}}$, has the general form (see, e.g., Kittel, 1968) $\mathcal{D}_{\text{e}} \approx c_V v_{\text{e}} \lambda/3$, where c_V is the specific heat at constant volume of the degenerate electrons, v_{e} is some typical (or relevant) electron velocity, and λ is an electron collisional mean free path. In the following, we shall derive the diffusion coefficient for nonrelativistic electrons.

The specific heat of a nearly completely degenerate electron gas was given by (3.114) in the last chapter. The momentum parameter x_f in that equation is very much less than unity for nonrelativistic electrons so that

$$c_V \approx \frac{8\pi^3 m_e^2 c}{3h^3} k^2 T x_f \quad \text{erg cm}^{-3} \text{ K}^{-1} \tag{4.71}$$

where a factor of $1/\rho$ has been deleted from (3.114) to convert to the indicated units. Since $x \propto (\rho/\mu_{\text{e}})^{1/3}$ from (3.51) is a good approximation to x_f at low temperatures, it is easy to see that $c_V \propto (\rho/\mu_{\text{e}})^{1/3} T$.

For v_{e} and λ, we must recall an important fact of degenerate life: any collisional process involving a degenerate electron cannot result in that electron being scattered into an already filled energy state. What this means is that only electrons near the top of the Fermi sea can participate effectively in the conduction process. Thus the velocity v_{e} should satisfy $m_{\text{e}} v_{\text{e}} \approx p_F \propto x \propto (\rho/\mu_{\text{e}})^{1/3}$. The most efficient means of scattering these electrons is via Coulomb interactions with the surrounding ion gas. Thus, write $\lambda = 1/(\sigma_{\text{C}} n_{\text{I}})$ where n_{I} is the ion number density and σ_{C} is the Coulomb scattering cross section. A typical way to estimate σ_{C} is to consider what electron–ion impact parameters result in a "significant" degree of scattering. Following arguments similar to those in Spitzer (1962), we can see that an encounter in which the electron kinetic energy is about the same as the electron–ion electrostatic potential will result in a significant scatter. Thus consider electrons for which $m_{\text{e}} v_{\text{e}}^2 \approx Z_c e^2/s$. The significant impact parameter is then $s \propto 1/v_{\text{e}}^2 \propto (\rho/\mu_{\text{e}})^{-2/3}$. The cross section, in simplest terms, is $\sigma_{\text{C}} \approx \pi s^2 \propto (\rho/\mu_{\text{e}})^{-4/3}$. Thus $\lambda \propto (\rho/\mu_{\text{e}})^{4/3}/n_{\text{I}}$ or, after introducing the ion mean molecular weight μ_{I}, $\lambda \propto (\rho/\mu_{\text{e}})^{4/3} (\mu_{\text{I}}/\rho)$, and $\mathcal{D}_{\text{e}} \propto \left(\mu_{\text{I}}/\mu_{\text{e}}^2\right) \rho T$.

Inserting these results into (4.67) and accounting for all the numerical factors previously ignored, we find that the conductive opacity is

$$\kappa_{\text{cond}} \approx 4 \times 10^{-8} \frac{\mu_{\text{e}}^2}{\mu_{\text{I}}} Z_c^2 \left(\frac{T}{\rho}\right)^2 \quad \text{cm}^2 \text{ g}^{-1} \tag{4.72}$$

(and remember that Z_c is the ion charge). As crude as this derivation has been, the final result is not so bad when compared to accurate calculations. The temperature and density exponents are about right ($s \approx n \approx -2$) and the coefficient is correct to within an order of magnitude (or so).

As an example of where conductive opacities are important, consider the deep interior of a typical cool white dwarf with $\rho \approx 10^6$ g cm^{-3}, $T \approx 10^7$ K, and a composition of carbon (which is close enough). The results of the last chapter imply that the gas is certainly degenerate and the material pressure ionized. This implies that the radiative opacity is electron scattering with $\kappa_{\text{e}} \approx 0.2$ cm^2 g^{-1}. Equation (4.72) yields $\kappa_{\text{cond}} \approx 5 \times 10^{-5}$ cm^2 g^{-1} with $\mu_{\text{e}} = 2$, $\mu_{\text{I}} = 12$, and $Z_c = 6$. Because $\kappa_{\text{cond}} \ll \kappa_{\text{rad}}$, the total opacity is $\kappa_{\text{tot}} \approx \kappa_{\text{cond}}$ after applying (4.69). Thus the radiative opacity is of no consequence.

4.6 Tabulated Opacities

As has been emphasized repeatedly here, modern stellar structure and evolution studies never use the simple kinds of expressions quoted here for opacities except, perhaps, for pedagogic purposes. In practice, extensive tables or, sometimes, analytic fits to these tables[3] are used that give radiative and conductive opacities over wide ranges of temperature and density for various compositions of interest. Usually a specific opacity is obtained by a multidimensional interpolation in tables; for example, interpolation is density, temperature, and X, Y, and Z.

The earlier efforts at large-scale computer calculations of opacities were due to the group at the Los Alamos National Laboratory (LANL) starting in the middle 1960s as a spin-off of nuclear weapons diagnostics.[4] Figures 4.2 and 4.3 show two sets of these older radiative LANL opacities (both from Cox and Tabor, 1976) plotted as functions of temperature and density for two different compositions.

The first is the "King IVa" set in which the composition is $X = 0.70$, $Y = 0.28$, and $Z = 0.02$ with a solar mix of metals (see, e.g., Figs. 1.2

[3] One such fit is given by Stellingwerf (1975).

[4] LANL was formerly known as LASL (for Los Alamos Scientific Laboratory) and is referred to as such in the older compilations. The `FORTRAN` program "csotesr" found on the CD-ROM on the endcover of this text yields opacities based on the older LANL (Cox and Stewart) tables. This code also computes conductive opacities.

and 2.19). The name "King IVa" given to the tabulated set of opacities from which this figure was generated means that it was the IV$^{\rm th}$ (sub a) set requested by David King. This kind of nomenclature is often found in the older published LANL tables. (These opacities were used by King in his, and his collaborators', studies of Cepheid variables.) The most obvious feature is the pronounced hump around $T \approx 10^4$ K. As density increases, the location of the peak of the hump moves out to slightly higher temperatures. This behavior reflects the temperature versus density relation of the Saha equation for the half-ionization point of hydrogen as given by (3.41) and that relation is indicated on the figure. (The relation should actually be modified because the mixture used for the figure is not pure hydrogen but it is close enough.) The sharp drop in opacity to the left of $T \approx 10^4$ K signals the demise of free–free and bound–free transitions as hydrogen becomes neutral and the radiation field cools to lower energies, but H^- prevents the opacity from disappearing altogether (see Fig. 4.4).

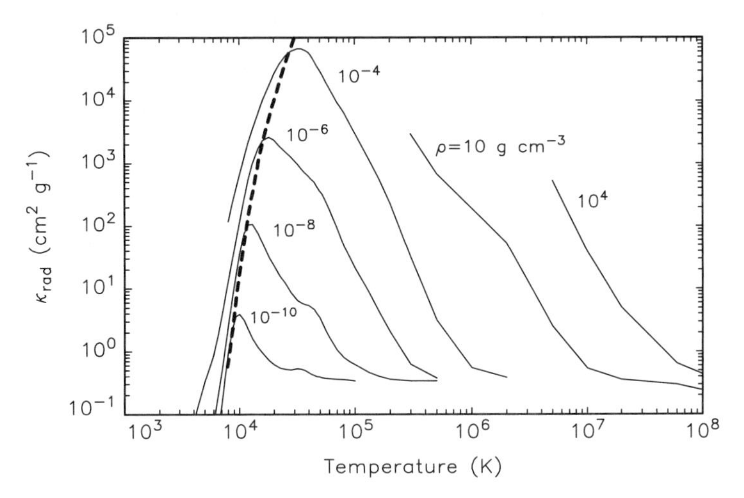

Fig. 4.2. Plots of the LANL radiative opacities for the King IVa mixture $X = 0.7$, $Y = 0.28$, and $Z = 0.02$. The mix of metals comprising Z corresponds to those seen in the solar atmosphere. Material for this figure comes from the tabulations of Cox and Tabor (1976). The dashed line shows the half-ionization curve for pure hydrogen.

Features at higher temperatures include the effects of first and second helium ionization, which can be detected as mild increases in opacity at temperatures a little over 10^4 K and near 10^5 K at the lower densities. These features, although seemingly minor, are important for many variable stars, which are "driven" by helium ionization. (See Chap. 8, and, in this case, the devil is in the details!) Apart from such irregularities, the opacities roughly follow a Kramers' law and fall off in temperature until high temperatures are reached, whereupon electron scattering takes over ($\kappa_e \approx 0.34$). At the highest temperatures, the opacity dips below the Thompson scattering level and this is due to relativistic effects.

Figure 4.2 also illustrates some problems with using tabulated opacities. First of all, they do not completely cover the temperature–density plane (that would be impossible) but rather include just enough information to be of use for modeling certain classes of stars. If you wish to study stars whose properties are very different from those for which the given table was computed, then you have to extend the table or make a new one. Never extrapolate off a table (if possible). Secondly, you will note that the lines in the figure do not always look smooth—they are not—and we have made no attempt at smoothing but have just connected the tabulated points by straight lines. This is where intelligent interpolation is needed.

The second figure shows the results for an almost pure helium mix (with $X = 0$, $Y = 0.97$, $Z = 0.03$) opacity set requested by Morris Aizenman. (This is the Aizenman IV table and it was used in modeling the deep interiors of evolved stars.) Here the first and second helium ionization stages are well marked by the double-humped peaks (and see Ex. 4.6). Also note that the opacities are about an order of magnitude lower (for a given T and ρ) than the hydrogen-rich mixture before the electron scattering threshold is reached.

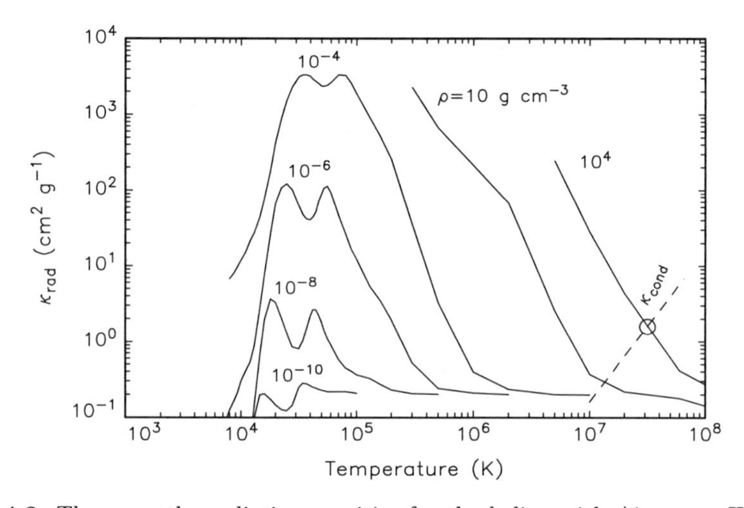

Fig. 4.3. These are the radiative opacities for the helium-rich Aizenman IV mix $X = 0$, $Y = 0.97$, and $Z = 0.03$ from Cox and Tabor (1976). Also shown (as a dashed line) is the conductive opacity for $\log \rho$ of $+4$ from (4.72).

In Fig. 4.3 we also show the conductive opacity for pure helium at a density of 10^4 g cm^{-3} (from 4.72). The intersection of this opacity with the radiative opacity at the same density and temperature is indicated by a circle. If density is kept fixed, then a reduction in temperature causes $\kappa_{\rm cond}$ to decrease as T^2 (see 4.72). At the same time the radiative opacity increases (see figure) so that $\kappa_{\rm cond} < \kappa_{\rm rad}$. This means, in accordance with our previous arguments, that the total opacity becomes more like the conductive opacity and the radiative opacity begins not to count. Conversely, a rise in temperature makes $\kappa_{\rm rad}$ more important. Thus if the *total* opacity were plotted on the figure, the opacity contours would be very different in some regions of ρ and T and especially where densities are high and temperatures are low.

Figure 4.4 shows the opacities for two mixtures from Cox and Tabor (1976). The one labeled "Pop II" is for a typical metal-rich mixture ($Z = 0.02$), whereas "Pop III" has no metals at all. (The density is 10^{-6} g cm^{-3} for both mixtures.) What is apparent is the precipitous drop-off in opacity for temperatures below a few thousand degrees for the Pop III mixture compared to Pop II. This is due to the virtual absence of H$^-$ opacity, which needs metals to provide electrons. At 1,500 K the two cases differ by over three orders of magnitude. Were you to make zero-age main sequence models for Pop III objects you would find that their structures are very different than normal stars with metals. This is no idle observation because, according to Big Bang cosmologies, the first stars were of Pop III variety and their evolution must have been very different than succeeding generations, which were enriched in metals.[5]

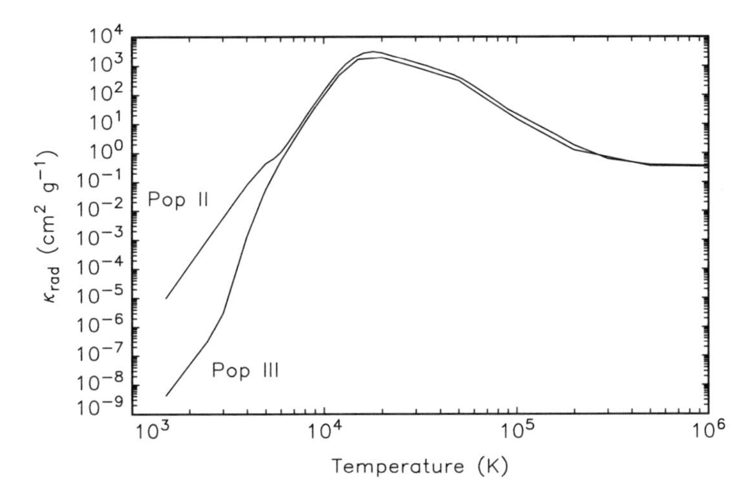

Fig. 4.4. These opacities are also from Cox and Tabor (1976) and show the effect of the H$^-$ opacity at low temperatures. "Pop II" has $Z = 0.02$, whereas "Pop III" has no metals. The density is 10^{-6} g cm^{-3}.

[5] No Pop III stars have been found but the star HE0107–5240 (a 16th mag giant in Phoenix) comes close. Its surface iron abundance is a mere 1/200,000 that of the sun's. See Tytell, D. 2003, Sky&Tel, 105, 20.

More recently, since roughly 1990, two other groups have been actively engaged in computing opacities using improved physics. The older group, formed in 1984, consists of an international consortium of atomic physicists and astrophysicists. They go under the catchy name of "The Opacity Project" (OP). The second effort involves investigators at the Lawrence Livermore National Laboratory (LLNL) in Livermore, CA. Their opacity code is called OPAL. Both groups make their opacity tables available (most conveniently on the World Wide Web from which they may be downloaded) and, upon request, can usually provide tables for new mixtures. In addition, both groups make available sophisticated interpolation routines to yield smooth and consistent results. References to pertinent publications are given at the end of this chapter, including some from groups other than LLNL or OP.

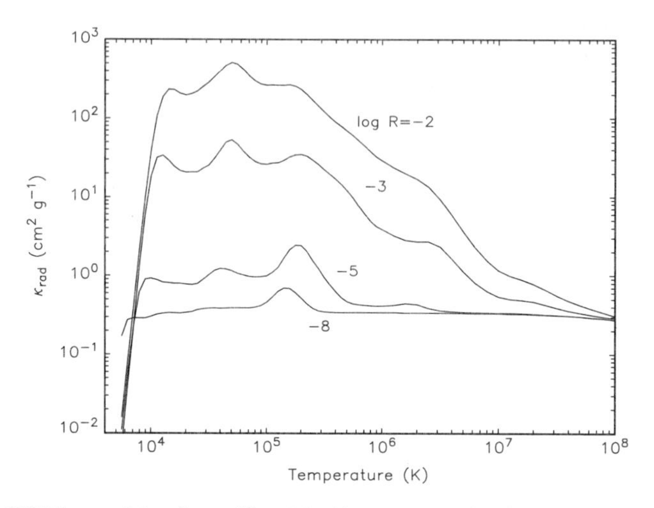

Fig. 4.5. OPAL opacities for a $X = 0.7$, $Z = 0.02$, and solar metals mixture. The parameter $R = \rho/T_6^3$. See text.

To indicate the level of physics required for the calculations of OP and LLNL, we quote from the LLNL Website (as of February 1999):

> Briefly, the calculations [for OPAL] are based on a physical picture approach that carries out a many-body expansion of the grand canonical partition function. The method includes electron degeneracy and the leading quantum diffraction term as well as systematic corrections necessary for strongly-coupled plasma regimes. The atomic data are obtained from a parametric potential that is fast enough for in-line calculations while achieving an accuracy comparable to single configuration Dirac-Fock results. The calculations use detailed term accounting; for example, the bound–bound transitions are treated in full intermediate or pure LS coupling depending on the element. Degeneracy and plasma collective effects are included in inverse bremsstrahlung and Thomson scattering. Most line

broadening is treated with a Voigt profile that accounts for Doppler, natural width, electron impacts, and for neutral and singly ionized metals broadening by H and He atoms. The exceptions are one-, two-, and three-electron systems where linear Stark broadening by the ions is included.

This is why very fast computers are needed. Also, if you read between the lines, you should realize that equations of state are computed as part of the program. (Both LLNL and OP provide these.) However, the question arises, "How comparable are the opacity results from LLNL and OP?" Were you simply to plot the results, the naked eye would have a difficult time seeing any differences between them. But differences of, in some cases, as much as 30% do occur. It would be worth your while to look into Iglesias and Rogers (1996) for some comparisons. This is all very difficult stuff.

To illustrate what is available, and in what form it is made available, Fig. 4.5 shows OPAL results for the mixture $X = 0.7$ and $Z = 0.02$ with a solar atmosphere mix of metals. Plotted is opacity as a function of temperature for four values of the parameter $R \equiv \rho/T_6^3$, where ρ is in g cm^{-3} and T_6 is the temperature in units of 10^6 K. To translate this to κ versus ρ and T takes a little work, but the numbers for this figure were taken directly from OPAL and it shows the range of ρ and T covered by the tables.

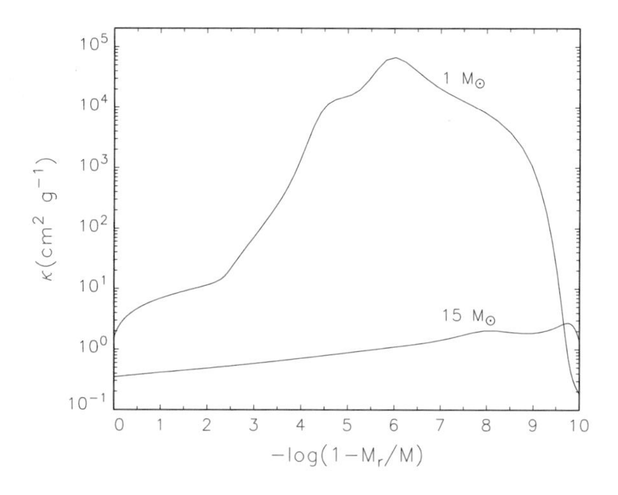

Fig. 4.6. This is to show how dramatically opacities differ between main sequence stars. The opacities are plotted for 1 $\mathcal{M}_\odot$ and 15 $\mathcal{M}_\odot$ ZAMS models on the same vertical scale. The mass scale, $-\log(1 - \mathcal{M}_r/\mathcal{M})$, emphasizes the outer layers.

What do opacities look like in actual stars (well, in models, at least)? Figure 4.6 shows the run of opacity in two ZAMS models. (The mass scale is the same as that used in Fig. 3.11.) For 1 $\mathcal{M}_\odot$, the opacity begins to rise at a $\mathcal{M}_r/\mathcal{M}$ of about $1 - 10^{-3}$, which corresponds to a temperature of a few$\times 10^5$ K. This agrees with Fig. 4.2, where opacity begins to take off in that range.

The peak opacity is reached at $\mathcal{M}_r/\mathcal{M} \approx 1 - 10^{-6}$, where $T \approx 5 \times 10^4$ K and $\rho \approx 10^{-4}$ g cm^3 (as it does in Fig. 4.2). What is striking is the mountain that radiation has to surmount in the very outer layers (by radius, not mass) in the model. If radiation had to carry all the flux, temperature gradients would have to be very high. At roughly the same time (or place), $\nabla_{\rm ad}$, as shown in Fig. 3.11, shows a deep trough. This combination is a double whammy, and convection takes over the task of moving most of the power through the star.

The 15 $\mathcal{M}_\odot$ model is entirely different. Temperatures begin to drop below a few$\times 10^5$ K where the $\mathcal{M}_r/\mathcal{M}$ is only $1 - 10^{-7}$ with densities around a low 10^{-6} g cm^{-3}. Looking at Fig. 4.2, we see this combination of T and ρ means a relatively small opacity. And so it goes. There is a minor convection zone in the envelope, but it is very near the surface. The central regions are convective, but this is due to vigorous nuclear burning concentrated around the stellar center (which is a story for the next chapter).

Finally, Fig. 4.7, from Hayashi, Hōshi, and Sugimoto (1962), shows what regions of the $\log \rho$–$\log T$ plane are dominated by various kinds of opacity. (The composition is typical Population I.) The line labeled $\psi = 0$ denotes the onset of degeneracy and, as you may verify, corresponds roughly to the transition line where conduction takes over.

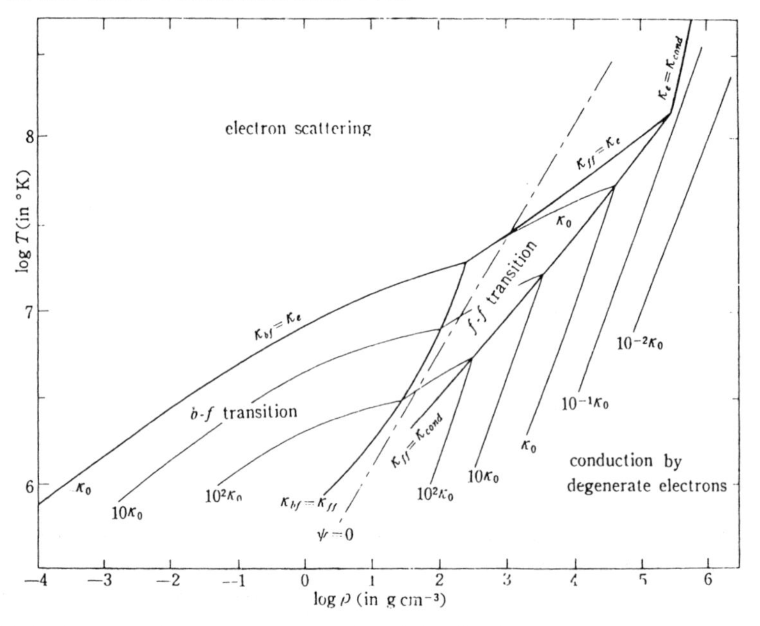

Fig. 4.7. This figure, from Hayashi et al. (1962), illustrates where various opacities are most important as functions of temperature and density. The mixture is Population I. The opacity nomenclature is almost the same as in the text except that the lines are labeled in units of the electron scattering opacity here denoted by $\kappa_0 (= \kappa_e) = 0.2(1 + X)$. Reproduced with permission.

4.7 Some Observed Spectra

In a text such as this, saturated as it is with theory, we believe it wise to introduce some observational material having to do with real atmospheres. In particular, we shall discuss briefly how some stars are classified. This is not "just" taxonomy because it draws on the physics of atmospheres and the intricate problems of observation itself. The subject is further complicated by the bewildering number of kinds of stars, each kind with its own peculiarities. The last come about from differences in atmospheric temperature, pressure, density, local gravity, fluid flows, surface composition, presence of magnetic fields, and even external influences such as companions or incident radiation fields.

For a short tour consider Fig. 4.8, which shows the spectra of "normal" (i.e., no oddball) main sequence (dwarf) stars. As reviewed briefly in Appendix A, such stars belong to luminosity class V with stars of decreasing effective temperature given the spectral class labels O, B, A, F, G, K, and M in that order (plus two other new classes to be discussed separately below). Further subdivision is gained by appending, for examples, a numeral 0, 1, $\cdots$, in order of decreasing temperature (and sometimes a 0.5 appears). The sun is a G2V star. Stars of a given spectral class with large appended numbers are called "late," whereas "early," and hotter, stars have small numbers (for now irrelevant historical reasons). Thus the sun can be called an "early G dwarf." This classification scheme, which evolved over a number of years, is variously called the "MKK" (for Morgan, Keenan, and Kellman, 1943) or the "Yerkes" system (for the observatory where MKK did their work). An excellent short review of the development of this scheme is given by Jaschek and Jaschek (1987) in their Chapter 3.

Shown in the figure are the spectra of 16 dwarf stars spanning the classes O–M with some subclasses combined (e.g., O7V–B9V) because, at this resolution, intermediate spectra would not be distinguishable. (We shall, despite the combining, refer to a particular curve as if it represents only one class of star.) The acquisition and treatment of the data is discussed in Silva and Cornell (1992).[6] The wavelength coverage is from 351 to 893 nm at 11Å resolution (with 1 nm=10Å) with a binning interval of 5Å. This means that features of roughly 10Å or less are not distinguishable in the figure. The vertical separation between each spectrum has been designed for visual clarity and has nothing to do with intrinsic luminosity.

The ordering of the curves is that the star with the highest effective temperature starts with the topmost curve (O5V) in the left-hand panel. The spectra of successively cooler stars then proceeds downward. The right-hand panel continues the sequence from top to bottom. The first overall impression is that the hotter stars emit photons predominantly in the blue (and, most likely, the ultraviolet) parts of the spectrum, and the coolest stars

[6] The data for this figure were downloaded from the WWW. See the references.

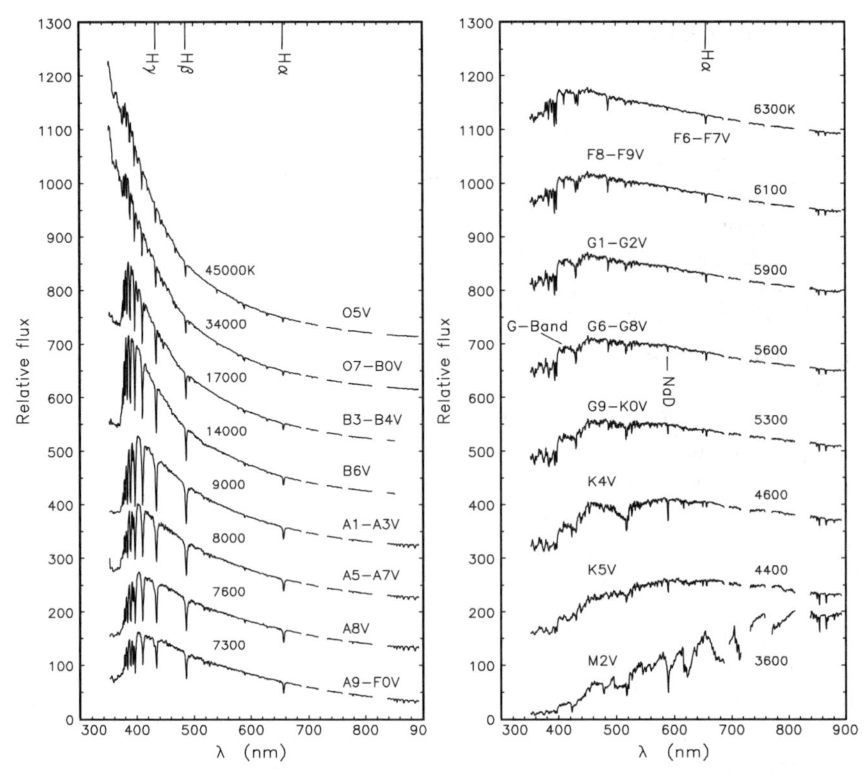

Fig. 4.8. Spectra of main sequence (luminosity class V) stars of spectral classes O–M derived from the work of Silva and Cornell (1992).

shine mostly in the red and near-infrared bands. (For orientation, Table 4.1, adapted from Jaschek and Jaschek, 1987, gives the names and rough wavelength bands for various parts of the electromagnetic spectrum.) Were these stars to shine as blackbodies—and note that they do so only approximately—the association of hot atmosphere with short wavelengths and cool with long wavelengths is obviously correct. The devil, however, is in the details. But devil aside, we also show in the figure effective temperatures for a single class or some average when classes are combined (quoted from Lang, 1991).

All the spectra show sharp absorption lines (the dips) and these are crucial in spectral classification. We have labeled some of these lines. Many others are there, but, in most cases, the resolution is too coarse to show them. The O– and B–stars in the sequence have strong lines—barely discernible here—due to HeIλ4471 and HeIIλ4541, where the numbers are the wavelengths in Å. The relative depth of these lines primarily determines the class. (We will have a bit more to say about spectral lines in the next section.) For the cooler stars in these two classes, the HeII lines diminish in strength because the radiation field and/or collisions are not vigorous enough to produce the

Table 4.1. Wavelength Regions

Name	Wavelength interval unit
Extreme ultraviolet	<1,000 Å
Ultraviolet	1,000–3,000 Å
Classical	3,000–4,900 Å
Visual	4,900–7,000 Å
Near infrared	7,000–10^5 Å
Far infrared	1–10^3 μm
Radio	0.1–10^4 cm

necessary first-ionized helium. Also present, and indicated in the figure, are the hydrogen Balmer lines H$\alpha\lambda$6562, H$\beta\lambda$4861, and H$\gamma\lambda$4330. These represent photons absorbed by the first excited state of hydrogen ($n = 2$), which is populated primarily by collisional processes. The early (i.e., hotter) A–stars show maximum strength in these lines.

As we enter the F- and G-stars, hydrogen lines weaken but metal lines appear. Among these are the H and K lines of CaII (at 3,968Å and 3,933Å), the NaDλ5889 line (indicated in the figure), and numerous iron lines. The calcium lines may be picked out by the "cliff" near 4,000Å while the depression known as the "G-band" (indicated) is primarily due to Fe.

The spectra of the cooler K-stars and class M-stars are dominated by metallic lines and molecular bands from, for example, TiO (which makes a decent white paint). These stars are obviously cool enough that such molecules can escape being torn apart by vigorous collisions or the radiation field. From our perspective, only the brave tread on ground such as that shown for the M2V star in the figure.

But observational astronomers are brave—especially since new tools for observing faint objects in the near-infrared have become available. In §2.2.2 we briefly discussed brown dwarfs, which, almost by definition, must be cooler than spectral class M dwarfs and we suspect (correctly) that there are intermediate objects. The next cooler spectral class after M, called the class "L" dwarfs, is characterized by the replacement of metal oxide bands (e.g., TiO in class M) by those of metallic hydrides and neutral alkali metals.[7] Descriptions of the classification of these very cool stars are given in Martín et al. (1999) and Kirkpatrick et al. (1999)—the 1999 date giving a clue to what new stuff this is. Subdivisions (thus far) are L0, L1 $\cdots$ L8 with a corresponding range of $T_{\rm eff}$ from about 2,200 down to 1,500 K or so. Over 100 of these objects have been discovered and they are a mix of (real) stars and brown dwarfs.

[7] It would have been nicer to follow M by N but class N had been preempted by the "carbon stars," which are late luminosity class giant stars with strong bands of carbon compounds but no metallic oxide bands.

But even cooler objects have been observed in the near infrared. At L7V–L8V there is evidence that weak features may be due to the emergence of methane (CH_4) in the spectra while the hydrides become less conspicuous. This introduces the spectral class T, which is the newest class proposed. Here, methane and H_2O absorption bands become progressively more conspicuous as temperature decreases. The subdivision sequence is from T0V to T8V where, at T8V, the effective temperature is around 900 K, and here also is where Gliese 229B lives, as discussed in §2.2.2. Recent (as of this writing) references to spectral class T objects and their classification are McLean et al. (2001), and Burgasser et al. (2002). By the time this text is printed, there will be much more to talk about. So keep your infrared-sensitive eyes out.

4.8 Line Profiles and the Curve of Growth

For the professional astronomer whose specialty is stellar atmospheres, the spectral lines in Fig. 4.8 tell a lot about the particular star. Among other things (such as the physical state of the atmosphere), elemental abundances may come through loud and clear—with, of course, a lot of work done beforehand.

This section will briefly explore some aspects of line formation and how, in a simple model, abundances can, in principle, be determined. We will first discuss the cross section for absorption of radiation by a classical charged oscillator.

Think of the oscillator as an electron in an excited state of an atom. The electron will decay to a lower energy level within some time, τ, determined by quantum mechanics.[8] As examples, the lifetime for the transition 2P–1S (Lyα) in hydrogen is $\tau = 1.6 \times 10^{-9}$ s. But, since the lifetime is not infinite, this means, by the uncertainty principle, that there is an uncertainty in the energy of the level given by $\Delta E \times \tau = \hbar$. Since we will deal in frequencies (as in $\Delta E = \hbar 2\pi\nu$), all the above implies an uncertainty in frequency of $\gamma = 1/\tau$. This "gamma" (Oh, no! Not another γ?!) is called the "damping constant."

4.8.1 The Lorentz Profile

Now back to the absorption of radiation by a classical oscillator. What we just discussed had to do with emission of radiation, but, by detailed balance (and see Ex. 4.9), emission and absorption are, in a sense, mirror images of one another and quantities such as γ will crop up. So, without further ado, the following gives the absorption cross section for radiation incident on a *stationary* atom in some state where ν_0 is the photon frequency necessary to

[8] If the electron is in a high quantum state, the lifetime can be estimated using classical arguments because of the Bohr correspondence principle. See, for example, Problem 14.21 in Jackson (1999, §16.8).

promote an electron to a particular higher state (see Jackson, 1999; §9–1 et seq. of Mihalas, 1978; or §4.2 of Rose, 1998, for derivations):

$$\sigma_a(\text{Lorentz}) = \frac{e^2}{mc} f \frac{(\gamma/4\pi)}{(\nu - \nu_0)^2 + (\gamma/4\pi)^2} . \tag{4.73}$$

In (4.73), γ is the sum of the damping constants for the two levels involved, and f is the *oscillator strength*. We sneak in the latter because it contains all the quantum mechanics that were not in the classical model. For the 2P–1S transition in hydrogen it is equal to 0.416. (A simplification has been made here by assuming the frequency ν is fairly close to ν_0.) The shape of (4.73) is a *Lorentz profile* or *Lorentzian* (named after H.A. Lorentz, who not only studied the fundamental properties of electromagnetism in pre-quantum days, but who also helped give us Lorentz length contraction). Looking ahead, a sample Lorentzian is shown in Fig. 4.9. You can check (4.73) to find that the half-width of the profile (in frequency) at half-maximum is $\gamma/4\pi$ (and measure it in Fig. 4.9 where it was arbitrarily set to unity). Incidentally, we shall see something very similar to (4.73) when we discuss resonant nuclear cross sections in Chapter 6.

In real life, however, atoms are not standing still while light bathes them—especially in stars.

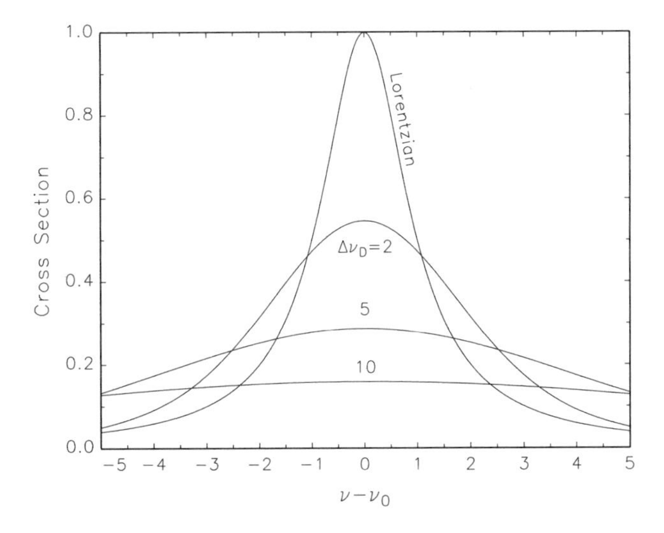

Fig. 4.9. Shown is the behavior of the absorption cross section from pure Lorentzian ($\Delta\nu_D = 0$, $T = 0$) to lines well-broadened by increasing temperature. The natural width is taken so that $\gamma/4\pi = 1$. See text for details.

4.8.2 Doppler Broadening

Because atoms are in motion, the arriving photons are Doppler shifted in frequency as seen by those atoms. This does not change the basic physics of (4.73), but if we consider an ensemble of moving atoms, then the total ensemble average absorption cross section need not look like a Lorentzian.

Assume that the gas is ideal so that we know the distribution of velocities that an individual photon may see; that is, the Maxwell–Boltzmann distribution discussed in §3.3. Equation 3.25 is the one we want after we realize that we must back off a step and put in some angular information. A factor of 2π comes from integration over the azimuthal angle, ϕ, (in spherical coordinates). We can always choose the z–axis to lie along a line connecting the incoming flux of photons and the gas (assumed to be at the origin) with the axis pointed in the direction of the incoming photons. In that case, all is well by symmetry and we do not have to remove the 2π. This leaves the co-latitude angle ϑ. This introduces the factor $\sin\vartheta\, d\vartheta$, which, when integrated over angle, gives a factor of two that must be removed. If we set $\mu = \cos\vartheta$, then you may easily verify that in velocity space (3.25) expands out to

$$\frac{dn(v,\mu)}{n} = (2\pi)^{-1/2}\left(\frac{m}{kT}\right)^{3/2} e^{-mv^2/2kT}\, v^2\, dv\, d\mu\ . \tag{4.74}$$

If the photon flux is all at frequency ν, then a particular atom sees a Doppler-shifted frequency of $\nu(1 - v\mu/c)$. Thus if, for example, the atom is headed toward the photons (at $\vartheta = \pi$ or $\mu = -1$), the atom sees the photon as being blue-shifted. Thus the denominator of (4.73) becomes

$$\frac{1}{[\nu(1 - v\mu/c) - \nu_0]^2 + (\gamma/4\pi)^2}\,.$$

The next step is to, in effect, undo the Doppler shift on ν and adjust ν_0. Note that

$$\begin{aligned}\nu\left(1 - \frac{v}{c}\mu\right) - \nu_0 &= \nu - \nu_0\left(1 + \frac{v}{c}\mu\right) - (\nu - \nu_0)\frac{v}{c}\mu \\ &\approx \nu - \nu_0\left(1 + \frac{v}{c}\mu\right)\end{aligned} \tag{4.75}$$

where the last term in the first equation is dropped because it is the product of two (presumably) small terms.

We now fold in the Doppler-shifted cross section with the velocity distribution to get an ensemble average for the cross section; that is,

$$\begin{aligned}\sigma_a(\nu,\nu_0,T) &= \frac{1}{(2\pi)^{1/2}}\left(\frac{m}{kT}\right)^{3/2}\frac{e^2}{mc}\,f\,\frac{\gamma}{4\pi}\,\times \\ &\times \int_0^\infty\int_{-1}^{+1}\frac{e^{-mv^2/2kT}v^2\,dv\,d\mu}{\left(\nu - \nu_0 - \nu_0 v\mu/c\right)^2 + (\gamma/4\pi)^2}\,.\end{aligned} \tag{4.76}$$

In mathematical terms, this is a convolution of a Gaussian (the Maxwellian) and a Lorentzian. Thus the shape of σ_a will turn out to be a hybrid of the two.

The next series of steps are tedious and not particularly enlightening. They consist of a series of substitutions plus the recognition that the denominator is of the form $x^2 + a^2$, which, when integrated over dx (i.e., ν), yields some arctangents. As a guide, we recommend §5.4 of Rose (1998), which you will probably need to read to do Ex. 4.12, where you are to derive the following expression for the cross section:

$$\sigma_a = \frac{e^2 f}{mc}\, \pi^{1/2}\, \frac{1}{\Delta\nu_D}\, H(a, \Delta\nu/\Delta\nu_D)\,. \tag{4.77}$$

The various new quantities here are

$$\Delta\nu_D = \nu_0 \left(\frac{2kT}{mc^2}\right)^{1/2} \tag{4.78}$$

which is the *Doppler width* that measures the half-width at half-maximum of the cross section if Doppler broadening dominates over the natural line width γ;

$$a = \frac{\gamma}{4\pi}\, \frac{1}{\Delta\nu_D} \tag{4.79}$$

which compares the two widths;

$$\Delta\nu = \nu - \nu_0 \tag{4.80}$$

as the new frequency variable; and, finally, the *Voigt function*,

$$H\left(a, u = \frac{\Delta\nu}{\Delta\nu_D}\right) = \frac{a}{\pi} \int_{-\infty}^{\infty} \frac{e^{-y^2}\, dy}{a^2 + (u-y)^2}\,. \tag{4.81}$$

So it all boils down to what the well-studied Voigt function looks like. Mihalas (1978, §9–2) gives a series expansion (which unfortunately contains even more integrals) for $a \ll 1$ corresponding to the usual case that Doppler broadening overwhelms the natural width γ. In making our figures we have used the FORTRAN programs (on CD) in Thompson (1997, §19.6) for $H(a,u)$, which he also calls the plasma dispersion function. If you wish to reproduce some of our results, be aware that Thompson does not give limits on how well the program works when u is large (with a still fairly small). But, for $u^2 \gg 1$, Mihalas gives $H(a,u) \approx a/u^2\pi^{1/2}$. This is the behavior of σ_a in the *wings* of the absorption line far from line center ($\nu = \nu_0$) in frequency.

Figure 4.9 shows the shape of the absorption cross section (sans multiplicative constants) for various values of $\Delta\nu_D$. The natural width is fixed by $\gamma/4\pi = 1$ so that all frequency units may be easily scaled. The curve labeled "Lorentzian" is self-explanatory and corresponds to $\Delta\nu_D = 0$. Note that the cross section near the center of the line decreases in height as $\Delta\nu_D$ increases (i.e., as temperature increases). What is happening is that more and more photons are being absorbed at frequencies further removed from line center and are not, in effect, available near line center.

4.8.3 Curve of Growth

If we have a hot macroscopic sample of absorbing atoms with radiation passing through the sample, we expect that the radiation will be attenuated around ν_0 as viewed by an observer peering at the sample in the direction of the source of radiation. The more absorbing atoms in the path, the more the attenuation; that is, the radiation acts as a probe of both the abundance of absorbing atoms and their temperature.

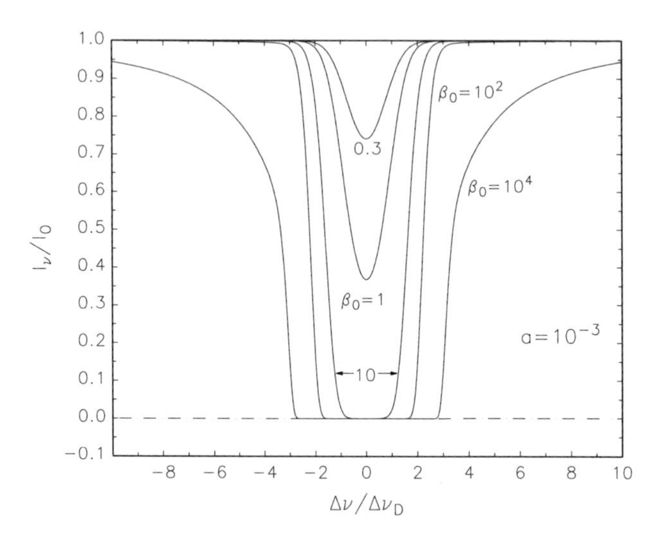

Fig. 4.10. The evolution of a spectral absorption line with increasing number density of absorbers is shown for sample values of β_0 (see text). The ratio $a = \gamma/4\pi\Delta\nu_D$ is fixed at 10^{-3}.

In a very simple model, let radiation of uniform intensity $I_\nu = I_0$ (for all ν) be incident on a slab of thickness Δr in which the number density of absorbers is n_a. If there were no absorbers, then an observer would see a flat spectrum with intensity I_0. Call this spectrum the *continuum.* We seek deviations from the continuum. Treating this as straight attenuation (no scattering, no angular problems, no emission—stimulated or not—etc.), (4.4) and the discussion preceding it states that

$$\frac{I_\nu}{I_0} = e^{-\tau\Delta r} = e^{-\kappa_\nu\rho\Delta r} = e^{-n_a\sigma_{a,\nu}\Delta r} = e^{-\beta_0} . \tag{4.82}$$

If we know the temperature of the sample (as a big "if") plus the microscopic physics and Δr (another big "if"), than the only variable is n_a. The resulting spectrum, I_ν/I_0, for a series of different $\beta_0 \equiv n_a \sigma_{a,\nu} \Delta r$ is shown in Fig. 4.10.

For small β_0, where few absorbers are in the line of sight, there is only a modest dip in I_ν near line center—as a "weak line." As the number of absorbers increases the line becomes deeper until the line becomes *saturated*; that is, I_ν flattens out to zero near line center. "Saturation" is apt because there are no photons left with frequencies near ν_0 to be absorbed by the time the beam leaves the slab. As β_0 increases further, all that is left are the wings of the line. If we had done a more realistic calculation, the line shapes would have been a little different but the overall effect would have been very similar (as in Fig. 10–1 of Mihalas, 1978).

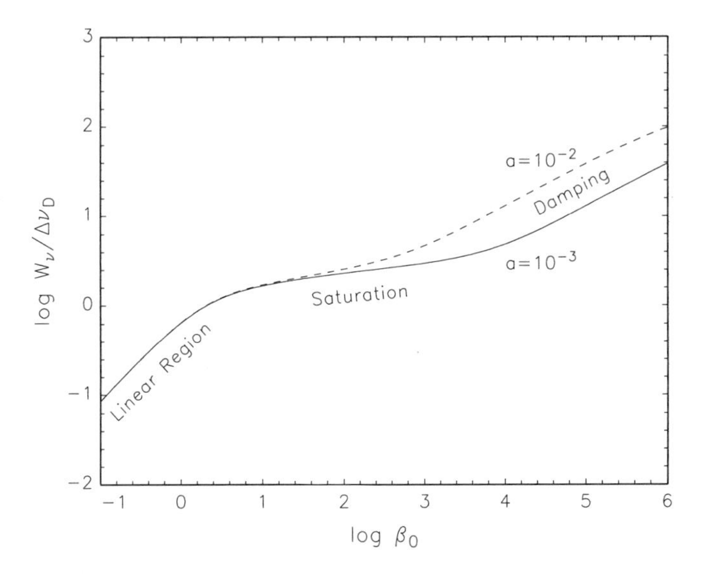

Fig. 4.11. The solid curve is the curve of growth corresponding to the line profiles of Fig. 4.10 with $a = 10^{-3}$ ($a \propto 1/\Delta\nu_D$). For $a = 10^{-2}$ (dashed line) the damping portion of the curve begins sooner because $\Delta\nu_D$ has been reduced.

Observing spectral lines with high resolution is not always possible. What is often done is to measure the *equivalent width*, defined as

$$W_\nu = \int_0^\infty \left(1 - \frac{I_\nu}{I_0}\right) d\nu = \int_0^\infty \left(1 - e^{-\beta_0}\right) d\nu \,. \tag{4.83}$$

Thus as β_0 decreases toward zero, so does W_ν. In this respect, W_ν is a measure of β_0 and hence n_a (all other things being equal). An example of what W_ν looks like is shown in Fig. 4.11 for $a = 10^{-3}$ ($a \propto 1/\Delta\nu_D$) and so corresponds to integrating the line profiles of Fig. 4.10 through a range of β_0. Such a curve is called a "curve of growth."

For small β_0 (weak line), $W_\nu \approx \beta_0$ so that on a log–log plot we have a straight line. (See Mihalas, 1978, §10–3 for more details.) Hence the "linear region" in the figure. In the saturated section, $W_\nu \sim \sqrt{\ln \beta_0}$, which is a mild dependence on β_0 and thus the curve rises very slowly. For large β_0, $W_\nu \sim \beta_0^{1/2}$ and we have a straight line again, but not as steep as the linear region.

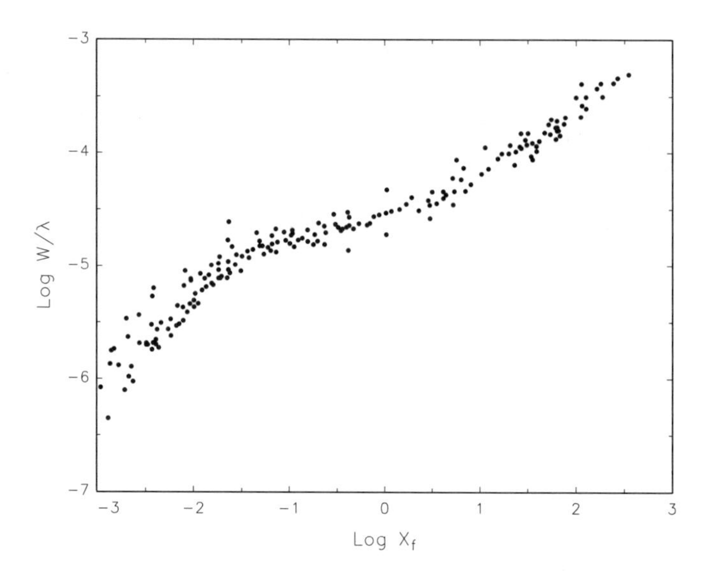

Fig. 4.12. The composite curve of growth for some 200 lines of iron (Fe I) and titanium (Ti I) in the sun. The quantity X_f is effectively our β_0 and W/λ is the equivalent width divided by the central wavelength of the line. See Wright (1948).

What we have shown thus far is, obviously, pure theory. Real lines are broadened by other processes such as turbulence (very difficult to model), collisional processes, etc. Each of these may overwhelm the effects of Doppler broadening. Yet, the curve of growth is a valuable diagnostic when applied to many lines using more sophisticated atmospheric calculations. An example is shown in Fig. 4.12 (originally due to Wright, 1948, and see Fig. 10–3 of Mihalas, 1978) for the sun. Here 75 lines of Fe I and 137 of Ti I have been observed to determine temperatures and abundances. We cannot go into the details but it sure has the general features of our Fig. 4.11 if you try to fit (by eye) a curve of growth to the observations.

4.9 Exercises

Exercise 4.1. We introduced the source function, $S(\tau)$, in §4.1 but, for the most part, had little to say about it. To rectify this partially, consider the emergent intensity at $\tau = 0$ (from 4.12)

$$I(0, \mu \geq 0) = \int_0^\infty e^{-t/\mu} \frac{S(t)}{\mu}\, dt$$

where we assume that the source function is a linear function of depth; that is,

$$S(\tau) = S_0 + S_1 \tau$$

with S_0 and S_1 constant. Integrate and show that $I(0, \mu \geq 0) = S_0 + \mu S_1$. (This is the Eddington–Barbier relation.) Obviously this may be written as $I(0, \mu \geq 0) = S(\tau = \mu)$. Thus the source function is directly a "source" for the emergent intensity. Interpret this more fully; for example, what does $\tau = \mu$ imply? (And, beware, the Eddington–Barbier relation is of limited usefulness. See Mihalas, 1978, for comments.)

Exercise 4.2. In deriving the diffusion approximation, we computed the flux in (4.17). It should be clear that the ratio $\left[\partial B(\tau)/\partial\tau\right]/B(\tau)$ is a relative measure of the anisotropy in the radiation field. Show that, aside from order unity factors, this ratio is approximately $\left(T_{\text{eff}}/T\right)^4$. Give a simple interpretation of this result.

Exercise 4.3. Suppose you wanted to "improve" upon the diffusion equation by adding a second-order term to the series expansion of the source function given by (4.14). Let that term be $(t-\tau)^2\left(\partial^2 B/\partial\tau^2\right)/2$. Show, by explicitly finding corresponding additional terms to the intensity (of Eqs. 4.15–4.16) for large τ and integrating (4.3) for the flux, that your efforts have been in vain; that is, show that the second-order term contributes nothing to the flux. You must go to third-order to find anything new!

Exercise 4.4. Consult the references and download opacities from the WWW and plot some (as in Figs. 4.2–4.5). Try a variety of mixes.

Exercise 4.5. Reproduce κ_{cond} of (4.72) by putting in all the numerical factors left out of the discussion leading up to that equation. In doing so it is worthwhile getting a numerical expression for the electron mean free path λ. You should find it is rather long compared to those associated with atomic processes (e.g., absorption, etc.). The point here is that conduction is efficient because the electrons can travel relatively long distances. It's the exclusion principle working again.

Exercise 4.6. Go back to Exercise 3.1 and find the half-ionization temperatures for the first and second stages of pure helium at densities 10^{-4}, 10^{-6}, and 10^{-8} g cm^{-3}. Plot these three results on Fig. 4.3 to verify that the bumps in that figure do correspond to helium ionization stages.

Exercise 4.7. Plot κ_{tot} on Fig. 4.3 for densities 10^4 and 10 g cm^{-3} using κ_{cond} of (4.72) and the parallel circuit result (4.69) to show how radically that figure can change when conduction is taken into account. (You can go to the Cox and Tabor, 1976, tables directly to get the numbers used in Fig. 4.3.)

Exercise 4.8. Using the approximations (4.62–4.65) for bound–free, free–free, and H^- opacities, try to duplicate the tabulated opacities shown in Fig. 4.2 for densities 10^{-4} through 10^{-8} g cm^{-3}. Recall that the range of (supposed) temperature validity for these approximations depends on the state of hydrogen ionization. And don't be too disappointed if the comparison doesn't look that good. These are only order of magnitude, or so, approximations.

Exercise 4.9. Consider an atom having only two levels, labeled j and i, with energies $E_j > E_i$. The energy difference is $E_j - E_i = h\nu_{ji}$, where ν_{ji} is the frequency of a photon transition from level j to i. For simplicity, the states are assumed to be nondegenerate in the sense that only one state in the atom has energy E_j or E_i. Thus g_j and g_i of (3.9) are both unity. In real life, the two levels have a small but nonzero width due to relative Doppler motions in a mixture of such atoms and to the intrinsic lifetime of the levels. Here we ignore these effects and assume that a line in emission is infinitely narrow and only photons of precise frequency ν_{ji} can be absorbed or emitted. What we discuss here are the Einstein coefficients that relate the rates at which photons are emitted from j by spontaneous emission or by "induced" ("stimulated") emission to absorption of photons on level i. Therefore define the Einstein coefficient A_{ji} such that $n_j A_{ji}$ is the rate at which states j in a mixture spontaneously decay to i where n_j is the number density of atoms in state j. Spontaneous emission, however, is not the only means of decay. The ambient radiation field can "induce" state j to decay due to interaction of photons of energy $h\nu_{ji}$ with j. The rate at which this occurs should be proportional to the field intensity I_ν at frequency ν_{ji}. Thus define B_{ji} so that the induced rate is $n_j B_{ji} I_\nu$. The last coefficient, B_{ij}, describes the absorption rate from i to j as $n_i B_{ij} I_\nu$ and it is, obviously, proportional to the intensity of the radiation field. In thermodynamic equilibrium the rates up and down must balance (as in "detailed balance"). Thus

$$n_j(A_{ji} + B_{ji} I_\nu) = n_i B_{ij} I_\nu$$

and now for the problem.

1. Assume thermodynamic equilibrium so that $I_\nu = B_\nu$, and we have the ideal gas Boltzmann result

$$\frac{n_j}{n_i} = \exp\left(-h\nu_{ji}/kT\right) .$$

Show that the Einstein coefficients are related by

$$B_{ij} = B_{ji} \quad \text{and} \quad A_{ji} = B_{ji}\frac{2h\nu_{ji}^3}{c^2} .$$

Thus induced emission takes place as long as we have spontaneous emission or absorption. Otherwise the laser would be impossible.

2. Show that

$$\frac{\text{rate of induced emission}}{\text{total rate of emission}} = \exp\left(-h\nu_{ji}/kT\right).$$

3. Compute the above ratio for the Lyman-α and 21 cm lines of hydrogen at a temperature of 10^4 K.

Exercise 4.10. The Eddington result for the run of temperature with optical depth was given by (4.44); i.e.,

$$T^4(\tau) = \frac{1}{2} T_{\text{eff}}^4 \left(1 + \frac{3}{2}\tau\right).$$

We now use this to examine what this implies for convection. Looking ahead to Chapter 5, convection takes place if $\nabla > \nabla_{\text{ad}}$ where

$$\nabla = \frac{d\ln T}{d\ln P} \quad \text{(Eq. 4.28)} \quad \text{and} \quad \nabla_{\text{ad}} = \frac{\Gamma_2 - 1}{\Gamma_2} \quad \text{(Eq. 3.94)}.$$

We wish to rephrase this as a condition on Γ_2 at large optical depths by finding what are the derivatives in ∇.

1. Show that

$$\frac{d\ln T}{d\tau} = \frac{3}{8 + 12\tau}.$$

2. After integrating (4.46) with constant opacity, combine that result with hydrostatic equilibrium of (4.45), and then use the definition of optical depth of (4.8), to show that

$$\frac{d\ln P}{d\tau} = \frac{1}{\tau}.$$

3. For large optical depths use the above to show that $\Gamma_2 < 4/3$ implies convection; i.e., the "magic" 4/3 strikes again.

Exercise 4.11. Show that $\partial B_\nu/\partial T$ in the definition of the Rosseland mean opacity (4.22) most heavily weights those photons with frequencies $\nu \approx kT/h$.

Exercise 4.12. Derive (4.77) for σ_a in terms of the Voigt function.

Exercise 4.13. Show that (4.39) for P_{rad} gives the usual result $P_{\text{rad}} = aT^4/3$ in LTE where $I = B(T)$.

Exercise 4.14. Use the `FORTRAN` code "csotest" on the CD-ROM to reproduce Fig. 4.2. The correspondence will not be exact because we have not told you what the mix of metals is.

4.10 References and Suggested Readings

Introductory Remarks and §4.1–§4.2: Radiative Transfer & The Diffusion Equation

We recommend the texts

- ▹ Mihalas, D. 1978, *Stellar Atmospheres*, 2d ed. (San Francisco: Freeman)
- ▹ Mihalas, D., & Mihalas, B.W. 1984, *Foundations of Radiative Hydrodynamics* (Oxford: Oxford University Press).

The emphasis of these two is different, but complementary, and both contain modern and practical material. Chapters 4–8 of

- ▹ Cox, J.P. 1968, *Principles of Stellar Structure*, in two volumes (New York: Gordon & Breach)

discusses stellar atmospheres more from the viewpoint of applications to stellar interiors than do the Mihalas references.

- ▹ Rybicki, G.B., & Lightman, A.P. 1979, *Radiative Processes in Astrophysics* (New York: Wiley & Sons)

and Chapters 4 and 5 of

- ▹ Rose, W.K. 1998, *Stellar Astrophysics* (Cambridge: Cambridge University Press)

also contain useful material. A little less intensive, but clear, is

- ▹ Böhm-Vitense, E. 1989, *Introduction to Stellar Astrophysics, Vol. 2, Stellar Atmospheres* (Cambridge: Cambridge University Press).

§4.3: A Simple Atmosphere

Mihalas (1978) discusses many simplified atmospheric calculations that we do not attempt. Various applications of the Eddington limit are discussed in

- ▹ Shapiro, S.L., & Teukolsky, S.A. 1983, *Black Holes, White Dwarfs, and Neutron Stars* (New York: Wiley Interscience).

The use of interpolation among atmospheres in making stellar models is reviewed in

- ▹ Kippenhahn, R., Weigert, A., & Hofmeister, E. 1967, MethCompPhys, 7, 53.

§4.4: Radiative Opacity Sources

The material in

- ▹ Clayton, D.D. 1968 *Principles of Stellar Evolution and Nucleosynthesis*, (New York: McGraw-Hill)

is presented from a physicist's point of view and we recommend it highly. Cox, J.P. 1968, *Principles of Stellar Structure*, bases his exposition primarily on the LANL method of calculating opacities. The LANL method (in an older but still good discussion) is given by

- ▹ Cox, A.N. 1965, in Chapter 3 of *Stellar Structure*, Eds. L.H. Aller & D.B. McLaughlin (Chicago: University of Chicago Press).

The reference to

▷ Landau, L.D. & Lifshitz, E.M. 1971, *Classical Theory of Fields* (Oxford: Pergamon Press)

can be found, as with other volumes in this classic series, in more recent editions. Also see (the now classic)

▷ Jackson, J.D. 1999, *Classical Electrodynamics*, 3rd ed. (New York: John wiley & Sons).

The monograph by

▷ Spitzer, L. 1962, *Physics of Fully Ionized Gases*, 2nd ed. (New York: Interscience)

contains much of interest for the astrophysicist. Another work of his is

▷ Spitzer, L. Jr. 1978, *Physical Processes in the Interstellar Medium* (New York: Wiley & Sons).

The text by

▷ Schwarzschild, M. 1958, *Structure and Evolution of the Stars* (Princeton: Princeton University Press)

is counted as the first modern work describing how stars evolve. It is now out of date but still worth perusing.

§4.5: Heat Transfer by Conduction

Several undergraduate solid-state (now "condensed matter") texts give the basic material on thermal conduction by electrons. The text by

▷ Kittel, C. 1968, *Introduction to Solid State Physics* (New York: Wiley & Sons)

(or later editions) is particularly clear.

Conductive opacities are discussed in

▷ Hubbard, W.B., & Lampe, M. 1969, ApJS, 18, 297

▷ Lamb, D.Q., & Van Horn, H.M. 1975, ApJ, 200, 306

▷ Itoh, N., Mitake, S., Iyetomi, H., & Ichimaru, S. 1983, ApJ, 273, 774

▷ Itoh, N., Kahyama, Y. Matsumoto, N., & Seki, M. 1984, ApJ, 285, 758.

The reference to Spitzer (1962) is given above.

§4.6: Tabulated Opacities

A relatively simple fit to opacities is given by

▷ Stellingwerf, R.F. 1975, ApJ, 195, 441

with a footnote correction in

▷ Ibid. 1975, ApJ, 199, 705.

The range of composition is somewhat limited to $0.6 < X < 0.8$, $0.2 < Y < 0.4$, and $0.001 < Z < 0.02$. To implement his prescription, however, you must supply the electron pressure. See also

▷ Iben, I. Jr. 1975, ApJ, 196, 525.

Extensive tabulations of radiative opacities from LANL may be found in

▷ Cox, A.N., & Stewart, J.N. 1970, ApJS, 19, pp. 243, 261

▷ Cox, A.N., & Tabor, J.E. 1976, ApJS, 31, 271

▷ Weiss, A., Keady, J.J., & Magee, N.H. Jr. 1990, ADNDT, 45, 209.

For seminal papers of the Opacity Project (OP), see

▷ Mihalas, D., Hummer, D.G., Mihalas, B.W., & Däppen, W. 1990, ApJ, 350, 300

▷ Hummer, D.G., & Mihalas, D. 1988, ApJ, 331, 794

▷ Seaton, M.J. 1987, JPhysB, 20, 6363

▷ Seaton, M.J., Yan, Y., Mihalas, D., & Pradhan, A.K. 1994, MNRAS, 266, 805

and

▷ Seaton, M.J. 1995, ed. *The Opacity Project, Vol. 1* (Bristol: Institute of Physics Publishing).

Sample papers from LLNL for the OPAL code are

▷ Rogers, F.J., & Iglesias, C.A. 1992, ApJS, 79, 507

▷ Ibid. 1993, ApJ, 401, 361, & ApJ, 412, 712

▷ Ibid. 1994, Science, 263, 50.

The latest paper, which contains comparisons to OP and LANL, is

▷ Iglesias, C.A., & Rogers, F.J. 1996, ApJ, 464, 943,

and, for information about the corresponding equation of state, see

▷ Rogers, R.F., Swenson, F., & Iglesias, C.A. 1996, ApJ, 456, 902.

The penultimate paper also makes comparisons to the opacities of

▷ Alexander, D.R., & Ferguson, J.W. 1994, ApJ, 437, 879

which we have not discussed.

The following are the WWW addresses for LLNL (i.e., OPAL) and OP as of February 1999. They will probably change at some time:

http://www-phys.llnl.gov/V_Div/OPAL/
http://vizier.u-strasbg.fr/OP.html

We shall have other occasions to refer to the classic article by

▷ Hayashi, C., Hōshi, R., & Sugimoto, D. 1962, PTPJS, Vol. 22.

It is now outdated by modern standards but contains a particularly clear development of the ingredients of stellar structure.

§4.7: Some Observed Spectra

For a comprehensive review of how spectra are used to classify stars, we recommend

▷ Jaschek C., & Jaschek, M. 1987, *The Classification of Stars*, (Cambridge: Cambridge University Press).

The MKK system is described in

▷ Morgan, W.W., Keenan, P.C., & Kellman, E. 1943, *An Atlas of Stellar Spectra with an Outline of Spectral Classification* (Chicago: University of Chicago Press).

Figure 4.8 derives from the work of

▷ Silva, D.R., & Cornell, M.E. 1992, ApJS, 81,865

using data downloaded from

http://zebu.uoregon.edu/spectra.html

which, if still active, should be looked into because data are also available for other luminosity classes. The effective temperatures shown in Fig. 4.8 are from

▷ Lang, K.R. 1991, *Astrophysical Data: Planets and Stars* (Berlin: Springer-Verlag).

Spectral class L dwarf stars and brown dwarfs are discussed in

▷ Martín, E.L., Delfosse, X., Basri, G., Goldman, B., Forveille, T., Zapatero, O., & Maria, R. 1999, AJ, 118, 2466

and

▷ Kirkpatrick, J.D., et al. 1999, ApJ, 519, 802

(the latter article having ten authors). Going yet further into the class T objects, we recommend

▷ McLean, I.S., Prato, L., Sungsoo, S.K., Wilcox, M.K., Kirkpatrick, J.D., & Burgasser, A. 2001, ApJ, 561, L115

and

▷ Burgasser, A.J., et al. 2002, ApJ, 564, 421.

§4.8: Line Profiles and the Curve of Growth

For those of you taking graduate courses,

▷ Jackson, J.D. 1999, *Classical Electrodynamics*, 3rd ed. (New York: John Wiley & Sons)

is the place to go for reading about the interaction of radiation with matter. The classic text to consult is Mihalas (1978), while

▷ Rose, W.K. 1998, *Stellar Astrophysics* (Cambridge: Cambridge University Press)

fills in some derivations.

▷ Thompson, W.J. 1997, *Atlas for Computing Mathematical Functions* (New York: Wiley-Interscience)

contains many `FORTRAN 90` and `Mathematica` programs in the text and on a compact disk. We used his program to compute the Voigt function, $H(a, u)$. Watch out, however. The program gives nonsensical results for very large u.

Figure 4.12 is based on

▷ Wright, K. 1948, *Publications of the Dominion Astrophysical Observatory, Victoria*, 8, 1.

This is a classic paper that was almost ready to be published in 1940 before WWII intervened. It is worth reading to see, among other things, how difficult pre-computer astronomy was compared to how observations and analyses are done today.

5 Heat Transfer by Convection

"Double, double toil and trouble;
Fire burn and cauldron bubble."
— W. Shakespeare (Macbeth)
That about sums it up.

The major portion of this chapter will be devoted to a discussion of the "mixing length theory," or "MLT," of convective heat transport in stars. Although this theory has many faults, it has served as a useful phenomenological model for a description of stellar convection for more than 40 years and most numerical simulations of stellar evolution use it in one guise or another. Near the end of the chapter we shall discuss alternatives to the MLT and why a realistic description of convection is so difficult.

Our discussion of the MLT will partly parallel that of Cox (1968), where details of the usual derivation of the MLT are the most completely laid out in the textbook literature. (We also recommend the paper by Gough and Weiss, 1976, which is a still-not-outdated review of calibrations of the MLT.) The first edition of this text (Hansen and Kawaler, 1994) went to considerable effort to explore the explicit and implicit assumptions behind the MLT—a task not attempted in standard texts. We still like that approach but many of our colleagues told us that it just didn't work in the classroom (and it made the chapter too long). OK, friends and colleagues, we gave in—mostly because we agreed with you. For those of you who would like to see the "complete" treatment, please consult the first edition.

5.1 The Mixing Length Theory

The mixing length theory was originally formulated in its "stellar" form by Biermann (1951), Vitense (1953), and Böhm–Vitense (1958) based on earlier 1925 work of Prandtl (see Prandtl, 1952). Since then, it has been elaborated on and modified in many ways and one should no longer call it just the MLT without citing exactly which version is being referred to. A "classic" derivation of one version of the theory may be found in Cox (1968, Chap. 14).

The general idea behind the theory is to imagine that the stellar fluid is composed of readily identifiable "eddies," "parcels," or "elements" (or, in more colloquial terms, "bubbles" and "blobs"), that can move from regions of high heat content to regions of lower heat content, or conversely; that is, they are capable of transporting or convecting heat through the fluid. These parcels arise from unspecified instabilities in the fluid but have properties not drastically different from their surroundings. If conditions are ripe, then buoyancy effects cause the parcels to, say, rise in the star through some characteristic distance ℓ, the *mixing length*, before they lose their identity as separate parcels and break up and merge with the surrounding fluid. As they rise, they maintain pressure equilibrium with their surroundings. Since these particular parcels start their rise in an environment having a higher heat content (higher temperature) than where they break up, heat is thereby transported from the starting position up to the level at the additional height ℓ. To complicate matters, the parcel may radiatively release heat to its surroundings as it rises. At the same time, cool parcels at a higher level sink a distance ℓ, and they too break up. The net effect is heat transport directed outward in the star. The rate of transfer is established by the parcel formation rate, velocity (w) of rise, ℓ, the heat content of the star as a function of depth, and by how radiatively "leaky" the parcels are as they rise.

This sounds relatively simple and, in fact, it is—in the context of MLT. We shall also see that most formulations of the theory have a major virtue for computation: all that matters is that temperature, density, and other stellar quantities be known at a single radius of interest. If so, then a convective heat flux may be computed at that point. The MLT is thus a *local* theory.

The sequence we shall follow in discussing the MLT is first to derive the criterion for buoyancy, and then to estimate the heat leakage from a parcel. This will give us the equations of motion. Finally, we shall find expressions for the convective flux in the limit of "efficient" convection and discuss how they are used.

5.1.1 Criteria for Convection

First of all, certain general assumptions are made that should be explicitly set forth. Besides neglecting magnetic fields, rotation, and the like, we assume the following (with comments).

1. A readily identifiable parcel has a characteristic dimension of the same order of size as the mixing length ℓ.
2. The mixing length is much shorter than any scale length associated with the structure of the star. Examples of such lengths are the pressure scale height, λ_P of (3.1), and similar scale heights for temperature and density.
3. The parcel always has the same internal pressure as that of its surroundings. This means that however the convective processes work, the time scales associated with them are always long enough that pressure equilibrium is maintained. Thus, for example, if v_s is the local sound speed in the parcel, then the sound traversal time across the parcel, ℓ/v_s, is short compared to, say, the ascent or descent time of the parcel through the distance ℓ.
4. Acoustic phenomena may be ignored altogether, as may shocks, etc.
5. Temperatures and densities within and outside a parcel differ by only a small amount.

The combination of these assumptions constitutes the "Boussinesq" approximation. What it implies is that the fluid is *almost* incompressible and that density variations (which may give rise to buoyancy effects) and temperature variations in the fluid are very small. The Boussinesq approximation usually works very well in the laboratory, where scale heights are large compared to container sizes (which roughly set the maximum size of a convective cell). In its application to stars, however, we shall see that the mixing length must be near the size of λ_P or one of the other scale heights for reasonable results to be obtained. Thus, in practice, the MLT will turn out to violate one of its internal assumptions. Furthermore, it is unfortunate that laboratory-derived constraints on the MLT are essentially nonexistent because of the following. The dimensionless Rayleigh number (see any text in fluid dynamics and Ex. 5.4) associated with laboratory fluids is usually less than 10^{11} but stellar convection is characterized by high values, 10^{20}—give or take a few orders of magnitude. The same situation applies to the Prandtl number where, in stars, it is around 10^{-9}, but in the lab it is of order unity (and see Ex. 5.3). Note also, in passing, the troublesome consequences of (1) in the above: how can the parcel get very far if its dimensions are of the same order as the distance it travels (ℓ)?

With the above alerts in mind, consider a plane parallel fluid under the influence of gravity, where z measures the height up through the fluid. Inside a typical parcel created by some unspecified process, denote the interior temperature, pressure, and density by T', P, and ρ', respectively. Outside the parcel, the corresponding quantities are denoted by T, P, and ρ. Note that the pressures inside and out are the same by virtue of assumption (3). Suppose that $T' > T$ (but not by much) so that the parcel is hotter than its surroundings. Normally this implies that $\rho' < \rho$ because of the interior versus exterior pressure equilibration. If the volume of the parcel is $V \sim \ell^3$, then Archimedes' principle states that the parcel will experience a net upward buoyancy force of

$$\rho V g - \rho' V g \tag{5.1}$$

where g is the local gravity. Note that we have not specified exactly what the volume of the parcel is in terms of ℓ. It could be spherical ($4\pi[\ell/2]^3/3$), a cube (ℓ^3), or what have you. These fine distinctions involving constants of order unity give rise to some of the variants in mixing length theory and we shall ignore them. In any case, the parcel now commences to rise.

We must eventually determine what is the mean velocity of the parcel as it rises through the mixing length distance, ℓ, and what its temperature is compared to the ambient temperature when it merges into the surrounding fluid. The latter comparison will tell us how much energy the parcel will release when it loses its identity. For all this we shall first need information about temperature gradients.

We denote β to be the *negative* of the ambient temperature gradient

$$\beta = -\frac{dT}{dz} \tag{5.2}$$

where (almost always) $\beta > 0$. This gradient is assumed to be known despite the fact that heat transported by rising and descending parcels may very well establish just what that gradient is. We can relate β to other known quantities by observing that

$$\frac{dT}{dz} = \frac{dT}{dP}\frac{dP}{dz} = T\frac{d\ln T}{d\ln P}\frac{d\ln P}{dz} = -\frac{T}{\lambda_P}\nabla = -\beta \tag{5.3}$$

where ∇ is the "actual del" introduced in the preceding chapter as (4.28). The pressure scale height in the above may be recast in terms of the local sound speed with the aid of (1.38):

$$\lambda_P = -\left(\frac{d\ln P}{dz}\right)^{-1} = \frac{P}{g\rho} = \frac{v_s^2}{g\Gamma_1} \tag{5.4}$$

where Γ_1 is the adiabatic exponent defined by (3.93). Thus

$$\beta = \frac{T}{\lambda_P}\nabla = \frac{g\Gamma_1 T}{v_s^2}\nabla\ . \tag{5.5}$$

To describe how the temperature inside the parcel varies as the parcel rises, first write

$$\frac{dT'}{dz} = T'\frac{d\ln T'}{d\ln P'}\frac{d\ln P'}{dz}\ .$$

If we assume, as a start, that the rising parcel exchanges no heat with its surroundings, then the term $d\ln T'/d\ln P'$ must describe adiabatic variations of temperature with pressure. This is the thermodynamic derivative $\nabla_{\rm ad} = (d\ln T/d\ln P)_{\rm ad}$ introduced earlier (as in 3.94 and 3.96). Because all fluctuations are assumed to be small, it is appropriate to replace the lone factor of T' by T in the right-hand side of the above. (We can't do the same with the temperature gradients because they drive the motions.) In addition, we replace P' with P, because of pressure equilibration, so that the last factor may be turned into a pressure scale height. Finally we may append an "ad" subscript to dT'/dz because of the adiabaticity assumption and write

$$\left(\frac{dT'}{dz}\right)_{\rm ad} = -\frac{T}{\lambda_P}\nabla_{\rm ad} = -\beta_{\rm ad}\ . \tag{5.6}$$

Thus,

$$\beta - \beta_{\rm ad} = \frac{T}{\lambda_P}\left(\nabla - \nabla_{\rm ad}\right)\ . \tag{5.7}$$

The question is now whether the parcel, once having commenced to rise adiabatically, will continue to rise. It may well be that as the parcel rises to greater heights and its internal pressure drops, its interior temperature may also have decreased adiabatically to a level where the parcel is cooler than its surroundings. In that case it has negative buoyancy and it tends to sink back down before traversing a mixing length. It is the other possibility, in which the parcel's temperature continues to exceed that of the surroundings, that is of major interest. Here, the fluid is said to be *convectively unstable* and the perturbation that causes the parcel to rise takes place in an environment that encourages further rising until a mixing length is traversed. This latter condition may be expressed as follows.

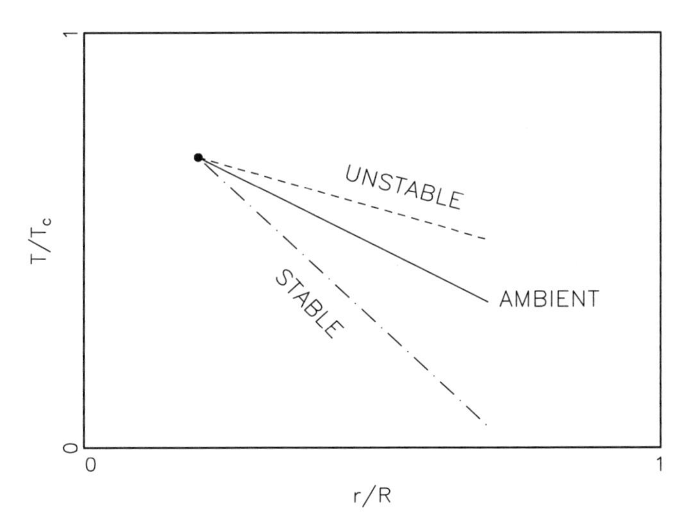

Fig. 5.1. A schematic run of temperature (normalized to some common temperature T_c) versus radius for: (a) the ambient medium (solid line); (b) a convectively unstable parcel (dashed line) with $(dT'/dr)_{\rm ad} > (dT/dr)$; (c) a stable parcel (dashed-dotted line) with $(dT'/dr)_{\rm ad} < (dT/dr)$.

First observe that both dT'/dz and dT/dz are assumed to be *negative.* Thus the condition that T' decreases *more slowly* than T with height is expressed as

$$\left(\frac{dT'}{dz}\right)_{\rm ad} > \left(\frac{dT}{dz}\right) \quad \text{(convectively unstable)}\,. \tag{5.8}$$

This convectively unstable situation is illustrated in Fig. 5.1. Another way to express this is to use (5.2), (5.5), and (5.6) to write

$$\beta > \beta_{\rm ad} \quad \text{(convectively unstable), or} \tag{5.9}$$

$$\nabla > \nabla_{\rm ad} \quad \text{(convectively unstable)}\,. \tag{5.10}$$

The three conditions (5.8–5.10) are equivalent to and, in the stellar context, are often called the *Schwarzschild criteria* (K. Schwarzschild, 1906). These are *local* criteria and thus require information from only the one height (or radius) of interest in the star.

The criteria (5.8–5.10) are also equivalent to the statement that if entropy decreases outward at some point ($dS/dr < 0$), then the fluid is convectively unstable (Cox, 1968, §13.4, or our §7.3.3). Put another way, convection does not take place in hydrostatic stars where the entropy increases outward. It will turn out that in regions where convection is very efficient, ∇ is only very slightly greater than $\nabla_{\rm ad}$. In such regions the entropy is very nearly constant with height. (Ex. 5.1 asks you to examine the role of entropy.)

We finally begin to see what role the "dels" play in convection. If, by some mischance, adiabatic perturbations arise where the local run of temperature versus pressure is such that the local value of ∇ is greater than the local thermodynamic equivalent, $\nabla_{\rm ad}$, then convection should be present. However, if this is really so, then convection must change the thermal structure and, hence, ∇, and so on.

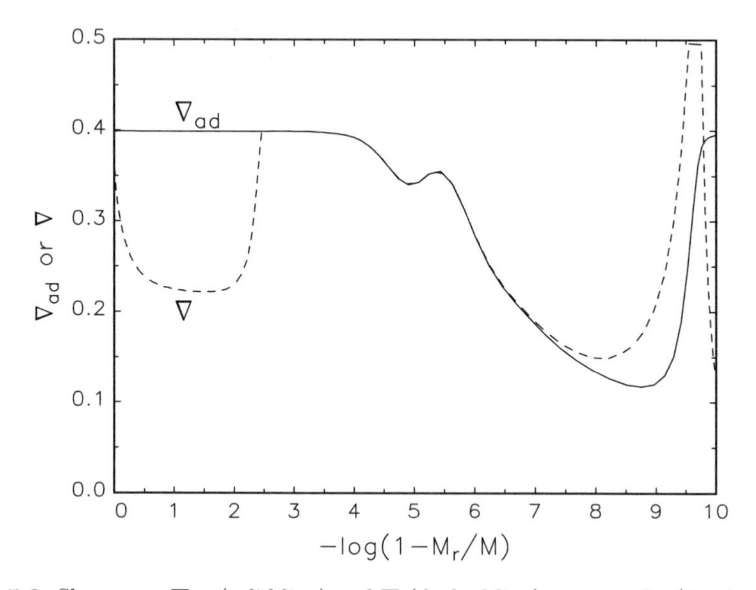

Fig. 5.2. Shown are $\nabla_{\rm ad}$ (solid line) and ∇ (dashed line) versus $-\log(1-\mathcal{M}_r/\mathcal{M})$ for a model ZAMS sun. See also Fig. 3.11.

As an example of how ∇ and $\nabla_{\rm ad}$ behave in a typical star, Fig. 5.2 shows their run in a ZAMS sun. The abscissa is $-\log(1-\mathcal{M}_r/\mathcal{M})$ and it emphasizes the outer layers (as in Fig. 3.11, which showed only $\nabla_{\rm ad}$ for the same model). It is apparent that the model is radiative (i.e., transports heat solely by radiative transfer) from the center, at $-\log(1-\mathcal{M}_r/\mathcal{M}) = 0$, to $\mathcal{M}_r/\mathcal{M} \approx 0.997$ because $\nabla < \nabla_{\rm ad}$ throughout that region. (The latter mass point corresponds to roughly $r/R = 0.8$.) It is impossible to detect in this figure but past that point (to $1 - \mathcal{M}_r/\mathcal{M} \approx 10^{-7}$) ∇ is *very* slightly larger than $\nabla_{\rm ad}$; that is, that region is convective and, as it turns out, efficiently so. The actual gradient ∇ then clearly exceeds $\nabla_{\rm ad}$ until very nearly the model surface after which radiative transfer takes over again. (A different, and perhaps clearer, view of this figure will be shown as Fig. 5.3.) The version of the MLT used to construct this model is almost what we shall describe in the next few pages, but it also included nonadiabatic (heat leakage) effects, which will eventually be ignored in this chapter. The ZAMS code, found on the CD–ROM at the end of this text, was used to make this model and that code includes heat leakage.

5.1.2 Radiative Leakage

Real life is never adiabatic. Because our parcel is either cooler or hotter than its surroundings, heat must be exchanged between the two. Thus consider the energy equation

$$\frac{dQ'}{dt} = -\boldsymbol{\nabla} \bullet \boldsymbol{\mathcal{F}}_{\rm rad} \tag{5.11}$$

where Q' is the heat content per unit volume in the parcel and $\boldsymbol{\mathcal{F}}_{\rm rad}$ is the radiant flux from the parcel out to the ambient medium. (Don't confuse the gradient or divergence operators used here and the various "dels.") Now recall that the Boussinesq conditions of no acoustic phenomena and small fluctuations imply that the only time density perturbations are to be taken into account is when they are coupled to gravity to cause buoyancy. Thus in considering heat balance, $P\,dV$ work terms are neglected so that (5.11) reduces to

$$\left(\frac{\partial T'}{\partial t}\right) = -\frac{1}{\rho c_P} \boldsymbol{\nabla} \bullet \boldsymbol{\mathcal{F}}_{\rm rad} \tag{5.12}$$

where c_P is in erg g^{-1} K^{-1}.

For the radiative flux we again choose a diffusion approximation, assuming the medium to be optically thick, and write

$$\boldsymbol{\mathcal{F}}_{\rm rad} = -\mathcal{K}\, \boldsymbol{\nabla} T \tag{5.13}$$

where the diffusion constant $\mathcal{K}$ is assumed constant over the mixing length ℓ. The gradient term on the righthand side of this equation and subsequent righthand sides of the energy equation will be taken to be some linear combination of T and T' (as will be done shortly). We now introduce the opacity by using a relation identical to that used in the calculation of conductive opacities (see 4.67); namely,

$$\mathcal{K} = \frac{4acT^3}{3\kappa\rho} \,. \tag{5.14}$$

Equation (5.12) then becomes

$$\left(\frac{\partial T'}{\partial t}\right) = \frac{\mathcal{K}}{\rho c_P}\nabla^2 T = \frac{4acT^3}{3\kappa\rho^2 c_P}\nabla^2 T\,. \tag{5.15}$$

The ratio $\mathcal{K}/(\rho c_P)$ is the thermal diffusivity (or conductivity), ν_T, with

$$\nu_T = \frac{4acT^3}{3\kappa\rho^2 c_P} \tag{5.16}$$

and it has the units of $\mathrm{cm^2\ s^{-1}}$. For the characteristic length in these units choose ℓ itself. Therefore a characteristic radiative cooling time associated with radiation from, or into, the parcel is ℓ^2/ν_T. The energy equation then becomes

$$\left(\frac{\partial T'}{\partial t}\right) = \nu_T\,\nabla^2 T \tag{5.17}$$

and, in this form, it is sometimes known as Fourier's equation (see Landau and Lifshitz, 1959, §50).

Since we have no way to determine the precise structure of our parcel, we should model $\nabla^2 T$ as simply as we can. The obvious way is to replace it by $(T - T')/\ell^2$. The order of the temperatures is correct because, for example, if $T' > T$ then $\nabla^2 T$ and $(\partial T'/\partial t)$ are both negative as they should be: the parcel loses heat because it is hotter than its surroundings.

At this point we have to amend the energy equation somewhat because, as it stands, it is in Eulerian form and it describes the change in heat content at a fixed position. To follow what happens in the parcel as it moves, convert to Lagrangian coordinates by the following well-known transformation, which is discussed in any text in hydrodynamics:

$$\frac{DT'}{Dt} = \left(\frac{\partial T'}{\partial t}\right) + \mathbf{w} \bullet \boldsymbol{\nabla} T' \tag{5.18}$$

where D denotes the Lagrangian (or Stokes) operator and $\mathbf{w}$ is the (vector) parcel velocity. Since only vertical movement is contemplated here, $\mathbf{w} \bullet \boldsymbol{\nabla} T' = w\,(\partial T'/\partial z)$. The term $\partial T'/\partial t$ in (5.18) takes care of the instantaneous heat loss so that the remaining advective term, $w\,(\partial T'/\partial z)$, should describe how T' behaves without such loses. Thus we identify the derivative in the last term as

$$\left(\frac{\partial T'}{\partial z}\right) = \left(\frac{dT'}{dz}\right)_{\mathrm{ad}} = -\beta_{\mathrm{ad}}$$

from (5.6), and (5.18) becomes

$$\frac{DT'}{Dt} = \frac{\nu_T}{\ell^2}(T - T') - \beta_{\mathrm{ad}} w \tag{5.19}$$

after replacing the Laplacian in (5.17) by its numerical difference analogue.

Now compare T to T' by introducing

$$\Delta T = T' - T \tag{5.20}$$

which depends on z and t. By combining (5.19) and (5.20) and realizing that the vertical rate of change of the ambient temperature as seen by the moving parcel is

$$\frac{DT}{Dt} = \frac{dT}{dz}\frac{dz}{dt} = -\beta w$$

we arrive at

$$\frac{D\Delta T}{Dt} = (\beta - \beta_{\text{ad}})\, w - \frac{\nu_T}{\ell^2}\Delta T\,. \tag{5.21}$$

This equation describes the time-dependent temperature contrast between the parcel and its immediate surroundings as the parcel moves.

5.1.3 The Equation of Motion

Can we now say something about w? Because buoyancy forces are responsible for the motion of the parcel,

$$\frac{dw}{dt} = \frac{(\rho - \rho')}{\rho}\, g \tag{5.22}$$

in a first approximation to the acceleration implicit in (5.1). Note that this expression ignores any viscous effects that might impede the flow of the parcel (but see Hansen and Kawaler, 1994, and Ex. 5.3).

The small relative density contrast $(\rho - \rho')/\rho$ is related to that in temperature through the coefficient of thermal expansion, $-\mathcal{Q}$, taken at constant pressure; that is,

$$-\mathcal{Q} = \left(\frac{d\,\ln\rho}{d\,\ln T}\right)_P\,.$$

(The constraint is required by pressure equilibrium.) If the density is written as a function of temperature, pressure, and composition (denoted by μ) with $\rho = \rho(T, P, \mu)$, then $-\mathcal{Q}$ is

$$-\mathcal{Q} = \left(\frac{\partial\ln\rho}{\partial\ln T}\right)_{\mu,P} + \left(\frac{\partial\ln\rho}{\partial\ln\mu}\right)_{P,T}\left(\frac{\partial\ln\mu}{\partial\ln T}\right)_P\,. \tag{5.23}$$

It follows that

$$\frac{(\rho - \rho')}{\rho} = -\mathcal{Q}\frac{(T - T')}{T} = \frac{\mathcal{Q}}{T}\Delta T \tag{5.24}$$

is the desired relation.

The coefficient $\mathcal{Q}$ is generally clumsy to compute because of the composition dependence, which should be kept in to allow for composition changes either within the parcel (such as ionization) or in the surroundings. If composition changes may be neglected, then it is easy to show that

$$\mathcal{Q} = \frac{\chi_T}{\chi_\rho} \tag{5.25}$$

where χ_T and χ_ρ were discussed in §3.7.1. For a mixture of ideal gas and radiation, $\mathcal{Q} = (4-3\beta)/\beta$, where β is the ratio of gas to total pressure (as in 3.106 and 3.107).

The equation of motion is then

$$\frac{dw}{dt} = \frac{\mathcal{Q}g}{T}\Delta T \tag{5.26}$$

which is to be considered along with the energy equation (5.21).

5.1.4 Convective Efficiencies and Time Scales

Consider the differential equations for ΔT and w given as (5.21) and (5.26). If we suppose that all the coefficients in those equations are constant not only over the distance ℓ but for all time, then it is an easy matter to solve for ΔT and w as functions of time. The solutions will be of the form ΔT or $w \propto \exp(\sigma t)$, where σ is a complex angular frequency. We shall find that $1/\Re(\sigma)$ defines a characteristic time scale for the growth (decay) of convection if $\nabla > \nabla_{\text{ad}}$ ("$\Re$" means "real part of").

To carry out this analysis, combine the two differential equations into a single second-order equation for either w or ΔT, substitute $\exp(\sigma t)$, and find that σ must satisfy the characteristic equation

$$\sigma^2 + \sigma\frac{\nu_T}{\ell^2} - \frac{\mathcal{Q}g}{T}(\beta - \beta_{\text{ad}}) = 0 \, . \tag{5.27}$$

The last term in (5.27) is important not only here but in the theory of variable stars (see Chap. 8), planetary atmospheres, and other fields, and it deserves a name. Define the *Brunt-Väisälä frequency*, N, by

$$N^2 = -\frac{\mathcal{Q}}{T}g(\beta - \beta_{\text{ad}}) \tag{5.28}$$

or, if composition gradients are not present (see 5.25),

$$N^2 = -\frac{\chi_T}{\chi_\rho}(\nabla - \nabla_{\text{ad}})\, g\lambda_P \, . \tag{5.29}$$

If there is a gradient in the mean molecular weight because the composition of nuclear species changes with height, then the parenthesis should contain an additional term

$$+\frac{\chi_\mu}{\chi_T}\frac{d\ln\mu}{d\ln P} \quad \text{with} \quad \chi_\mu = \left(\frac{\partial \ln P}{\partial \ln \mu}\right)_{\rho,T} \, .$$

Note that the effects of ionization on μ are already automatically accounted for through the equation of state and (5.29) should not be modified if μ changes solely due to those effects. (See, Cox 1968, §13.3 for an analysis.)

Since $(\beta - \beta_{\rm ad}) > 0$ means the fluid is convective, so does $N^2 < 0$, as yet another way to express the criterion for instability. Equation (5.27) then becomes

$$\sigma^2 + \sigma \frac{\nu_T}{\ell^2} + N^2 = 0 \,. \tag{5.30}$$

Looking at the roots of σ, it is obvious that $N^2 < 0$ means that σ is real with one positive root; that is, both w and ΔT may increases exponentially. What limits the motion in the context of the MLT is the ultimate breakup of the parcel. If the radiative cooling time, ℓ^2/ν_T, is very long compared to $1/|N|$, then $\sigma = |N|$. This is the case of *efficient convection* because the parcel loses essentially no heat during its travels until it breaks up. For $\ell^2/\nu_T \ll 1$ (short cooling times) and $N^2 < 0$, $\sigma = -N^2\ell^2/\nu_T \ll 1$ so w and ΔT increase, but slowly. From this point on, we will consider the limit $\nu_T \to 0$ so that no heat is lost. This case of *adiabatic convection* simplifies the analysis enormously although, when we present numerical results, we will have allowed for heat loss in our calculations (as discussed more fully in Hansen & Kawaler 1994).

Now in the adiabatic limit but with $N^2 > 0$ (no convection), σ is pure imaginary and the parcel oscillates around some mean value. Inclusion of radiative losses would cause the motion to gradually damp out.

Related to N^2 is the *Schwarzschild discriminant*,

$$A_{\rm s}(r) = \frac{d\ln\rho}{dr} - \frac{1}{\Gamma_1}\frac{d\ln P}{dr} \tag{5.31}$$

where the logarithmic derivatives are over the actual run of ambient pressure and density in the star. With a little effort, this can be converted into

$$A_{\rm s} = \frac{\chi_T}{\chi_\rho}\left(\nabla - \nabla_{\rm ad}\right)\frac{1}{\lambda_P} \tag{5.32}$$

if composition gradients are again not present. It should be clear from (5.10) that either $N^2 < 0$ or $A_{\rm s} > 0$ implies convective instability. The relation between N^2 and $A_{\rm s}$ is

$$N^2 = -A_{\rm s}g \,. \tag{5.33}$$

Note that the difference $\nabla - \nabla_{\rm ad}$ will appear numerous times in this chapter. It is frequently called the *superadiabatic gradient* because it measures how much larger (or smaller) the actual gradient is compared to the adiabatic gradient. Many authors denote the superadiabatic gradient by $\Delta\nabla$, but we shall not: we have enough deltas as it stands.

Figure 5.3 shows the run of $N = \sqrt{|N^2|}$ versus $-\log(1 - \mathcal{M}_r/\mathcal{M})$ for the same ZAMS sun considered in Fig. 5.2. Because we cannot plot N for negative (convectively unstable) N^2 on such a plot, radiative and convective regions are indicated (although the tiny radiative region just near the surface is not labeled). Here, as contrasted with Fig. 5.2, we clearly see the effects of

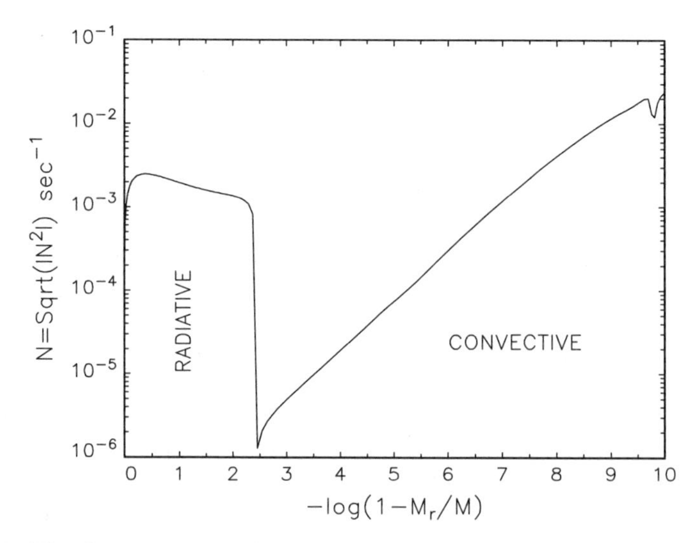

Fig. 5.3. This figure corresponds to Fig. 5.2 and is the same ZAMS sun model but shows how $N = \sqrt{|N^2|}$ varies with $-\log(1 - \mathcal{M}_r/\mathcal{M})$. See text.

the sign of $\nabla - \nabla_{\text{ad}}$.[1] Since N is a frequency, then the time scale $1/N$ should measure either how quickly potentially convective parcels are suppressed in their motion in radiative zones or what turnover times are for parcels in convective zones. For the radiative core in this model suppression times are about 15 minutes. In the convection zone turnover times range from a sluggish week or so to a vigorous few hundred seconds. However, these time scales do not tell us how efficient the convection is.

Finally, we estimate the magnitudes of w and ΔT. It is most convenient to measure these two quantities in terms of σ since both $w = dz/dt$ and ΔT vary as $\exp(\sigma t)$. From this we have $w = \sigma z$. Taking $z = \ell$, a characteristic (or, perhaps, terminal) velocity of a convecting parcel is

$$w = \sigma \ell \, . \tag{5.34}$$

Similarly (and leaving in ν_T for clarity),

$$\frac{d\Delta T}{dt} = \sigma \Delta T = (\beta - \beta_{\text{ad}})\, w - \frac{\nu_T}{\ell^2} \Delta T$$

or, after substituting for w and solving for ΔT, the temperature contrast between the parcel and its surroundings is (in various guises and see 5.27)

$$\Delta T = \frac{\sigma \ell \, (\beta - \beta_{\text{ad}})}{\sigma + \nu_T/\ell^2} = \frac{\sigma^2 \ell T}{\mathcal{Q} g} \, . \tag{5.35}$$

[1] Note that the cusp at abscissa value 2.5 should be much sharper and should extend down to zero as N^2 changes sign. The lack of resolution in the figure is due to the coarseness of the output from the numerical stellar model.

5.1.5 Convective Fluxes

We are now prepared to find convective fluxes in this version of MLT. We have assumed that a convectively unstable parcel rises a total distance ℓ with a characteristic velocity w, while the typical temperature contrast between the parcel and its surroundings is ΔT. The parcel merges into the surroundings at ℓ. Realize that we have assumed that various ambient quantities in the star, such as those contained in the coefficients of (5.21) and (5.26), are constant over the vertical distance ℓ. This assumption is only reasonable if assumption (1) we started out with really is satisfied—namely, if ℓ is smaller than all other scale heights. Granting this, then the heat released by the parcel upon dissolution is $\rho c_P \Delta T$ erg cm^{-3}, where c_P is used here because there is still pressure equilibration. The *rate* at which heat is released is then $\rho w c_P \Delta T$ erg cm^{-2} s^{-1} and this is the convective flux. Thus,

$$\mathcal{F}_{\rm conv} = \rho w c_P \Delta T = \frac{\rho c_P T \sigma^3 \ell^2}{\mathcal{Q} g} \tag{5.36}$$

in a few of many forms. The convective luminosity is then

$$\mathcal{L}_{\rm conv}(r) = 4\pi r^2 \mathcal{F}_{\rm conv}(r) \ . \tag{5.37}$$

For adiabatic convection with no radiative losses, $\sigma = \sqrt{-N^2}$, so that (using 5.29)

$$\mathcal{F}_{\rm conv} = \frac{\rho c_P \ell^2 T g^{1/2} \mathcal{Q}^{1/2} \left(\nabla - \nabla_{\rm ad}\right)^{3/2}}{\lambda_P^{3/2}} \ . \tag{5.38}$$

The above result is the same as that given by Cox (1968, Eq. 14.122) for adiabatic convection to within factors of order unity.

The flux $\mathcal{F}_{\rm conv}$ may also be phrased to contain the Mach number of the convective parcels. Using (5.36) and $\sigma = w/\ell$,

$$\mathcal{F}_{\rm conv} = \frac{\rho c_P T}{\ell \mathcal{Q} g} v_s^3 \left(\frac{w}{v_s}\right)^3 \ . \tag{5.39}$$

The Mach number, w/v_s, comes in as a high power in $\mathcal{F}_{\rm conv}$ and it spells possible trouble. If w/v_s approaches unity, then assumption (4) of the MLT has clearly been violated because acoustic effects are no longer ignorable. What is usually done in practice when the Mach number approaches unity is to limit the convective flux by setting w/v_s in the above to, say, 1/2. What this has to do with reality remains a mystery.

Finally, what to do about the mixing length ℓ? Here is where you can get all kinds of advice, usually contradictory. The most popular tack is to set it equal to some fraction α (usually called the "mixing length parameter"), perhaps something like one-half, of the pressure scale height, thereby violating assumption (2) that we started our discussion with. Or, better yet, the fraction is determined by making evolutionary models and comparing them with observations. The fraction that works best is then the fraction *du jour*. What is unfortunate is that models for other classes of stars may give different fractions of the pressure scale height (or for whatever stellar scale height that is used). What is fortunate, however, is that it doesn't make much of a difference if the convection is efficient—as discussed below.

5.1.6 Calculations in the MLT

Thus far we have assumed that ∇ (or β) and ℓ are known and that, from this, we can then compute σ, etc., and, finally, $\mathcal{F}_{\text{conv}}$. There are three practical difficulties with this. The first is that there is no guarantee the system is consistent; that is, $\mathcal{F}_{\text{conv}}$, if not zero, must surely help determine the structure of the star and, hence, ∇, ∇_{ad}, β, and so on. Some sort of iterative process is then required. The second difficulty, and we will see this below, is that ∇ is very close in value to ∇_{ad} for efficient convection and the difference $(\nabla-\nabla_{\text{ad}})$ in (5.38) is not well determined. Thirdly, in stellar structure calculations it is very often true that it is the total flux that is relatively well determined at some stage in the calculation and not ∇ (or, worse yet, $\nabla - \nabla_{\text{ad}}$). This situation is common when examining the outer stellar regions where the luminosity is nearly constant and is primarily determined by nuclear burning deeper within the star. The issue is also complicated by the fact that the total flux is compounded from the convective flux (if present) and the radiative flux (which is always there). How is this last situation handled?

Well, for adiabatic convection with no radiative leakage, things are relatively easy. First make believe that all the flux is carried by radiation and compute ∇_{rad} of (4.30, in §4.2.1)—viz.,

$$\nabla_{\text{rad}} = \frac{3}{4acG}\frac{r^2 P\kappa}{T^4 \mathcal{M}_r}\mathcal{F}_{\text{tot}} \; . \tag{5.40}$$

If $\nabla = \nabla_{\text{rad}}$, then the flux is carried solely by radiation and you're done. But if $\nabla_{\text{rad}} > \nabla$, then convection must contribute with

$$\mathcal{F}_{\text{tot}}(r) = \mathcal{F}_{\text{conv}}(r) + \mathcal{F}_{\text{rad}}(r) \tag{5.41}$$

where $\mathcal{F}_{\text{conv}}$ is given by (5.38). Using (5.2), (5.5), (5.13), (5.14), and (5.16), find

$$\mathcal{F}_{\text{rad}}(r) = \frac{\rho c_P T}{\lambda_P}\nu_T \nabla \tag{5.42}$$

so that, after taking out common factors,

$$\mathcal{F}_{\text{tot}} = \frac{\rho c_P T \ell^2}{\lambda_P}\left[\left(\frac{Qg}{\lambda_P}\right)^{1/2}(\nabla - \nabla_{\text{ad}})^{3/2} + \frac{\nu_T}{\ell^2}\nabla\right] \; . \tag{5.43}$$

(Note that ν_T appears here because of the radiative flux, not because it has been included in leakage from parcels, which it hasn't.) Since $\mathcal{F}_{\text{tot}}$ is known and we have provisional values for everything else, we can solve for $(\nabla - \nabla_{\rm ad})$ in (5.43) and see if that is consistent with the value of ∇ that the model thinks is correct. If not, we had better iterate yet again.

5.1.7 Numeric Examples

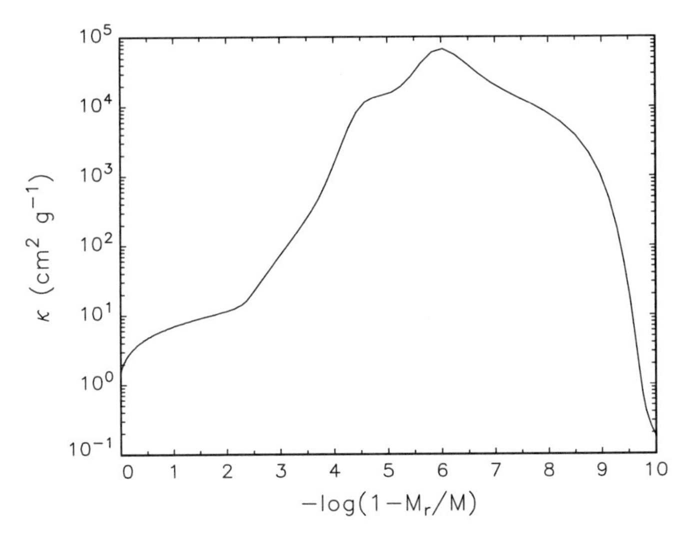

Fig. 5.4. This plot of opacity versus mass is to be compared to Figs. 3.11, 5.2, and 5.3 for a ZAMS sun.

An illustration of what happens to ∇ and $\nabla_{\rm ad}$ was shown in Fig. 5.2 for the solar ZAMS model. The core is radiative (as shown in Fig. 5.3) but from a mass level of about $\mathcal{M}_r/\mathcal{M} = 1 - 10^{-2.5}$ to $1 - 10^{-6}$ convection is present and $\nabla = \nabla_{\rm ad}$ to a few decimal places. Note that the convection continues past this level but ∇ is greater than $\nabla_{\rm ad}$ by a bit. This is because radiative leakage has been included in the model and the convection must be pushed by a larger ∇ to make up for it.

Why is the envelope of our ZAMS sun convective? One contributing factor was shown in Fig. 3.11 where we plotted $\nabla_{\rm ad}$. Referring back to that figure you see that $\nabla_{\rm ad}$ drops to low values (from 0.4 to less than 0.2) because of ionization. This means that it is "easier" for ∇ to exceed $\nabla_{\rm ad}$, and thus convection is more likely. The second factor is the opacity, shown in Fig. 5.4 (and see Fig. 4.6). Note that the opacity (on a logarithmic scale) increases rapidly at the same mass point that convection starts in Fig. 5.3. This is no accident, since $\nabla_{\rm rad}$ of (5.40) is proportional to opacity. What happens is that the increasing opacity acts as a dam to radiation, which, if convection were not ready to take over, would require ∇ to increase to keep the radiative flux up to the required level. But increasing ∇ past a critical point (i.e., past $\nabla_{\rm ad}$) triggers convection. Things may now settle down, so to speak, as $\nabla \to \nabla_{\rm ad}$ and radiative transfer just goes along for the ride.

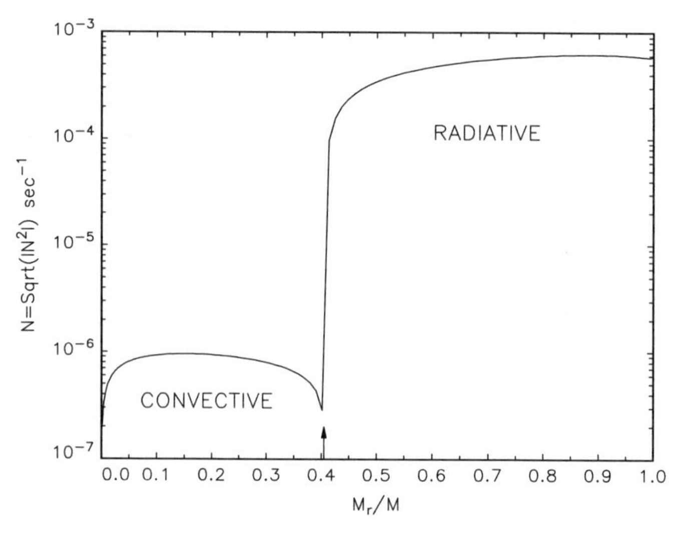

Fig. 5.5. Compare this figure for a 15 $\mathcal{M}_\odot$ ZAMS model with Fig. 5.3.

Now for the other side of the main sequence coin. Figure 5.5 shows $|N|$ for a 15 $\mathcal{M}_\odot$ upper ZAMS model. Compared to Fig. 5.3, the convective and radiative zones have been switched. (Note that $\mathcal{M}_r/\mathcal{M}$ is now the independent variable so that the outer regions are no longer emphasized as they were in Fig. 5.3.) We now have a convective core and radiative envelope. (There is a minute convection zone way out near the surface but it doesn't show here.) Because the core is so hot, the stellar fluid is essentially completely ionized, so $\nabla_{\rm ad}$ should not be as small as it was in the envelope of the 1 $\mathcal{M}_\odot$ model. It is equal to 0.32 at the center—not 0.4—because radiation pressure tends to depress $\nabla_{\rm ad}$ (as Eq. 3.111 indicates), as shown in Fig. 5.6. A high opacity doesn't trigger convection as it did before because electron scattering dominates and κ is a mere 0.35 cm^2 g^{-1}, and, from Fig. 4.6, we know that envelope opacities aren't large either. So, what's up now?

Again looking back to the expression for $\nabla_{\rm rad}$ of (4.30), we find the ratio $\mathcal{L}_r/\mathcal{M}_r$. Can this be the culprit? From (1.58), $d\mathcal{L}_r/d\mathcal{M}_r = \varepsilon$, and you can use this to show (see Chap. 7) that $\mathcal{L}_r/\mathcal{M}_r \to \varepsilon$ as $r \to 0$. So what is ε deep down? The model gives $\varepsilon(r=0) = 7 \times 10^4$ erg s^{-1}. Putting in the rest of the numbers and taking ℓ as the radius of the model or perhaps the size of the convection zone itself (since λ_P is infinite at the center), we find $\nabla_{\rm rad} = 2.5$, which is nearly ten times larger than $\nabla_{\rm ad}$ and convection rules. The reason that upper main sequence stars are prone to convect in their cores rather than less massive ones is that the CNO cycle is so temperature dependent ($\varepsilon \propto T^{15}$, as in Table 1.1). This causes a steep increase in ε as the center is approached in these stars. Because so much power is concentrated in a small volume, ∇_{rad} would have to be large if radiation were called upon to carry all the flux. This triggers convection, which can then do the job.

The radiative cooling time ℓ^2/ν_T can also be computed to be about 10^6 years, whereas a characteristic convective transport time, $|N|^{-1}$, is only about 10^6 s (see Fig. 5.5). Comparing these two numbers implies that the convection is indeed adiabatic—and see Fig. 5.6, where $\nabla = \nabla_{\text{ad}}$ in the convection zone. With $w = |N|\ell$, we find that the convective velocity is only 10^5 cm s^{-1}, which is only a little faster than the supersonic Concorde. Considering how far fluid must travel, this is relatively sluggish. The fluid need not travel very swiftly, however, because the dense, high heat content interior can convect a lot of power without much effort. The convection is not only adiabatic but also efficient.

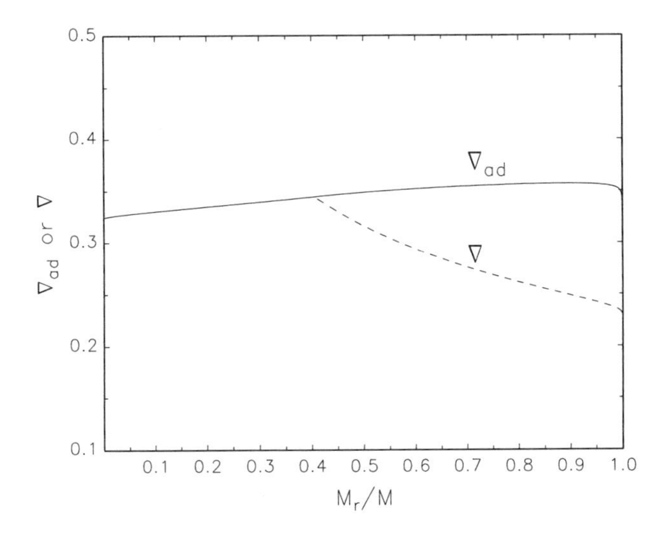

Fig. 5.6. Compare this figure for a 15 $\mathcal{M}_\odot$ ZAMS model with Fig. 5.2.

From what we have seen thus far, it appears that stars with low (on some scale) effective temperatures tend to have convective envelopes, whereas hot stars do not. Figure 5.7 shows the HR diagram divided into two regions, where, for hotter stars on the left, convection is inefficient or hardly present in the outer stellar layers, whereas the cooler stars have active and efficient envelope convection zones. The primary reason for this is that the ionization zones in the cooler stars lie deeper in the star where opacities are high (not electron scattering as in the hot deep interior of upper main sequence stars) and ∇_{ad} is relatively small. The analysis leading to Fig. 5.7 may be found in Cox (1968, §20.5), but it takes a bit of work. Note that the dividing line crosses the main sequence at about spectral class F0 ($T_{\rm eff} \approx 7,500$ K). This corresponds to a mass near 1.5 $\mathcal{M}_\odot$, which is in accord with our "Mass Cut Table" of Fig. 2.4.

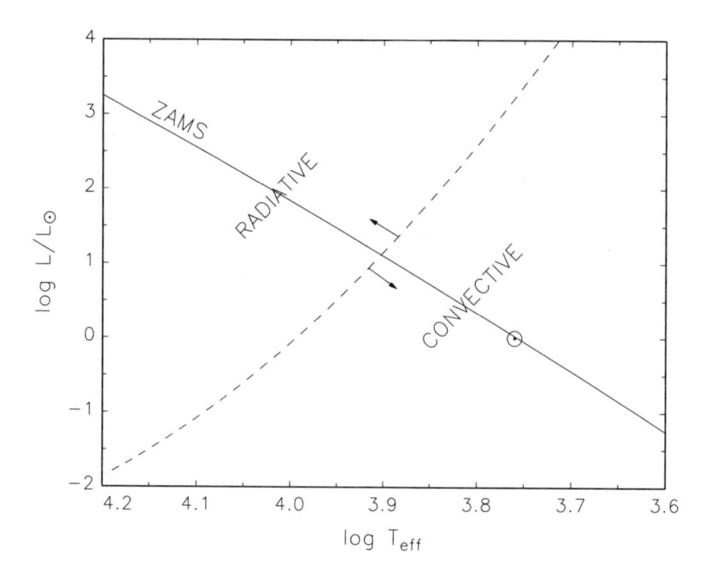

Fig. 5.7. The dashed line in this schematic HR diagram divides those stars with active and efficient outer convection zones (to the right of the line) from those (to the left) that have feeble and inefficient convection. There is, of course, a gradual transition across the line. The sun is indicated by the $\odot$ sign. Adapted from Cox (1968).

5.2 Variations on the MLT

This chapter has mentioned attempts to improve the modeling of parcels and their motions by introducing geometrical factors, changing the method of averaging various quantities, etc. We shall not describe these attempts because all they tend to do is change the efficiency of convection somewhat. As a practical matter, one has to be very careful when reading the literature to figure out precisely what version (or version of a version) is being used. Besides simple geometrical factors, the choice of scale height used to construct the mixing length differs among authors. Variations here include computing the mixing length using the pressure scale height unless that scale is longer than the distance from the stellar position in question to the upper boundary of the convection zone. In that case the latter distance is used (as in Böhm and Cassinelli, 1971). Each of the variations sounds reasonable but they are still carried out in the framework of the MLT. Recent studies that explore the effect on stellar evolution of varying mixing length parameters are Pedersen et al. (1990) and Stothers and Chin (1997).

Some attempts at modeling stellar convection that are certainly not part of generic formulations of the MLT include the following. We have imagined our parcel rising (for example) at some characteristic velocity w and then merging with the ambient medium and releasing heat once ℓ has been traversed. At the top of the convection zone, where ∇ finally becomes smaller than $\nabla_{\rm ad}$, all these parcels are supposed to be able to stop in their tracks and not penetrate into the stable layer above. This does not even sound reasonable. We expect some fluid elements to *overshoot* into the overlying stably stratified medium. Below the convection zone we might also expect downward flowing elements to penetrate into the underlying stable medium to some extent. The question is how to model this behavior without really working too hard. Descriptions of how this is done in practice by some researchers may be found in the references, and see the reviews by Trimble (1992) and Trimble and Aschwanden (1998).

One way to attack this problem, however, is to consider a rising parcel having velocity w_0 at the top of a convection zone at radius r_0. Assuming that the parcel does not come to a complete halt at r_0 as it begins to enter the convectively stable region above r_0, it must still experience a deceleration because of negative buoyancy. The work done against the parcel by the negative buoyancy may be estimated by combining (5.1), (5.20), and (5.24). Consideration of the kinetic energy lost by the parcel then yields the velocity at some position $r \geq r_0$

$$w^2(r) = w_0^2 + 2 \int_{r_0}^{r} \frac{gQ\Delta T}{T} \, dr' \tag{5.44}$$

where viscous dissipation has been neglected and the fluid is assumed to be essentially incompressible over some distance less than a density scale height. The temperature contrast $\Delta T = T' - T$ is negative in the stable fluid and thus the parcel must eventually slow down and stop (over a distance less than ℓ). Because of the integral, this makes the theory nonlocal. Thus, unlike the standard MLT, information from some range of radii is necessary to make a statement about some point further removed. This makes the computation (of various things) a good deal more complicated.

The effects of overshooting are twofold: it may mix matter of varying composition past the convective interface and it may transport heat. To calculate these effects requires even more work than we have implied. To be fully consistent, the term ΔT must be handled correctly. It is the difference in temperature between the parcel's interior and the ambient medium. Presumably T is known but T' is another matter. If we are going to treat ΔT as something that can vary with height for a *particular* parcel, then consider the following. Two parcels start rising from different positions, r_1 and r_2, within the convection zone. When they reach $r_3 > r_1, r_2$, their interior temperatures will not be the same in general. What then is the meaning of $T' - T$ at r_3? A little thought reveals that to do this sort of thing also requires a nonlocal treatment of handling temperatures, etc.

There are different implementations of the above and not all give the same kind of answer for the same kind of stellar model. For an upper main sequence model of the sort discussed in the last section, it appears that typical overshooting lengths are a little bit less than a pressure scale height. How much less (zero to 70% of a scale height?) is a matter of debate and the consequences are important. Overshooting in the convective cores of massive stars affects the lifetime (by mixing fuel) of that stage, and may easily cause the products of nuclear burning to emerge at the stellar surface where they might become visible (as in Wolf–Rayet stars). The caution here is that all these modifications, as well intentioned as they may be, are still based primarily on the MLT with all its potential faults.

Another consideration is whether the Boussinesq assumption of no acoustic effects is reasonable. We have mentioned the Reynolds number only briefly (see Hansen and Kawaler, 1994) and it can easily be estimated. For the 15 $\mathcal{M}_\odot$ ZAMS core it is about 10^{13} and it is even larger for red supergiant envelopes. Because laboratory values of the Reynolds number exceeding $\sim$100 imply a transition from smooth to turbulent flow, we expect stellar convection to be characterized by turbulent eddies of many size scales. (What does a parcel size of the order of a "mixing length" mean in this context?) It has been suggested that the energy flux carried by acoustic noise due to turbulence in some convection zones may actually be comparable to that carried by mean flows. One model, due to Lighthill (1952, 1954), and Proudman (1952), yields an estimate for this flux of $\mathcal{F}_{\text{turb}} \approx a \int \left(\rho w^3 M^5/\ell\right) dr$, where the integration is to be carried out over the extent of the convection zone, M is the Mach number, and a is a number of order 10. If this result is really applicable to stars, then the computation of convective fluxes for high–velocity flows must be far more difficult than what has been outlined in this chapter.

Brave attempts to model turbulence—without doing full-scale hydrodynamic calculations—include Canuto & Mazzitelli (1991), Canuto & Dubovikov (1998) (and see Kupka 1999), and references therein. These attempts are worthwhile in that they point out the defects (and virtues) of the MLT. As we will briefly discuss shortly, the state of the art (as this text is written) in convective modeling is in transition. We are on the verge (in how many years?) of incorporating realistic three dimensional hydrodynamics into stellar evolution calculations—but we aren't there yet. For now, and for the most part, we have to make due with the MLT and its variations.

5.2.1 Beyond the MLT

Any attempt to model convection in all its glory in realistic stellar models must be, a priori, highly nonlocal and nonlinear because the full equations of hydrodynamics, including turbulence, must be considered. This may turn out to be an impossible program for the near future (but see §5.3). There has, however, been some progress on a somewhat less ambitious scale, which we now discuss very briefly. An excellent introduction to what follows may be found in the first few chapters of Chandrasekhar (1981). References to the works of those involved with this program in the stellar context are listed at the end of this chapter.

The analytical and computational thrust of this research consists in considering the equations of hydrodynamics, but where relevant variables are separated into two parts. One corresponds to a horizontal mean of a variable while the other part deals with fluctuations. For example, the spatial and temporal behavior of the density (in plane parallel geometry) is $\rho(z,t) = \langle\rho(z,t)\rangle + \rho'(z,t)$. It is the second, and fluctuating, term on the right that will describe how mass, energy, etc., are transported. The next step is to introduce the anelastic approximation in which $\partial\rho'/\partial t$ is neglected in the equation of continuity

$$\left(\frac{\partial \rho}{\partial t}\right) + \nabla \bullet (\rho \mathbf{v}) = 0 \,.$$

The effect of this is to filter out acoustic waves (as in the MLT). The reason this is done is that acoustic waves have higher frequencies than gravity waves and deleting the former allows the time evolution of the convection to be followed more efficiently on the computer. It also assumes that acoustic fluxes are relatively unimportant in transporting energy. This may well restrict the validity of the model to those envelopes in which convection is not too vigorous.

The next step is to expand the horizontal structure of fluctuating quantities into a finite number of horizontal "planforms" or "modes." These planforms may be of various shapes. For example, a hexagonal planform might be used where fluid may rise in the center of the hexagon and sink at the edges (or the other way around). Such structures are seen in the laboratory under the right conditions in Bénard cells, or on the surface of the sun (as a direct consequence of the underlying convective layer). Another form might be that of a "roll." In any case, such planforms of various shapes and sizes may be added together to model the convective motions.

Before we get further (but not very far) into hydrodynamic calculations, consider a more classic problem in stellar structure.

5.2.2 Semiconvection

This important topic is related in some respects to overshooting. Briefly, the process of semiconvection comes about as indicated in the following example.

In a pioneering evolutionary study, Schwarzschild and Härm (1958) found that the convective cores of massive ($\mathcal{M} > 10\mathcal{M}_\odot$) stars behaved in a curious fashion. It turns out that the convective core of such a star tends to be larger than the region of active CNO cycle burning and, because of the contribution of radiation pressure to the equation of state, the convective core tends to move outward as evolution proceeds. This means that if we assume convective motions can efficiently mix material in such cores, then snapshots of the hydrogen content of the star as it evolves might resemble that shown in Fig. 5.8. The evolutionary stages are labeled as discontinuous "steps" in hydrogen mass fraction starting from step zero on the main sequence when the star is homogeneous and has an initial hydrogen mass fraction $X = 0.7$. (Incidently, all special effects such as mass loss or simple overshooting are ignored here.) What is the effect of these discontinuities on the model and are they consistent with the equations of stellar structure developed thus far? Consider the following.

If the standard MLT has been used to describe convection, then, as discussed previously, $\nabla = \nabla_{\text{ad}}$ to high precision in the core. Outside the core, $\nabla = \nabla_{\text{rad}}$. Now ∇_{ad} should be roughly continuous across the outer edge of the composition discontinuity because Γ_2 is roughly independent of composition provided that ionization is complete. We expect the latter to be true since we are deep within the star. What about ∇_{rad}? Is it continuous? That "del" is given by equation (5.40) and, in that expression, P cannot be discontinuous because that would introduce infinite radial derivatives of pressure and, hence, infinite forces. Similarly, temperature cannot be discontinuous because the radiative luminosity is proportional to its gradient. Lastly, $\mathcal{M}_r$ and $\mathcal{L}(r)$ are continuous provided that density doesn't do something bizarre and if we stay outside the energy generation region. This leaves the opacity.

In the deep interior $\kappa \approx 0.2(1 + X)$ from electron scattering. Since X is discontinuous, so is κ and, thus, ∇_{rad}. However, we do see that the ratio $\nabla_{\text{rad}}/\kappa$ is continuous.

If quantities interior to the composition discontinuity are indicated by an "i" subscript, and those exterior by an "e," then we must have

$$\frac{\nabla_{\text{rad,i}}}{\kappa_i} = \frac{\nabla_{\text{rad,e}}}{\kappa_{\text{e}}} \ .$$

For the hydrogen profile evolution shown in Fig. 5.8, $X_{\text{i}} \leq X_{\text{e}}$ implies $\kappa_{\text{i}} \leq \kappa_{\text{e}}$, or

$$\nabla_{\text{rad,e}} \geq \nabla_{\text{rad,i}}$$

across the discontinuity in X. An illustrative run of the ratio $\nabla_{\text{rad}}/\nabla_{\text{ad}}$ with radius is shown in Fig. 5.9 for the ZAMS stage and one later stage. Note that ∇_{rad} must be greater than ∇_{ad} for points within the convection zone (otherwise there is no convection). Because of the discontinuity in ∇_{rad}, however, there is a small region outside the convection zone that is not radiative; that is, this region must also be convective. If this is really so, then we have

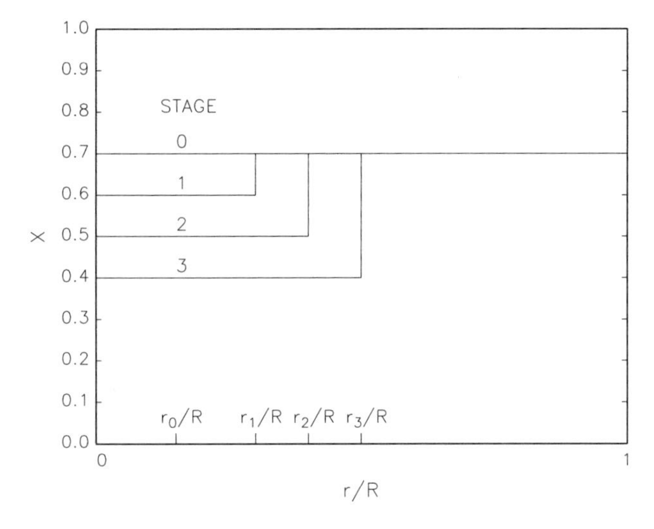

Fig. 5.8. Schematic profiles of the hydrogen mass fraction, X, for a massive star as evolution proceeds from the ZAMS (step 0). As the star evolves (steps 1 through 3), the convection zone moves outward and mixing causes the hydrogen, which is gradually being depleted by nuclear burning, to have constant abundance out to larger radii. The convection zone is assumed to extend out to the discontinuity in X. The radii r_0, r_1, etc., are the outermost radii of the convection zones and hydrogen discontinuities in the various evolutionary stages.

a contradictory situation. Just what happens in this region is still a matter of some debate. It is supposed that some mixing takes place in this region so that some composition gradients are smoothed out. Exactly how this is accomplished has not been established. But, however it is done, it is referred to as "semiconvective mixing" and it may have a strong effect on the later stages of evolution when shell sources are effective. Brief summaries of this general problem may be found in Chiosi and Maeder (1986) in the context of massive stars, in the general review by Trimble (1992), and in Hansen (1978) for other evolved objects where chemical discontinuities play a role in structure.

5.3 Hydrodynamic Calculations

Here we are talking numerical modeling on a huge scale. Computers keep getting faster, can store more information, and can strain the numerical techniques of the most proficient investigators. And the end is still not in sight. The "end" being, for stellar evolutionists, the incorporation of truly realistic three dimensional fluid dynamic codes into their comparatively puny evolution codes. And yet there has been a lot of progress.

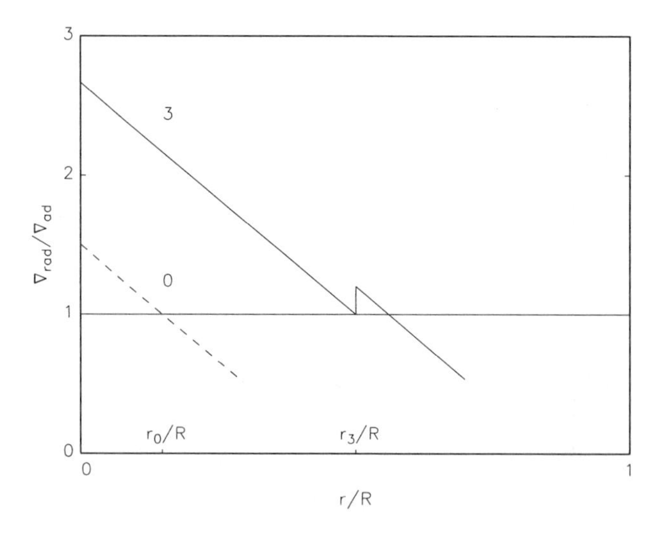

Fig. 5.9. An illustration of how the ratio $\nabla_{\text{rad}}/\nabla_{\text{ad}}$ might vary as a function of radius for two of the evolutionary stages pictured in Fig. 5.8. The radii r_0 and r_3 are the same as in that figure.

One area of current interest is the calibration of stellar models and, in particular, models of the sun. We will return to *helioseismology* in chapters 8 and 9, but the sun "rings" in many modes of oscillation (see §2.10). Many of these modes penetrate (are active) to large depths and how they ring tells us a good deal about that structure at depth. (The sun is not alone in this respect, but that's a story for later.) Thus if a model of the present-day sun does not give the correct oscillation modes, something is wrong with the model. With over a million modes to use, you shouldn't be surprised that some remarkable results have been forthcoming, and there may be no better way to plumb the solar interior than using those modes.

One result concerns the structure of the convective envelope and radiative core of the sun as they respond to the sun's rotation. Or, put more properly perhaps, how does the sun rotate inside? It is easy to observe the sun's surface rotate and it does so differentially with the equatorial regions having a shorter rotation period than do higher latitudes. (That is, the equator slips ahead of the higher latitudes.) This, however, tells us nothing about what happens deeper down. Recent analysis of the helioseismological data has established that the convective envelope also rotates differentially but the radiative core rotates like a solid body down to a considerable depth. Between the two zones is a thin transition region (the "tachocline"). A reasonable question to ask is "What is the transition region like and how does rotation affect it?" And answering this question requires modern supercomputers. (Of course one decade's supercomputer is the next decade's desktop.)

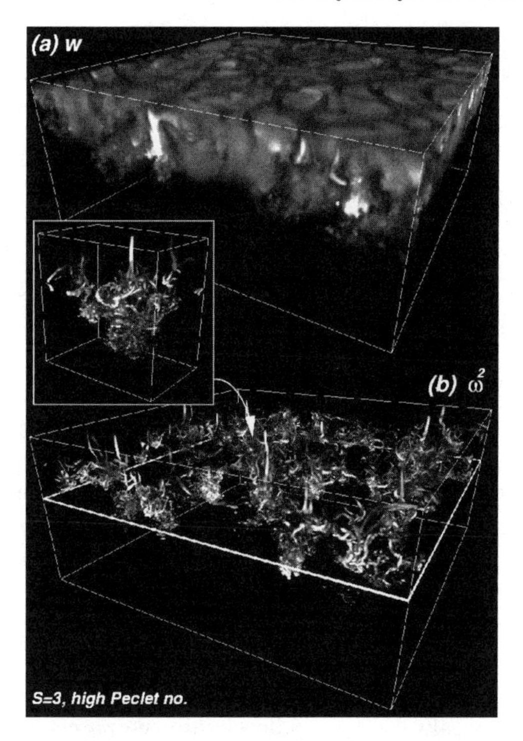

Fig. 5.10. This illustrates the complexity of modern 3D hydrodynamic modeling of fluid motions in stars. (See text.) A full color version of this figure appeared in Brummell et al. (2002) as their Fig. 11. Courtesy of Nic Brummell, and reproduced with permission.

Figure 5.10 shows the results from one set of calculations (Brummel et al., 2002; and see the beautiful color graphics in that paper and in some of the references therein). The smudgy or lacelike structures in the two larger boxes represent convective motions on various scales as they overlie a stable radiative layer.[2] Box (a) shows downflows (in white) and upflows (in darker tones) as the material overturns. Material that flows down into the stable layer, mixes with it, and alters the background state of the stable layer, is said to be undergoing "penetration." "Overshooting," as we discussed earlier, means that inertia has carried material into the stable layer but doesn't change the properties of the layer in any marked way. Box (b) shows the "enstrophy" density of the flow, which is defined as the square of the vorticity; that is $\omega^2 = |\boldsymbol{\nabla}\times\mathbf{u}|^2$, where $\mathbf{u}$ is the velocity. (You may think of ω^2 as a measure of the "twistyness" of the fluid around a point—although there's more to it than that.) White shows strong values of ω^2 while darker tones are less so. The insert box shows ω^2 for a downflow.

[2] Some of these simulations had as many as 1.5×10^8 mesh points!

We are going to have to refer you to the original paper to see what the authors conclude, but it, and papers of similar intent, are worth perusing to see what state-of-the-art stellar hydrodynamics calculations can do.

5.4 Exercises

Exercise 5.1. Verify, in detail, that $dS/dr < 0$ implies local convective instability in the same sense as does $\nabla > \nabla_{\rm ad}$ (as in the discussion following Eq. 5.10). This takes a bit of thermodynamic manipulation.

Exercise 5.2. Verify (5.25); that is, show that the coefficient of thermal expansion, neglecting composition changes, is $\mathcal{Q} = \chi_T/\chi_\rho$.

Exercise 5.3. In the Navier–Stokes equation of motion for an incompressible fluid (which is consistent with the Boussinesq approximation) we find a drag term $\nu\nabla^2\mathbf{w}$, where ν is the kinematic viscosity having the same units as ν_T. (See, for example, Landau and Lifshitz, 1959, §15.) We replace the Laplacian by w/ℓ^2 and amend the equation of motion (5.22) to read

$$\frac{dw}{dt} = \frac{(\rho - \rho')}{\rho} g - \frac{\nu}{\ell^2} w \, .$$

The additional term always acts to decelerate the parcel. An estimate for ν is from Chapter 5 of

▷ Spitzer, L. 1962, *Physics of Fully Ionized Gases*, 2nd ed. (New York: Interscience)

and is

$$\nu \approx \frac{2 \times 10^{-15}}{\rho} \frac{T^{5/2} A^{1/2}}{Z^4 \ln \Lambda_c} \quad \text{cm}^2 \text{ s}^{-1}$$

where

$$\Lambda_c \approx 10^4 \frac{T^{3/2}}{n_{\rm e}^{1/2}}$$

and A and Z are the atomic weight and charge of the ions. Use this to make an estimate of the Prandtl number

$$\mathbf{Pr} = \frac{\nu}{\nu_T}$$

for a typical point in the sun. Note that laboratory experiments work around $\mathbf{Pr} \sim 1$. Comments?

Exercise 5.4. The dimensionless Rayleigh number, **Ra**, is a measure of how well the driving of convection (as in $\nabla - \nabla_{\rm ad}$ terms) compares to damping processes (ν_T and ν). It is defined by

$$\mathbf{Ra} = \frac{\mathcal{Q} g}{\lambda_P} \left(\nabla - \nabla_{\rm ad}\right) \frac{\ell^4}{\nu_T \nu} \, .$$

Compute a sample value (say in a solar radiative zone). Laboratory experiments have **Ra** of about 10^{11} or less. Comments? Also show (using Eq. 5.20, 5.21, 5.24, and the amended equation of motion given in Ex. 5.3) that $\mathbf{Ra} > 1$ implies that w and $D\Delta T/Dt$ have exponentially growing solutions. Laboratory convection usually sets in at about $\mathbf{Ra} \sim 10^3$.

5.5 References and Suggested Readings

§5.1: The Mixing Length Theory

Chapter 14 of

▹ Cox, J.P. 1968, *Principles of Stellar Structure*, in two volumes (New York: Gordon & Breach)

contains a good description of the MLT as it is usually applied and should be read after our chapter, but it isn't easy. See also the review by

▹ Gough, D.O., & Weiss, N.O. 1976, MNRAS, 176, 589.

The first edition of this text, which discussed the MLT in more detail, is

▹ Hansen, C.J., & Kawaler, S.D. 1994, *Stellar Interiors: Physical Principles, Structure, and Evolution*, 1st ed. (New York: Springer-Verlag).

The development of the MLT may be traced through the following papers:

▹ Schwarzschild, K. 1906, GottNach, 1, 41

▹ Biermann, L. 1951, ZsAp, 28, 304

▹ Prandtl, L. 1952, *Essentials of Fluid Dynamics* (London: Blakie)

▹ Vitense, E. 1953, ZsAp, 32, 135

▹ Böhm-Vitense, E. 1958, ZsAp, 46, 108.

An accessible and modern commentary on Boussinesq convection and the MLT may be found in

▹ Spiegel, E.A. 1971, ARA&A, 9, 323.

As usual, we recommend

▹ Landau, L.D., & Lifshitz, E.M. 1959, *Fluid Mechanics* (Reading: Addison–Wesley)

(or later editions) for their physical style.

§5.2: Variations on the MLT

▹ Chan, K.L., & Sofia, S. 1987, Science, 235, 465

give an interesting discussion on some validity tests for the MLT in deep convection zones. The paper by

▹ Böhm, K.H., & Cassinelli, J. 1971 A&A, 12, 21

concerns the convective envelopes of white dwarfs.

▹ Pedersen, B.B., VandenBerg, D.A., & Irwin, A.W. 1990, ApJ, 352, 279 have constructed series of evolutionary models with different mixing length parameters. You may wish to consult this paper to see how these series differ from one to the other. An example of another attempt to manipulate the mixing length is

▷ Stothers, R.B., & Chin, C-W. 1997, ApJ, 478, L103.

Early estimates of the rate of production of acoustic noise due to turbulence may be found in

▷ Lighthill, M.J. 1952, PRSocL, A221, 564

▷ Ibid. 1954, PRSocL, A222, 1

▷ Proudman, J. 1952, PRSocL, A214, 119.

More modern attempts include

▷ Canuto, V.M., & Mazzitelli, I. 1991, ApJ, 370, 295,

▷ Canuto, V.M., & Dubovikov, M.S. 1998, ApJ, 493, 827,

▷ Kupka, F. 1999, ApJ, 526, L45.

The following papers by A. Maeder and collaborators describe one model for overshooting in a version of the MLT and give results for the evolution of upper main sequence stars:

▷ Maeder, A. 1975, A&A, 40, 303

▷ Ibid. 1975, A&A, 43, 61

▷ Ibid. 1976, A&A, 47, 389

▷ Ibid. 1982, A&A, 105, 149

▷ Maeder, A., & Bouvier, P. 1976, A&A, 50, 309

▷ Maeder, A., & Mermilliod, J.C. 1981, A&A, 93, 136.

Additional material, including a brief appraisal of methods used in handling semiconvection, is given by

▷ Chiosi, C., & Maeder, A. 1986, ARA&A, 24, 329.

A recent compilation of evolutionary results that summarizes 40,000 models is

▷ Maeder, A. 1990, A&AS, 84, 139.

Marked differences in structure may result from varying degrees of discontinuities in models. Some results are discussed in

▷ Hansen, C.J. 1978, ARA&A, 16, 15.

A short review containing up-to-date material on semiconvection and overshoot is due to

▷ Trimble, V. 1992, PASP, 104, 1.

Other relevant material is in

▷ Trimble, V., & Aschwanden, M. 1998, PASP, 111, 385.

The text by

▷ Chandrasekhar, S. 1981, *Hydrodynamic and Hydromagnetic Stability* (New York: Dover)

should be on every theorists' bookshelf. It can be heavily mathematical at times, but it is an indispensable introduction to modern developments in convection theory.

Planform methods, with applications, are given in the following series of papers involving many of the same names as collaborators:

▷ Gough, D.O., Spiegel, E.A., & Toomre, J. 1975, JFlMech, 68, 695

▷ Latour, J., Spiegel, E.A., Toomre, J., & Zahn, J.-P. 1976, ApJ, 207, 233

▷ Toomre, J., Zahn, J.-P., Latour, J., & Spiegel, E.A. 1976, ApJ, 207, 545

▷ Latour, J., Toomre, J., & Zahn, J.-P. 1981, ApJ, 248, 1081

▷ Latour, J., Toomre, J., & Zahn, J.-P. 1983, SolPhys, 82, 387

▷ Nordlund, Å, & Stein, R.F. 1990, CompPhysC, 59, 119.

Other examples of dealing with real convection can be found in

▷ Lydon, T.J., Fox, P.A., & Sofia, S. 1992, ApJ, 397, 701

▷ Ibid. 1993, ApJ, 403, L79.

The upper main sequence calculations, which led to our discussion of stellar semiconvection, are those of

▷ Schwarzschild, M., & Härm, R. 1958, ApJ, 128, 348.

§5.3: Hydrodynamic Calculations

Figure 5.10 is a black & white version of Fig. 11 of

▷ Brummell, N.H., Clune, T.L., & Toomre, J. 2002, ApJ., 570, 825

produced especially for us by Nic Brummell. This paper is but one in a long series (and see the references in that paper). If your interests lay more along the lines of observables on the solar surface, then

▷ Stein, R.F., & Nordlund, Å. 1998, ApJ, 499, 914

is a prime place to look. Their aim is to use hydrodynamic calculations to model solar granulation with, in our estimation, considerable success.

6 Stellar Energy Sources

"Don't get smart alecksy
with the galaxy
leave the atom alone."
— *E.Y. Harburg (on the "Bomb")*

"We never met a nucleus we didn't like."
— *We made that one up.*

Now that we have explored how energy is transported in the stellar interior, let's backtrack and see how to generate that energy. This chapter will discuss energy production by the conversion of gravitational energy into internal energy and by thermonuclear processes. One section will deal with energy loss by neutrinos.

6.1 Gravitational Energy Sources

In the first chapter (§1.3.2) we described how the virial theorem could be used to estimate the Kelvin–Helmholtz time scale for maintaining the luminosity of a star by means of gravitational contraction. What we obtained was a global property of the system because the virial deals only with integrated quantities. We now examine how this works on a local scale.

The local condition for thermal balance was discussed earlier (in §1.5). It described how the local rate of thermonuclear energy generation, ε, is balanced by the mass divergence of luminosity with

$$\frac{\partial \mathcal{L}_r}{\partial \mathcal{M}_r} = \varepsilon \tag{6.1}$$

for a given gram of material (with ε in erg g^{-1} s^{-1}). If this equality doesn't hold, then the energy content decreases (increases) for $\partial \mathcal{L}_r / \partial \mathcal{M}_r$ greater (less) than ε. The difference $\varepsilon - (\partial \mathcal{L}_r / \partial \mathcal{M}_r)$ is then the rate at which heat is added to, or removed from, each gram. This difference is just dQ/dt, where Q is the heat content in erg g^{-1}. Combining this with the first law of thermodynamics yields

$$\frac{dQ}{dt} = \frac{\partial E}{\partial t} + P \frac{\partial}{\partial t}\left(\frac{1}{\rho}\right) = \varepsilon - \frac{\partial \mathcal{L}_r}{\partial \mathcal{M}_r} \tag{6.2}$$

where it is understood that the partial time derivatives are applied in the Lagrangian mode so that a particular gram of matter is followed in time.

Now rewrite (6.2) by defining the *gravitational* energy generation rate

$$\varepsilon_{\rm grav} = -\left[\frac{\partial E}{\partial t} + P\frac{\partial}{\partial t}\left(\frac{1}{\rho}\right)\right] \tag{6.3}$$

so that (6.2) becomes

$$\frac{\partial \mathcal{L}_r}{\partial \mathcal{M}_r} = \varepsilon + \varepsilon_{\rm grav}\,. \tag{6.4}$$

Note that $\varepsilon_{\rm grav}$ may be positive, negative, or zero.

Another way of expressing $\varepsilon_{\rm grav}$ is to cast it in a more useful form containing time derivatives of density and pressure rather the internal energy. This involves a little work using thermodynamic identities and some of the thermodynamic derivatives discussed in Chapter 3 (see, e.g., Cox, 1968, §17.6). The result we shall use is

$$\varepsilon_{\rm grav} = -\frac{P}{\rho\,(\Gamma_3 - 1)}\left[\frac{\partial \ln P}{\partial t} - \Gamma_1\frac{\partial \ln \rho}{\partial t}\right] \tag{6.5}$$

where, for simplicity, we assume Γ_1 is constant in time. Note that $\varepsilon_{\rm grav}$ is zero for adiabatic processes where $P \propto \rho^{\Gamma_1}$. Energy release in gravitational sources thus arises from departures from adiabaticity during contraction or expansion. This may also be seen by rewriting (6.5) in the form

$$\varepsilon_{\rm grav} = -\frac{P}{\rho\,(\Gamma_3 - 1)}\frac{\partial}{\partial t}\left[\ln\left(P/\rho^{\Gamma_1}\right)\right]. \tag{6.6}$$

Thus, for example, if the pressure rises less rapidly than adiabatic upon compression so that

$$P \sim \rho^{\Gamma_1 - \delta}$$

with δ small but positive, then

$$\varepsilon_{\rm grav} = \delta\frac{P}{\rho\,(\Gamma_3 - 1)}\frac{\partial \ln \rho}{\partial t} > 0$$

where the greater than sign is used for compression. This result is reasonable because a less than adiabatic rise of pressure upon compression implies that energy is being released by $P\,dV$ work.

We now improve on our earlier estimate of the Kelvin–Helmholtz time scale. As in the discussion of §1.3.2, we ignore thermonuclear energy sources so that (6.4) becomes $(\partial \mathcal{L}_r/\partial \mathcal{M}_r) = \varepsilon_{\rm grav}$. If we assume homologous contraction, which is often fairly close to the truth, then the results of §1.6, with mass held constant, yield

$$\frac{\rho}{\rho_0} = \left(\frac{\mathcal{R}}{\mathcal{R}_0}\right)^{-3} \tag{6.7}$$

and

$$\frac{P}{P_0} = \left(\frac{\mathcal{R}}{\mathcal{R}_0}\right)^{-4} . \tag{6.8}$$

Hydrostatic equilibrium has been used in the derivation of these relations, which requires that time scales must be long compared to dynamic times.

If we take the "0" star to be some initial configuration at time zero, then the time-dependent source $\varepsilon_{\rm grav}$ may be found by applying (6.5) and then expressing the result in terms of the initial values P_0, ρ_0, the adiabatic exponents Γ_1 and Γ_3, and the rate of change of $\mathcal{R}$. The total luminosity is then found as a function of time by integrating over all mass. This sequence yields

$$\mathcal{L}(t) = -\int_{\mathcal{M}} \frac{P_0}{\rho_0} \frac{\mathcal{R}_0}{\mathcal{R}^2(t)} \frac{d\mathcal{R}}{dt} \left[\frac{3\Gamma_1 - 4}{\Gamma_3 - 1}\right] d\mathcal{M}_r . \tag{6.9}$$

The time–independent factors in the integral may be reexpressed using the virial result (1.23) with $\ddot{I} = 0$. Thus, for example, eliminate the initial pressure and density using

$$-\Omega_0 = 3 \int_{\mathcal{M}} \frac{P_0}{\rho_0} \, d\mathcal{M}_r .$$

If the adiabatic exponents are constant in space as well as time, then

$$\mathcal{L}(t) = \frac{1}{3} \frac{\Omega_0 \mathcal{R}_0}{\mathcal{R}^2} \frac{d\mathcal{R}}{dt} \left[\frac{3\Gamma_1 - 4}{\Gamma_3 - 1}\right] .$$

Simplify this further by using (1.8) where, as you may recall, q is a dimensionless constant of order unity with $\Omega_0 = -q\left(G\mathcal{M}^2/\mathcal{R}_0\right)$. Thus

$$\mathcal{L}(t) = -q \frac{G\mathcal{M}^2}{\mathcal{R}^2} \frac{d\mathcal{R}}{dt} \left[\frac{\Gamma_1 - 4/3}{\Gamma_3 - 1}\right] . \tag{6.10}$$

If $\Gamma_1 = \Gamma_3 = 5/3$, then we regain the luminosity relation of (1.30). Note that here, however, we have specified a bit more carefully just which Γs are involved in the luminosity. Before it was a generic γ from a γ–law equation of state. Also—perhaps as expected from our discussion of the virial theorem—if $\Gamma_1 = 4/3$, then expansion or contraction supplies no power to provide luminosity.

Real stars sometimes contract in a near-homologous fashion. It is more usual, however, for evolving stars to contract in some regions while other regions expand. The most familiar example of the latter is during post-main sequence evolution where the core contracts and the outer regions expand, as discussed in Chapter 2. In those real situations, $\varepsilon_{\rm grav}$ must be treated as the local quantity it is.

6.2 Thermonuclear Energy Sources

The major text references for this section (and most of those following) are the excellent books by Don Clayton (1968), who emphasizes nuclear astrophysics, and that of Dave Arnett (1996) on supernovae and nucleosynthesis. We advise referring to those texts for many of the details we shall be forced to leave out in the following.

After we first consider how reaction rates are found using experimental data supplemented by theory, we shall examine hydrogen, helium, and advanced stages of thermonuclear burning.

6.2.1 Preliminaries

We shall concern ourselves, for the moment at least, only with reactions initiated by charged particles. Neutron–induced reactions, which are of great importance for nucleosynthesis, will be discussed later.

Most thermonuclear reactions in stars proceed through an intermediate nuclear state called the *compound nucleus*. That is, if α represents some *projectile* (say a proton or α-particle), and X is a *target* nucleus, and these react to give rise finally to nuclear products β and Y, then the compound nucleus Z^* is the intermediary state. In reaction equation language this statement reads

$$\alpha + X \rightarrow Z^* \rightarrow Y + \beta \qquad \text{or} \qquad X(\alpha, \beta)Y \tag{6.11}$$

where the last expression is shorthand for the net reaction. The "$*$" appended to Z implies that the compound nucleus is (almost always) in an excited state.

A basic assumption here is that Z^* forgets how it was formed and may decay or break up by any means consistent with conservation laws and selection rules. (And see Ex. 6.4 for application of some selection rules.) One permitted breakup "channel" is that consisting of the particles that originally formed Z^*—in our example, $\alpha + X$ (the "entrance channel"). As an example, suppose we had produced an excited state of ^{12}C by way of proton capture on ^{11}B with

$$\mathrm{p} + {}^{11}\mathrm{B} \rightarrow {}^{12}\mathrm{C}^* .$$

The compound state $^{12}\mathrm{C}^*$ may then break up in a number of ways ("exit channels") such as

$$\begin{aligned} {}^{12}\mathrm{C}^* &\rightarrow {}^{12}\mathrm{C}^{**} + \gamma \\ &\rightarrow {}^{11}\mathrm{B} + \mathrm{p} \\ &\rightarrow {}^{11}\mathrm{C} + \mathrm{n} \\ &\rightarrow {}^{12}\mathrm{N} + \mathrm{e}^- + \overline{\nu}_\mathrm{e} \\ &\rightarrow {}^{8}\mathrm{Be} + \alpha, \quad \text{etc.} \end{aligned}$$

Here, a "$**$" means another excited (or ground) state of ^{12}C and the other symbols, γ, p, n, e^-, $\overline{\nu}_\mathrm{e}$, and α refer, respectively, to gamma-ray photon, proton, neutron, electron, electron antineutrino, and alpha particle (^{4}He nucleus).

The state $^{12}\mathrm{C}^*$ may be thought of as some combination of the states of all the possible exit channels or decay modes. With each of these modes we can associate a mean-life for decay, τ_i, and, through the uncertainty principle, a width in energy, Γ_i, where

$$\Gamma_i \tau_i = \hbar \,. \tag{6.12}$$

Here, $\hbar$ is Planck's constant divided by 2π. (Excuse us for even more Γs, but we're just using the normal nomenclature!) Thus the more long-lived the state is, the less likely it is that the decay will take place and the smaller is the width in energy. The probability that $^{12}\mathrm{C}^*$ will decay through the channel i is measured by the comparison of τ_i to the sum of all the other lifetimes. Put more precisely, the probability $\mathcal{P}_i$ is given by

$$\mathcal{P}_i = \frac{1/\tau_i}{\sum_j (1/\tau_j)} = \frac{\tau}{\tau_i} \tag{6.13}$$

where τ, the total mean-life of $^{12}\mathrm{C}^*$, is

$$\tau = \left(\sum_j \frac{1}{\tau_j} \right)^{-1} \,. \tag{6.14}$$

If Γ is defined as the *total energy width*, $\sum_j \Gamma_j$, then

$$\mathcal{P}_i = \frac{\Gamma_i}{\Gamma} \,. \tag{6.15}$$

Among the possible exit channels in our example is the channel by which $^{12}\mathrm{C}^*$ was formed in the first place—that is, the entrance channel. Denote its width by Γ_{entr}. If we could turn the clock backward then, in a crude sense, the ratio $\Gamma_{\mathrm{entr}}/\Gamma$ should be a measure of forming $^{12}\mathrm{C}^*$ through that channel.

For charged particle channels the Γ_i are broken down into two factors. There is the intrinsic probability that the state Z^* is composed of the charged constituents of the ith channel in a common nuclear potential. Here is where the nuclear physics really comes in. We shall just give it a name and a symbol: that factor is the "reduced width" γ_i^2. The second factor describes the probability that the separate constituents of the compound state ($\alpha + X$ or $Y + \beta$ of 6.11) can dissociate from one another and become distinct particles far–removed from one another. For charged particles this means that not only must their combined energy within the nuclear potential well be positive relative to the total energy (including rest mass energy) of the separated particles but, further, that they be able to overcome the Coulomb barrier between them. This last probability will be denoted by P_ℓ, the *Coulomb penetration factor*. To clarify what this means, we first digress into some nuclear energetics.

6.2.2 Nuclear Energetics

The total binding energy, $B_{\mathcal{E}}$, of a nucleus is defined as the energy required to break up and disperse to infinity all the constituent nucleons (protons and neutrons) in that nucleus. Using "mc^2" arguments, this is

$$B_{\mathcal{E}} = (\text{mass of constituent nucleons} - \text{mass of bound nucleus})\, c^2, \quad (6.16)$$

where $B_{\mathcal{E}}$ is usually expressed in units of MeV. (Note that we do not worry here about some refinements concerning just how "mass" is defined, such as whether electronic binding energies are included or not.) The *average binding energy per nucleon*, $B_{\mathcal{E}}/A$, is defined as $B_{\mathcal{E}}$ divided by the total nucleon mass number, A (in integer amu). It is a measure of the energy required to remove the most energetic nucleon from a given nucleus in its ground state. Figure 6.1 shows a schematic of the experimentally derived $B_{\mathcal{E}}/A$ for the most stable isobar of nuclei with atomic mass number A.

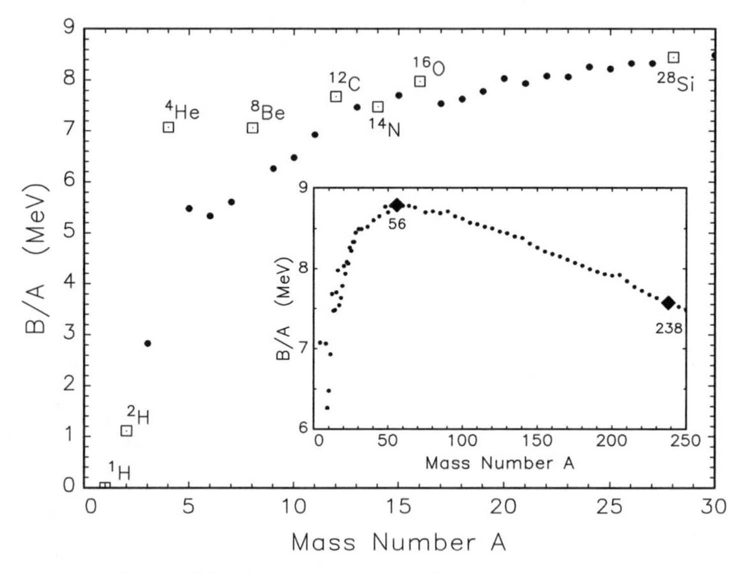

Fig. 6.1. Plotted is the binding energy per nucleon, $B_{\mathcal{E}}/A$, as a function of atomic mass number A for the most stable isobar of A. The main area is for the lighter nuclei, whereas the insert is for most of the rest. Some nuclei discussed in the text are given special emphasis. Data from Wapstra et al. (1988).

The important feature of the binding energy per nucleon curve is its rapid rise from low mass number nuclei to a plateau around A of 60, and then a gradual decline thereafter. The plateau around $A \approx 60$ includes the relatively common element iron, and thus that region is commonly referred to as the "iron-peak" region. The significance of this curve is well known: it, unfortunately, makes fission and fusion weaponry possible. The fusion branch of the

$\mathcal{B}_{\mathcal{E}}/A$ curve is that region below $A \approx 60$. Thus if we fuse two light nuclei on that branch, energy is released. In particular, the requirement for energy release in fusion is that $\mathcal{B}_{\mathcal{E}}(1) + \mathcal{B}_{\mathcal{E}}(2) < \mathcal{B}_{\mathcal{E}}(3)$ where nuclei 1 and 2 combine to give nucleus 3. That is, if we tear the first two nuclei apart and then reconstitute the remains to form the third, then energy is left over.

As an example, consider the reaction that fuses three ^{4}He nuclei (or α-particles) to form one nucleus of ^{12}C. Reading roughly from Fig. 6.1, $\mathcal{B}_{\mathcal{E}}/A$ for ^{4}He is ≈ 7.1 MeV per nucleon. For carbon, $\mathcal{B}_{\mathcal{E}}/A \approx 7.7$. (More exact numbers are 7.074 and 7.680 MeV.) The total binding energies are then $4 \times 7.1 \approx 28.4$, and $12 \times 7.7 \approx 92.4$ MeV, respectively. Taking apart three α-particles then requires $3 \times 28.4 \approx 85.2$ MeV. Reconstituting the resulting protons and neutrons into ^{12}C gives back 92.4 MeV leaving an excess, or "Q-value," for the reaction of about 7.2 MeV (really 7.275 MeV). In pseudo-equation form this reads

$$
\begin{aligned}
3 \times {}^4\text{He} &\longrightarrow {}^{12}\text{C} \\
3 \times {}^4\text{He (in MeV)} - {}^{12}\text{C (in MeV)} &= \; ? \\
3 \times [4 \times (-7.1\,\text{MeV})] - 12 \times (-7.7\,\text{MeV}) &= \; ? \\
-85.2\,\text{MeV} + 92.4\,\text{MeV} = 7.2\,\text{MeV} &= \text{Q-value}\,.
\end{aligned}
$$

Since a gram of pure ^{4}He consists of $N_{\text{A}}/4 \approx 1.5 \times 10^{23}$ atoms and it requires three ^{4}He nuclei per reaction, then a complete conversion to ^{12}C yields approximately 6×10^{17} erg g^{-1} after the MeV have been converted to ergs (by multiplying by 1.602×10^{-6} erg MeV^{-1}). In practice, nuclear mass or, better yet, mass excess ($\Delta = [M - A]c^2$) tables are used to find the Q-values of reactions (as in Wapstra et al., 1988).

Fission on the branch with A greater than about 60 achieves the same end as the above. What this means is that nuclei around the iron peak, which are the most tightly bound of all nuclei (per constituent nucleon), are not of much use as an energy source. Thus any star that ends up with nuclei in the iron peak has lost potential fuel and this is a matter of grave consequence for the star.

Once enough nucleons have been packed into a nucleus, $\mathcal{B}_{\mathcal{E}}/A$ approaches its saturation value of around 8 MeV. To remove one nucleon in a high-lying state from such a nucleus then requires about 8 MeV. How much deeper is the nuclear well below 8 MeV? We can estimate this from the observation that nucleons in the nuclear well behave, to zeroth order, as independent fermions of spin 1/2 in a zero-temperature sea. That is, the nuclear protons (or neutrons) are stacked in energy in pairs from the bottom of the potential well on up. The Fermi energy of the most energetic proton may be found from the number density of protons in the nucleus as we did with electrons in a Fermi sea in Chapter 3. For this, assume that the nucleus is spherical with radius R. A perfectly good estimate for the nuclear radius is

$$R \approx 1.4 \times 10^{-13} A^{1/3} \;\; \text{cm}\,. \tag{6.17}$$

With A nucleons (each with a mass of about 1.67×10^{-24} g) packed into the sphere, this yields a nuclear density of around 1.5×10^{14} g cm^{-3}.

Since the proton and neutron rest mass energies are near 940 MeV, and all other nuclear energies are in the low MeV range, the nucleons are non-relativistic. If the protons and neutrons are considered independently, and if $Z \approx N \approx A/2$, then a simple calculation yields (see §3.5.1), for either neutrons or protons, a Fermi momentum of $p_F \approx 1.15 \times 10^{-14}$ g cm s^{-1} and a Fermi energy $\mathcal{E}_F = p_F^2/2m$ of close to 25 MeV. Thus the potential depth of the typical massive nucleus is about $25 + 8 \approx 30$ MeV.

For a charged particle seeking either to leave or enter a nucleus, there is another important energy—the maximum height of the Coulomb barrier between the interacting nuclei. If, for example, the charges of the target and projectile are Z_X and Z_α, then that height is $B_\mathrm{C} = Z_\alpha Z_X e^2/R$, where R is now taken to be the minimum interparticle separation

$$R = 1.4 \left(A_\alpha^{1/3} + A_X^{1/3} \right) \quad \mathrm{fm} \tag{6.18}$$

and "fm" denotes "femtometer" or "fermi" and is 10^{-13} cm. The Coulomb barrier height is then

$$B_\mathrm{C} \approx 1.44 \, \frac{Z_\alpha Z_X}{R} \quad \mathrm{MeV} \tag{6.19}$$

with R in fm. For typical fusioning nuclei with, say, $Z_\alpha \approx Z_X \sim 2$, and $A_\alpha \approx A_X \sim 4$, $B_\mathrm{C} \sim$ MeV. The potential well energetics just described are summarized in Fig. 6.2.

The barrier penetration factor mentioned previously is the probability that a particle may quantum mechanically tunnel through the Coulomb barrier shown in Fig. 6.2. Two situations are of interest here. A projectile, such as α of reaction (6.11), with initial kinetic energy $\mathcal{E}$, must tunnel through the Coulomb barrier to reach the target X or, in the exit channel, the compound state Z^* breaks up internally into $Y + \beta$ and β must have enough energy to tunnel its way out to freedom from Y. It is usually the case in stellar charged particle reactions that particles β of the exit channel have energies comparable to, or greater than, B_C, whereas entrance channel particles α have much lower energies than B_C. The "IN" and "OUT" in Fig. 6.2 is our cartoon version of these statements.

The relation between the barrier penetration factor P_ℓ, reduced width (γ_i^2), and Γ_i is

$$\Gamma_i = 2 P_\ell \gamma_i^2 \tag{6.20}$$

where, as we shall see shortly, P_ℓ can depend sensitively on the ratio of kinetic energy to Coulomb barrier height. The subscript "ℓ" refers to the angular momentum quantum number ℓ and it implies that the effectiveness of barrier penetration also depends on the relative angular momenta of the particles involved in the reaction.

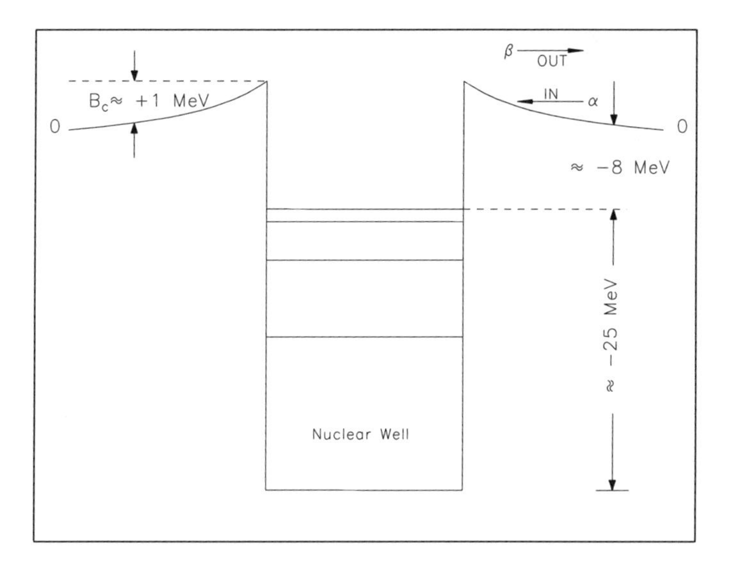

Fig. 6.2. A sketch of a nuclear potential well including the $1/r$ Coulomb barrier between the target nucleus and a hypothetical charged projectile. (Not to scale.) "OUT" indicates an exit channel particle leaving the final nucleus above the Coulomb barrier, while "IN" is the entrance channel particle trying to enter the target nucleus below the Coulomb barrier.

6.2.3 Astrophysical Thermonuclear Cross Sections and Reaction Rates

To derive a rate for a particular reaction, we need the cross section for that reaction. For the moment, imagine that we are in the laboratory and the target nucleus X is stationary while being bombarded by a beam of monoenergetic particles, α, of velocity v. Calling $\sigma_{\alpha\beta}$ the cross section for the model reaction of (6.11), where the subscripts tell us who is doing what to whom, we have (as in the definition 4.57 used in electron scattering)

$$\sigma_{\alpha\beta}(v) = \frac{\text{number of reactions per unit time per target X}}{\text{incident flux of projectiles } \alpha} \quad \text{cm}^2. \tag{6.21}$$

The incident projectile flux is $n_\alpha v$, where n_α is the number density of projectiles in the beam so that the reaction rate per target nucleus is $n_\alpha v\, \sigma_{\alpha\beta}$. The *total* reaction rate (in number of events per unit time per unit volume of target) is

$$r_{\alpha\beta}(v) = n_\alpha n_X\, \sigma_{\alpha\beta}(v)\, v \quad \text{cm}^{-3}\ \text{s}^{-1} \tag{6.22}$$

where n_X is the target number density. If α and X are the same species of particle (as in the proton-proton reaction or the first stage of the triple-α reaction), then care must be taken when applying (6.22) to avoid double counting. In that situation, (6.22) *must be divided by 2.*

The above expression for the reaction rate does not, of course, apply to the stellar environment. In almost all situations of astrophysical interest both targets and projectiles are in Maxwell–Boltzmann energy distributions. We shall not go through the details here (see Clayton, Chap. 4, or Cox §17.12) but it is easy to show that the distribution of the product $n_\alpha n_X$ is also Maxwell–Boltzmann but it *must be expressed in the center-of-mass system*. The result of integrating (6.22) over all particles in their respective distributions is the rate per unit volume, $r_{\alpha\beta}$, phrased in terms of a suitably averaged product of cross section and velocity:

$$r_{\alpha\beta} = n_\alpha n_X \langle\sigma v\rangle_{\alpha\beta} \quad \text{cm}^{-3}\ \text{s}^{-1} . \tag{6.23}$$

The total number densities appear in this expression as n_α and n_X, and $\langle\sigma v\rangle_{\alpha\beta}$ is constructed by the weighting of cross section and velocity with the differential Maxwell–Boltzmann distribution, $\Psi(\mathcal{E})$, in energy space. The latter was given in Chapter 3 as (3.26): that is,

$$\Psi(\mathcal{E}) = \frac{2}{\pi^{1/2}} \frac{1}{(kT)^{3/2}} e^{-\mathcal{E}/kT} \mathcal{E}^{1/2} . \tag{6.24}$$

Note that here, and hereafter in this section (unless warned beforehand), $\mathcal{E}$ will be the center-of-mass energy of an α-X pair. With this weighting $\langle\sigma v\rangle_{\alpha\beta}$ is given by

$$\langle\sigma v\rangle_{\alpha\beta} = \frac{\int_0^\infty \Psi(\mathcal{E})\, \sigma_{\alpha\beta}(\mathcal{E})\, v\, d\mathcal{E}}{\int_0^\infty \Psi(\mathcal{E})\, d\mathcal{E}} \tag{6.25}$$

where the velocity is in the center-of-mass. Note that the integral in the denominator is unity because of normalization.

Putting in Ψ explicitly and using $v = (2\mathcal{E}/m)^{1/2}$, $\langle\sigma v\rangle_{\alpha\beta}$ becomes

$$\langle\sigma v\rangle_{\alpha\beta} = \left(\frac{8}{\pi m}\right)^{1/2} (kT)^{-3/2} \int_0^\infty \sigma_{\alpha\beta}(\mathcal{E})\, e^{-\mathcal{E}/kT}\, \mathcal{E}\, d\mathcal{E} \quad \text{cm}^3\ \text{s}^{-1}. \tag{6.26}$$

The mass, m, in this expression is the *reduced mass*,

$$m = m_\alpha m_X / (m_\alpha + m_X) . \tag{6.27}$$

The problem is now reduced to finding $\sigma_{\alpha\beta}(\mathcal{E})$ in the center-of-mass. It turns out that the cross section can almost always be written in the form

$$\sigma_{\alpha\beta}(\mathcal{E}) = \pi \lambda\!\!\!^{-2} g\, (2\ell + 1) \frac{\Gamma_\alpha \Gamma_\beta}{\Gamma^2} f(\mathcal{E}) . \tag{6.28}$$

Here the widths are as defined previously, with Γ_α being the width of the entrance channel ($\alpha + X$), Γ_β is that of the exit channel ($\beta + Y$), and Γ the total width. All are functions of energy. The entrance channel angular momentum, $\mathbf{L}$ (and this does not include the spins of the particles), makes its appearance in the quantum number ℓ ($\ell = 0, 1, \cdots, |L|$). The wavelength $\lambdabar$ is the reduced DeBroglie wavelength,

$$\pi\lambdabar^2 = \frac{\pi\hbar^2}{2\mathcal{E}m} = \frac{0.657}{\mathcal{E}(\mathrm{MeV})}\frac{1}{\mu} \quad \text{barns} \quad (1\,\text{barn} = 10^{-24}\ \text{cm}^2) \tag{6.29}$$

where $\mathcal{E}$ on the right-hand side is in MeV. The quantity μ (not to be confused with other μs we have defined) is the reduced mass in amu units (with one amu equal to 931.494 MeV in μc^2 units) and it is given by $\mu = A_\alpha A_X/(A_\alpha + A_X)$. The factor g is statistical and contains information on the spins of the target, projectile, and compound nucleus (and you are invited to derive it in Ex. 6.5). It is just a number of order unity for our purposes. Finally, the factor $f(\mathcal{E})$; we shall refer to it as a "shape factor."[1]

The various important pieces of the cross section come about as follows. For the factor containing the DeBroglie wavelength: suppose we have a nucleus that is a perfect absorber. If a particle α with linear momentum p comes within an impact parameter distance s, then its quantized angular momentum is $sp = \ell\hbar$. Thus each ℓ has associated with it an impact parameter s_ℓ. For angular momenta between ℓ and $\ell+1$, the target area, or fractional cross section, of a ring bounded by s_ℓ and $s_{\ell+1}$ is $\sigma_{\ell,\ell+1} = \pi\left(s_{\ell+1}^2 - s_\ell^2\right) = \pi\lambdabar^2(2\ell+1)$, where $\lambdabar = \hbar/p$.

The factor $\Gamma_\alpha\Gamma_\beta/\Gamma\,\Gamma$ is the joint probability of forming $\alpha+X$ and then $\beta+Y$ through the compound state Z^*, as was indicated by our earlier discussion.

The shape factor, $f(\mathcal{E})$, hides many physical effects that we can only allude to. We shall give it one of two forms: either the "resonant" or "nonresonant" $f(\mathcal{E})$. The first form varies rapidly with energy over some interesting energy range whereas the second form is always slowly varying.

1. The *resonant* form of $f(\mathcal{E})$ is that of an isolated Breit–Wigner resonance (in its most simple guise, and see Arnett, 1996, for a sample derivation) whereby

$$f(\mathcal{E}) = \frac{\Gamma^2}{(\mathcal{E} - \mathcal{E}_r)^2 + (\Gamma/2)^2}\,. \tag{6.30}$$

This form of $f(\mathcal{E})$ is strongly peaked at, or near, the *resonance energy* $\mathcal{E}_r$ and reflects the fact that the compound nucleus has a discrete state at an energy $\mathcal{E}^*$ corresponding to

$$m_{Z^*}\,c^2 + \mathcal{E}^* = m_\alpha c^2 + m_X c^2 + \mathcal{E}_r\,. \tag{6.31}$$

If Γ does not vary appreciably over the interval $\mathcal{E}_r - \Gamma/2 \le \mathcal{E} \le \mathcal{E}_r + \Gamma/2$, then the width of the state at half maximum is Γ. The Breit–Wigner resonant cross section is then

[1] Some of §6.2.3 appeared years ago (c. 1964) in a series of lecture notes derived from lectures given by A.G.W. Cameron at Yale University. The senior author of this text (C.J.H.) was responsible for preparing the portions that finally appear here. This is where I (C.J.H.) learned his nuclear astrophysics. Thanks, Al.

$$\sigma_{\alpha,\beta} = \pi \lambda\!\!\!^{-2} g(2\ell+1) \frac{\Gamma_\alpha \Gamma_\beta}{(\mathcal{E}-\mathcal{E}_r)^2 + (\Gamma/2)^2} \, . \tag{6.32}$$

For energies near $\mathcal{E}_r$, the last factor dominates and the cross section has a sharp peak around $\mathcal{E}_r$.

2. The *nonresonant* case arises when $f(\mathcal{E})$ is a constant or is slowly varying compared to other factors in the cross section. This situation commonly arises when $\mathcal{E}$ is far removed from $\mathcal{E}_r$, that is, when the reaction takes place far in the tail of a resonance. It may also arise when the reaction is intrinsically nonresonant (as we shall see in the example of the proton-proton reaction). This case will be treated first because it contains features common to many types of charged particle reactions.

6.2.4 Nonresonant Reaction Rates

These rates require further discussion of the Coulomb barrier penetration factor because it is part of the widths contained in $\sigma_{\alpha\beta}$ (and see 6.20). We first note that nuclear reactions of major astrophysical interest are exothermic: they produce energy and the Q-value is positive in the nuclear energy equation

$$m_\alpha c^2 + m_X c^2 = m_\beta c^2 + m_Y c^2 + \mathrm{Q} \, . \tag{6.33}$$

The Q-value is usually of order MeV and that is shared as kinetic energy among the exit channel particles if they are in their ground states. In a real reaction, the entrance channel kinetic energy is added on to the left-hand side of (6.33) but reappears in the exit channel and hence does not change what energy is added to the system—namely, Q per reaction. The entrance channel energy is, however, very important for the cross section. In terms of kT, entrance channel energies are typically $kT = 8.6174 \times 10^{-8}\, T(\text{in K})$ keV (see Eq. 3.37). The total temperature range spanning the hydrogen-, helium-, and carbon-burning stages is about $10^7 \lesssim T(\text{K}) \lesssim 10^9$ under normal circumstances. In terms of kT this is $1 \lesssim kT \lesssim 100$ keV. Thus input channel energies are usually considerably less than those of the exit channel.

The Coulomb barrier height, B_{C}, of (6.19) is also very large in comparison to input channel energies. From a classical perspective, this means that the target and projectile can never combine under these circumstances because the Coulomb barrier of Fig. 6.2 cannot be penetrated. On the quantum level, on the other hand, this is possible. As in other such problems involving tunneling through a relatively high barrier, however, we expect the barrier penetrability factor, P_ℓ, to be very sensitive to energy. We again refer the reader to Clayton (or almost any text on nuclear physics) where the following form for P_ℓ in the entrance channel is derived using WKBJ methods and which is valid for $B_{\text{C}} \gg \mathcal{E}$:

$$P_\ell(\mathcal{E}) \propto e^{-2\pi\eta} \tag{6.34}$$

where η is the dimensionless Sommerfeld factor

$$\eta = \frac{Z_\alpha Z_X e^2}{\hbar v} = 0.1574 Z_\alpha Z_X \left(\frac{\mu}{\mathcal{E}}\right)^{1/2} . \tag{6.35}$$

Here, $\mathcal{E}$ is the entrance channel kinetic energy (CM) in MeV and μ is the reduced mass of (6.27) expressed in amus. For $\mathcal{E} \sim kT \lesssim 100$ keV, $\mu \sim 1$, and $Z_\alpha Z_X \gtrsim 2$, find that $2\pi\eta \gtrsim 12$. With this factor in the exponential of P_ℓ it is clear that the latter is a very sensitive function of $\mathcal{E}$. Other energy- and angular momentum-dependent terms enter into a complete formulation of P_ℓ; but, unless energies become comparable to the barrier height, these terms are not nearly as important as the exponential. Thus, for example, there is a factor of $1/\mathcal{E}^{1/2}$ that belongs in (6.34), but this is usually ignored. The effect of $\ell > 0$ (non-head-on collisions) in (6.35) is to decrease P_ℓ, but this can usually be accommodated by modifying other factors.[2] (Collisions with $\ell > 0$ mean that the incoming particle is arriving off-center, so to speak, and it's tougher to catch things flying off to the side.) For energies small compared to B_C, we shall use (6.34) because it is more than adequate. Any proportionality factors we have left out of that expression will be absorbed in various elements of the cross section.

The situation for the exit channel is different. Here, for charged particle channels, stellar exothermic reactions with Q-values in excess of a couple of MeV mean that the Coulomb barrier is relatively easy to tunnel through near or above its top and P_ℓ becomes insensitive to $\mathcal{E}$. This is also true when the exit channel is electromagnetic and a γ–ray is produced along with a charged ion or when the exiting particle is a neutron. Thus the exit channel width Γ_β in expression (6.28) for the cross section is taken to be a constant independent of the entrance channel energy $\mathcal{E}$ or, at worst, it may vary slowly with energy. The total width, Γ, is also assumed to vary slowly with $\mathcal{E}$ by the same arguments.

Putting the above elements together, and remembering that $\lambda\!\!\!^{-2}$ goes as $1/\mathcal{E}$, the nonresonant form for the cross section of (6.28) becomes

$$\sigma_{\alpha\beta}(\mathcal{E}) = \frac{S(\mathcal{E})}{\mathcal{E}}\, e^{-2\pi\eta} . \tag{6.36}$$

Here $S(\mathcal{E})$ is a slowly varying function of $\mathcal{E}$ and contains all the energy dependencies not contained in $\lambda\!\!\!^{-}$ or in the exponential of P_ℓ.

Since the form of η is known, a common procedure is to extract $\sigma_{\alpha\beta}$ at experimentally accessible laboratory energies (usually, and unfortunately, at energies not much below 100 keV) and plot $S(\mathcal{E}) = \sigma_{\alpha\beta}\, \mathcal{E}\, e^{2\pi\eta}$ as a function of $\mathcal{E}$. If necessary $S(\mathcal{E})$ is then extrapolated, with perhaps some help from theory, down to astrophysically interesting energies. We shall give an example shortly.

---

[2] To find the exact dependence of the barrier penetrability on ℓ and $\mathcal{E}$ requires calculating the Coulomb wave functions. These are discussed in most nuclear physics texts and in some mathematical physics texts.

Some of the major uncertainties in reaction rates derive from this procedure because low experimental energies usually involve low cross sections and, hence, relatively large experimental errors.

Assuming that $\sigma_{\alpha\beta}(\mathcal{E})$ is known either from experiment and/or by extrapolation of $S(\mathcal{E})$, we can now introduce (6.36) into the expression (6.26) for $\langle\sigma v\rangle_{\alpha\beta}$ and find

$$\langle\sigma v\rangle_{\alpha\beta} = \left(\frac{8}{\pi m}\right)^{1/2} (kT)^{-3/2} \int_0^\infty S(\mathcal{E}) \exp\left[-\left(\frac{\mathcal{E}}{kT} + \frac{b}{\mathcal{E}^{1/2}}\right)\right] d\mathcal{E}\,. \tag{6.37}$$

Here, m is the reduced mass in grams, S is in erg cm^2, and $b/\mathcal{E}^{1/2}$ replaces $2\pi\eta$ with

$$b = 0.99 Z_\alpha Z_X \mu^{1/2} \quad (\mathrm{MeV})^{1/2} \tag{6.38}$$

and μ is in amu. As a first step in evaluating the integral in (6.37) $S(\mathcal{E})$ is either evaluated at some typical energy where most reactions take place, or it is extrapolated to zero energy, yielding $S(0)$, and a first-order constant derivative, $dS/d\mathcal{E}|_0$, is added on to that. These are refinements (albeit often necessary) and, for our purposes, it will be sufficient to assume that S is some experimentally determined constant that may be taken out from inside the integral.

With S assumed constant, the numerical form of $\langle\sigma v\rangle_{\alpha\beta}$ is then

$$\langle\sigma v\rangle_{\alpha\beta} = \frac{1.6\times10^{-15}}{\mu^{1/2}(kT)^{3/2}}\, S \int_0^\infty \exp\left[-\left(\frac{\mathcal{E}}{kT} + \frac{b}{\mathcal{E}^{1/2}}\right)\right] d\mathcal{E} \;\; \mathrm{cm}^3\ \mathrm{s}^{-1} \tag{6.39}$$

where S is now in MeV-barns and $\mathcal{E}$ and kT are in MeV.[3]

The structure of the integrand in (6.39) reflects the combination of two strongly competing factors. The barrier penetration factor contributes the factor $\exp\left(-b/\mathcal{E}^{1/2}\right)$, which increases rapidly with increasing energy, whereas the Maxwell–Boltzmann exponential decreases rapidly as energy increases. The integrand thus increases as energy increases because the Coulomb barrier becomes more penetrable but, to offset that, the number of pairs of particles available for the reaction decreases in the exponential tail of the distribution. What results is a compromise between the two competing factors. This is illustrated in Fig. 6.3, where the integrand of (6.39) is plotted for two temperatures. The reactants are protons and ^{12}C nuclei.

The integrand of (6.39) is aptly called the "Gamow peak" because of its shape and in honor of George Gamow, who early on investigated the problems of quantum mechanical transmission though barriers (and who made many contributions to astrophysics). It is easy to show that the summit of the peak lies at an energy of

$$\mathcal{E}_0 = 1.22\left(Z_\alpha^2 Z_X^2 \mu T_6^2\right)^{1/3} \text{ keV}. \tag{6.40}$$

[3] Our continual switching of units is done not to be capricious: you will see all these combinations in the literature.

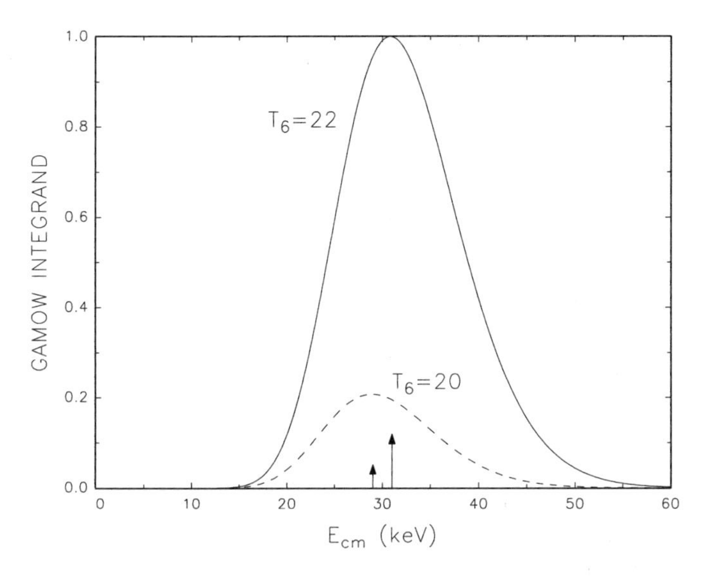

Fig. 6.3. The integrand of (6.39) is plotted against center-of-mass energy (in keV) for the temperatures $T_6 = 20$ and $T_6 = 22$. The input channel is protons on ^{12}C. The figure has been scaled by dividing both integrands by 6.6×10^{-22}, which is the maximum (prescaled) of the integrand for $T_6 = 22$. The short (tall) vertical arrow indicates the summit of the Gamow peak for $T_6 = 20$ ($T_6 = 22$) (see text).

Here, T_6 is the temperature in units of 10^6 K. Thus, for example, at a temperature of 2.2×10^7 K ($T_6 = 22$), $kT = 1.896$ keV and, for the $p+{}^{12}$C reaction, $\mathcal{E}_0 \approx 30.8$ keV. (The $p+{}^{12}$C reaction will be used as a prototype reaction for both resonant and nonresonant rates here.) The location of the summit in energy for these conditions is indicated by the large arrow in Fig. 6.3. Note that this result depends only on the reaction being nonresonant; the details of the nuclear physics are almost irrelevant. Also shown in Fig. 6.3 is the integrand for a slightly lower temperature of $T_6 = 20$, where the height of the peak is lower by a factor of almost five compared to the higher temperature. This is characteristic of low-energy charged particle reactions where the rate is a sensitive function of temperature. If you compute logarithmic derivatives of these numbers, you will find that the height of the peak varies as roughly the 17th power of the temperature. This should come as no surprise because we are dealing with one of the reactions in the CNO cycles (see §1.5).

The approximate full width of the Gamow peak (at $1/e$ of maximum) is

$$\Delta \approx 2.3 \left(\mathcal{E}_0 \, kT\right)^{1/2} \tag{6.41}$$

in whatever common units are used for $\mathcal{E}_0$ and kT. In the above example with $T_6 = 22$, $\Delta \approx 17.6$ keV, so that roughly half of the reactions arise from reactant pairs with energies $22 \lesssim \mathcal{E} \lesssim 40$ keV $>> kT \sim 2$ keV. (You can almost eyeball this from Fig. 6.3.)

A closed expression for the integral in (6.39) does not exist, but a perfectly useful approximation may be derived that follows from replacing the integrand with a Gaussian of the same height and curvature at maximum. A full description of the procedure is given in Clayton (1968, §4–3). A simple integration over the resulting Gaussian then yields

$$\langle \sigma v \rangle_{\alpha\beta} = \frac{0.72 \times 10^{-18} S}{\mu Z_\alpha Z_X} \, e^{-\tau} \, \tau^2 \ \ \text{cm}^3 \ \text{s}^{-1} \tag{6.42}$$

where S is now in keV-barns, and

$$\tau = \frac{3\mathcal{E}_0}{kT} \, . \tag{6.43}$$

Correction terms that improve on the Gaussian consist of multiplying (6.42) by $1+(5/12\tau)+\cdots$. Because $\mathcal{E}_0$ is usually much greater than kT, we shall neglect this correction (and while we have not included anything fancy for possible slow energy variations in S, it can be expressed as a Taylor series with knowledge of $dS/d\mathcal{E}$, as in Clayton, §4–3).

We can express (6.42) in terms of temperature by unwinding τ to find

$$\langle \sigma v \rangle_{\alpha\beta} = \frac{0.72 \times 10^{-18} S a^2}{\mu Z_\alpha Z_X} \, \frac{e^{-a T_6^{-1/3}}}{T_6^{2/3}} \ \ \text{cm}^3 \ \text{s}^{-1} \tag{6.44}$$

where $a = 42.49 \left(Z_\alpha^2 Z_X^2 \mu \right)^{1/3}$ and S is in keV-barns. The temperature exponents in the exponential and the denominator are characteristic of nonresonant reactions.

Example: The ^{12}C (p, γ) ^{13}N Reaction

To give an example of how the above results are applied, consider the well-studied reaction $^{12}\text{C}\,(\text{p},\gamma)\,^{13}\text{N}$. (Clayton also uses this reaction as a prototype. The experiments done for this reaction in the astrophysical context are classics, as you will see.) At typical hydrogen-burning temperatures on the main sequence, this reaction proceeds primarily through the low-energy tail of a resonance in ^{13}N (at 2.37 MeV) which, in the laboratory frame, is directly accessed by a proton with an energy of 0.46 MeV (and remember, the ^{12}C nuclei are stationary in the laboratory frame). The cross section for this reaction is shown in Fig. 6.4 (taken from Fowler et al., 1967) where the abscissa is the laboratory energy of the proton.

The laboratory to center-of-mass conversion of the total kinetic energy of a projectile-target pair is $\mathcal{E}(\text{CM}) = \mathcal{E}(\text{Lab})[m_X/(m_\alpha + m_X)]$, where, in our example, α refers to the proton and X to ^{12}C. For $T_6 = 22$, the Gamow peak (center-of-mass) energy of $\mathcal{E}_0 = 30.8$ keV corresponds to a proton laboratory energy of 33.4 keV = 0.0334 MeV. It is clear that the information contained in Fig. 6.4 must be extrapolated down to energies well below the

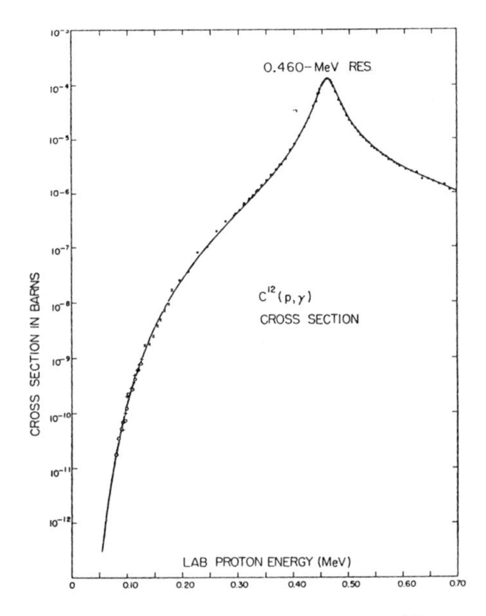

Fig. 6.4. The experimental cross section for the reaction $^{12}C\,(p,\gamma)\,^{13}N$ from Fowler et al. (1967). The solid curve is based on theory. (See also Fig. 6.5.) Reproduced, with permission, from the *Annual Review of Astronomy and Astrophysics*, Vol. 5, ©1967 by Annual Reviews, Inc.

experimental data in order that $\langle\sigma v\rangle$ be computed. This is done, as discussed earlier, by removing the penetration factor and $1/\mathcal{E}$ dependence (of 6.36) from the data and then plotting $S(\mathcal{E})$. The results of this procedure are shown in Fig. 6.5 (also from Fowler et al., 1967). Extrapolation to low energies yields $S(\mathcal{E}=0)=1.4$ keV-barns for the reaction. The rate may now be computed using (6.42) and ancillary equations.

The result we now quote for the nonresonant reaction $^{12}C(p,\gamma)^{13}N$ is from Fowler et al. (1975) where all their corrections for the energy dependence in $S(\mathcal{E})$ and adjustments to the Gaussian approximation to the shape of the Gamow peak are included for completeness. (Note that in their notation $\langle\sigma v\rangle_{p\gamma}$ becomes $\langle^{12}C\,p\rangle_{\gamma}$.) Their result is

$$\begin{aligned}\langle\sigma v\rangle_{p\gamma} &= 3.39\times10^{-17}(1+0.0304\,T_9^{1/3}+1.19\,T_9^{2/3}+0.254\,T_9+\\ &+\ 2.06\,T_9^{4/3}+1.12\,T_9^{5/3})\times T_9^{-2/3}\times\\ &\times\ \exp\left[-13.69/T_9^{1/3}-(T_9/1.5)^2\right]\ \mathrm{cm^3\ s^{-1}}.\end{aligned}\qquad(6.45)$$

The correction terms alluded to in the comments following equation (6.43) are those in the parenthesis following the unit term and in the T_9^2 term of the exponential. You may easily verify that the rest of the expression follows from the information already given. Note that the units of temperature are in

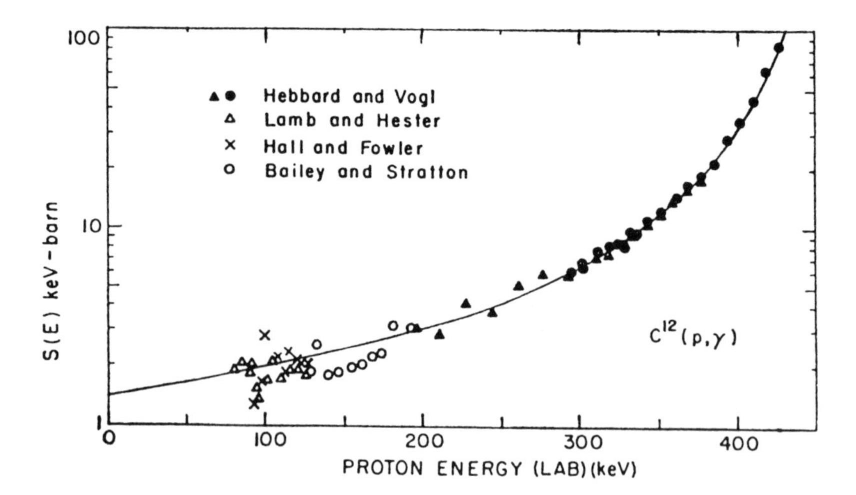

Fig. 6.5. The nonresonant factor $S(\mathcal{E})$ for the reaction ${}^{12}\mathrm{C}\,(\mathrm{p},\gamma)\,{}^{13}\mathrm{N}$ with an extrapolation to low energies (from Fowler et al., 1967). Reproduced, with permission, from the *Annual Review of Astronomy and Astrophysics*, Vol. 5, ©1967 by Annual Reviews, Inc.

billions of degrees, which is generally the case for the reaction rate material published by Fowler and his coauthors (for a listing of those papers, see the references at the end of this chapter). The quoted limits of applicability of (6.45) are $0 \leq T_9 \leq 0.55$ for reasons to be made apparent shortly.

Note: From now on in this chapter we shall usually not include all correction terms or information on all resonances for a reaction (nor electron screening terms—see later). Instead, only those terms necessary for "quick and dirty" calculations will be given.

Once $\langle \sigma v \rangle$ has been found, then the total reaction rate per unit volume is given by (see 6.23 and 1.45)

$$r_{\alpha\beta} = \rho^2 N_A^2 \frac{X_\alpha X_X}{A_\alpha A_X} \langle \sigma v \rangle_{\alpha\beta} \quad \mathrm{cm}^{-3}\ \mathrm{s}^{-1} \tag{6.46}$$

where the X_i are the mass fractions of particles α and X.[4]

The energy generation rate per gram, $\varepsilon_{\alpha\beta}$, is simply the reaction rate multiplied by the Q-value (in ergs) of the reaction divided by the density. The units of $\varepsilon_{\alpha\beta}$ are then erg g^{-1} s^{-1} in accordance with the discussion in §1.5. Thus

$$\varepsilon_{\alpha\beta} = \frac{r_{\alpha\beta}\mathrm{Q}}{\rho} \quad \mathrm{erg\ g^{-1}\ s^{-1}}. \tag{6.47}$$

[4] If α and X are the same, then the right-hand side should be divided by 2 as indicated in the discussion following (6.22).

A little care must be exercised if a neutrino is produced as a result of the reaction. Under all but the most unusual circumstances, matter is essentially transparent to these particles and thus the energy associated with the neutrino is lost from the star. (In any case, the neutrino is almost never captured where is was emitted.) Therefore, that energy must then be subtracted from Q. Neutrinos play no role in the $^{12}\text{C}\,(\text{p},\gamma)\,^{13}\text{N}$ reaction and its Q-value is 1.944 MeV=3.115×10^{-6} ergs.

The functional dependence of ε on temperature is, of course, the same as that of the reaction rate: both go as $\exp\left(-aT^{-1/3}\right)/T^{2/3}$ using (6.44). The density dependence for ε is obviously linear (once the ρ^2 term in $r_{\alpha\beta}$ is included). From these considerations it is easy to derive the temperature and density exponents used in previous chapters. With

$$\varepsilon = \varepsilon_0 \rho^\lambda T^\nu \tag{6.48}$$

the logarithmic derivatives ν and λ for nonresonant reactions are

$$\lambda = \left(\frac{\partial \ln \varepsilon}{\partial \ln \rho}\right)_T = 1 \tag{6.49}$$

$$\nu = \left(\frac{\partial \ln \varepsilon}{\partial \ln T}\right)_\rho = \frac{a}{3T_6^{1/3}} - \tfrac{2}{3}\,. \tag{6.50}$$

Note again that these expressions do not include any of the correction terms discussed above (and given in 6.45). In addition, it is clear that ν depends on temperature. For $^{12}\text{C}\,(\text{p},\gamma)\,^{13}\text{N}$ at $T_6 = 20$, $\nu \approx 16$. This is a number characteristic of the CNO cycles, as was stated in Chapter 1 and deduced from Fig. 6.3.

What happens if we now systematically raise the temperature of the gas containing protons and ^{12}C? At some point the Gamow peak will begin to encroach upon the resonance at 0.424 MeV (in the center-of-mass) and our notions about how the peak is formed will break down. This temperature is easy to estimate. If $\mathcal{E}_0 + \Delta/2 \approx \mathcal{E}_r$, then the peak begins to overlap the resonance. For this reaction $\mathcal{E}_0 + \Delta/2 = 0.393T_9^{2/3} + 0.213T_9^{5/6}$, which equals 0.42 MeV when $T_9 \approx 0.6$. This is the upper temperature limit quoted for (6.45). (The width of the resonance must also be taken into account and that can lower the temperature calculated for the limit.) Therefore, on to resonant reactions.

6.2.5 Resonant Reaction Rates

The form of the cross section for a resonant reaction is dominated by the factor $\left[(\mathcal{E} - \mathcal{E}_r)^2 + (\Gamma/2)^2\right]^{-1}$ of (6.32). Because $\Psi(\mathcal{E})$, the Maxwell–Boltzmann distribution function given by (6.24), and the Γs vary slowly over a resonance (at least as long as $\mathcal{E}_r \gg \Gamma/2$), what is usually done is to evaluate $\Psi(\mathcal{E})$ and

Γ_α at $\mathcal{E}_r$, thereby letting the resonant form act like a delta function.[5] Using this approximation $\langle \sigma v \rangle_{\alpha\beta}$ becomes

$$\begin{aligned}\langle \sigma v \rangle_{\alpha\beta} &= \frac{\pi \hbar^2 g(2\ell+1)}{2m}\left(\frac{8}{\pi m}\right)^{1/2} (kT)^{-3/2}\, e^{-\mathcal{E}_r/kT}\, \Gamma_\alpha(\mathcal{E}_r)\Gamma_\beta(\mathcal{E}_r) \times \\ &\times \int_0^\infty \frac{d\mathcal{E}}{(\mathcal{E}-\mathcal{E}_r)^2 + (\Gamma/2)^2} \end{aligned} \tag{6.51}$$

where we have used (6.26), (6.29), and (6.32). All that is left to evaluate is the integral over the resonance denominator. Because the integrand peaks sharply at $\mathcal{E}_r$ and nowhere else, including negative energies, it is customary to extend the lower limit of the integral to $-\infty$. The integral is elementary if Γ is taken constant and yields

$$\langle \sigma v \rangle_{\alpha\beta} = \hbar^2 \left(\frac{2\pi}{mkT}\right)^{3/2} g(2\ell+1)\frac{\Gamma_\alpha \Gamma_\beta}{\Gamma}\, e^{-\mathcal{E}_r/kT}\,. \tag{6.52}$$

This form is particularly useful because sometimes a low resolution experiment only yields an integrated cross section, $\int_{\text{res}} \sigma(\mathcal{E})\, d\mathcal{E}$, where the integral is only over the resonance. The same sort of delta-function trick used above may be used here to yield

$$\int_{\text{res}} \sigma(\mathcal{E})\, d\mathcal{E} = \frac{\hbar^2 \pi^2}{m\mathcal{E}_r}\left\{ g(2\ell+1)\frac{\Gamma_\alpha \Gamma_\beta}{\Gamma}\right\}. \tag{6.53}$$

The term in braces is called $(\omega\gamma)_r$ by Fowler et al. (1967) and these are tabulated for many reactions by them. In these terms,

$$\begin{aligned}\langle \sigma v \rangle_{\alpha\beta} &= \left(\frac{2\pi\hbar^2}{mkT}\right)^{3/2} \frac{(\omega\gamma)_r}{\hbar}\, e^{-\mathcal{E}_r/kT} \\ &= 2.56 \times 10^{-13} \frac{(\omega\gamma)_r}{(\mu T_9)^{3/2}}\, e^{-11.605\,\mathcal{E}_r/T_9} \end{aligned} \tag{6.54}$$

where $(\omega\gamma)_r$ and $\mathcal{E}_r$ are in MeV.

Fowler et al. (1967, 1975) give $(\omega\gamma)_r = 6.29 \times 10^{-7}$ MeV, $\Gamma(\mathcal{E}_r) = 0.0325$ MeV, and $\mathcal{E}_r = 0.424$ MeV for the 0.46 MeV (lab) state in ^{13}N. When inserted into (6.54) these yield

$$\langle \sigma v \rangle_{p\gamma} = \frac{1.8 \times 10^{-19}}{T_9^{3/2}}\, e^{-4.925/T_9} \quad \text{cm}^3\ \text{s}^{-1} \tag{6.55}$$

[5] Note that this is one way to do it. Often, in practice, $\langle \sigma v \rangle_{\alpha\beta}$ is numerically integrated over the resonance using the experimental data. This is especially true if the resonance is broad or $\mathcal{E}_r$ is at low energy or at a negative energy with respect to where the input channel produces the compound state.

for the resonant contribution to the $^{12}C(p,\gamma)^{13}N$ reaction. This result is applicable for $0.25 \le T_9 \le 7$ and is to be added onto the nonresonant expression (6.45). You will also want to consult Caughlan et al. (1985, 1988) for numerical tabulations of this rate (and others). The temperature exponent for a resonant rate of the form (6.52 or 6.54) is

$$\nu = \frac{11.61\mathcal{E}_r}{T_9} - \frac{3}{2} \tag{6.56}$$

with $\mathcal{E}_r$ in MeV and, of course, $\lambda = 1$.

Finally, Fig. 6.6 shows $N_A\langle\sigma v\rangle_{p\gamma}$ as the total of (6.45) and (6.55) (multiplied by N_A). Note the shallow dip near $T9 \approx 0.3$ as the nonresonant rate gives way to the resonance. What is remarkable is that the range of temperature shown ($0.015 \lesssim T9 \lesssim 3$) is of astrophysical interest and, over that range, the rate varies by some 19 orders of magnitude. Amazing.

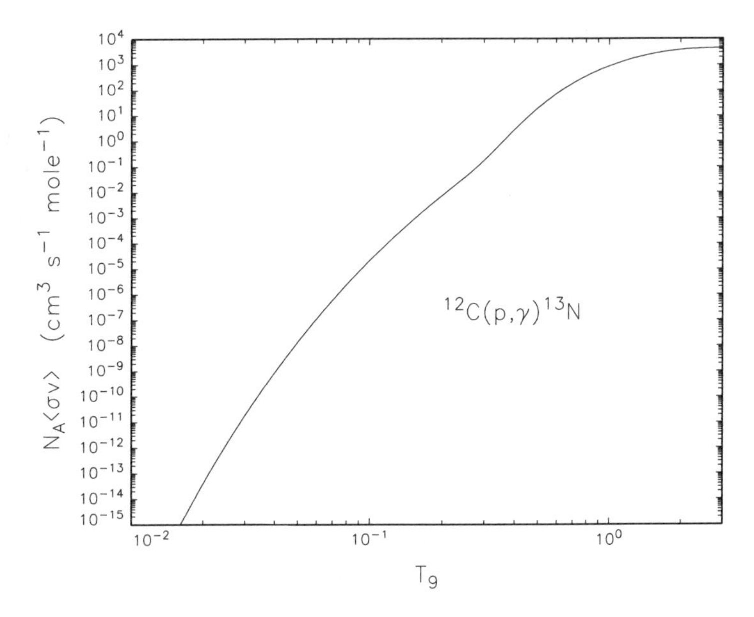

Fig. 6.6. The rate $N_A\langle\sigma v\rangle_{p\gamma}$ is plotted versus T_9 for the $^{12}C(p,\gamma)^{13}N$ reaction. The units are cm^3 s^{-1} mole^{-1}. See text for details.

6.2.6 Other Forms of Reaction Rates

Thus far we have only considered reactions initiated by protons (in particular) or charged nuclei such as alpha particles (to be dealt with in detail later). We now consider other types of reactions that either do not fall into this category or require special handling (such as the proton-proton reaction).

Neutron Capture and the S-Process

It is believed that most of the elements with $A \gtrsim 60$ were formed as the result of successions of neutron capture reactions and electron decays, as we have already discussed in §2.8.1—and here we expand on some details. These reactions can take place during some of the more normal stages of evolution or in supernovae. An example of the former in helium burning is where neutrons are formed by the reaction $^{13}C(\alpha, n)^{16}O$ and these neutrons are then captured on "seed" nuclei in the iron range of elements. This is an example of "s-process" nucleosynthesis. The rapid "r-process" is usually associated with the fast time scale of supernovae, where a myriad of reactions take place involving many nuclei. In neither process does the production of very heavy nuclei represent an energy source—and thus has little direct effect on evolution—but both are important for our understanding of heavy element abundances found in nature.

The experimental determination of neutron capture cross sections is difficult to come by because neutral particles are hard to control and the neutron has a relatively short lifetime. (The determination by Mampe et al., 1989, gives $\tau = 887.6 \pm 3$ s for the e-folding life). It is fortunate that the form of the cross section is relatively simple and, for low energies, varies as v^{-1} (and see Clayton 1968, §7–3). Because they are unaffected by a Coulomb barrier, there are no strong energy dependencies. Thus, at the lowest level of approximation, $\langle \sigma v \rangle$ is constant for any given reaction. For a summary of experiment versus theory see Käppeler et al. (1989, 1998).

We have already shown (in Figs. 2.19 and 2.20) the abundance of nuclides formed by the s- and r-processes as found in the solar system. These are presumably the byproducts from many stars over many generations. In each star, moreover, there may have been more than one episode of s-processing, each having its individual time of processing and intensity of neutron exposure. A typical neutron capture reaction has a rate $n_n n_A \langle \sigma v \rangle_A$ where n_n is the neutron number density, $\langle \sigma v \rangle_A$ is the average "$\langle \sigma v \rangle$" for capture of a neutron by a nucleus of mass A with number density n_A. This rate changes that number by $-dn_A/dt$. At the same time, ignoring β-decays, that nuclide is fed by neutron capture on nuclei with mass $A-1$. Therefore, all else aside,

$$\frac{dn_A}{dt} = -n_n n_A \langle \sigma v \rangle_A + n_n n_{A-1} \langle \sigma v \rangle_{A-1} . \tag{6.57}$$

We now convert the time variable to a "neutron exposure time," τ, by $d\tau = v n_n(t)\, dt$ so that

$$\frac{dn_A}{d\tau} = -\sigma_A n_A + \sigma_{A-1} n_{A-1} . \tag{6.58}$$

Here v is the thermal velocity and σ_A is the cross section for that velocity.

The time scales for the s-process are usually relatively long so the reactions tend to equilibrate (but never quite make it). If we assume that the reaction chain does equilibrate, then $dn_A/d\tau$ should tend to zero. If true, then the products $\sigma_A n_A$ should not differ significantly one from the other or, at least, they should be smooth over extended ranges of A. This may seem a bit surprising since the abundances in Fig. 2.21 are shown on a logarithmic scale and they do go up and down (especially around the neutron magic numbers, where the $\sigma_A n_A$ might show some action).

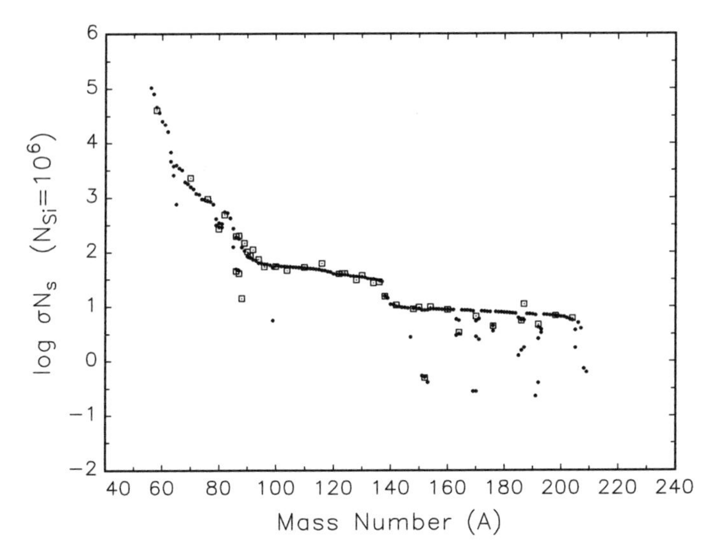

Fig. 6.7. The products of neutron capture cross section and s-process number densities are shown plotted versus atomic mass number A. The little boxes represent known abundances while the dots follow from calculations.

Figure 6.7, adapted from Käppeler et al. (1989), shows σn_s versus A where n_s is a s-process number density and σ is the neutron capture cross section for that nucleus. The little boxes show the product for nuclei whose abundances are known and the dots are for those whose abundances have not been determined. (Note that some 200 neutron capture cross sections have been fairly well-determined, which is a much larger number than abundance determinations.) What has been done here is to choose a reasonable set of exposure times and neutron fluxes so as to eventually match the known abundances (the boxes). The dotted results then follow from the calculation. In any case, the curve is smooth (except for some outliers and around magic neutron numbers, see Fig. 2.21) and, in principle, the stellar environment leading to Fig. 6.7 might be identified. As an example, see Käppeler et al. (1990), where helium shell burning in low-mass stars yield s-process abundances that closely match what is seen in the solar system.

Weak Interactions

As an aftermath of the reaction ^{12}C(p,γ)^{13}N, the nucleus ^{13}N is left in the ground state after emission of the γ–ray. Several things may now happen to that nucleus. One possibility is the reaction ^{13}N(p,γ)^{14}O (Q=4.628 MeV). The $\langle \sigma v \rangle$ for this reaction is

$$\begin{aligned} \langle \sigma v \rangle_{\mathrm{p}\gamma} &= \frac{6.71 \times 10^{-17}}{T_9^{2/3}} \, e^{-15.202/T_9^{1/3}} + \\ &+ \frac{4.04 \times 10^{-19}}{T_9^{3/2}} \, e^{-6.348/T_9} \ \mathrm{cm^3\ s^{-1}} \end{aligned} \tag{6.59}$$

for temperatures appropriate to main sequence CNO cycling. You should be able to identify both nonresonant and resonant contributions in this expression. The resonant term is only important at higher temperatures.

The time scale for destruction of ^{13}N is clearly the number density of that nucleus divided by the rate of the reaction (i.e., the rate of destruction of ^{13}N). Thus define

$$\tau_{\mathrm{p}\gamma} = \frac{n(^{13}\mathrm{N})}{r_{\mathrm{p},\gamma}} = \frac{1}{n_{\mathrm{p}} \langle \sigma v \rangle_{\mathrm{p}\gamma}} \quad \mathrm{s} \tag{6.60}$$

where n_{p} is the proton number density. (We assume, in these sorts of arguments, that volume is constant with time so that number density does not change for that reason. You could better describe number density as number per gram.) The number density of protons is $n_{\mathrm{p}} = N_{\mathrm{A}} \rho X$, where X is the hydrogen mass fraction. For $T_6 = 20$, $X \approx 1$, and $\rho \approx 10$, find that $n_{\mathrm{p}} \approx 6 \times 10^{24}$ cm^{-3}, and $\langle \sigma v \rangle_{\mathrm{p}\gamma} \approx 4 \times 10^{-40}$ cm^3 s^{-1}. This yields a time scale $\tau_{\mathrm{p}\gamma} \approx 10^7$ years.

There is, however, a complication here because ^{13}N is an unstable nucleus even in its ground state and positron decays into ^{13}C with a half–life of only 10 minutes.[6] The reaction is

$$^{13}\mathrm{N} \longrightarrow {}^{13}\mathrm{C} + \mathrm{e}^+ + \nu_{\mathrm{e}}, \ \ \tau_{1/2} = 10 \ \mathrm{min} \tag{6.61}$$

(and see Table 6.2). We thus encounter a typical situation in nuclear astrophysics: a choice must always be made regarding what reactions are important in any given situation. Here it appears that the (p,γ) reaction may safely be ignored because the ^{13}N nuclei are whisked away by positron decay before the protons can get at them. On the other hand, were the temperature and

[6] The Q-value for the reaction is 2.22 MeV, but the neutrino carries away an average of 0.71 MeV, leaving 1.51 MeV for the positron. This last energy is eventually returned to the stellar gas when the positron annihilates with an ambient electron.

density higher, say, 2.5×10^8 K and 10^3 g cm^{-3}, then the resonant contribution reduces $\tau_{p\gamma}$ to about one minute so that the capture reaction competes with the beta-decay. Such is the case in the explosive burning of hydrogen near the surface of a classical nova (as discussed in §2.11.1).

Another interesting feature arises because of the rapidity of the positron decay of ^{13}N. Before the start of hydrogen-burning in a Pop I star the abundance of ^{12}C is about 0.5% by mass of the total but the concentration of ^{13}N is, of course, zero because it is unstable. When burning does commence, the concentration of ^{13}N begins to be built up by the $^{12}\mathrm{C}(\mathrm{p},\gamma)^{13}\mathrm{N}$ reaction. If temperatures are not too high, proton captures on ^{13}N can be ignored (as in the above) and the rate of change of abundance for ^{13}N is given by

$$\frac{d^{13}\mathrm{N}}{dt} = n_{\mathrm{p}}\,{}^{12}\mathrm{C}\,\langle\sigma v\rangle_{\mathrm{p}\gamma} - \lambda\,{}^{13}\mathrm{N} \quad \mathrm{cm}^{-3}\ \mathrm{s}^{-1} \tag{6.62}$$

where ^{13}N and ^{12}C represent the number densities of the respective nuclei. The beta-decay constant, λ, is related to the half-life by $\lambda = 0.693/\tau_{1/2}$. The time development of ^{13}N under conditions of constant temperature and density and the assumption that elapsed times are sufficiently short that n_{p} and ^{12}C also remain constant is

$$^{13}\mathrm{N}(t) = \frac{n_{\mathrm{p}}{}^{12}\mathrm{C}\,\langle\sigma v\rangle_{\mathrm{p}\gamma}}{\lambda}\left(1 - e^{-\lambda t}\right) \quad \mathrm{cm}^{-3}\,. \tag{6.63}$$

This means that the concentration of ^{13}N rapidly approaches an equilibrium value and that it can just as well be computed by setting the time derivative of ^{13}N in (6.62) to zero and solving for ^{13}N. This situation is also common in nuclear astrophysics where the concentration of a nuclide involved in a comparatively rapid reaction may often easily be computed. Other examples arise in reaction chains where other considerations apply (as in the proton-proton chains to be discussed shortly).

Another kind of reaction, which results in the emission of an electron neutrino, is the capture by a nucleus of either a free electron or one in an atomic orbital. An example from the proton-proton chains is

$$\mathrm{e}^- + {}^7\mathrm{Be} \longrightarrow {}^7\mathrm{Li} + \nu_{\mathrm{e}}\,. \tag{6.64}$$

In the laboratory, neutral atoms of ^{7}Be capture atomic K–shell electrons with a half–life of about 53 days for the capture. In the hydrogen-burning stellar interior, however, temperatures are high enough to completely ionize essentially all of the ^{7}Be present and the reaction must proceed using the free electrons in the stellar plasma. The rate of the reaction is determined, in effect, by how well the wave functions of the electrons overlap the nucleus and by the intrinsic strength of the weak interaction process (see Chiu, 1968, Chap. 6, for example). An effective $\langle\sigma v\rangle$ is

$$\langle\sigma v\rangle_{\mathrm{e}^-\,\nu_{\mathrm{e}}} = \frac{2.23 \times 10^{-34}}{T_9^{1/2}} \quad \mathrm{cm}^3\ \mathrm{s}^{-1} \tag{6.65}$$

exclusive of some correction terms (with restrictions for $T_9 \leq 3$, see Caughlin et al. 1988). This expression has a form entirely different from those found thus far.

To find the rate for $^7\text{Be}(\text{e}^-, \nu_\text{e})^7\text{Li}$, (6.65) has to be multiplied by n_e, the free electron number density, and the number density of ^{7}Be. The energy generation rate follows, as usual, by dividing by the density and multiplying by the Q-value. In this reaction the total Q is 0.862 MeV but the neutrinos (of two energies depending on what state of ^{7}Li is produced) carry away all but 0.046 MeV of that figure. Even though this eventually means that the reaction is a minor direct contributor to the energy generation rate in the proton-proton chains, the neutrinos so produced are among those seen by neutrino detectors "focused" on the sun (of which more in Chap. 9).

Electron captures are also important in high-density situations, where electron Fermi energies range into the MeVs—such as in at least one kind of supernova—and we shall touch upon this in §6.8 when we discuss neutrino emission mechanisms. Pertinent references are Fuller et al. (1982, 1985).

The Proton–Proton Reaction

The proton-proton chains are initiated by the reaction

$$^1\text{H} + {}^1\text{H} \rightarrow {}^2\text{H} + \text{e}^+ + \nu_\text{e} \tag{6.66}$$

where ^{2}H is a deuteron (often given the designation ^{2}D). This crucial but, as it turns out, unlikely reaction requires that two protons form a coupled system (the "diproton") while flashing past one another and, at practically that same instant, one of these protons must undergo a weak decay by emitting a positron and electron neutrino. The two remaining massive particles, proton and neutron, are then left together as the rather fragile deuteron (2.22 MeV binding energy). This sequence of events is so unlikely that it probably will never be measurable with any certainty in the laboratory. However, the theory—for once—appears to be quite reliable.

We shall not derive the rate for this reaction here (see Chiu, 1968; Clayton, 1968; and, further back in time, the pioneering work of Bethe, Critchfield, and Salpeter listed in the references). It turns out that one of the major uncertainties is the beta-decay lifetime for the neutron, which is needed to compute the reverse process of proton decay. (Other problems may arise because of unusual conditions in the stellar plasma but these are not of a fundamental nature.)

The reaction is nonresonant and the energy dependence of the cross section arises mostly from the Coulomb barrier between the initial proton pair. The Q-value is 1.192 MeV if the energy carried away by the electron neutrino is discarded. From Caughlan and Fowler (1988) we find that for the $^1\text{H}(^1\text{H}, \text{e}^+ + \nu_\text{e})\,^2\text{H}$ reaction

$$\langle\sigma v\rangle_{\rm pp} = \frac{6.34\times10^{-39}}{T_9^{2/3}}\left(1+0.123T_9^{1/3}+1.09T_9^{2/3}+0.938T_9\right)$$
$$\times\ \exp\left(-3.380/T_9^{1/3}\right)\ {\rm cm^3\ s^{-1}}\ . \qquad (6.67)$$

(New measurements of the neutron lifetime and other corrections result in an increase of the multiplicative factor by a small number of percent. We have not included these here. See Gould and Guessoum, 1990.)

The reaction rate is obtained by multiplying by $n_{\rm p}^2/2$, where the factor of $1/2$ comes about because of the double-counting problem for identical initial particles discussed earlier. The result, excluding correction terms, is

$$r_{\rm pp} = \frac{1.15\times10^9}{T_9^{2/3}}X^2\rho^2\,e^{-3.380/T_9^{1/3}}\ \ {\rm cm^{-3}\ s^{-1}} \qquad (6.68)$$

where X is the hydrogen mass fraction. The temperature exponent for the reaction rate and energy generation rate is

$$\nu_{\rm pp} = \frac{11.3}{T_6^{1/3}} - \frac{2}{3} \qquad (6.69)$$

which, for a solar center temperature of about $T_6 = 15$, is $\nu\approx4$.

It is easy to compute the mean life of a proton against destruction by the pp-reaction—namely,

$$\tau_{\rm p} = -\frac{n_{\rm p}}{dn_{\rm p}/dt} = \frac{n_{\rm p}}{2r_{\rm pp}}\ . \qquad (6.70)$$

(Note that a factor of 2 appears because each reaction destroys two protons.) For $T_6\approx15$, $\rho\approx100$ g cm^{-3}, and $X\approx0.7$, find that $\tau_{\rm p}\approx6\times10^9$ years. That this time scale is close to the nuclear time scale given by (1.91) is no accident: the pp-reaction is so slow that it effectively controls the rate at which the pp-chains operate as a whole—as will be discussed further in the next section, and see Table 6.1.

6.2.7 Special Effects

A major modification to normal reaction rates discussed above has to do with alterations to the Coulomb potential between reactants due to the presence of intervening electrons. This is the problem of "electron screening." Here we only treat the regime where the effects are "weak" and, even then, only approximately. The case of "strong" screening is beyond the scope of this text and, in any event, many questions regarding this regime have not been satisfactorily resolved. For a very readable first paper on the subject, see Salpeter (1954).

Consider two completely ionized identical nuclear reactants of nuclear charge Z. It is assumed that the medium consists solely of these species and free electrons. We introduced the Wigner–Seitz radius, a, in Chapter 3, where it was defined by $(4\pi a^3/3) = (1/n_{\rm I})$, where $n_{\rm I}$ was the ion number density. If $Z^2e^2/a \ll kT$, then a simple exercise in Debye–Hückel theory yields the following expression for the electrostatic potential of one ion surrounded by a cloud of electrons (see, for example, Landau and Lifshitz, 1958, §74):

$$\phi(r) = \frac{Ze}{r}\, e^{-\kappa_d r} \tag{6.71}$$

where κ_d, the inverse of the Debye radius, is

$$\kappa_d = \left[\frac{4\pi e^2}{kT}\left(Z^2 n_{\rm I} + n_{\rm e}\right)\right]^{1/2} . \tag{6.72}$$

The net effect of the exponential is to reduce the potential barrier below its pure Coulomb value of Ze/r at a given radius. In other words, the electrons screen the ions from one another to some extent.

Since we are interested in how this modified potential affects the barrier penetrability, the radii of interest for nuclear reactions are those roughly equal to, or less than, the classical turning point of the motion which, for zero angular momentum, is given by $r_t = Z^2e^2/\mathcal{E}$ where $\mathcal{E}$ is the kinetic energy at infinite separation. For $\mathcal{E} \approx \mathcal{E}_0 \sim 10$ keV at the Gamow peak, and Z of unity, the turning point radius is about 10^{-11} cm. We can then approximate $\phi(r)$ for $r \leq r_t$ by

$$\phi(r) \approx \frac{Ze}{r}(1 - \kappa_d r) \tag{6.73}$$

if $\kappa_d r_t \ll 1$ or, in this numerical example, if $(n_{\rm I}/kT) \ll 10^{39}$ cm^{-3} erg^{-1}. For a solar type main sequence star with $\rho_{\rm c} \approx 100$ and $T_{\rm c} \approx 10^7$ K, $n_{\rm I}/kT \approx 10^{34}$, which seems safe enough.

The above implies that the electrostatic potential energy, $U = Ze\phi$, has been reduced by an amount $U_0 \approx Z^2e^2\kappa_d$ because of the screening presence of the electron cloud surrounding the ions. This, in turn, implies that the interacting charged particles effectively have their center-of-mass kinetic energy enhanced by an amount U_0; that is, they do not have to use up as much kinetic energy in approaching one another because the Coulomb barrier has, in effect, been reduced in height. Therefore, as a first go at seeing what this means, replace $\sigma_{\alpha\beta}(\mathcal{E})$ by $\sigma_{\alpha\beta}(\mathcal{E}+U_0)$ in expression (6.26). The other kinetic energies appearing in that expression are not altered in this approximation. We then transform the variable of integration in (6.26) from $\mathcal{E}$ to $\mathcal{E}-U_0$ with the result

$$\langle \sigma v \rangle_{\alpha\beta} \propto \int_{U_0}^{\infty} (\mathcal{E} - U_0)\sigma_{\alpha\beta}(\mathcal{E})\, e^{-\mathcal{E}/kT}\, e^{U_0/kT}\, d\mathcal{E} \, .$$

Because the dominant contribution to the rate occurs for $\mathcal{E}$ equal to either $\mathcal{E}_0$ or $\mathcal{E}_r$ (in the nonresonant and resonant forms, respectively), and both of these are greater than U_0 in cases where all these approximations will apply, then notice that the major change made in the transform of the integral is the introduction of the factor $e^{U_0/kT}$. The lower limit of the integral is now U_0 but this may be replaced by zero since very low energies (compared to $\mathcal{E}_0$ or $\mathcal{E}_r$) contribute little to the total and, similarly, the linear term in energy may be replaced by $\mathcal{E}$. The result is an integral identical to the original except for the factor $e^{U_0/kT}$. In other words,

$$\langle \sigma v \rangle_{\alpha\beta} \text{ (with screening)} = \langle \sigma v \rangle_{\alpha\beta} \text{ (unscreened)} \times e^{U_0/kT} \tag{6.74}$$

with $e^{U_0/kT} \geq 1$. Thus the rate is increased by the screening.

If this is done consistently, with due account made for differences of charge of the reactants, etc. (see Clayton, 1968, §4–8; or Cox, 1968, §17.15), then, as an example, for protons on ^{12}C in a Pop I mix at $T_6 = 20$ and $\rho = 100$ g cm^{-3}, you will find that $\exp(U_0/kT) \approx 1.25$.[7] The effect of screening may thus be significant even at relatively low densities and high temperatures. The above formalism falls apart badly, however, when U_0/kT approaches anything like unity and other steps must be taken. Some of this will be brought up again later in the context of helium-burning reactions. For now, note the curious quantum mechanical fact that reactions may occur even at very low or "zero" temperatures because of zero-point energy vibrations in a lattice as *the* extreme in screening.

We shall not explicitly indicate that screening corrections should be applied to many reactions discussed in this chapter but keep them in mind because nuclear burning at the higher densities may be effected strongly by these corrections.

6.3 The Proton–Proton Chains

The major reaction sequences in the proton-proton chains are given in Table 6.1. By "major" we mean that some minor reactions have been left out of this tabulation and that the reactions given are those appropriate to hydrogen-burning at normal main sequence temperatures.

There are three "chains," denoted by "PP–I," "PP–II," and "PP–III," and these are accessed by alternative reaction paths as indicated by the downward pointing arrows (before the third and fifth reactions). The end products of each chain are ^{4}He nuclei. Generally speaking, these chains become more important in the order I, II, and III as temperature increases. Starting at the pp-reaction itself and going to the end of any of these chains eventually involves using four protons to make each α-particle. To do so, two of the

[7] In doing this simple calculation you will find that our U_0 is of the opposite sign from that used by Clayton but is consistent with Cox. This is merely a pedagogical preference.

Table 6.1. The Proton-Proton Chains

$$\text{PP–I}\quad\left\{\begin{array}{rcl} ^1\text{H}+{}^1\text{H} & \longrightarrow & ^2\text{H}+\text{e}^+ +\nu_\text{e} \\ ^2\text{H}+{}^1\text{H} & \longrightarrow & ^3\text{He}+\gamma \\ \Downarrow {}^3\text{He}+{}^3\text{He} & \longrightarrow & ^4\text{He}+{}^1\text{H}+{}^1\text{H} \\ & \text{–or–} & \end{array}\right.$$

$$\text{PP–II}\quad\left\{\begin{array}{rcl} ^3\text{He}+{}^4\text{He} & \longrightarrow & ^7\text{Be}+\gamma \\ \Downarrow {}^7\text{Be}+\text{e}^- & \longrightarrow & ^7\text{Li}+\nu_\text{e}(+\gamma) \\ ^7\text{Li}+{}^1\text{H} & \longrightarrow & ^4\text{He}+{}^4\text{He} \\ & \text{–or–} & \end{array}\right.$$

$$\text{PP–III}\quad\left\{\begin{array}{rcl} ^7\text{Be}+{}^1\text{H} & \longrightarrow & ^8\text{B}+\gamma \\ ^8\text{B} & \longrightarrow & ^8\text{Be}+\text{e}^+ +\nu_\text{e} \\ ^8\text{Be} & \longrightarrow & ^4\text{He}+{}^4\text{He} \end{array}\right.$$

protons must be converted to neutrons and this is done by means of some combination of positron decays or electron captures.

Another "view" of Table 6.1 is shown in Fig. 6.8, where the paths for the three chains (on an essentially charge versus mass number plot) are indicated by arrows, one type per chain. If you follow the arrows, they all end up at ^{4}He, as promised.

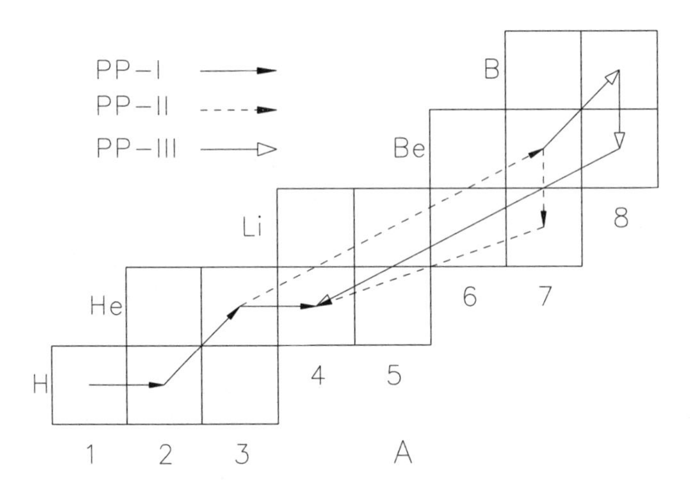

Fig. 6.8. Starting from ^{1}H, the arrows for the reaction sequences in the three pp-chains all end up at ^{4}He. Adapted from Käppeler et al. (1998).

The various reactions in the pp-chains may proceed at wildly different rates, and this is illustrated in Fig. 6.9, where $N_{\mathrm{A}}\langle\sigma v\rangle$ is plotted versus temperature for all capture reactions. The intrinsically slowest is the pp-reaction, and its run of $N_{\mathrm{A}}\langle\sigma v\rangle$ has been multiplied by 10^{18} just so it could appear in the figure. The next reaction in the PP–I chain, $^2\mathrm{H}(\mathrm{p},\gamma)^3\mathrm{He}$, is so fast that the abundance of the fragile nucleus ^{2}H is kept at a very low level, which may be computed using an equilibrium argument similar to that discussed for ^{13}N. The last reaction in the PP–I chain operates much more slowly then the preceding, but, again, it is very fast compared to the pp-reaction and a long-term equilibrium abundance of ^{3}He may be calculated (see Clayton, 1968, §§5-2, 5-3, for a full discussion of equilibration and see our Ex. 6.9).

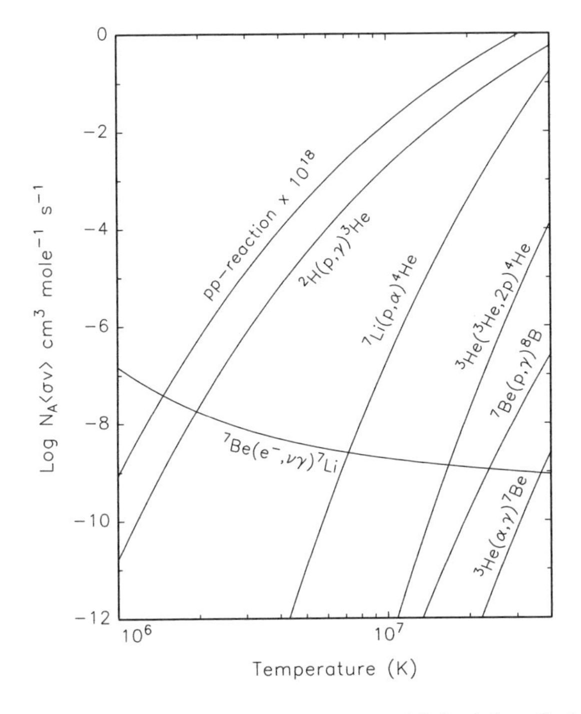

Fig. 6.9. Shown as a function of temperature are $N_{\mathrm{A}}\langle\sigma v\rangle$ for all the capture reactions in the proton-proton chains of Table 6.1. Note that the curve for the proton-proton reaction has been multiplied by a factor of 10^{18}. Material for this figure came from Caughlan and Fowler (1988).

As the temperature is raised, the equilibrium abundance of ^{3}He decreases until the first reaction in the PP–II chain begins to compete. Note that ^{4}He is always present in main sequence hydrogen-burning because of its production in Big Bang nucleosynthesis and the rate for the ^{3}He + ^{3}He reaction depends on the square of the number density $n(^3\text{He})$.

The PP–II chain continues with an electron capture on ^{7}Be. Compared to ion capture reactions its rate is comparatively constant with temperature (see 6.65 and recall that computation of its rate requires multiplying $\langle\sigma v\rangle$ by both electron and ^{7}Be number densities). The alternative reaction in PP–III is proton capture on ^{7}Be. Since the number density of electrons is roughly the same as that of protons, the crossing point of the curves for $^7\text{Be}(\text{e}^-,\nu)^7\text{Li}$ and $^7\text{Be}(\text{p},\gamma)^8\text{B}$ yields the temperature at which the PP–III chain begins to compete with PP–II. This is at a temperature of around $T_6 \approx 24$. The ^{8}B nucleus produced by the proton capture is unstable to positron decay with a half–life of about 0.8 s. The accompanying neutrino, along with that from the ^{7}Be electron capture, is being detected in solar neutrino experiments (see Chap. 9).

The final nuclear event in the PP–III chain, which is the decay of ^{8}Be into two α-particles, is not only of great importance because it terminates the pp-chains, but it, and its inverse, is also one of the key reactions in helium-burning (see §6.5). The ^{8}Be nucleus is spectacularly unstable with a mean lifetime of about 10^{-16} s.

Because the slow pp-reaction starts off the pp-chains, the rate of processing to helium is controlled by that reaction. Thus the energy generation rate for the pp-chains must be proportional to $\langle\sigma v\rangle_{\text{pp}}$ of (6.67). The overall Q-value for the chains depends, however, on the weighted contributions of the three subchains to the rate of processing. Each of these contributes differently to the energy release because of the quantities and energies of the neutrinos lost among the chains. An overall effective Q-value may be estimated and used to compute the energy generation rate. From Fowler et al. (1975), this effective Q-value is

$$\text{Q}_{\text{eff}}(\text{pp-chains}) = 13.116\left[1 + 1.412\times 10^8(1/X - 1)\,e^{-4.998/T_9^{1/3}}\right]\ \text{MeV}, \tag{6.75}$$

where X is the hydrogen mass fraction. This may be used in conjunction with (6.67) or (6.68) to form the effective energy generation rate

$$\begin{aligned} \varepsilon_{\text{eff}}(\text{pp-chains}) &= r_{\text{pp}}\,\text{Q}_{\text{eff}}/\rho \\ &\approx \frac{2.4\times 10^4 \rho X^2}{T_9^{2/3}}\,e^{-3.380/T_9^{1/3}} \quad \text{erg g}^{-1}\ \text{s}^{-1} \end{aligned} \tag{6.76}$$

where only the leading term for Q_{eff} has been used. Note that since the dominant temperature dependence is still in the exponential of $\langle\sigma v\rangle_{\text{pp}}$, the temperature exponent for the energy generation rate for the combined pp-chains is again given by (6.69).

6.3.1 Deuterium and Lithium Burning

We include this short section on deuterium and lithium not only because their burning plays a special role in some stars but also because of the cosmological implications of their abundances. For reviews, see Boesgaard and Steigman (1985) and Steigman (1985). A complete compilation of cosmological results may be found in Yang et al. (1984), for example.

Deuterium is produced in "standard models" of the Big Bang by the reaction $p + n \rightarrow {}^2H + \gamma$. If temperatures are still very high, however, the reverse reaction destroys ^{2}H as rapidly as it is formed. Only when universal expansion has sufficiently cooled the radiation field does ^{2}H persist and pp-reactions can process nuclei to ^{4}He. The amount of ^{4}He left after this stage is done (expressed as a mass fraction) is $0.24 < Y_{\rm prim} < 0.26$ and this may later be incorporated into stars. Observations of metal-poor (and, hence, old) galaxies indicate a mass fraction at the lower end of this range. The amount of ^{2}H and ^{3}He left over from the Big Bang is $1 < 10^5 \left[\left({}^2H + {}^3He\right)/{}^1H\right] < 20$ where the nuclear designations refer to number densities. Since ^{2}H is such a fragile nucleus, it is readily burned in stars and, in particular, in pre–main sequence evolution if temperatures exceed $T \gtrsim 6 \times 10^5$ K. It can then serve as an energy source (for a short time) to supplement gravitational contraction.

Lithium, as ^{7}Li, is produced and destroyed in the early universe by the same reactions given for the pp-chains. The final primordial amount left is $0.8 \lesssim 10^{10}({}^7Li/{}^1H) \lesssim 10$ from standard models. It too can be processed in stars by burning, mixing, etc. (and through cosmic rays), and we expect to see varying amounts in stellar atmospheres and the interstellar medium. Pop I stars and their associated gas show a maximum abundance ratio by number density of ${}^7Li/{}^1H < 10^{-9}$, whereas Pop II stars generally have ${}^7Li/{}^1H \sim 10^{-10}$. Among the many puzzles in nucleosynthesis and stellar evolution, however, is the following—and it has to do with the sun. Among the oldest objects in the solar system are the meteorites. The abundance of ^{7}Li has been measured in one class of these (the Type I Chondrites) to be ${}^7Li/{}^1H \sim 10^{-9}$ and this is consistent with the sun's being a Pop I star. The "lithium problem" for the sun, however, is that the solar surface abundance of lithium is only $\left({}^7Li/{}^1H\right) \sim 10^{-11}$, which is down by two orders of magnitude from what we expect. Standard evolutionary models for the sun cannot explain this and we raise this as a warning flag because the sun is our standard among stars.

6.4 The Carbon–Nitrogen–Oxygen Cycles

The major reactions comprising the CNO cycles at normally occurring hydrogen-burning temperatures are given in Table 6.2.

The general structure of the CNO cycles (or you can call them the CNOF cycles because the last involves fluorine—but few do) consists of a series of proton captures on isotopes of CNO interspersed with positron decays, and

Table 6.2. The Carbon-Nitrogen-Oxygen cycles

$$
\begin{aligned}
{}^{12}\mathrm{C} + {}^{1}\mathrm{H} &\longrightarrow {}^{13}\mathrm{N} + \gamma \\
{}^{13}\mathrm{N} &\longrightarrow {}^{13}\mathrm{C} + \mathrm{e}^{+} + \nu_{\mathrm{e}} \\
{}^{13}\mathrm{C} + {}^{1}\mathrm{H} &\longrightarrow {}^{14}\mathrm{N} + \gamma \\
{}^{14}\mathrm{N} + {}^{1}\mathrm{H} &\longrightarrow {}^{15}\mathrm{O} + \gamma \\
{}^{15}\mathrm{O} &\longrightarrow {}^{15}\mathrm{N} + \mathrm{e}^{+} + \nu_{\mathrm{e}} \\
\Downarrow {}^{15}\mathrm{N} + {}^{1}\mathrm{H} &\longrightarrow {}^{12}\mathrm{C} + {}^{4}\mathrm{He} \\
&\text{–or–} \\
{}^{15}\mathrm{N} + {}^{1}\mathrm{H} &\longrightarrow {}^{16}\mathrm{O} + \gamma \\
{}^{16}\mathrm{O} + {}^{1}\mathrm{H} &\longrightarrow {}^{17}\mathrm{F} + \gamma \\
{}^{17}\mathrm{F} &\longrightarrow {}^{17}\mathrm{O} + \mathrm{e}^{+} + \nu_{\mathrm{e}} \\
\Downarrow {}^{17}\mathrm{O} + {}^{1}\mathrm{H} &\longrightarrow {}^{14}\mathrm{N} + {}^{4}\mathrm{He} \\
&\text{–or–} \\
{}^{17}\mathrm{O} + {}^{1}\mathrm{H} &\longrightarrow {}^{18}\mathrm{F} + \gamma \\
{}^{18}\mathrm{F} + \mathrm{e}^{-} &\longrightarrow {}^{18}\mathrm{O} + \nu_{\mathrm{e}} \\
{}^{18}\mathrm{O} + {}^{1}\mathrm{H} &\longrightarrow {}^{19}\mathrm{F} + \gamma \\
{}^{19}\mathrm{F} + {}^{1}\mathrm{H} &\longrightarrow {}^{16}\mathrm{O} + {}^{4}\mathrm{He}
\end{aligned}
$$

ending with a proton capture reaction yielding ^{4}He. The first set of reactions listed in Table 6.2 is called the CN cycle and the isotopes of carbon and nitrogen act as catalysts; that is, you can start almost anywhere in the cycle, destroy one of these isotopes, and, by looping around the cycle, eventually find a reaction that makes the same isotope. This does not mean that the concentrations will remain constant through time because that depends primarily on the relative rates of the reactions in the cycles taken as a whole.

The second set of reactions in Table 6.2, when combined with the first, constitute the CNO cycle (sometimes called a tricycle). It arises from a combination of two factors: either ^{16}O is (very likely) in the stellar mixture in the first place or, in any case, it will eventually be made by the reaction ${}^{15}\mathrm{N}(\mathrm{p},\gamma){}^{16}\mathrm{O}$. Note that the final reaction in the CNO cycle sends ^{14}N right back into the CN cycle.

The branching to the third segment of the cycle is somewhat uncertain because the rate for the reaction ${}^{17}\mathrm{O}(\mathrm{p},\alpha){}^{14}\mathrm{N}$ is not well determined. We include that branch for completeness only.

An exact description of just how the CNO cycles operate in hydrogen-burning is not a trivial matter because of the intricate cycling of isotopes. Both the rate of energy generation and the detailed abundances of all the isotopes depend on the initial concentrations of the catalytic nuclei, the mean lifetimes for the individual reactions as they depend on temperature, and how long the processing has been going on. It was recognized early on, however, that a key reaction in the cycles was ${}^{14}\mathrm{N}(\mathrm{p},\gamma){}^{15}\mathrm{O}$. It is relatively slow and involves the isotope ^{14}N, which appears in the first two cycles (see Caughlan and Fowler, 1962). As we saw in the last section, a slow reaction in a chain of reactions often sets the pace for the whole. We cannot go through the complete analysis here (see Clayton 1968, §5-4) but the important result is that if temperatures are high enough to initiate CN or CNO hydrogen-burning in main sequence stars, then the most abundant nucleus will end up being ^{14}N after long enough periods of time have elapsed. This means that the cycles eventually convert almost all of the original CN or CNO nuclei, depending on temperatures and time scales, to ^{14}N. It is thought that virtually all ^{14}N seen in nature has been produced in this way.

If not enough time is available to allow the CN or CNO cycles to reach equilibrium, then the above results have to be reevaluated. This means that the differential equations that govern the creation and destruction of individual isotopes must be followed explicitly in time. This is also usually necessary when detailed isotope ratios are desired. For example, many highly evolved stars show abundance ratios between ^{12}C, ^{13}C, and ^{14}N in their spectra that are anomalous compared to some sort of cosmic standard. It is highly likely that what is being seen here is the effect of CNO processed material having being brought to the stellar surface by mixing perhaps coupled with mass loss (in, e.g., red giant or asymptotic giant phases). Thus we see directly the products of nuclear burning. For a review of this important topic see, for example, Iben and Renzini (1984).

An estimate for the energy generation rate for the CN and CNO cycles may be obtained as follows. If enough time has elapsed so that the cycles are in equilibrium, then the reaction rate for the cycles is set by the rate of $^{14}\mathrm{N}(\mathrm{p},\gamma)^{15}\mathrm{O}$. To find the energy generation rate we then need an overall Q-value. Fowler et al. (1975) recommend 24.97 MeV per proton capture on ^{14}N. We also need the number densities of protons (that is easy) and ^{14}N. The last is tricky because, if the cycles are in equilibrium, $n(^{14}\mathrm{N})$ is the sum of the number densities of the original (before burning) CN or CNO nuclei. But, as discussed above, this depends on details of temperature and time scale history.

To get an idea of what errors might arise from making the wrong choice between CN and CNO, we should look at typical abundances of these nuclei in nature and a good place to look is the sun. Bahcall and Ulrich (1988), in a review on the status of solar models, quote the following relative abundances (from L. Aller) for C, N, and O in the solar atmosphere: $n(\mathrm{C}) = 0.28$, $n(\mathrm{N}) = 0.059$, and $n(\mathrm{O}) = 0.498$. These number densities are normalized so that the total number density of all metals (i.e, all elements except hydrogen and helium) is unity. Adding these up we find that CNO constitutes about 84% of all metals by number and that O makes up about 60% of CNO by number. Delving a little deeper into the tables in the review by Bahcall and Ulrich and multiplying number densities by atomic weights also reveals that the mass fraction of CNO is $X_{\mathrm{CNO}} \approx 0.74Z$, where Z is the metal mass fraction (see §1.4). Thus CNO elements comprise the majority of metals in the solar atmosphere and, by extension, of the atmospheres of other normal Pop I stars. In addition, we find that $X_{\rm O} \approx 0.67 X_{\rm CNO}$. What this all means is that considerable caution should be exercised in choosing CN or CNO to represent ^{14}N in the CNO energy generation rate. As a compromise, and after reviewing the above figures, a reasonable choice is to set the CN or CNO mass fraction to $Z/2$ with a possible error of about 25%.

Putting this together—and after consulting Fowler et al. (1975) for the reaction rate of $^{14}{\rm N}({\rm p},\gamma)^{15}{\rm O}$—we find a very useful estimate for the energy generation rate for the CN or CNO cycles of

$$\varepsilon_{\rm CNO} \approx \frac{4.4 \times 10^{25} \rho X\, Z}{T_9^{2/3}}\, e^{-15.228/T_9^{1/3}} \quad {\rm erg\ g^{-1}\ s^{-1}}. \tag{6.77}$$

Detailed evolutionary calculations require more than this but it should suffice for making simple ZAMS models.

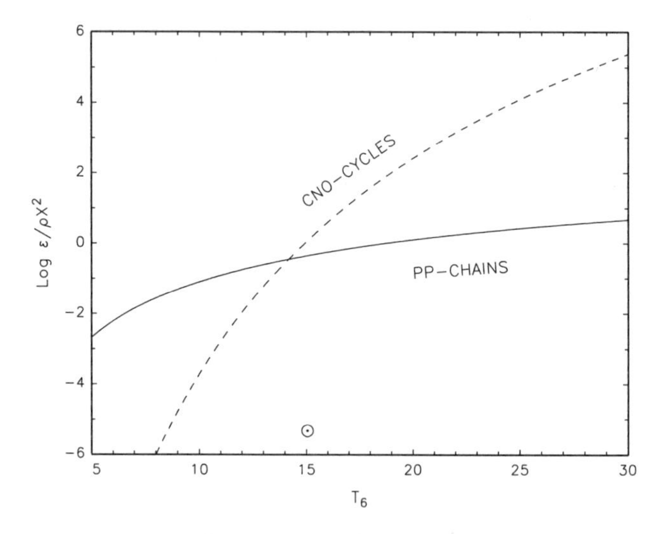

Fig. 6.10. Plots of $\varepsilon_{\rm pp}/\rho X^2$ and $\varepsilon_{\rm CNO}/\rho XZ$ as a function of temperature. (The legend on the ordinate is generic: "$\varepsilon/\rho X^2$" refers to either depending on context.) To obtain the energy generation rates, you must multiply by the density and the appropriate mass fractions. The temperature of the present-day solar center is indicated by the sun sign.

Figure 6.10 shows the pp-chain and CNO cycle energy generation rates derived from (6.76) and (6.77) with density and either factors of X^2 (for the pp-chains) or XZ (for CNO) removed. For a solar central temperature of $T_6 \approx 15$, $X = 0.7$, and $Z = 0.02$, find that $\varepsilon_{\rm pp} \approx 10 \times \varepsilon_{\rm CNO}$, so that the CNO contribution to the total energy generation rate is roughly 10%. However, it does not take much more massive a star on the main sequence with a higher than solar central temperature before the greater temperature sensitivity of the CNO cycles wins out and the pp-chains lose their dominance. Note that the temperature exponent for the CNO cycles is

$$\nu(\text{CNO}) = \frac{50.8}{T_6^{1/3}} - \tfrac{2}{3} \tag{6.78}$$

which is about 18 for $T_6 = 20$.

6.5 Helium-Burning Reactions

This section will deal with the triple-α and subsequent reactions in helium-burning. For the most part, the stellar environment is assumed to correspond to that of the cores of normal post-main sequence stars where temperatures do not greatly exceed 10^8 K. The primary reaction sequence considered here is

$$\alpha + \alpha \longrightarrow {}^8\text{Be}(\alpha,\gamma)^{12}\text{C}(\alpha,\gamma)^{16}\text{O} \, .$$

For an excellent historical review of the subject see Fowler (1986).

Helium-burning begins with the inverse of the ${}^8\text{Be} \rightarrow 2\,{}^4\text{He}$ decay that terminates the PP–III chain; that is, the first reaction is ${}^4\text{He} + {}^4\text{He} \rightarrow {}^8\text{Be}$, which is endothermic (energy absorbing) by 91.78 keV. We remarked earlier that ${}^8\text{Be}$ has a lifetime of only 10^{-16} s. Thus to produce ${}^8\text{Be}$ in any quantity whatsoever, the α-particles must have sufficient energy to gain access to the ground state of ${}^8\text{Be}$ and the formation rate of ${}^8\text{Be}$ must be sufficiently rapid to make up for its short lifetime. Since the ground state of ${}^8\text{Be}$ has a finite width ($\Gamma_{\alpha\alpha} \sim 7$ eV), we may ask at what temperature the Gamow peak begins to encroach upon that resonance. In other words, if the reaction does not begin to look resonant, then the reaction rate for production may not catch up with the inverse decay. From (6.40), the location of the peak is at $\mathcal{E}_0 = 3.9 T_6^{2/3}$ keV ($Z_\alpha = 2$, $\mu = 2$) and it equals 92 keV when $T = 1.2 \times 10^8$ K. If the effects of electron screening in high-density situations are ignored (for the present) then this roughly sets the minimum temperature for helium-burning.

Assume then that temperatures exceed 10^8 K and the ${}^8\text{Be}$ producing reaction proceeds rapidly. If rapid enough, the formation rate of ${}^8\text{Be}$ should begin to match the rate at which it is destroyed by decays; that is, the concentration of ${}^8\text{Be}$ should approach equilibrium. (This should, and can, be justified—as it is in the references.) One way to find the equilibrium concentration is to compute the rate of production by equating $n_\alpha^2 \langle \sigma v \rangle_{\alpha\alpha}/2$ (remember the factor of two for like particles) to $\lambda\, n({}^8\text{Be})$, where λ is the decay constant for ${}^8\text{Be}$. This unfortunately requires knowing $\langle \sigma v \rangle_{\alpha\alpha}$. But there is an easier and more illuminating way to go about it. We may assume chemical equilibrium and use the Saha equation (3.35) except that we now have nuclei and not atoms, ions, and electrons as was the case for the hydrogen ionization reaction (3.31).

The equilibrium reaction we are talking about is

$$\alpha + \alpha \Longleftrightarrow {}^8\text{Be}$$

and several easy modifications must be made to the Saha equation of (3.35). The first is to replace the number densities by $(n^+, n_e) \rightarrow n_\alpha$ and $n^0 \rightarrow n(^8\text{Be})$. The statistical factors, g, are unity for both ^{4}He and the ground state of ^{8}Be because both have zero spin. Instead of an ionization potential χ_H we now have the Q-value, which is –91.78 keV. Finally, the mass, m_e, is replaced by $m_\alpha^2/m(^8\text{Be}) \approx m_\alpha/2$ as the reduced mass. The "nuclear" Saha equation is then

$$\begin{aligned} \frac{n_\alpha^2}{n(^8\text{Be})} &= \left(\frac{\pi m_\alpha kT}{h^2}\right)^{3/2} e^{-\text{Q}/kT} \\ &= 1.69 \times 10^{34} T_9^{3/2}\, e^{1.065/T_9} . \end{aligned} \tag{6.79}$$

For typical conditions at, say, the start of the helium flash in lower mass stars where $\rho \approx 10^6$ g cm^{-3} ($n_\alpha \approx 1.5 \times 10^{29}$ cm^{-3} if the flash starts with pure helium) and $T_9 \approx 0.1$, find that the equilibrium concentration of ^{8}Be is about 10^{21} cm^{-3} or $n(^8\text{Be})/n_\alpha$ is only 7×10^{-9}.

With a seed of ^{8}Be nuclei now in place, however, the second stage of the triple-α reaction may now continue with the capture reaction $^8\text{Be}(\alpha,\gamma)^{12}\text{C}$. This is an exothermic resonant reaction, with Q=7.367 MeV, which proceeds through an excited state $^{12}\text{C}^*$ with zero spin at 7.654 MeV. The emission of a γ-ray photon by $^{12}\text{C}^*$ does not come easily because once the compound excited state is formed it almost always decays right back to ^{8}Be and an α-particle. Yet, as in the first step of the triple-α described above, the forward reaction is sufficiently rapid (assuming a high enough temperature) that a small pool of ^{12}C nuclei in the excited state is built up and, again, the nuclear Saha equation may be used to find the concentration in the pool. It is not difficult to do this and it should be obvious that it finally results in an expression for $n(^{12}\text{C}^*)/n_\alpha^3$ as a function of temperature after (6.79) is applied.

Having found $n(^{12}\text{C}^*)$, we can then determine the net rate of decay of $^{12}\text{C}^*$ by γ-ray cascade (or electron–positron pair emission) rather than by an α-particle: $n(^{12}\text{C}^*) \times \Gamma_\text{rad}/\hbar$, where the combination $\Gamma_\text{rad}/\hbar$ is the decay rate, λ_rad, through the uncertainty relation. (The value of Γ_rad is only 3.67 meV.) The overall sequence of the triple-α is illustrated in Fig. 6.11. (The entries to the right of the ^{12}C levels are the spins and parities of the levels.) It should be clear that the overall rate of the triple-α reaction is the same as the formation rate of the ground state of ^{12}C.

The above contains all the elements for computing the energy generation rate of the triple-α sequence. The final result we quote is taken from Harris et al. (1983, in their Table 1) where the quantity $N_\text{A}^2 \langle \alpha\alpha\alpha \rangle$ is to be found. This is multiplied by $\rho^2 Y^3 N_\text{A} \text{Q}/6A_\alpha^3$ and Q=7.367–0.0918=7.275 MeV to yield

$$\varepsilon_{\alpha\alpha\alpha} = \varepsilon_{3\alpha} = \frac{5.1 \times 10^8 \rho^2 Y^3}{T_9^3}\, e^{-4.4027/T_9} \ \text{erg g}^{-1}\ \text{s}^{-1} . \tag{6.80}$$

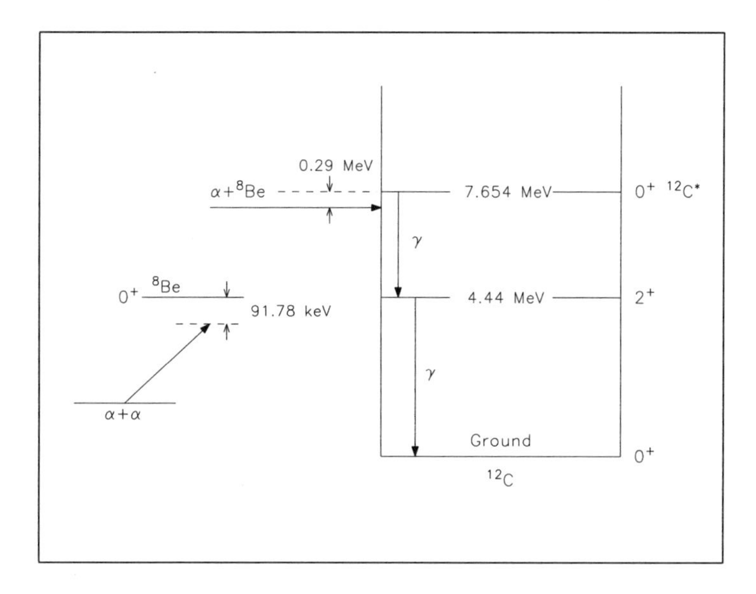

Fig. 6.11. The level diagrams and energetics of the two reactions composing the triple-α reaction (not to scale). The final result is the nucleus ^{12}C.

Here, $A_\alpha = 4$ (as part of n_α), and the division by 6 in the multiplying factor comes about because of triple counting of α-particles (as in dividing by 2 for double counting of protons in the pp chains). To verify (6.80) requires searching through some of the papers already referenced and we suggest you try to reproduce it to gain experience in how to use these references.[8] If intermediate rates are fast enough to satisfy the Saha equation, then the uncertainty in (6.80) is estimated to be only 15% (Fowler, 1986).

It is easy to show from (6.80) that the temperature and density exponents for the triple-α reaction are

$$\lambda_{3\alpha} = 2, \quad \text{and} \quad \nu_{3\alpha} = \frac{4.4}{T_9} - 3\,. \tag{6.81}$$

For $T_8 = 1$, $\nu_{3\alpha} \approx 40$, which is considerably larger than the corresponding exponent for hydrogen-burning. This means that the helium fuel is potentially more explosive than hydrogen—a fact of considerable interest for the helium flash, as discussed in §2.5.

The effects of screening are difficult to assess for a reaction such as the triple-α and we shall not attempt to do so here. In addition, the above analysis is inappropriate for temperatures much below 10^8 K because ^{4}He(α)^{8}Be no longer samples the resonance in ^{8}Be strongly. For an attempt to combine these

[8] In particular, you will need Fowler et al. (1967, 1975) and Harris et al. (1983). High-temperature correction factors and individual rates for the two parts of the triple-α may be found in Caughlan and Fowler (1988).

various elements see Fushiki and Lamb (1987), who give general expressions for the energy generation rate, including effects of weak and strong screening and, in the very-high-density limit, "pycnonuclear" effects (a term coined by Cameron, 1959; see also Ichimaru et al., 1992). These corrections can be very important.

The next step in helium-burning is α capture on ^{12}C to form ^{16}O. In the $^{12}\mathrm{C}(\alpha,\gamma)^{16}\mathrm{O}$ reaction, the $^{12}\mathrm{C}+\alpha$ pair, at zero initial energy, enters ^{16}O at 7.12 MeV (which is the Q-value). The nearest resonance in ^{16}O, however, lies some 45 keV below that energy. Hence the reaction proceeds only in the upper tail of the resonance at temperatures near 10^8 K. Unfortunately the nuclear parameters for this resonance and the detailed behavior of the resonance tail are hard to come by experimentally: a direct measurement of the cross section fails by a few orders of magnitude with present capabilities. To complicate matters, there are two (at least) levels well above the entry point of ^{4}He that can contribute (either constructively or by distructive interference) to the rate. A (hopefully) outdated quote by Fowler (1985) reads: "If users find that their results in a given study are sensitive to the rate of the $^{12}\mathrm{C}(\alpha,\gamma)^{16}\mathrm{O}$ reaction, then they should repeat their calculations with 0.5 times and 2 times the values recommended here." That is, give a factor of two either way. (And see Imbriani et al., 2001, for an example of this philosophy applied to evolutionary models.)

That situation may have been remedied by Kunz et al. (2002), who have used older experimental data plus new results of their own to calculate (using **R**–matrix theory) a new rate that they claim should be accurate to $\pm 30\%$. Whether this is an optimistic appraisal or not, their rate is (are you ready?)

$$\begin{aligned} N_{\mathrm{A}}\langle\sigma v\rangle(\alpha,{}^{12}\mathrm{C}) = {} & \frac{a_0}{T_9^2\left(1+a_1T_9^{-2/3}\right)^2}\exp\left[-a_2T_9^{-1/3}-(T_9/a_3)^2\right]+ \\ & +\frac{a_4}{T_9^2\left(1+a_5T_9^{-2/3}\right)^2}\exp\left[-a_2T_9^{-1/3}\right]+ \\ & +\frac{\tilde{a}_9}{T_9^{1/3}}\exp\left[-a_{11}T_9^{-1/3}\right] \end{aligned} \tag{6.82}$$

in the units of cm^3 s^{-1} mole^{-1}. The various constants (in Kunz et al. notation) are $a_0 = 1.21\times10^8$, $a_1 = 6.06\times10^{-2}$, $a_2 = 32.12$, $a_3 = 1.7$, $a_4 = 7.4\times10^8$, $a_5 = 0.47$, $a_{11} = 38.534$, and $\tilde{a}_9 = 3.06\times10^{10}$. The temperature range is $0.02 \lesssim T_9 \lesssim 10$. This may all seem picayune, but nature has found a way to produce a ratio $^{12}\mathrm{C}/^{16}\mathrm{O}$ that seems to fit our needs, so pay attention! Furthermore, the amounts of ^{16}O made also control to a large extent the amounts of heavier elements made in later burning stages.

The next reaction in the helium-burning sequence, $^{16}\mathrm{O}(\alpha,\gamma)^{20}\mathrm{Ne}$, is rather slow at normal helium-burning temperatures because no appropriate resonance in ^{20}Ne is available nearby where the α enters ^{20}Ne. (^{20}Ne is one of those even-even nuclei that have a very low density of levels.) Thus the competition between how fast ^{12}C is produced by the triple-α and how quickly it is converted to ^{16}O primarily determines the final relative abundances of these two nuclei. For the later evolutionary stages of lower-mass stars, this may determine whether the final core, as in a white dwarf, is mostly carbon or oxygen. For your reference, the energy generation rate for the $^{16}\mathrm{O}(\alpha,\gamma)^{20}\mathrm{Ne}$ reaction (Q=4.734 MeV) is, from Caughlan and Fowler (1988),

$$\begin{aligned}\varepsilon(\alpha, {}^{16}\mathrm{O}) &= \frac{6.69 \times 10^{26} Y\, X_{16}\, \rho}{T_9^{2/3}} \times \\ &\times \exp\left[-39.757 T_9^{-1/3} - (0.631 T_9)^2\right] \ \mathrm{erg\ g^{-1}\ s^{-1}} \quad (6.83)\end{aligned}$$

for not overly high temperatures.

Other capture reactions using α-particles that are of some importance to nucleosynthesis are those on various C, N, and O isotopes, where one of the exit channel particles is a neutron—and we have discussed these briefly before.

6.6 Carbon, Neon, and Oxygen Burning

Once α-particles have been used up in helium-burning and if temperatures can rise to $T_9 \sim 0.5$–1, carbon burning commences and, at yet higher temperatures ($T_9 \gtrsim 1$), oxygen burning. Intermediate between these two burning stages is neon burning, which uses high-energy photons to break down ^{20}Ne by "photodisintegration" (see below) via $^{20}\mathrm{Ne}(\gamma,\alpha)^{16}\mathrm{O}$.

The important branches of the reactions $^{12}\mathrm{C}+^{12}\mathrm{C}$ and $^{16}\mathrm{O}+^{16}\mathrm{O}$ are given in Table 6.3, where "yield" is the percentage of time the reaction results in the particular products on the right-hand side. The yield depends weakly on temperature and we ignore minor branches.

Table 6.3. Carbon- and Oxygen-Burning Reactions

Reaction	Yield	Q (MeV)
$^{12}\mathrm{C} + {}^{12}\mathrm{C} \rightarrow {}^{20}\mathrm{Ne} + \alpha$	44%	4.621
$^{12}\mathrm{C} + {}^{12}\mathrm{C} \rightarrow {}^{23}\mathrm{Na} + \mathrm{p}$	56%	2.242
$^{16}\mathrm{O} + {}^{16}\mathrm{O} \rightarrow {}^{28}\mathrm{Si} + \alpha$	21%	9.593
$^{16}\mathrm{O} + {}^{16}\mathrm{O} \rightarrow {}^{31}\mathrm{P} + \mathrm{p}$	61%	7.678
$^{16}\mathrm{O} + {}^{16}\mathrm{O} \rightarrow {}^{31}\mathrm{S} + \mathrm{n}$	18%	1.500

The $^{12}\mathrm{C}+{}^{12}\mathrm{C}$ reactions are followed by $^{23}\mathrm{Na}(\mathrm{p},\alpha)^{20}\mathrm{Ne}$ (Q=2.379 MeV), and $^{23}\mathrm{Na}(\mathrm{p},\gamma)^{24}\mathrm{Mg}$ (Q=11.691 MeV) using the protons released from the second reaction in Table 6.3. The α-particles can then be used on ^{16}O to form ^{20}Ne or on ^{20}Ne to yield ^{24}Mg. Depending somewhat on temperature and density, the net result of this chain of reactions is the formation of ^{20}Ne followed by lesser amounts of ^{23}Na and ^{24}Mg for quiescent carbon burning. (For a good review of quiescent heavy ion burning see Thielemann and Arnett, 1985.) The rate of energy generation for the two branches of the $^{12}\mathrm{C} + {}^{12}\mathrm{C}$ reaction is, from Caughlan and Fowler (1988),

$$\varepsilon(^{12}\mathrm{C} + {}^{12}\mathrm{C}) = \frac{1.43 \times 10^{42} \mathrm{Q}\, \eta \rho X_{12}^2}{T_9^{3/2}} \, e^{-84.165 T_9^{-1/3}} \quad \mathrm{erg\ g^{-1}\ s^{-1}} \tag{6.84}$$

where the proper Q-value (in MeV) is to be used and η is the yield of Table 6.3 multiplied by 10^{-2}. If, as a convenience, you wish to make believe that the reaction ends up as ^{24}Mg, then use $\eta=1$ and Q=13.933 MeV. Note, however, that the $^{12}\mathrm{C} + {}^{12}\mathrm{C}$ rate is not very well-determined for temperatures below $T_9 \lesssim 1$.

You may easily check that the temperature and density exponents for (6.84) are $\nu = 28/T_9^{1/3} - 1.5$ and $\lambda = 1$. The large temperature exponent is, as usual, due to the large nuclear charge of the reactants. These heavy ion reactions are especially susceptible to electron screening effects (and often take place in dense environments) so take care if you require accurate rates.

Intermediate between carbon and oxygen burning are a set of reactions that use up the neon just produced and constitute the neon burning stage. The first of this set has been referred to briefly before and is a result of the intensity of the radiation field as temperatures exceed $T_9 \gtrsim 1$. A temperature of T_9 of unity is about 0.1 MeV and there are substantial numbers of photons with energies exceeding that figure in the tail of the Planck distribution. These energies are in the range of those of low-lying nuclear states for some nuclei and it is now possible to excite those unstable states. The result is often the emission of particles from the nucleus in a "photodisintegration" reaction, which is the analogue of ionization in atoms (and see Ex. 6.10). The relevant reaction for neon burning is $^{20}\mathrm{Ne}(\gamma,\alpha)^{16}\mathrm{O}$, which is the inverse of the last reaction in helium-burning. And, as in helium-burning, the α-particle produced can be captured right back by a ^{20}Ne nucleus, but, more to the point, temperatures are sufficiently high to allow the sequence $^{20}\mathrm{Ne}(\alpha,\gamma)^{24}\mathrm{Mg}(\alpha,\gamma)^{28}\mathrm{Si}$. (Note how, in the later burning stages, the reaction sequences get more convoluted with nucleons and α-particles being tossed around to make a great variety of heavy nuclei.) The net result is a pool of ^{16}O, ^{24}Mg, and ^{28}Si (see Fig. 2.30).

The next stage is the burning of oxygen by $^{16}\mathrm{O}+{}^{16}\mathrm{O}$. (Note that $^{12}\mathrm{C}+{}^{16}\mathrm{O}$ is, in principle, possible at some point, but ^{12}C is rapidly used up by carbon-burning and the rate of $^{12}\mathrm{C} + {}^{16}\mathrm{O}$ is intrinsically slow.) The three main reactions and their yields and Q-values are given in Table 6.3 with an energy generation rate of

$$\varepsilon(^{16}\mathrm{O} + {}^{16}\mathrm{O}) = \frac{1.3 \times 10^{52} \mathrm{Q}\, \eta \rho X_{16}^2}{T_9^{2/3}} \, e^{-135.93 T_9^{-1/3}} \times$$

$$\times \; e^{\left[-0.629T_9^{2/3} - 0.445T_9^{4/3} + 0.0103T_9^2\right]} \; . \qquad (6.85)$$

If all the reactions were somehow to proceed to ^{32}S, the total Q would be 16.542 MeV.

Many reactions are possible after the last three in Table 6.3. Examples are $^{31}\text{S} \rightarrow {}^{31}\text{P} + \text{e}^+ + \nu_\text{e}$, $^{31}\text{P}(\text{p}, \alpha)^{28}\text{Si}(\alpha, \gamma)^{32}\text{S}$, etc. Completion of this stage of burning results in ^{28}Si, ^{30}Si, ^{32}S and, depending on conditions of temperature and density, ^{42}Ca and ^{46}Ti.

6.7 Silicon "Burning"

When temperatures begin to exceed some 3×10^9 K, a bewildering number of reactions are possible. We pointed out previously the effects of photodisintegration during neon burning where the radiation field was capable of "ionizing" nucleons from nuclei, which could then be used to build even more massive nuclei. As a relevant example consider $^{28}\text{Si}(\gamma, \alpha)^{24}\text{Mg}$ followed by $^{24}\text{Mg}(\alpha, \text{p})^{27}\text{Al}$ and, finally, $^{27}\text{Al}(\alpha, \text{p})^{30}\text{Si}$. Here nucleons have been recycled with the aid of photons effectively to add two neutrons to ^{28}Si to produce ^{30}Si. Amplify this to include many reactions that eventually lead up to nuclei in the iron peak—as those with the highest binding energy per nucleon—and you have the essentials of silicon burning (or perhaps "melting").

To follow all these reactions in detail is a daunting task and one that was first carried out in the pioneering calculations of Truran et al. (1966). What is required is consideration of many nuclei and the reactions that connect them (plus the cross sections for the reactions). One such reaction "network," along with the possible types of reactions, is shown in Fig. 6.12. As the burning accelerates, the reactions proceed sufficiently rapidly that a state of "quasi-static equilibrium" begins to take hold. By this we mean that photodisintegration and particle capture reactions are nearly in equilibrium but with a bias toward the production of nuclei in the iron peak. Because of the rapidity of the reactions, the abundances of most nuclei may be approximated by a nuclear version of the Saha equation where, instead of ionization potentials, nuclear masses and energies are used. (See, for example, Bodansky, Clayton, and Fowler, 1968.)

The result of silicon burning is production of nuclei in the iron peak. If enough time is allowed to elapse (as in quiescent burning), then the most abundant of these is ^{56}Fe. If, on the other hand, the time scales are short—as in a supernova—and electron or positron decays and electron captures do not have enough time to go to completion, then ^{56}Ni is the most abundant. This is crucial to our understanding of supernovae and was discussed in Chapter 2.

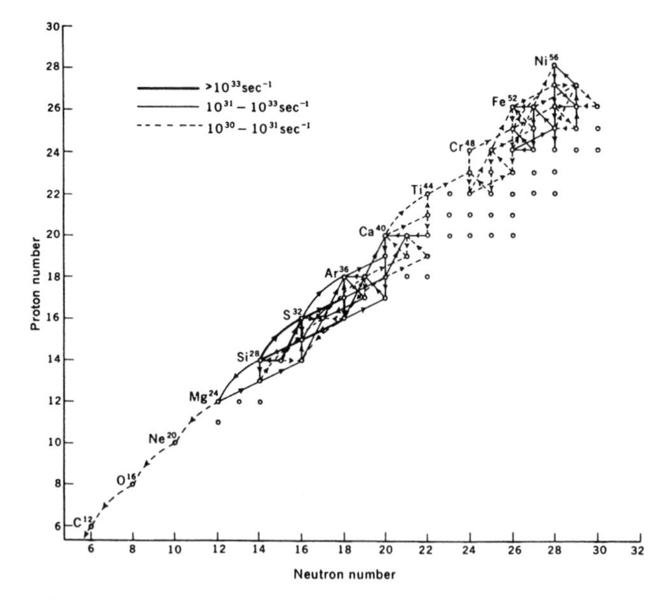

Fig. 6.12. A sample reaction network for silicon burning that also shows the reactions possible between nuclei in the network. Adapted by Clayton (1968) from Truran et al. (1966). Copyright ©1968, 1983 by D.D. Clayton and reproduced by permission.

6.8 Neutrino Emission Mechanisms

Matter at ordinary temperatures and densities is extraordinarily transparent to neutrinos. Thus if neutrinos are produced in normal stellar interiors they are an energy-loss mechanism because they carry off energy to space. But there are situations where neutrinos may be intercepted by stellar material and either be absorbed or scattered. To get an idea of how extreme those conditions must be, consider a typical neutrino capture cross section of $\sigma_\nu \sim 10^{-44}\mathcal{E}_\nu^2$ cm^2, where $\mathcal{E}_\nu$ is the neutrino energy in MeV. The target for capture is unspecified here.[9] The mean free path is then $\lambda = 1/n\sigma_\nu$, where n is the number density of targets which, if they have a mean molecular weight near unity, yields $\lambda \sim 10^{20}\mathcal{E}_\nu^{-2}/\rho$ cm. To get short mean free paths we need very high densities or neutrino energies. These conditions for stars are met, as far as we know, only in the cores of supernovae. In the core collapse phase of Type II supernovae densities approach and then exceed those of nuclear densities (as discussed in Chap. 2). If a typical density is $\rho \sim 10^{14}$ g cm^{-3}, then typical nucleon kinetic energies are in the 20 MeV range if we choose Fermi energies as being representative (as in §6.2.2). If neutrinos are produced with these

[9] Much of this is discussed in Shapiro and Teukolsky (1983, Chap. 18), Bethe (1990), and references therein. We make no attempt to give derivations of this material or that which follows.

kinds of energies then a crude estimate of λ_ν is 25 meters. Thus, neutrinos in this sort of environment are effectively "trapped" and whatever energy they carry with them may be deposited in the collapsing core. This energy transfer mechanism is one ingredient in some supernovae calculations. (For an example of the difficulties in computing how neutrinos are transported, see Burrows et al., 2000.)

Aside from such extreme environments, we should consider neutrinos as an energy drain for stars. If ε_ν represents the specific energy rate (in erg $\text{g}^{-1}\ \text{s}^{-1}$) at which neutrinos are produced, then the energy equation, excluding other factors, is $d\mathcal{L}_r/d\mathcal{M}_r = -\varepsilon_\nu$ where the minus sign reminds us that ε_ν is a power drain. What we shall briefly explore here are some important neutrino-producing mechanisms in the later stages of evolution.

The most familiar of these are electron (or positron) decay and electron capture involving nuclei. We have seen examples of these reactions in hydrogen-burning and more advanced stages. Usually the associated neutrino losses are rather modest. This is not the case, however, for the stages immediately prior to Type II supernova core collapse. The reason is that the Fermi energies of electrons in very dense environments are sufficiently high that electrons near the top of the Fermi sea are capable of being captured on protons in most nuclei. The result is not only copious neutrino production but also a shift to neutron–rich nuclei. This process can be a first step in the transformation of ordinary matter to neutron star matter. A brief overview is given by Bethe (1990). More esoteric mechanisms include the following.

Pair Annihilation Neutrinos

These neutrinos come about by the annihilation of an electron by a positron in the reaction

$$\mathrm{e}^- + \mathrm{e}^+ \longrightarrow \nu_\mathrm{e} + \overline{\nu}_\mathrm{e}\,. \tag{6.86}$$

But where can we get sufficient numbers of positrons to make this reaction at all interesting? This is not that difficult if temperatures are high enough ($kT \sim 2m_e c^2$) so that some fraction of ambient photons are capable of pair creation via

$$\gamma + \gamma \Longleftrightarrow \mathrm{e}^- + \mathrm{e}^+. \tag{6.87}$$

If this reaction goes rapidly enough then the equilibrium number densities of both electrons and positrons can be calculated from the condition on their chemical potentials, $\mu(\mathrm{e}^-) + \mu(\mathrm{e}^+) = 0$ (see §3.1). Thus, in chemical equilibrium, $\mu(\mathrm{e}^-) = -\mu(\mathrm{e}^+)$. A further requirement is that the hot gas be electrically neutral, which leads to the condition on the number densities

$$n_{\mathrm{e}^-}(\text{total}) = n_{\mathrm{e}^-}(\text{free}) + n_{\mathrm{e}^+}$$

where "free" refers to those electrons that would normally be associated with nuclei were no pairs created and "total" refers to the total of free plus pair–created electrons. The number density $n_{\mathrm{e}^-}(\text{free})$ can be calculated from the density and composition of the material and is a "given." Equation (3.44) for fermion number densities as a function of chemical potential and temperature is true for both electrons and positrons provided that $-\mu(e^-)$ is used for the positrons. This information is sufficient and both n_{e^-} and n_{e^+} may be found although the calculation is not easy. The rate of neutrino emission then follows from application of the Weinberg–Salam–Glashow theory of electroweak interactions.

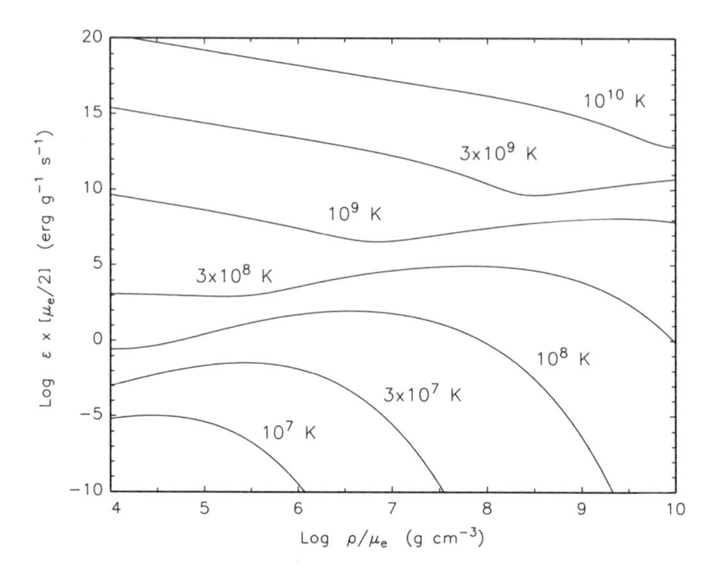

Fig. 6.13. The combined neutrino loss rates (in erg g^{-1} s^{-1}) for pair annihilation, photo-, and plasma neutrinos versus ρ/μ_e and temperature. Adapted from the calculations of Itoh and collaborators.

Photoneutrinos and Bremsstrahlung Neutrinos

The first of these is the analogue of electron–photon scattering except that instead of a final photon we get a neutrino–antineutrino pair. That is,

$$e^- + \gamma \longrightarrow e^- + \nu_e + \overline{\nu}_e \ . \tag{6.88}$$

The rule seems to be that if you can get an exiting photon, then it is also possible to get a ν_e–$\overline{\nu}_e$ pair. Thus ordinary bremsstrahlung, which yields a photon when an electron is scattered off an ion, is a likely candidate. This is an important energy loss mechanism for hot white dwarfs.

Plasma Neutrinos

From elementary physics we know that a free photon cannot create an electron–positron pair because energy and momentum cannot both be conserved in the process. In a very dense plasma, however, electromagnetic waves can be quantized in such a way that they behave like relativistic Bose particles with finite mass, "plasmons," that can decay into either e^-–e^+ or ν_e–$\overline{\nu}_e$ pairs. You might look upon plasmons as heavy photons created especially to cause trouble for some stars.

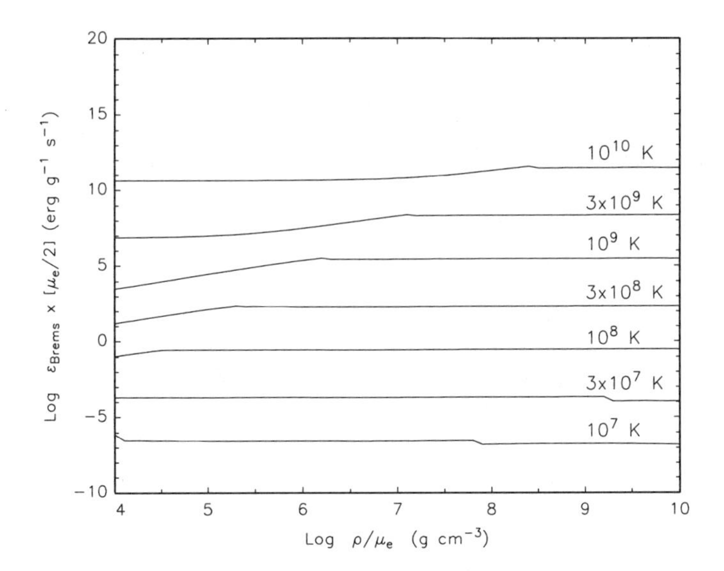

Fig. 6.14. Bremsstrahlung neutrino loss rates for pure ^{12}C. The kinks in the curves are due to overlap in fitting formulas. Adapted from the calculations of Itoh and collaborators.

Figures 6.13 and 6.14 summarize the neutrino power generated by the above reactions as functions of temperature and density. (Fig. 6.14 is for the case of pure ^{12}C.) In these figures μ_e is the usual mean molecular weight for electrons.[10] As an application of Fig. 6.13, consider SN1987A discussed in Chapter 2 and, in particular Fig. 2.30, which gave an overview of the evolutionary stages leading to explosion. The next-to-last stage, lasting some two days, consists of the building up of a $1\,\mathcal{M}_\odot$ iron peak core with a central temperature and density of $T_c \approx 3.7\times10^9$ K, $\rho_c \approx 4.9\times10^7$ g cm^{-3}. Whether the

[10] The calculations done to construct these figures were based on the work of Naoki Itoh and his collaborators (see references). We used their analytic fits to neutrino rates but they also include tables. If you wish to duplicate the figures using these fits, we warn you that some of their expressions contain obvious errors.

core is silicon or iron, μ_e is still about two. An eyeball estimate from Fig. 6.13 of the neutrino power under these conditions is $\varepsilon_\nu \sim 10^{13}$ erg g^{-1} s^{-1}. If the whole core released this specific power (and this is an overestimate), then the total neutrino luminosity is $\mathcal{L}_\nu \sim 5 \times 10^{12}\ \mathcal{L}_\odot$ (!), which is not too far distant from the value quoted in Fig. 2.30. This extraordinary luminosity loss in neutrinos is primarily due to the pair annihilation process but, in any event, it shows how important these elusive particles can sometimes be for stars.

6.9 Exercises

Exercise 6.1. Derive $\varepsilon_{\rm grav}$ of (6.5).

Exercise 6.2. Suppose the sun's thermonuclear energy source had turned itself off at the time of the hominid "Lucy" (A. Afarensis) some 3 Myr B.P. ("Before Present") ago but its luminosity has remained unchanged since that time. If so, estimate the sun's radius and effective temperature during Lucy's time. (We won't ask whether she cared one way or the other.)

Exercise 6.3. Verify Eqs. (6.40)–(6.42).

Exercise 6.4. A nuclear reaction involving only nuclei (including protons and neutrons) must obey some selection rules: otherwise the reaction can't proceed. This exercise explores rules pertaining to intrinsic spin, parity, and angular momentum. You really need some quantum mechanics to understand this fully but we will give the rules as recipes (e.g., see Clayton, 1968, §4–4). So, imagine you have some combination of two nucleons and/or nuclei combining to make a compound state of a nucleus. The entrance channel reactants have intrinsic vector spin and parity ($\Pi = \pm 1$) $\mathbf{S}_i^{\Pi_i}$ (where i labels the reactant) and their relative angular momentum vector is $\mathbf{L}$. The compound nucleus has a total angular momentum vector $\mathbf{J}_c^{\Pi_c}$ made up of all the things that go on inside the compound state. The total spin of the reactants is

$$\mathbf{S} = \mathbf{S}_1^{\Pi_1} + \mathbf{S}_2^{\Pi_2}\,.$$

By the addition rules of QM the scalar value of $\mathbf{S}$ ranges between

$$|s_1 - s_2| \le s \le |s_1 + s_2|$$

where the s_i are either integers or half-integers. The total angular momentum of the entrance channel is $\mathbf{J} = \mathbf{S} + \mathrm{L}$ with permitted scalar values

$$|\ell - s| \le J \le |\ell + s|$$

where $\ell = 0, 1, 2, \cdots, |L|$. J must equal $|J_c|$. The parity, telling how things look when viewed in a mirror, is either + or − and "adds" up as

$$\Pi(J) = \Pi(s_1)\Pi(s_2)(-1)^\ell\,.$$

To conserve parity in strong nuclear reactions (no electrons or neutrinos involved) requires $\Pi(J_c) = \Pi(J)$. Given all this, determine the following:

1. Our favorite CNO reaction, $^{12}C(p, \gamma)^{13}N$, forms the compound states in ^{13}N with J^{Π} of (in order of increasing energy) $(1/2)^+$, and $(3/2)^-$. What are the permitted values of ℓ of the entrance channel to form each of these states? (The spin and parity of protons and neutrons is $1/2^+$. The ground state of ^{12}C is 0^+.)
2. For $^{13}C(p, \gamma)^{14}N$ there are states in ^{14}N with 2^- and 1^-. If ^{13}C ground is $(1/2)^-$, what are the permissible values of ℓ?

Exercise 6.5. The "statistical factor" g that appears in (6.28) and subsequent expressions for $\langle\sigma v\rangle$ needs some explaining. For each spin and angular momentum in Ex. 6.4, there are, for example, $(2s_1 + 1)$ possible values (i.e, orientations) of s_1. If the entrance channel particles are not polarized, then all of these may participate in the interaction. The factor g is such that we sum over all possible compound state angular momenta and average over all possible incoming angular momenta. From this show that the following replacement holds:

$$g(2\ell + 1) \longrightarrow \frac{(2J + 1)}{(2s_1 + 1)(2s_2 + 1)} .$$

Exercise 6.6. As far as we know, helium is of no use for the metabolism of creatures such as us despite its being the second most abundant element in the universe. But carbon and oxygen are essential. Therefore, let's burn helium (as ^{4}He) and convert it to ^{12}C and ^{16}O. We shall use the triple-α to make ^{12}C, then add on another ^{4}He to make ^{16}O and, further, see how much ^{16}O is destroyed to make ^{20}Ne in one additional reaction. The imagined site for the burning is in the core of an intermediate mass star, where $\rho = 10^4$ g cm^{-3} and $T_9 = 0.15$. (These are typical figures for nondegenerate helium-burning in such a star.) We denote the number densities (in cm^{-3}) of the various elements by n_4, n_{12}, n_{16}, and n_{20}, and the corresponding mass fractions by X_4, X_{12}, X_{16}, and X_{20}. The average of the cross section times velocity for the reactions are denoted by $\langle 3\alpha\rangle$ for the 3-α; $\langle\alpha 12\rangle$ for $^{12}C(\alpha, \gamma)^{16}O$; $\langle\alpha 16\rangle$ for $^{16}O(\alpha, \gamma)^{20}Ne$. The $\langle\sigma v\rangle$s are to be deduced from equations (6.80), (6.82), (6.83), and accompanying material.

1. Show that the reaction rate equations that govern the creation and destruction of the nuclei are given by

$$\begin{aligned}
\frac{dn_4}{dt} &= -\frac{3n_4^3}{6}\langle 3\alpha\rangle - n_4 n_{12}\langle\alpha 12\rangle - n_4 n_{16}\langle\alpha 16\rangle \\
\frac{dn_{12}}{dt} &= \frac{n_4^3\langle 3\alpha\rangle}{6} - n_4 n_{12}\langle\alpha 12\rangle \\
\frac{dn_{16}}{dt} &= n_4 n_{12}\langle\alpha 12\rangle - n_4 n_{16}\langle\alpha 16\rangle \\
\frac{dn_{20}}{dt} &= n_4 n_{16}\langle\alpha 16\rangle .
\end{aligned}$$

2. Now integrate these equations as a function of time where the initial condition is that of pure helium at the temperature and density given above. You are to keep both of the latter fixed for all time. Note that this will involve numerical integration of four simultaneous first-order differential equations. The best way to do this is to first convert the number densities in the rate equations to mass fractions (by $n = \rho X N_{\mathrm{A}}/A$) so that the dependent variables are now the mass fractions. This will be convenient for two reasons. First off, they all range between zero and unity, so the scale of the variables is nice. Secondly, we know that

$$X_4 + X_{12} + X_{16} + X_{20} = 1$$

where, initially, $X_4 = 1$ and the others are zero. This provides a conservation check on the computations so that if the sum is *not* unity at some stage you are in trouble. Before you integrate the dX_i/dt first try to estimate the kind of time scales you are up against. The 3α reaction basically controls the flow of nuclear processing. Once you run out of αs, you're done. Therefore solve the first rate equation analytically, keeping only the 3α reaction. Then find the time (starting from time zero) when half of the αs are used up using that solution. (The time will be about 10^4 years.) Now you have a rough idea of what time steps to use at the beginning of the time integrations and you can numerically integrate the full set of equations until the αs are essentially used up.
3. Plot your results for the X_i as functions of time and see how much carbon and oxygen you are left with. You may wish to redo the calculations by changing the problematical $\langle\alpha 12\rangle$ by a $\pm 30\%$ either way to see what happens.

Exercise 6.7. In the normal course of evolution of a massive star, the end products of nuclear burning are elements in the iron region of nucleon number. From our previous discussions, we know this is a disaster because the star continues to contract and heat up. If the temperatures get high enough, the radiation field is capable of initiating photodisintegration reactions and all the iron peak elements end up as a puddle of nucleons. This can happen on such rapid time scales that the abundances of nuclei (as functions of temperature and density) can be calculated approximately as if the gas were in chemical equilibrium. This sounds like the Saha equation but the folks who do high-temperature nuclear astrophysics call it "Nuclear Statistical Equilibrium" or NSE for short. To look at this in a very simplified version consider a gas composed only of ^{56}Ni and ^{4}He where the "chemical reaction" between them is

$$14\,{}^{4}\mathrm{He} \Longleftrightarrow {}^{56}\mathrm{Ni} + \mathrm{Q}\,.$$

You may compute the Q-value for the reaction from the mass excesses

$$\begin{aligned} [M - A]\,c^2({}^{4}\mathrm{He}) &= 2.42494 \quad \mathrm{MeV} \\ [M - A]\,c^2({}^{56}\mathrm{Ni}) &= -53.902 \quad \mathrm{MeV}\,. \end{aligned}$$

1. Set up the "Saha equation" for the reaction, making believe that you are dealing with atoms and ions and assume that both nuclei are in their ground states. Let the statistical weights be equal to one (and this is all right because the spins of the ground states are zero).
2. Convert your Saha equation so that the unknowns are mass fractions X_4 and X_{56} where $X_4 + X_{56} = 1$.
3. Fix the physical density to be $\rho = 10^7$ g cm^{-3} and solve for X_4 and X_{56} for temperatures in the range $4.5 \leq T_9 \leq 6.5$.
4. At what temperature is $X_4 = X_{56}$?
5. And, yes, plot up your results for the Xs versus T_9.

Exercise 6.8. We have alluded to the nefarious helium core flash several times, but here is your chance to actually do something about it. Suppose you have a gram of pure helium (as ^{4}He) in the center of a pre-helium flash red supergiant. The density and temperature of the gram are, respectively, $\rho = 2 \times 10^5$ g cm^{-3} and $T = 1.5 \times 10^8$ K. This is hot enough to burn helium by the triple-α reaction—which is the only reaction you will use. The energy generation rate for the reaction is given by (6.80).

You are now to follow the time evolution of the gram as helium-burning proceeds by computing the temperature, $T(t)$, as a function of time. Start the clock running at time zero at the conditions stated. Assume that the density remains constant for all time, that no heat is allowed to leave the gram, and find $T(t)$ until that time when the material begins to become nondegenerate. For this use the nonrelativistic demarcation line $\rho/\mu_e = 6 \times 10^{-9} T^{3/2}$ (see 3.70). Remember to use a specific heat that is the sum of the electron specific heat (Eq. 3.115) and the ideal gas specific heat for pure helium. (Assume that x may be found from $\rho/\mu_e = Bx^3$ of 3.51.)

Just so this problem doesn't become too difficult, assume that the helium concentration does *not* change with time. Also, plot T versus time in days.

We warn you, this is not an easy problem and you may find yourself in trouble if you are not careful since a good solution requires solving a tough differential equation numerically. You will be able to recognize the flash when it happens because the temperature will suddenly skyrocket after not too many days of burning.

Exercise 6.9. (The next two exercises were suggested by Ellen Zweibel.) This has to do with the PP–I chain of Table 6.1. If we denote ^{1}H by the subscript "1," ^{2}H by "2," ^{3}He by "3," ^{4}He by "4," and let $\langle\sigma v\rangle_{ij}$ be the average cross section times velocity connecting nuclei i and j, then show that the differential equations for creation and destruction of the nuclei in Table 6.1 are

$$\frac{dn_1}{dt} = -\frac{2}{2}\langle\sigma v\rangle_{11} n_1^2 + \frac{2}{2}\langle\sigma v\rangle_{33} n_3^2 - \langle\sigma v\rangle_{12} n_1 n_2$$

$$\frac{dn_2}{dt} = \frac{1}{2}\langle\sigma v\rangle_{11} n_1^2 - \langle\sigma v\rangle_{12} n_1 n_2$$

$$\frac{dn_3}{dt} = -\frac{2}{2}\langle\sigma v\rangle_{33}n_3^2 + \langle\sigma v\rangle_{12}n_1n_2$$
$$\frac{dn_4}{dt} = \frac{1}{2}\langle\sigma v\rangle_{33}n_3^2 \,.$$

According to the discussion in §6.3 and the numbers given in Fig. 6.9, the ^{2}H(p, γ)^{3}He and ^{3}He(^{3}He, 2p)^{4}He reactions are intrinsically much faster than the pp-reaction itself. This means that over short times (compared to the time scale set by the pp-reaction) ^{2}H and ^{3}He may be regarded as being in equilibrium (meaning their time derivatives may be taken to be zero). From this show that, to a good approximation,

$$\frac{dn_4}{dt} = \frac{1}{4}\langle\sigma v\rangle_{11}n_1^2 \,. \tag{6.89}$$

Argue why this is true (still assuming equilibrium) almost be inspection.

Exercise 6.10. Consider the two reactions $X(\alpha,\gamma)Y$ and $Y(\gamma,\alpha)X$, where the second reaction is the photodisintegration inverse of the first. The rates for these, in obvious notation, are proportional to $\langle\sigma v\rangle_{\alpha X}n_\alpha n_X$ and $\lambda_{\gamma Y}n_Y$. At low temperatures—e.g., keV—the radiation field is not intense enough to initiate the photodisintegration because MeVs are required. However, in silicon burning the two reactions may be in equilibrium (or quasi-equilibrium) so that

$$X + \alpha \Longleftrightarrow Y + \gamma$$

where the reactions proceed equally rapidly in both directions. If this is true, then the Saha equation can be used to find $\lambda_{\gamma Y}$. (This quantity is notoriously difficult to find experimentally.) Thus show that

$$\lambda_{\gamma Y} = \langle\sigma v\rangle_{\alpha X}\frac{g_\alpha g_X}{g_Y}\left(\frac{2\pi m_\alpha kT}{h^3}\right)^{3/2} e^{-Q/kT}$$

where Q is the Q-value for $X(\alpha,\gamma)Y$, m_α is the mass of α assuming that mass is much less than the mass of X, and the gs are the statistical weights. For the ^{24}Mg(α,γ)^{28}Si reaction the binding energies per nucleon for α, ^{24}Mg, and ^{28}Si are, respectively (and see Fig. 6.1), 7.074, 8.26, and 8.447 MeV. What is Q? What kind of temperatures would this imply for equilibrium? (This is a bit of a phony because at high temperatures excited states may be populated and these can partake in the reaction.)

6.10 References and Suggested Readings

§6.1: Gravitational Energy Sources

Sections 17.4–17.6 of

▷ Cox, J.P. 1968, *Principles of Stellar Structure*, in two volumes (New York: Gordon and Breach)
contains a fuller discussion of gravitational sources.

§6.2: Thermonuclear Energy Sources

Chapters 4 and 5 of
▷ Clayton, D.D. 1968, *Principles of Stellar Evolution and Nucleosynthesis* (New York: McGraw-Hill)
are still the most effective general textbook references for thermonuclear reactions and nucleosynthesis in stars. Some details have changed during the intervening years, but the overall picture he presents is still accurate.
▷ Arnett, D. 1996, *Supernovae and Nucleosynthesis* (Princeton: Princeton University Press)
has a slightly different slant. The two works complement each other very well. Cox (1968, Chap. 17) and
▷ Chiu, H.-Y. 1968, *Stellar Physics*, Vol. 1 (Waltham, MA: Blaisdell)
also contain useful material.

The source for mass excesses (Δ) of nuclei used in Fig. 6.1 is
▷ Wapstra, A.H., Audi, G., & Hoekstra, R. 1988, ADNDT, 39, 281.

The Breit–Wigner resonance cross section is derived in any number of nuclear physics texts. We refer to Arnett (1996) because his derivation avoids the messy details that the latter texts must go through. For those of you looking for more information and references on barrier penetration factors as applied to nuclear astrophysics, see
▷ Humbler, J., Fowler, W.A., & Zimmerman, B.A. 1987, A&A, 177, 317.

The major source for reaction rates (of many kinds) are the compilations and critical reviews of William A. Fowler and his collaborators. The following references should be consulted in sequence of publication because philosophy and nomenclature carry over to the later papers:
▷ Fowler, W.A., Caughlan, G.R., & Zimmerman, B.A. 1967, ARA&A, 5, 525

▷ Harris, M.J., Fowler, W.A., Caughlan, G.R., & Zimmerman, B.A. 1983, ARA&A, 21, 165

▷ Caughlan, G.R., Fowler, W.A., Harris, M.J., & Zimmerman, B.A. 1985, ADNDT 32, 197

▷ Caughlan, G.R., & Fowler, W.A. 1988, ADNDT, 40, 283.

Neutron capture cross sections are reviewed by
▷ Käppeler, F., Thielemann, F.-K, & Wiescher, M. 1998, ARN&PS, 48, 175
and references therein.
For a full description of electron capture and other weak reactions in nucleosynthesis see:
▷ Fuller, G.M., Fowler, W.A., & Newman, M.J. 1982, ApJS, 48, 279

▹ Ibid. 1985, ApJ, 293, 1.

The early developments of the physics of the pp-reaction are given in

▹ Bethe, H.A., & Critchfield, C.H. 1938, PhysRev, 54, 248

▹ Bethe, H.A. 1939, PhysRev, 55, p. 103 and 434

▹ Salpeter, E.E. 1952, PhysRev, 88, 547

▹ Ibid. 1952, ApJ, 115, 326.

Recent corrections are due to

▹ Gould, R.J., & Guessoum, N. 1990, ApJ, 359, L67,

and see

▹ Mampe, M., et al. 1989, PhysRevL, 63, 593

for an experimentally determined half-life of the neutron.

▹ Käppeler, F., Beer, H., & Wisshak, K. 1989, RepProgPhys, 52, 945

▹ Käppeler, F., Thielemann, F.-K., & Wiesher, M. 1998, ARN&PS, 48, 175

▹ Käppeler, F., Gallino, R., Busso, M., Picchio, G. & Raiteri, C.M. 1990, ApJ, 354, 630

have lots of material on the s-process.

The description of electron screening used in most modern works is based on the development of

▹ Salpeter, E.E. 1954, AustJPhys, 7, 373.

Debye–Hückel theory is discussed by

▹ Landau, L.D., & Lifshitz, E.M. 1958, *Statistical Physics* (London: Pergamon).

§6.3: The Proton–Proton Chains

The expression (6.75) for the average Q-value for the pp-chains is from

▹ Fowler, W.A., Caughlan, G.R., & Zimmerman, B.A. 1975, ARA&A, 13, 69.

An improved, but more complicated, estimate has been given by

▹ Mitalas, R. 1989, ApJ, 338, 308.

The papers cited on deuterium and lithium burning and their role in cosmology are

▹ Boesgaard, A.M., & Steigman, G. 1985, ARA&A, 23 319

▹ Steigman, G. 1985, in *Nucleosynthesis*, eds. W.D. Arnett and J.W. Truran (Chicago: University of Chicago Press), p. 48

▹ Yang, J., Turner, M.S., Steigman, G., Schramm, D.N., & Olive, K.A. 1984, ApJ, 281, 493.

See also

▹ Deliyannis, C., Demarque, P., Kawaler, S., Krauss, L., & Romanelli, P. 1989, PhysRevL, 62, 1583.

§6.4: The Carbon–Nitrogen–Oxygen Cycles

Many of the intricacies of the CNO cycles were worked out by

▹ Caughlan, G.R., & Fowler, W.A. 1962, ApJ, 136, 329.

See also

▹ Caughlan, G.R. 1965, ApJ, 141, 688.

"Fast" CNO cycles, which require more reactions to treat than are listed in Table 6.2, are reviewed in

▹ Starrfield, S., Sparks, W.M., & Truran, J.W. 1974, ApJS, 28, 247

in the context of classical novae. The papers referring to CNO abundances in the sun and anomalous abundances in red supergiants are

▹ Bahcall, J.N., & Ulrich, R.K. 1988, RevModPhys, 60, 297

▹ Iben, I. Jr., & Renzini, A. 1984, PhysRep, 105, 329.

§6.5: Helium-Burning Reactions

▹ Fowler, W.A. 1986, in *Highlights of Modern. Ap.*, eds. S.L. Shapiro and S.A. Teukolosky (New York: Wiley-Interscience), p. 1

is an excellent (and personal) introduction to helium-burning. Landmark papers include

▹ Salpeter, E.E. 1952, ApJ, 115, 326

▹ Ibid. 1953, ARNS, 2, 41

▹ Hoyle, F. 1954, ApJS, 1, 121.

Individual rates are given throughout the papers of Fowler and collaborators as listed above. The quote from Fowler (1985) may be found in

▹ Fowler, W.A. 1985, in *Nucleosynthesis*, eds. W.D. Arnett & J.W. Truran (Chicago: University of Chicago Press) p. 13.

See also

▹ Filippone, B.W. 1986, ARN&PS, 36, 717

for further comments on helium-burning.

▹ Buchmann, L. et al. 1993, PhysRevL, 70, 726

report some experimental results for part of the S–factor for $^{12}C(\alpha,\gamma)^{16}O$, and the nucleosynthetic calculations of

▹ Weaver, T.A., & Woosley, S.E. 1993, PhysRep, 227, 1

appear to be consistent with what can be inferred from the experiment. These results appear to have been superceded by

▹ Kunz, R., Fey, M., Jaeger, M., Mayer, A., Hammer, J.W., Staudt, G., Harissopulos, S., & Paradellis, T. 2002, ApJ, 567, 643.

For the results of varying the above rate, see

▹ Imbriani, G., Limongi, M., Gialanella, L., Terrasi, F., Straniero, O., & Chieffi, A. 2001, ApJ, 558, 903.

Screening for the triple-α is discussed in

▹ Fushiki, I., and Lamb, D.Q. 1987, ApJ, 317, 368

and references therein. "Pycnonuclear" screening was originally discussed by

▹ Cameron, A.G.W. 1959, ApJ, 130, 916

and a newer prespective is offered by

▹ Ichimaru, S., Ogata, S., & Van Horn, H.M. 1992, ApJ, 401, L35.

§6.6: Carbon and Oxygen Burning

▷ Thielemann, F.-K., & Arnett, W.D. 1985, in *Nucleosynthesis*, eds. W.D. Arnett and J.W. Truran (Chicago: Univ. of Chicago Press), p. 151

give an excellent review of the various late thermonuclear burning stages for hydrostatic stars. The rate quoted for the $^{12}C + ^{12}C$ reaction is from

▷ Caughlan, G.R., & Fowler, W.A. 1988, ADNDT, 40, 283.

And, of course, see Arnett (1996).

§6.7: Silicon "Burning"

The first network calculations for silicon burning were reported in

▷ Truran, J.W., Cameron, A.G.W., & Gilbert, A.A. 1966, CanJPhys, 44, 576.

▷ Bodansky, D., Clayton, D.D., & Fowler, W.A. 1968, ApJS, 16, 299

clarified the quasi-static equilibrium (QSE) nature of this burning and, for a more recent reference with details, see

▷ Woosley, S.E., Arnett, W.D., & Clayton, D.D. 1973, ApJS, 26, 231.

Another summary of advanced burning, but in the context of the radioactive dating of the elements, may be found in

▷ Cowan, J.J., Thielemann, K.-R., & Truran, J.W. 1991, ARA&A, 29, 447.

§6.8: Neutrino Emission Mechanisms

Chapter 18 of

▷ Shapiro, S.L., & Teukolsky, S.A. 1983, *White Dwarfs, Neutron Stars, and Black Holes* (New York: Wiley & Sons)

gives an overview of high-energy neutrino emission mechanisms.

▷ Bethe, H.A. 1990, RevModPhys, 62, 901

places some of these in the context of supernova explosions and includes a discussion of electron capture rates.

▷ Chiu H.-Y. 1968, *Stellar Physics* (Waltham: Blaisdell)

discusses weak interactions in considerable detail. He uses an outdated theory for these interactions (not the new unified electro-weak theory), but, for low-energy reactions, his results are perfectly acceptable. The neutrino rates shown in Figs. 6.13 and 6.14 are constructed from analytic fitting formulas from the following papers:

▷ Itoh, N., & Kohyama, Y. 1983, ApJ, 275, 858

▷ Itoh, N., Matsumoto, N., Seki, M., & Kohyama, Y. 1984, ApJ, 279, 413

▷ Itoh, N., Kohyama, Y., Matsumoto, N., & Seki, M. 1984 ApJ, 280, 787

▷ Itoh, N., Kohyama, Y., Matsumoto, N., & Seki, M. 1984, ApJ, 285, 304

▷ Munakata, M., Kohyama, Y., & Itoh, N. 1987, ApJ, 316, 708

▷ Itoh, N., Adachi, T., Nakagawa, M., Kohyama, Y., & Munakata, H. 1989, ApJ, 339, 354

▷ Itoh, N., Mutoh, H., Hikita, A., & Kohyama, Y. 1992, ApJ, 395, 622.

You should also check on errata for some of these papers.

▷ Burrows, A., Young, T., Pinto, P., Eastman, R., & Thompson, T.A. 2000, ApJ, 539, 865

illutrate how difficult it is to model neutrino transport.

7 Stellar Modeling

"To err is human,
but to really foul things up
requires a computer."
— Anonymous

"Every novel should have
a beginning, a muddle, and an end."
— Peter De Vries
We are now in the stellar muddling stage.

This chapter will end up having covered a diverse set of topics but all have the same underlying theme: what analytic and numeric techniques are used to model stars? Some of these techniques will yield approximate solutions to the equations of stellar structure, whereas others are designed for the exacting task of comparing model results to real stars. We start with some rather general considerations by reviewing the equations of stellar structure.

7.1 The Equations of Stellar Structure

We shall restrict ourselves, for the moment at least, to discussing what is necessary to model stars in hydrostatic equilibrium and thermal balance while neglecting complicating factors such as nonsphericity, magnetic fields, etc. The assumption of strict equilibrium implies that time-dependent (e.g., evolutionary) processes are ignored for now.

To construct an ab initio stellar model we must first specify the total stellar mass and the run of composition as a function of some coordinate such as radius or interior mass. What should come out at the end of the calculation is the run of mass versus radius (or the other way around), and the corresponding local values of pressure, density, temperature, and luminosity. To do this, we need the microscopic constituent physics implied in the following:

$$P = P(\rho, T, \mathbf{X}) \tag{7.1}$$

$$E = E(\rho, T, \mathbf{X}) \tag{7.2}$$

$$\kappa = \kappa(\rho, T, \mathbf{X}) \tag{7.3}$$

$$\varepsilon = \varepsilon(\rho, T, \mathbf{X}) \tag{7.4}$$

and various derivatives of these quantities. Here $\mathbf{X}$ is shorthand for composition (as in a specification of nuclear species). Thus given density, temperature, and composition, the four quantities should be available on demand by the model builder.

The structural and thermal differential relations to be satisfied include

$$\frac{dP}{dr} = -\frac{G\mathcal{M}_r}{r^2}\rho \quad \text{or} \quad \frac{dP}{d\mathcal{M}_r} = -\frac{G\mathcal{M}_r}{4\pi r^4} \tag{7.5}$$

$$\frac{d\mathcal{M}_r}{dr} = 4\pi r^2 \rho \quad \text{or} \quad \frac{dr}{d\mathcal{M}_r} = \frac{1}{4\pi r^2 \rho} \tag{7.6}$$

$$\frac{d\mathcal{L}_r}{dr} = 4\pi r^2 \varepsilon\rho \quad \text{or} \quad \frac{d\mathcal{L}_r}{d\mathcal{M}_r} = \varepsilon\ . \tag{7.7}$$

The righthand variants of these equations are set down because, as we shall see, it is often convenient to take $\mathcal{M}_r$ as the independent variable.

Accompanying the above are relations or criteria that establish what the modes of heat transfer are. In the simplest instance of allowing only local adiabatic convection—as in a mixing length theory—these are as follows. Compute (as 4.30)

$$\nabla_{\text{rad}} = \frac{3}{16\pi ac}\frac{P\kappa}{T^4}\frac{\mathcal{L}_r}{G\mathcal{M}_r} \tag{7.8}$$

and, as was done in Chap. 5, test to see if this exceeds ∇_{ad}. Then set

$$\nabla = \nabla_{\text{rad}} \quad \text{if} \quad \nabla_{\text{rad}} \le \nabla_{\text{ad}} \tag{7.9}$$

for pure diffusive radiative transfer or conduction, or

$$\nabla = \nabla_{\text{ad}} \quad \text{if} \quad \nabla_{\text{rad}} > \nabla_{\text{ad}} \tag{7.10}$$

when adiabatic convection is present locally. The quantity ∇ is given by (as, e.g., in 4.28)

$$\nabla = \frac{d\ln T}{d\ln P} = -\frac{r^2 P}{G\mathcal{M}_r\rho}\frac{1}{T}\frac{dT}{dr} = -\frac{4\pi r^4 P}{G\mathcal{M}_r T}\frac{dT}{d\mathcal{M}_r}\ . \tag{7.11}$$

These computations establish the local slope of temperature with respect to pressure.

Equations (7.5–7.11), when combined, are equivalent to a fourth-order differential equation in space or mass. Four boundary conditions are required to close the system. We could choose "zero" conditions which, with $\mathcal{M}_r$ as the independent variable, are

$$\text{at the center } (\mathcal{M}_r = 0), \quad r = \mathcal{L}_r = 0 \tag{7.12}$$

$$\text{and at the surface } (\mathcal{M}_r = \mathcal{M}), \quad \rho = T = 0\ . \tag{7.13}$$

We could also get fancy and apply some version of the photospheric boundary condition of §4.3 (or even something more sophisticated) but the above will do for now. Note that $\mathcal{M}$ was specified beforehand but other quantities such as total radius, $\mathcal{R}$, and total luminosity, $\mathcal{L}$, must be found as a result of the entire calculation.

In principle we may then solve for the structure. However, does a solution always exist, and, if so, is it unique? The answer to both these questions is "no" for many choices of total mass and composition. For the first, you can't make an equilibrium star out of any old thing. The second question is a little more subtle. It turns out that, for some combinations of total mass and composition, multiple solutions to the stellar structure equations are indeed possible (see Ex. 7.1). This is not obvious, but a hint is contained, for example, in the observation that the general equation of state for stellar material is exceedingly complicated and a given pressure may be generated at different temperatures and densities by some combination of ion ideal gas, degenerate electrons, radiation, etc., with each of these components having very different thermodynamic properties. The existence of multiple solutions contradicts what was long held to be a "theorem" in stellar astrophysics due to H. Vogt and H.N. Russell. However, the idea of uniqueness is still useful in that among a set of models all having the same mass and run of composition, usually only one seems to correspond to a real star or to have come from some realistic line of stellar evolution. The others are unstable in some fundamental way (as far as we know). Thus, for example, we now know that it is possible to construct "main sequences" for stars having burnable fuel where the sequence is double-valued with respect to mass.[1] As fascinating as these things may be to some theorists, we shall brush them aside and assume that if a stellar model of given mass and run of composition can be constructed, then it is the only one possible.

Before going into how the stellar structure equations are solved in practice, we introduce a simplification which, albeit restrictive, turns out to be of both practical and pedagogical value.

7.2 Polytropic Equations of State and Polytropes

The primary, and classic, reference for the beginning portions of this section is Chandrasekhar (1939). Similar material, although not as exhaustive, may be found in Cox (1968, §23.1), and Kippenhahn and Weigert (1990, §19).

We shall first discuss polytropes in a general way but then interrupt the narrative to consider how these approximations to stellar models and, to some extent, real stars are calculated in practice. This last may seem to take us far afield but, toward the end of the section, we shall return to polytropes for a discussion of how they are used.

---

[1] For a review of such problems, see Hansen (1978), where this topic and the notion of "secular stability" is discussed.

7.2.1 General Properties of Polytropes

In previous chapters we encountered equations of state where pressure was only a function of density (and, of course, composition). For example, the equation of state for a completely degenerate, nonrelativistic, electron gas was given by (3.65) as

$$P_e = 1.004 \times 10^{13} \left(\frac{\rho}{\mu_e}\right)^{5/3} \quad \text{dyne cm}^{-2} \tag{7.14}$$

which is a power law equation of state with $P \propto (\rho/\mu_e)^{5/3}$. We might then imagine a stellar model composed of a material for which μ_e is a constant throughout and in which *both* the equation of state *and* the actual run of pressure versus density satisfy (7.14). But, if this condition is imposed beforehand, it is likely to result in a conflict with the complete set of stellar structure equations and a self-consistent model would not be possible. *Polytropes* are pseudo-stellar models for which power law equations of pressure versus density such as (7.14) are assumed a priori but where no reference to heat transfer or thermal balance is made. Thus only the hydrostatic and mass equations are used and inconsistencies with respect to the complete set of stellar structure equations are avoided. This may seem to be a high price to pay for consistency, but the resulting polytropic structures have proven to be remarkably useful in the interpretation of many aspects of real stellar structure.

Another motivation for studying polytropes arises from consideration of the structure of certain types of adiabatic convection zones. In a region of efficient convection the actual "del" of (7.11) is given by $\nabla = \nabla_{\text{ad}} = 1 - 1/\Gamma_2$ as in (3.94). If Γ_2 is assumed constant, then integrating (7.11) yields

$$P(r) \propto T^{\Gamma_2/(\Gamma_2 - 1)}(r) \, . \tag{7.15}$$

If, in addition, the gas is ideal with $T \propto P/\rho$, then $P(r) \propto \rho^{\Gamma_2}(r)$ and we have the same situation as above: P obeys a power law relation with respect to density as a function of radius.

In particular, we define a polytropic stellar model to be one in which the pressure is given by

$$P(r) = K\rho^{1+1/n}(r) \tag{7.16}$$

where n, the *polytropic index*, is a constant as is the proportionality constant K.[2] Since the polytrope is to be in hydrostatic equilibrium, then the distribution of pressure and density must be consistent with both the equation of hydrostatic equilibrium and conservation of mass. To best see how this works, divide the hydrostatic equation by ρ, multiply by r^2, and then take the derivative with respect to r of both sides to find

[2] Be careful not to confuse this n with the n used as the power law density exponent of opacity.

$$\frac{d}{dr}\left(\frac{r^2}{\rho}\frac{dP}{dr}\right) = -G\frac{d\mathcal{M}_r}{dr} = -4\pi G r^2 \rho$$

where the mass equation has been used to obtain the final equality. Rewrite this as

$$\frac{1}{r^2}\frac{d}{dr}\left(\frac{r^2}{\rho}\frac{dP}{dr}\right) = -4\pi G\rho \tag{7.17}$$

which is Poisson's equation. The latter identification is clear if we define the potential Φ such that

$$g(r) = \frac{d\Phi}{dr} = \frac{G\mathcal{M}_r}{r^2} \tag{7.18}$$

eliminate the pressure derivative (using 7.5), and find

$$\nabla^2\Phi = 4\pi G\rho \tag{7.19}$$

in spherical coordinates.

We now perform a sequence of transformations with the intent of making (7.17) dimensionless. Define the dimensionless variable θ by

$$\rho(r) = \rho_c\,\theta^n(r) \tag{7.20}$$

where $\rho_c = \rho(r=0)$. The power law for pressure is then

$$P(r) = K\rho_c^{1+1/n}\,\theta^{n+1}(r) = P_c\,\theta^{1+n}(r)\ . \tag{7.21}$$

The central pressure, P_c, is clearly equal to

$$P_c = K\rho_c^{1+1/n}. \tag{7.22}$$

Now substitute these into Poisson's equation and find the second-order differential equation for $\theta(r)$

$$\frac{(n+1)P_c}{4\pi G\rho_c^2}\frac{1}{r^2}\frac{d}{dr}\left(r^2\frac{d\theta}{dr}\right) = -\theta^n. \tag{7.23}$$

Finally, introduce the new dimensionless radial coordinate, ξ, by

$$r = r_n\xi \tag{7.24}$$

where the scale length, r_n, is defined as

$$r_n^2 = \frac{(n+1)P_c}{4\pi G\rho_c^2}. \tag{7.25}$$

We append the subscript n on r_n to signal its association with the particular polytropic index n. We will (usually) do the same for θ_n for the same reason, as follows.

So, with the substitution (7.24), Poisson's equation becomes

$$\frac{1}{\xi^2}\frac{d}{d\xi}\left(\xi^2\frac{d\theta_n}{d\xi}\right) = -\theta_n^n \tag{7.26}$$

and is now called the *Lane–Emden equation.* Models corresponding to solutions of this equation for a chosen n are called "polytropes of index n" and the solutions themselves are "Lane–Emden solutions" and are denoted by $\theta_n(\xi)$.

Note that if the equation of state for the model material is an ideal gas with $P = \rho N_A kT/\mu$, then some easy manipulations yield

$$P(r) = K' T^{n+1}(r), \quad T(r) = T_c\,\theta_n(r) \quad \text{(ideal gas)} \tag{7.27}$$

with

$$K' = \left(\frac{N_A k}{\mu}\right)^{n+1} K^{-n}, \quad T_c = K\rho_c^{1/n}\left(\frac{N_A k}{\mu}\right)^{-1}. \tag{7.28}$$

Thus in a polytrope whose material equation of state is an ideal gas with constant μ, θ_n measures temperature. Finally, the radial scale factor in this case is

$$r_n^2 = \left(\frac{N_A k}{\mu}\right)^2 \frac{(n+1)T_c^2}{4\pi G P_c} = \frac{(n+1)K\rho_c^{1/n-1}}{4\pi G}. \tag{7.29}$$

To prepare complete polytropic models that might share some resemblance to stars, appropriate boundary conditions must be applied to the Lane–Emden equation. For a complete model, with center at $r = 0$ and a surface that has vanishing density, these boundary conditions are as follows. For ρ_c in (7.20) to really be the central density, we require that $\theta_n(\xi{=}0){=}1$. Furthermore, spherical symmetry at the center (dP/dr vanishing at $r = 0$) requires that $\theta_n' \equiv d\theta_n/d\xi = 0$ at $\xi = 0$. This last condition pins down the solution at the center so that divergent solutions of the second-order system are suppressed. The regular solutions are called "E-solutions."

If the surface is that place where $P = \rho = 0$, then we require that the solution θ_n vanish there also. More specifically, the surface is where the *first zero* of θ_n occurs as measured from the center outward. (We do not want the pressure to vanish both at the "surface" and at some interior point.) We denote the location of the first zero by ξ_1 and it depends on the value of the polytropic index n. To summarize, the boundary conditions for a whole model are

$$\theta_n(0) = 1, \quad \theta_n'(0) = 0 \quad \text{at } \xi = 0 \ \text{(the center)} \tag{7.30}$$

$$\theta_n(\xi_1) = 0 \quad \text{at } \xi = \xi_1 \ \text{(the surface)}. \tag{7.31}$$

Since ξ_1 is the location of the surface, then the total (dimensional) radius is at

$$\mathcal{R} = r_n\xi_1 = \left[\frac{(n+1)P_c}{4\pi G\rho_c^2}\right]^{1/2} \xi_1\,. \tag{7.32}$$

Thus specifying K, n, and either ρ_c or P_c, yields the radius $\mathcal{R}$.

Analytic E-solutions for θ_n are obtainable for $n = 0$, 1, and 5. Numerical methods must be used to obtain solutions to the Lane–Emden equation for general n.

1. The solution for $n = 0$ is the constant-density sphere discussed in earlier chapters with $\rho(r) = \rho_c$. You may easily verify that

$$\theta_0(\xi) = 1 - \frac{\xi^2}{6} \quad \text{with} \quad \xi_1 = \sqrt{6} \tag{7.33}$$

and $P(\xi) = P_c\theta(\xi) = P_c\left[1 - (\xi/\xi_1)^2\right]$. Except that we have not found P_c (which may be found once $\mathcal{M}$ and $\mathcal{R}$ are specified), this is the solution found for the constant–density sphere as given by (1.41). P_c is easily computed using (7.32) with $\xi_1 = \sqrt{6}$ to be $(3/8\pi)(G\mathcal{M}^2/\mathcal{R}^4)$ in accord with (1.40).

2. For $n = 1$, the solution θ_1 is the familiar "sinc" function

$$\theta_1(\xi) = \frac{\sin\xi}{\xi} \quad \text{with} \quad \xi_1 = \pi \,. \tag{7.34}$$

The pressure and density follow from $\rho = \rho_c\theta_1$ and $P = P_c\theta_1^2$.

3. The polytrope for $n = 5$ has a finite central density, but its radius is unbounded, with

$$\theta_5(\xi) = \left[1 + \xi^2/3\right]^{-1/2} \quad \text{and} \quad \xi_1 \to \infty \,. \tag{7.35}$$

Despite the infinite radius, this polytrope does has a finite amount of mass associated with it.

Complete and regular solutions with $n > 5$ are also infinite in extent but contain infinite mass. The range of n of interest to us for complete models is then $0 \le n \le 5$.

Given n and K, we can in principle find the dependence of P and ρ on ξ. However, we cannot obtain absolute physical numbers unless $\mathcal{R}$ and either ρ_c or P_c are first specified. This follows from (7.22) and (7.32). The main difficulty is that $\mathcal{R}$ is not known beforehand. But $\mathcal{M}$ is what we wish to specify and this turns out to be enough.

The mass contained in a sphere of radius r is found from (7.6) to be $\mathcal{M}_r = \int_0^r 4\pi r^2 \rho(r)\, dr$. In ξ-space this becomes

$$\mathcal{M}_\xi = 4\pi r_n^3 \rho_c \int_0^\xi \xi^2 \theta_n^n \, d\xi \,. \tag{7.36}$$

The integrand of this expression contains θ_n^n, but this is just the (negative) of the righthand side of the Lane–Emden equation (7.26). Therefore, make the replacement, notice that the factors of ξ^2 cancel, and what is left is a perfect differential under the integral. The result is

$$\mathcal{M}_\xi = 4\pi r_n^3 \rho_c \left(-\xi^2 \theta_n'\right)_\xi \tag{7.37}$$

where $\left(-\xi^2\theta_n'\right)_\xi$ means "evaluate $(-\xi^2\, d\theta_n/d\xi)$ at the point ξ." The total mass is given by $\mathcal{M} = \mathcal{M}(\xi_1)$. It should be clear that if $\mathcal{M}$ and $\mathcal{R}$ are specified in physical units, then all else follows.

In what comes next, the relations between $\mathcal{M}$, $\mathcal{R}$, etc., are given without derivation. For example,

$$\mathcal{M} = (4\pi)^{-1/2} \left(\frac{n+1}{G}\right)^{3/2} \frac{P_c^{3/2}}{\rho_c^2} \left(-\xi^2\theta_n'\right)_{\xi_1} \tag{7.38}$$

which, in conjunction with (7.22), gives ρ_c or P_c in terms of $\mathcal{M}$. A little more algebra yields

$$\begin{aligned} P_c &= \frac{1}{4\pi(n+1)\left(\theta_n'\right)^2_{\xi_1}} \frac{G\mathcal{M}^2}{\mathcal{R}^4} \\ &= \frac{8.952 \times 10^{14}}{(n+1)\left(\theta_n'\right)^2_{\xi_1}} \left(\frac{\mathcal{M}}{\mathcal{M}_\odot}\right)^2 \left(\frac{\mathcal{R}}{\mathcal{R}_\odot}\right)^{-4} \quad \text{dyne cm}^{-2}. \end{aligned} \tag{7.39}$$

Note that the last result requires n, but not K.

Another result that will prove useful follows from solving for K given n, $\mathcal{M}$, and $\mathcal{R}$:

$$K = \left[\frac{4\pi}{\xi^{n+1}\left(-\theta_n'\right)^{n-1}}\right]^{1/n}_{\xi_1} \frac{G}{n+1} \mathcal{M}^{1-1/n} \mathcal{R}^{-1+3/n}. \tag{7.40}$$

Note that if $n = 3$, K depends only on $\mathcal{M}$ or, turned around, $\mathcal{M}$ does not depend on $\mathcal{R}$ for any K if $n = 3$.

If the equation of state is that of an ideal gas, then the central temperature is given by

$$\begin{aligned} T_c &= \frac{1}{(n+1)\left(-\xi\theta_n'\right)_{\xi_1}} \frac{G\mu}{N_A k} \frac{\mathcal{M}}{\mathcal{R}} \quad \text{(ideal gas)} \\ &= \frac{2.293 \times 10^7}{(n+1)\left(-\xi\theta_n'\right)_{\xi_1}} \mu \left(\frac{\mathcal{M}}{\mathcal{M}_\odot}\right) \left(\frac{\mathcal{R}}{\mathcal{R}_\odot}\right)^{-1} \quad \text{K}. \end{aligned} \tag{7.41}$$

You may easily verify that T_c for the constant-density sphere ($n = 0$) is the same as given by the earlier result (1.56, and see Table 7.1 for the numbers you need).

A useful quantity that depends only on n is the ratio of central density to mean density. This is given by

$$\frac{\rho_c}{\langle\rho\rangle} = \frac{1}{3}\left(\frac{\xi}{-\theta_n'}\right)_{\xi_1}. \tag{7.42}$$

We will sometimes say "this stellar model looks like a polytrope of index so-and-so because its degree of central concentration is such-and-such." That is, comparison of central to mean density implies an n by way of (7.42, and see Table 7.1). This is often a useful way to look at things—if you know what you're doing.[3]

Finally, it is an easy matter to show that the gravitational potential energy of a polytrope is (see 1.7 and the discussion preceding that equation for a refresher on Ω)

$$\Omega = -\frac{3}{5-n}\frac{G\mathcal{M}^2}{\mathcal{R}} . \tag{7.43}$$

For the constant–density sphere the coefficient $3/(5-n)$ is just 3/5 and this is the value quoted for the quantity "q" after (1.8).

Now that some of the formalism is out of the way, what are interesting values for n?

1. The pressure of the completely degenerate but nonrelativistic electron gas goes as $\rho^{5/3}$. Hence, by the definition of the polytropic equation of state (7.16), n for this case is 1.5 (or "a three-halves polytrope").
2. The density exponent for the fully relativistic case is 4/3 and thus $n = 3$ (or "an n equal three polytrope").
3. Recall that $P \propto \rho^{5/3}$ in an ideal gas convection zone. If no ionization is taking place (almost a contradiction for a real convection zone) then $\Gamma_2 = 5/3$ and $n = 3/2$ again.

It will turn out that indices of 1.5 and 3 are the ones usually encountered in simple situations. How unfortunate it then is that neither of these values have analytic E-functions associated with them. Therefore, how *are* these nonanalytic cases computed? The following subsection looks into this question and serves as a brief introduction to how some stellar models are computed. After this, we shall use the results from the polytropic calculations.

7.2.2 Numerical Calculation of the Lane–Emden Functions

This section, and others like it, is not intended to be an introduction to numerical analysis, but rather a guide to some techniques used to make stellar models. The subject is obviously important because one practical end of theory is the computation of a number. Get that wrong and you may waste the valuable time of experimentalists and observers.

A primary reference for numerical techniques that work is *Numerical Recipes: The Art of Scientific Computing*, 2nd ed., by Press et al. (1992). (We are not biased just because the list of authors is heavily weighted by those practicing the art of astrophysics.)

The Lane–Emden equation we wish to solve is

[3] Check out Ex. 7.2, where you will check the density ratio for main sequence models.

$$\frac{1}{\xi^2}\frac{d}{d\xi}\left(\xi^2\frac{d\theta_n}{d\xi}\right) = -\theta_n^n \tag{7.44}$$

with boundary conditions for the complete E-solution, $\theta_n(0)=1$, $(d\theta_n/d\xi) = 0$ at $\xi = 0$, and the vanishing of θ_n at the surface.

Shooting for a Solution

The first, and most straightforward, method is to treat the system as an initial value problem by starting from the origin ($\xi = 0$), integrating outward, and then stopping when θ_n goes to zero at the initially undetermined surface. This is a version of the "shooting method," whereby one "shoots" from a starting point and hopes that the shot will end up at the right place. For this we need an "integrator." One of the most useful for this purpose is of the class called "Runge–Kutta" integrators, many of which are easy to program (even on a programmable handheld calculator—with some patience) and are accurate and stable for simple problems. Note, however, that some problems are not simple and special techniques are required. Runge–Kutta schemes involve evaluating a series of derivatives of the dependent variable, y, at a sequence of points in the interval starting at x in the independent variable and ending at $x+h$. The quantity h is called the "step size." These derivatives are then averaged together in a particular way to eventually find $y(x+h)$. As will be seen below, the solution is "leap-frogged" from x to $x+h$.

The most convenient way to pose the second-order Lane–Emden problem (and many others) for use in a Runge–Kutta scheme is to cast it in the form of two first-order equations. For notational convenience (and to make what follows better resemble what you will find in *Numerical Recipes*), introduce the new variables $x = \xi$, $y = \theta_n$, and $z = (d\theta_n/d\xi) = (dy/dx)$. The Lane–Emden equation now becomes

$$\begin{aligned} y' &= \frac{dy}{dx} = z, \\ z' &= \frac{dz}{dx} = -y^n - \frac{2}{x}z\,. \end{aligned} \tag{7.45}$$

Suppose we know the values of y and z at some point x_i. Call these values y_i and z_i. If h is some carefully chosen step size, then the goal is to find y_{i+1} and z_{i+1} at $x_{i+1} = x_i + h$. This is, of course, just what is meant by an initial value problem. The particular Runge–Kutta scheme we shall choose to illustrate the technique is the fourth-order Runge–Kutta integrator. (Lower- and higher-order schemes are available.) As promised before, y' and z' are evaluated in a series of steps leading up to x_{i+1} as follows.

Compute the quantities

$$\begin{aligned} k_1 &= h\,y'(x_i, y_i, z_i) \\ l_1 &= h\,z'(x_i, y_i, z_i) \end{aligned}$$

$$\begin{aligned}
k_2 &= h\,y'(x_i + h/2,\, y_i + k_1/2,\, z_i + l_1/2) \\
l_2 &= h\,z'(x_i + h/2,\, y_i + k_1/2,\, z_i + l_1/2) \\
k_3 &= h\,y'(x_i + h/2,\, y_i + k_2/2,\, z_i + l_2/2) \\
l_3 &= h\,z'(x_i + h/2,\, y_i + k_2/2,\, z_i + l_2/2) \\
k_4 &= h\,y'(x_i + h,\, y_i + k_3,\, z_i + l_3) \\
l_4 &= h\,z'(x_i + h,\, y_i + k_3,\, z_i + l_3)\,.
\end{aligned} \tag{7.46}$$

As you can see, these ks and ls are rough guesses of the changes in the values of the functions y and z at various steps along the way to x_{i+1}. These are then weighted and added to find

$$\begin{aligned}
y_{i+1} &= y_i + \frac{k_1}{6} + \frac{k_2}{3} + \frac{k_3}{3} + \frac{k_4}{6} \\
z_{i+1} &= z_i + \frac{l_1}{6} + \frac{l_2}{3} + \frac{l_3}{3} + \frac{l_4}{6}\,.
\end{aligned} \tag{7.47}$$

It looks a bit complicated but, once you get in the swing of it, it rolls right along. The error introduced in the integration by discrete steps in this scheme is of the order of the fifth power of the step size h; that is, in moving from x_i to x_{i+1}, y_{i+1} and z_{i+1} are good to $\mathcal{O}(h^4)$. The calculation is a trade-off between the computation time it takes to make many steps of size h and the accuracy desired. Even unlimited time on a computer may not give arbitrary precision, however, for the simple reason that no machine has unlimited precision in the way it represents numbers internally. One therefore always lives with some, albeit small, amount of error. More sophisticated versions of Runge–Kutta (and other) methods are available that automatically adjust step size to maintain some desired, but hopefully reasonable, level of accuracy.[4]

The polytrope calculation now marches from the origin to the surface. But the origin must be treated with care because the boundary condition $\theta_n'(0) = z(0) = 0$ means that equation (7.45) for z' is indeterminate at $x = 0$. Since the E-solutions are derived, in part, from the stellar structure equations, we should resolve what to do now before worrying about the same difficulties with more realistic models. The resolution of the problem is to expand $\theta_n(\xi)$ in the Lane–Emden equation in a series about the origin using the boundary conditions to establish constants in the expansion. This is not particularly difficult to carry out and we quote from Cox (1968, §23.1a, and see Ex. 7.3):

$$\theta_n(\xi) = 1 - \frac{1}{6}\xi^2 + \frac{n}{120}\xi^4 - \frac{n(8n-5)}{15120}\xi^6 + \cdots\,. \tag{7.48}$$

[4] An "error-correcting" Runge–Kutta scheme that is a favorite of ours is due to H.A. Watts and L.F. Shampine of Sandia Laboratories, Albuquerque, New Mexico. For a textbook reference, discussion, and annotated listing of these programs, see

▷ Forsythe, G.E., Malcolm, M.A., & Moler, C.B. 1977, *Computer Methods for Mathematical Computation* (Prentice Hall: N.J.), pp. 127–147.

This is the method we use in the ZAMS and PULS codes on the CD–ROM.

For $\xi \to 0$, find that $z' \to -1/3$, which may be used to start the integration if so desired. A better way is to start at some $0 < \xi \ll 1$, compute y, y', z, and z' from (7.48), and carry on from there. This is a better procedure because there is an irregular solution to the Lane–Emden equation that blows up at the origin, and numerical techniques, no matter how good they are claimed to be, sense this. Therefore, treat the origin as delicately as possible.

The outer surface of the polytrope is reached when $\theta_n = y$ crosses zero. To resolve just where the first zero lies usually entails adjusting the step size h to be smaller as θ_n begins to near zero and, perhaps, using some form of interpolation scheme.

Since now you know how to compute polytropes, you may check that Table 7.1 contains the correct values of θ_n and θ_n' evaluated at ξ_1, and $\rho_c/\langle\rho\rangle$ for a range of n. More complete tabulations of the entire functions and other material may be found in Appendix A.5 and Table 23.1 of Cox (1968).

Table 7.1. Some Polytropic Parameters

Index n	ξ_1	$-\theta_n'(\xi_1)$	$\rho_c/\langle\rho\rangle$
0	$\sqrt{6}$	$\sqrt{6}/3$	1
1.0	π	$1/\pi$	$\pi^2/3$
1.5	3.6538	0.20330	5.9907
2.0	4.3529	0.12725	11.402
3.0	6.8969	0.04243	54.183
4.0	14.972	0.00802	622.41

The Fitting Method

Another method for solving the Lane–Emden equation for E-solutions, and one that is used for real stellar models, is prompted by the following observation. In integrating from the center outward, it is possible that slight errors introduced within the deeper parts of the polytrope will be amplified as the low-density surface is approached. This is in the spirit of having the low-density surface being shaken like the tip of a whip when the heavy handle at the center is moved slightly. The same problems arise in complete stellar models where, because of the exceedingly large contrast between central and surface conditions of, for example, pressure and density, any inaccuracies in numerics near the center are felt manyfold by the time the surface is reached. One method for dealing with this is to integrate starting at *both* the center *and* the surface and see if the solutions join in some continuous way at a point between the two extremes. This is called the "fitting method" and it is a standard way to construct homogeneous, zero–age main sequence models.

The difficulty, as should be clear from what we have seen for polytropes, is that we do not know beforehand exactly where the surface ξ_1 is. Furthermore, since the system is second-order, we require two pieces of information at the surface to start an integration. The first is the requirement that $\theta_n(\xi)$ must be zero at ξ_1. As a second, we must make a *guess* at the first derivative of θ_n at ξ_1. Because pressure always decreases outward in a hydrostatic star we at least know that the sign of $\theta'_n(\xi_1)$ must be negative. Therefore, make reasonable guesses for ξ_1 and $\theta'_n(\xi_1)$, set $\theta_n(\xi_1)$ to zero, and let the integrator work its way inward from there. With your other hand, integrate from the center outward as before. At some interior point in the prospective polytrope (say near $\xi_1/2$) check if both θ_n and θ'_n (or the first-order variables y and z) of the inward and outward integrations match at that point—which we now call the "fitting point." They must fit for a complete E-solution because no discontinuities are lurking in the differential equations. If the solutions do not match, then one or both of ξ_1 and $\theta'_n(\xi_1)$ have been chosen incorrectly. To remedy this situation, the following strategy usually works (see, e.g., Press et al., 1992, §17.2).

Let x_s (in x-, y-, z-space, with subscript "s" denoting "surface") be the initial guess at the surface location (i.e., ξ_1). In addition, let x_f be the location of the fitting point ("f" for "fitting") and $y_{\rm o}(x_f)$, $z_{\rm o}(x_f)$, $y_{\rm i}(x_f)$, and $z_{\rm i}(x_f)$ be the values of y and z obtained at x_f by means of the outward (subscript "o") and inward (subscript "i") integrations. The situation is shown in Fig. 7.1. Note that in these variables $y(x_s)$ is always zero because it is the value of θ_n at the surface. What we wish to do is vary x_s and $z_s = z(x_s)$ and see what happens to $y_{\rm i}(x_f)$ and $z_{\rm i}(x_f)$. With this information, we then set up an algorithm to eventually match at the fitting point.

To implement the algorithm, first define

$$\begin{aligned} Y(x_s, z_s) &= y_{\rm i}(x_f) - y_{\rm o}(x_f) \\ Z(x_s, z_s) &= z_{\rm i}(x_f) - z_{\rm o}(x_f) \end{aligned} \tag{7.49}$$

and these are shown as the steps on Fig. 7.1 in a not-yet-converged solution. (Note that $y_{\rm o}(x_f)$ and $z_{\rm o}(x_f)$ are fixed from the outward integration.) We eventually want $Y = Z = 0$.

Now change x_s from its initial value to $x_s \to x_s + \delta x_s$, where δx_s is small compared to x_s, while keeping z_s at its original value, and integrate inward. Because x_s has changed, we expect Y and Z to both change. Denote these changes by δY_z and δZ_z where the "z" subscript means "only x_s has changed." Thus $Y \to Y + \delta Y_z$ and $Z \to Z + \delta Z_z$. Similarly, if x_s is fixed at its original value but $z_s \to z_s + \delta z_s$, then $Y \to Y + \delta Y_x$ and $Z \to Z + \delta Z_x$. It should be clear that $\delta Y_z / \delta x_s$ is the difference approximation to the partial derivative of Y with respect to x_s when z_s is kept fixed. In similar fashion, $\delta Y_x / \delta z_s$ represents $(\partial Y / \partial z)_{x_s}$, etc. (This is more difficult to explain in words than your just sitting down with a piece of paper and playing with it.)

The route to an E-solution is to compute the numeric derivatives, as in the above, and then imagine first-order Taylor expansions of the form

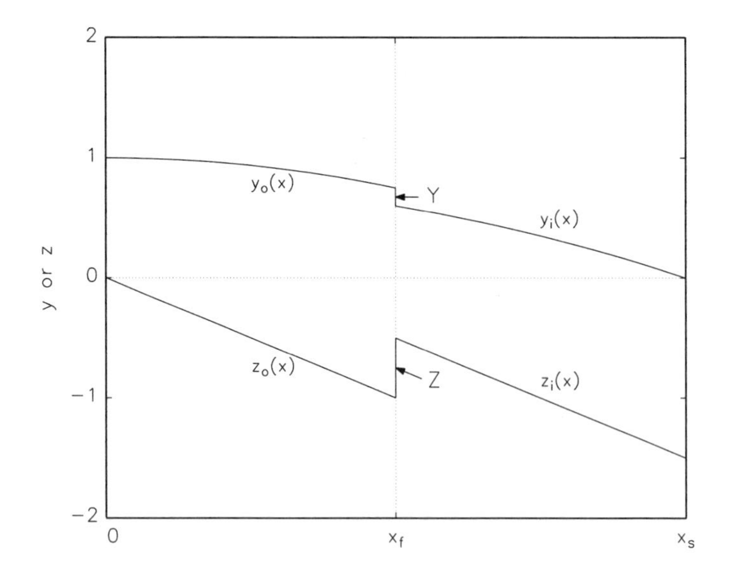

Fig. 7.1. An illustration of the fitting method for the two variables $y = \theta_n$ and $z = \theta'_n$. The solid curves indicate solutions that do not quite satisfy the conditions of continuity. Hence the "jumps" labeled Y and Z at x_f.

$$
\begin{aligned}
Y(x_s+\Delta x_s, z_s+\Delta z_s) &= Y(x_s, z_s) + \left(\frac{\delta Y_z}{\delta x_s}\right)\Delta x_s + \left(\frac{\delta Y_x}{\delta z_s}\right)\Delta z_s \\
Z(x_s+\Delta x_s, z_s+\Delta z_s) &= Z(x_s, z_s) + \left(\frac{\delta Z_z}{\delta x_s}\right)\Delta x_s + \left(\frac{\delta Z_x}{\delta z_s}\right)\Delta z_s
\end{aligned} \tag{7.50}
$$

where the partial derivatives of the expansion have been replaced by their known difference equivalents. The trick is to now find Δx_s and Δz_s such that the lefthand sides of both equations in (7.50) are zero. That is, we wish to find corrections to the initial guesses x_s and z_s such that Y and Z are both zero and thus complete the fitting at x_f. Thus solve the simultaneous, inhomogeneous, but *linear* system (which is why 7.50 is sometimes referred to as a "linearization" of the system)

$$
\begin{aligned}
\left(\frac{\delta Y_z}{\delta x_s}\right)\Delta x_s + \left(\frac{\delta Y_x}{\delta z_s}\right)\Delta z_s &= -Y(x_s, z_s) \\
\left(\frac{\delta Z_z}{\delta x_s}\right)\Delta x_s + \left(\frac{\delta Z_x}{\delta z_s}\right)\Delta z_s &= -Z(x_s, z_s)
\end{aligned} \tag{7.51}
$$

for the unknowns Δx_s and Δz_s. To first-order, these are the corrections needed to solve for the two roots of the combined equations $Y = 0$ and $Z = 0$.

Because the Lane–Emden equation is nonlinear, however, there is no guarantee that the new values of $x_s \to x_s + \Delta x_s$ and $z_s \to z_s + \Delta z_s$ will satisfy

$Y = Z = 0$. (If the equation were linear, you would be done.) This method is therefore iterated with successive corrections until both Δx_s and Δz_s become very small or, better yet, the E-functions approach some preassigned level of continuity at the fitting point. The method we have outlined for converting a nonlinear problem into one that is linear by Taylor expansions, and then solving for some small changes to find roots, is known by many as the "Newton–Raphson method." It comes up so frequently in numerical analysis that it is worth studying carefully—as in *Numerical Recipes* (§9.4), where it is called "Newton's rule."

In the case where the full fourth-order differential system for more realistic stellar models is to be used, the above fitting scheme must be generalized. It is clear that four quantities must be specified: two at the surface and two at the center of the desired model. At the center, two could be chosen out of the three nonzero quantities $P_{\rm c}$, $\rho_{\rm c}$, and $T_{\rm c}$. Given the composition, they are all connected through the equation of state. At the surface, two of $\mathcal{R}$, $\mathcal{L}$, and $T_{\rm eff}$ seem reasonable. Equations (7.5–7.11) are then integrated inward and outward and an attempt is made to match quantities that should be continuous at some interior point specified at some fixed $\mathcal{M}_r$. The variables to be fitted must be four in number and a convenient choice might be r, P, $\mathcal{L}_r$, and T. The same sort of algorithm is then used in which the quartet of surface and center quantities are varied independently and a Newton–Raphson root-finding scheme is employed to calculate corrections. This is more easily said than done but it is conceptually simple and is efficient when done properly. It is a standard method for constructing ZAMS models—as on the CD–ROM.

7.2.3 The U–V Plane

The notion of fitting continuous functions in a simplified form of a stellar model raises the question of what quantities in a star (or a model of one) are continuous. The coordinate r must be smooth. So must the pressure be continuous in a hydrostatic star because, otherwise, terrible things would happen to its radial derivative in the equation of motion $\rho\ddot{r} = -(dP/dr) - (G\mathcal{M}_r\rho/r^2)$, which would result in unbounded accelerations. Another continuous quantity is $\mathcal{M}_r$. A discontinuous interior mass would imply unbounded densities in the mass equation. Density itself, however, need not be continuous. (Air resting on lead is fine.) These statements may be recast by introducing two new dimensionless functions that, in effect, summarize the mass and hydrostatic equations. These are

$$\begin{aligned} \mathrm{U} &\equiv \frac{d\ln\mathcal{M}_r}{d\ln r} = \frac{4\pi r^3\rho}{\mathcal{M}_r} = 3\frac{\rho(r)}{\langle\rho(r)\rangle} \\ \mathrm{V} &\equiv -\frac{d\ln P}{d\ln r} = \frac{G\mathcal{M}_r\rho}{rP} = \frac{r}{\lambda_P}\,. \end{aligned} \tag{7.52}$$

The last part of the equation for U contains $\langle\rho(r)\rangle$, which is the average density interior to the point r. The last equation for V implies that V is the local radius measured in units of the local pressure scale height λ_P. From the preceding remarks it is clear that both U/ρ and V/ρ are continuous for static stars.

We shall only use these functions to illustrate some particular points in stellar structure and modeling but they were of great importance in the earlier history of these subjects. It is still worthwhile to see how U and V were used in the calculations described in Schwarzschild (1958), and Hayashi, Hōshi, and Sugimoto (1962). A useful modern reference is Kippenhahn and Weigert (1990, §23).

We shall need the values of U and V at the center and surface of a typical model for future reference. The limit as $r \to 0$ requires that we look at $\mathcal{M}_r$ in that limit. Later on we shall consider in detail how various stellar quantities behave near stellar center but, for now, we state that

$$\mathcal{M}_r \to (4\pi/3)\rho_c r^3 \quad \text{as} \quad r \to 0\,. \tag{7.53}$$

All that was done here was to realize that ρ has a zero gradient at the center (and, hence, its expansion must be in even powers of r) and to compute the mass in a tiny sphere of radius r of constant density. The surface presents a slightly more difficult problem. We assume here that both the density and pressure go to zero at the surface. Thus the ratio ρ/P in V might seem to be indeterminate but, if the gas is ideal, then ρ/P is just inversely proportional to temperature and, if temperature approaches zero at the surface, then V becomes unbounded. Gathering this together, the boundary conditions on U and V are

$$\begin{aligned} &\mathrm{U} \to 3, \quad \mathrm{V} \to 0 \quad \text{as} \quad r \to 0 \\ &\mathrm{U} \to 0, \quad \mathrm{V} \to \infty \quad \text{as} \quad r \to \mathcal{R}\,. \end{aligned} \tag{7.54}$$

The U–V variables in polytrope language are

$$\begin{aligned} \mathrm{U}(\xi) &= \frac{\xi\theta_n^n}{(-\theta_n')} \\ \mathrm{V}(\xi) &= (n+1)\frac{(-\xi\theta_n')}{\theta_n} \end{aligned} \tag{7.55}$$

and an E-solution phrased in terms of these for an $n = 3$ polytrope is illustrated in Fig. 7.2. This curve was produced by integrating the Lane–Emden equation from the surface inward starting with the proper values of ξ_1 and $\theta_n'(\xi_1) \equiv \theta_n'(\mathrm{E})$ and continuing on to the center. We also show the U–V path of a $5\,\mathcal{M}_\odot$ Pop–I ZAMS model. Its $\rho_c/\langle\rho\rangle$ is close to that of a $n = 3$ polytrope and it sits comfortably close to the E-solution.

It is instructive, however, to consider what happens if an *incorrect* value of $\theta_n'(\xi_1)$ is chosen at the start. If $\theta_n'(\xi_1)$ is chosen so that its magnitude is greater than $\theta_n'(\mathrm{E})$ (i.e., $|\theta_n'(\xi_1)| > |\theta_n'(\mathrm{E})|$), then—and we shall not go through the derivation—$\theta_n(\xi)$ (i.e., density) increases more rapidly than the

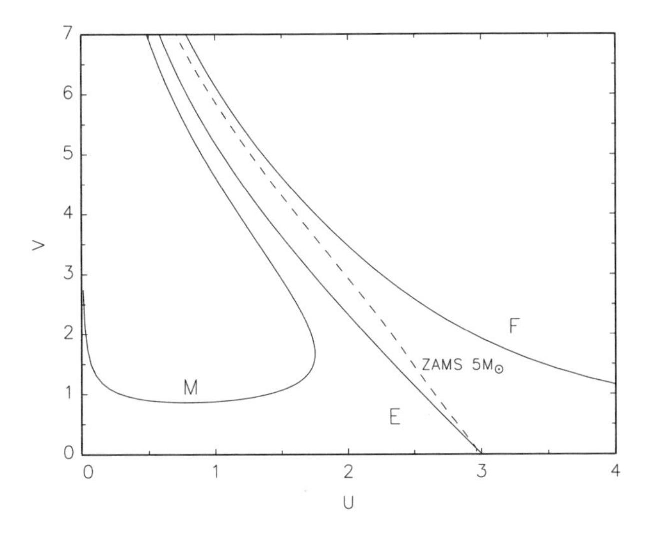

Fig. 7.2. E-(Lane–Emden), F-(collapsed), and M-(centrally condensed) solutions to the Lane–Emden equation for a polytrope of $n = 3$ (see text). The surface radius ξ_1 was chosen to be its nominal value $\xi_1 = 6.8968486$, while the derivatives of θ_3 at the surface were $\theta_3'(\mathrm{E}) = -0.0424298$, $\theta_3'(\mathrm{F}) = -0.051$, and $\theta_3'(\mathrm{M}) = -0.038$. The M-solution goes to $\mathrm{V} \to 4$ as $\mathrm{U} \to 0$ for the $n = 3$ polytrope. Also shown, as a dashed line, is what a $5\,\mathcal{M}_\odot$ ZAMS Pop–I model looks like on the U–V plane.

E-solution and the interior mass tends to zero before the center is reached (think of $d\mathcal{M}_r/dr$ in the mass equation). Thus $\mathrm{U} \to \infty$ as $\mathrm{V} \to 0$. These are the "F-solutions" (named after R.H. Fowler) or "solutions of the collapsed type" and one of these is denoted by "F" in Fig. 7.2.

If, on the other hand, $|\theta_n'(\xi)| < |\theta_n'(\mathrm{E})|$, then $\mathcal{M}_r$ remains finite but $\theta_n(\xi)$ blows up as the center is approached. These represent irregular solutions to the Lane–Emden equation and are called "centrally condensed" or "M-solutions" (after E.A. Milne); one is shown in the figure.

As they stand, the M- and F- solutions cannot represent complete stellar models because they are unphysical. If, however, polytropic equations of state hold for some portions of stellar interiors, then these solutions are still of some interest. They may, for example, represent pieces of the interiors of stars where other pieces (e.g., E-solutions) are to be tacked on to them in a continuous fashion.

7.2.4 Newton-Raphson or "Henyey" Methods

In integration by the fitting method the idea was to separate the model into two regions and then connect those two regions by continuity. In the Newton-Raphson or Henyey (Henyey, Forbes, and Gould, 1964) method, the "integration" of the stellar structure equations is performed over the model as a whole. This is best described by a simple example such as that discussed below.[5] This is a powerful technique (not really originated by astronomers) and is now the standard way to construct evolutionary models.

Consider the second-order system

$$\begin{aligned} \frac{dy}{dx} &= f(x,y,z) \\ \frac{dz}{dx} &= g(x,y,z) \end{aligned} \tag{7.56}$$

with boundary conditions on y and z specified at the endpoints of the interval $x_1 \leq x \leq x_N$. These boundary conditions may be represented by general functions b_1 and b_N such that

$$\begin{aligned} b_1(x_1,y_1,z_1) &= 0 \\ b_N(x_N,y_N,z_N) &= 0 \end{aligned} \tag{7.57}$$

where y_i, z_i are $y(x_i)$, $z(x_i)$. It is assumed that f, g, b_1, and b_N are well behaved and that there is no ambiguity about the location of x_1 and x_N (unlike the polytrope where ξ_1 was not known beforehand, but see §7.2.5). The differential equations are now cast in a "finite difference form" over a predetermined "mesh" in x. That is, we choose a sequence of points x_1, x_2, $\cdots$, x_N at which y and z are to be evaluated. (This choice must be made with some care for the sake of accuracy.) The question is how do we represent the differential equations. A perfectly respectable way is to replace the differentials with differences and to replace the righthand sides of (7.56) with an average of the derivatives over the interval covered by the differences. You may think of this as a differential representation of the trapezoidal rule of integration. It is also an example of "implicit" integration to be discussed later. We then have

$$\begin{aligned} \frac{y_{i+1}-y_i}{x_{i+1}-x_i} &= \frac{1}{2}(f_{i+1}+f_i) \\ \frac{z_{i+1}-z_i}{x_{i+1}-x_i} &= \frac{1}{2}(g_{i+1}+g_i) \end{aligned} \tag{7.58}$$

where f_{i+1} and g_{i+1} are shorthand for the functions $f(x_{i+1},y_{i+1},z_{i+1})$ and $g(x_{i+1},y_{i+1},z_{i+1})$, and $i=1,2,\cdots,N-1$. Expressions (7.58) then represent $2N-2$ equations but the two boundary conditions make up the difference so that all $2N$ variables y_i and z_i may, in principle, be found all at once. This is *not* an initial value problem.

To simplify the notation, we assume that the mesh in x is constant so that $x_{i+1}-x_i = \Delta x$ for all i between $i=1$ and $i=N-1$. (Real life is

[5] An excellent reference for these methods is the classic review by Kippenhahn, Weigert, and Hofmeister (1967) and, in more general language, §17.3 of Press et al. (1992), where they are called "relaxation methods."

hardly ever so well-behaved that this can be done without loss of accuracy.) The errors in y and z in this order difference scheme go as $|\Delta x|^3$.

The difficulty with this method is that values of y and z at i and $i+1$ are all mixed up in the difference equations and boundary conditions—but Newton-Raphson comes to the rescue. As in the example of fitting, this means that (7.57–7.58) are to be linearized to find the solution.

Suppose we have some notion of the run of y_i and z_i for all i. These "guesses" do not, in general, satisfy (7.57–7.58). We may imagine, however, that there are corrections Δy_i and Δz_i

$$\begin{aligned} y_i &\rightarrow y_i + \Delta y_i \\ z_i &\rightarrow z_i + \Delta z_i \end{aligned} \tag{7.59}$$

that lead to new values of y_i and z_i that might satisfy those equations. We now *estimate* the values of Δy_i and Δz_i for all i. If it turns out that these do not quite do the job, then we iterate the following procedure (or "relax" to a solution) until values are found that do.

A first-order estimate of Δy_i and Δz_i is obtained by introducing (7.59) into (7.57–7.58) and expanding those equations to first-order in a Taylor series around the original guesses. For example, the first equation in (7.58) becomes, to first-order in the Δs and after minor arrangement,

$$\begin{aligned} y_{i+1} - y_i - \frac{\Delta x}{2}\left(f_i + f_{i+1}\right) = & \\ + \left[\frac{\Delta x}{2}\left(\frac{\partial f}{\partial y}\right)_i + 1\right]\Delta y_i &+ \left[\frac{\Delta x}{2}\left(\frac{\partial f}{\partial y}\right)_{i+1} - 1\right]\Delta y_{i+1} + \\ + \left[\frac{\Delta x}{2}\left(\frac{\partial f}{\partial z}\right)_i\right]\Delta z_i &+ \left[\frac{\Delta x}{2}\left(\frac{\partial f}{\partial z}\right)_{i+1}\right]\Delta z_{i+1} \end{aligned} \tag{7.60}$$

where the partials are taken at constant y or z as the case may be. The corresponding equation for the z-derivative is obtained by replacing ys with zs and fs with gs. Note that the lefthand sides of these equations are zero when the difference equations are satisfied; that is, when the Δys and Δzs go to zero. The linearized forms of the boundary conditions are

$$b_{(1\text{ or }N)} + \left(\frac{\partial b}{\partial y}\right)_{(1\text{ or }N)} \Delta y_{(1\text{ or }N)} + \left(\frac{\partial b}{\partial z}\right)_{(1\text{ or }N)} \Delta z_{(1\text{ or }N)} = 0\,. \tag{7.61}$$

We now arrange all these equations in matrix form so that we shall ultimately solve an equation of the form

$$\mathbf{M} \cdot \mathbf{U} = \mathbf{R} \tag{7.62}$$

where the column vector $\mathbf{U}$ contains the unknown quantities Δy_i and Δz_i, $\mathbf{R}$ is the righthand side column vector, and $\mathbf{M}$ is the $N \times N$ coefficient matrix. The unknown vector is arranged like so:

$$\mathbf{U} \equiv (\Delta y_1, \Delta z_1, \Delta y_2, \Delta z_2, \cdots, \Delta y_N, \Delta z_N)^{\mathrm{T}} \tag{7.63}$$

where the superscript "T" indicates transpose; note the interlacing of the variables. Some of the elements in $\mathbf{R}$ we will denote by

$$\begin{aligned} Y_{i+1/2} &\equiv y_{i+1} - y_i - \frac{\Delta x}{2}(f_{i+1} + f_i) \\ Z_{i+1/2} &\equiv z_{i+1} - z_i - \frac{\Delta x}{2}(f_{i+1} + f_i) \end{aligned} \tag{7.64}$$

and come from terms such as given in (7.60). The "half-step" notation $i+1/2$ is meant to imply that a quantity is evaluated between i and $i+1$. The rest of $\mathbf{R}$ comes from the constant terms in the linearization of the boundary conditions (7.61), which are put in as the first and last elements in $\mathbf{R}$ to give

$$\mathbf{R} = (-b_1,\, Y_{3/2},\, Z_{3/2}, \cdots, Y_{N-1/2},\, Z_{N-1/2},\, -b_N)^{\mathrm{T}}. \tag{7.65}$$

Parts of the matrix elements in $\mathbf{M}$ come from (7.60) and are denoted by

$$\begin{aligned} A_i &\equiv \frac{\Delta x}{2}\left(\frac{\partial f}{\partial y}\right)_i, \quad C_i \equiv \frac{\Delta x}{2}\left(\frac{\partial g}{\partial y}\right)_i \\ B_i &\equiv \frac{\Delta x}{2}\left(\frac{\partial f}{\partial z}\right)_i, \quad D_i \equiv \frac{\Delta x}{2}\left(\frac{\partial g}{\partial z}\right)_i . \end{aligned} \tag{7.66}$$

Finally (and as is implied by 7.65), the order in which the equations are ranked in the matrix is (1) boundary condition at x_1; (2) the $2N-2$ difference equations; (3) the boundary condition at x_N. To conserve space, we give the following result for $\mathbf{M}$ for a mesh consisting of three points i =1,2,3. Once you construct this for yourself the entire scheme should become clear:

$$\begin{pmatrix} (\partial b/\partial y)_1 & (\partial b/\partial z)_1 & 0 & 0 & 0 & 0 \\ A_1+1 & B_1 & A_2-1 & B_2 & 0 & 0 \\ C_1 & D_1+1 & C_2 & D_2-1 & 0 & 0 \\ 0 & 0 & A_2+1 & B_2 & A_3-1 & B_3 \\ 0 & 0 & C_2 & D_2+1 & C_3 & D_3-1 \\ 0 & 0 & 0 & 0 & (\partial b/\partial y)_3 & (\partial b/\partial z)_3 \end{pmatrix} . \tag{7.67}$$

Note the particularly simple structure of this matrix in which no nonzero element is located further than two columns away from the diagonal (as an example of a "band diagonal" matrix). This fact makes the problem amenable to several accurate and efficient techniques for solving simultaneous linear equations and you are referred to §§2.4 and 17.3 of *Numerical Recipes* for more details.

Once the solution set $\mathbf{U}(\Delta y_i, \Delta z_i)$ is found, then new values of y_i and z_i are immediately obtained by adding Δy_i and Δz_i to the corresponding old guesses. If these new values are sufficiently close to the old values, then the solution has converged and the original difference equations and boundary conditions are presumably satisfied. If not, try again with the new values and iterate until they are. Suffice it to say, the novice (and professional) very often reaches this happy state only after many iterations in the multidimensional solution space of ys and zs. Convergence depends on many factors, the most important being a reasonably good initial guess and that is where experience and intuition count. Initial guesses that bear no resemblance to the desired solution may leave you stranded in a solution space of many dimensions and convergence may not be possible.[6] Even a good guess may not help, however, if the original differential equations are ill-behaved or if the mesh x_i is inappropriate. Another factor often comes in when constructing stellar models where tabulated equations of state, etc., are used. Very often such tables are "noisy" (usually from the introduction of incomplete physics at some temperature or density) and the difference equations see that noise. But, if all goes well, then it goes well indeed and the corrections Δy_i and Δz_i decrease as the square of their absolute values from one iteration to the next and convergence is swift.

With the above as an example, it should be clear how to extend the technique to differential systems of higher than second-order: the bookkeeping gets messier, but the principle remains the same. The stellar problem is fourth-order with two boundary conditions at each end as discussed in §7.1. For stellar evolution off the main sequence, what is usually done is first to construct a ZAMS model using the fitting method and then use that solution as the first guess for a Henyey model.

7.2.5 Eigenvalue Problems and the Henyey Method

The above scheme, as presented, cannot solve the simple problem of an E-solution polytrope. This is because we do not know the radius (ξ_1) beforehand. Thus, on the face of it, a grid in x cannot be established upon which the differences equations and boundary conditions are to be applied. Yet with some minor adjustments, this can be remedied. We note here that the following, and variations thereof, are calculational mainstays in some subareas of stellar astrophysics such as variable star analysis.

Recall (7.45), which is the Lane–Emden equation phrased in terms of $y = \theta_n$, $z = dy/dx$, with $x = \xi$:

$$\begin{aligned} y' &= \frac{dy}{dx} = z \\ z' &= \frac{dz}{dx} = -y^n - \frac{2}{x} z \,. \end{aligned} \tag{7.68}$$

[6] Of course what's even worse is if no solution exists at all! In that case, go back and examine the logic of your analysis. See the following section for an example of where this scheme doesn't work as presented.

The boundary conditions are $y = 1$ at $x = 0$, and $y = 0$ at some unknown $x = x_s = \xi_1$; in addition, it is required that $z = dy/dx = 0$ at $x = 0$ to obtain the regular E-solution.

We use the simple trick of rescaling x so that it lies within the closed interval $[0, 1]$ by letting

$$x \to x = \frac{\xi}{\xi_1} = \frac{\xi}{\lambda} . \tag{7.69}$$

Thus given $\xi_1 = \lambda$, the edge of the polytrope is at unity in the new x-coordinate. But, you say, we don't know what this λ is, so it looks as if we haven't gotten anywhere. This is true, so let's find λ as part of the overall problem in a relaxation method.

First transform the Lane–Emden equation into the new x-coordinate:

$$\begin{aligned} y' &= \frac{dy}{dx} = \lambda z \\ z' &= \frac{dz}{dx} = -\lambda y^n - \frac{2}{x} z . \end{aligned} \tag{7.70}$$

These equations are in the same form as we had previously but now the parameter λ appears; that is, the functions of (7.56) are now $f = f(x, y, z; \lambda)$ and $g = g(x, y, z; \lambda)$. It is no surprise that an additional variable appears because there are now an overabundance of conditions on the problem. The problem is still second-order but there are two boundary conditions at the center ($y = 1$, $z = 0$), a third at the surface ($y = 0$ at the new $x = 1$), and the system is overdetermined. With λ as an additional degree of freedom, however, the surplus boundary condition is no longer one too many. Here λ is an *eigenvalue* for the problem in much the same way that eigenvalues appear in other situations in physics. In this situation it yields the radius.

To solve this problem we proceed as in the previous subsection and assume we have a complete run of guesses for y and z over the mesh of $0 \le x \le 1$, but we also make a first guess for λ. We then linearize f and g, as before, by letting $y_i \to y_i + \Delta y_i$, etc., but also allow $\lambda \to \lambda + \Delta\lambda$. The scheme is to solve for the $2N + 1$ unknowns y_i, z_i, and λ. There are $2N - 2$ difference equations (as before) but now there are *three* boundary conditions yielding a total of $2N + 1$ equations. It is easy to see (by writing out all the linearized equations) that $\Delta\lambda$ comes in from such terms as the last term in

$$f \to f + \left(\frac{\partial f}{\partial y}\right)_{z_i,\lambda} \Delta y_i + \left(\frac{\partial f}{\partial z}\right)_{y_i,\lambda} \Delta z_i + \left(\frac{\partial f}{\partial \lambda}\right)_{y_i,z_i} \Delta\lambda .$$

Each difference equation thus contributes terms in $\Delta\lambda$ to a matrix algebra problem similar to the one encountered before except that there is now an extra row in the main matrix corresponding to the additional boundary condition and an extra column corresponding to the extra unknown, $\Delta\lambda$. This set of linear equations is somewhat more cumbersome to solve (because of the additional column of nonzero elements) but various techniques can be made to work.

In this specific problem the behavior of the functions near the center must be treated carefully as discussed previously. Either the differential equations are replaced by series solutions up to some point, or the derivatives appearing in the difference equations have to be calculated very accurately. Just how to incorporate series expansions into the Henyey scheme is left as an exercise, but note that all that is required is a redefinition of the boundary conditions so that the proper ones appear at a point slightly removed from the center. Once this has been mastered for polytropes, you will soon find that such eigenvalue problems hold no real terror—maybe.

7.2.6 Dynamic Problems

We can gain more insight into the difficulties of constructing stellar models and evolutionary sequences by examining how rapidly evolving models are handled. As will be pointed out later in this section, some of the techniques used here are also the stock-in-trade of those who compute models for slowly evolving stars.

By "rapidly" evolving we mean situations where evolutionary processes take place on dynamic time scales such as in supernovae, novae, and most variable stars. Somewhat special techniques are required for this. We shall follow Cox, Brownlee, and Eilers (1966) because it contains a reasonably complete exposition of the methods. You may also wish to consult Chapter 19 of *Numerical Recipes* for more general problems involving the numerical treatment of partial differential equations containing time. This discussion will be restricted to motions that are spherically symmetric.

Consider the equation of motion given earlier in (1.5) but expressed in the Lagrangian form

$$\ddot{r} = -4\pi r^2 \left(\frac{\partial P}{\partial \mathcal{M}_r} \right) - \frac{G\mathcal{M}_r}{r^2} \,. \tag{7.71}$$

As is usual in a Lagrangian description, the elements of interest in the star are tagged by the mass—in this instance by $\mathcal{M}_r$. To reduce (7.71) to a form suitable for numeric calculation, first divide the star into concentric shells containing masses labeled $\mathcal{M}_{i+1/2}$, where it is understood that $(i+1/2)$ means that the mass lies between radii r_{i+1} and r_i as illustrated in Fig. 7.3. The model center is at r_0. The spherical surfaces corresponding to these various radii will be referred to as "interfaces." The mass interior to the radius at interface r_{i+1} is then

$$\mathcal{M}_r \quad \text{at} \quad r_{i+1} \to \sum_{k=0}^{i} \mathcal{M}_{k+1/2} \,. \tag{7.72}$$

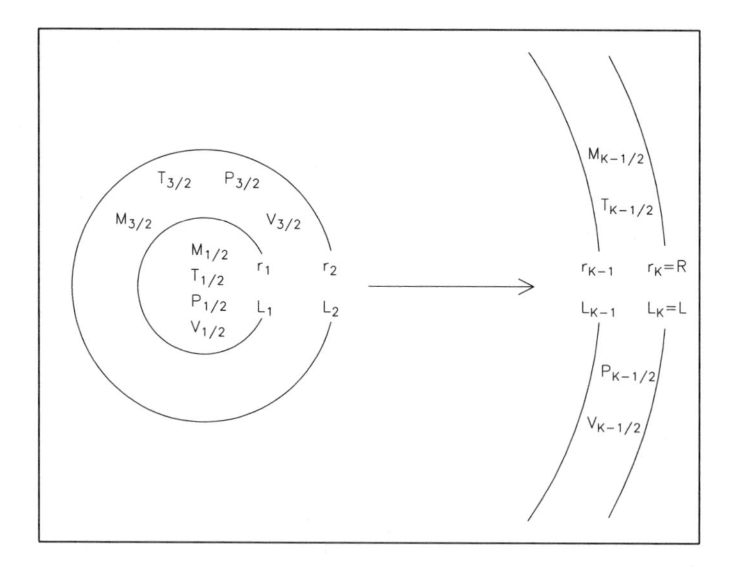

Fig. 7.3. This figure illustrates the shell and interface partitioning of a stellar model. Some variables are defined within the shells whereas others are defined at the interfaces (see text).

A natural way of expressing the mass equation is to define the density as

$$\rho_{i+1/2} = \frac{\mathcal{M}_{i+1/2}}{(4\pi/3)\left(r_{i+1}^3 - r_i^3\right)} = \frac{1}{V_{i+1/2}} \tag{7.73}$$

where, for later purposes, we introduce the specific volume, V. (Note that we are using a plain V here instead of our usual V_ρ.) The notation implies that the density is defined in shells and not on interfaces. Having made this choice, we can then reasonably define $T_{i+1/2}$ and $P_{i+1/2}$ as the temperature and pressure associated with the above density so that $P_{i+1/2}$ may be expressed in terms of the corresponding temperature, density, and composition within a shell.

The mass gradient of pressure in (7.71) is simply the difference of the pressures divided by the change in mass across an interface i. The change in mass is constructed as

$$\begin{aligned} \partial\mathcal{M}_r \ \text{ at } \ i &= \frac{1}{2}\left[\sum_{k=0}^{i}\mathcal{M}_{k+1/2} - \sum_{k=0}^{i-2}\mathcal{M}_{k+1/2}\right] \\ &= \frac{1}{2}\left[\mathcal{M}_{i+1/2} + \mathcal{M}_{i-1/2}\right] \end{aligned} \tag{7.74}$$

where $i = 0$ is not needed because the stellar center is never accelerated. Thus (7.71) is transcribed into difference form as

$$\ddot{r}_i = -8\pi r_i^2 \frac{P_{i+1/2} - P_{i-1/2}}{\mathcal{M}_{i+1/2} + \mathcal{M}_{i-1/2}} - \frac{G \sum_{k=0}^{i-1} \mathcal{M}_{k+1/2}}{r_i^2} . \tag{7.75}$$

A technical point now enters. What we are trying to do is to mock up the hydrodynamical time development of a continuous system using discrete time and space steps. The above equation of motion is fine for motions that are subsonic but has real problems if shocks develop.[7] The trouble is that shocks imply some sort of discontinuous behavior in, say, density. In practice, this means that mass shells in a numerical calculation tend to "rattle" around unacceptably near the shock front unless special steps are taken. To alleviate this, an artificial (or numerical, pseudo-, etc.) viscosity is often introduced that smooths out the shock profile and allows the computation to proceed without undue noise. One prescription, and there are quite a few, for an artificial viscosity is (see, for example, Richtmyer and Morton, 1967) $Q_{i+1/2} \sim \left(\dot{r}_{i+1} - \dot{r}_i\right)^2 \rho_{i+1/2}$ if a mass shell at $i+1/2$ is undergoing compression but is set to zero if the shell is expanding. This viscosity is added to the pressure in the difference form of the equation of motion. The net effect is to increase the entropy and spread the shock discontinuity over a few shells with a corresponding loss of resolution in the vicinity of the shock. In the following we shall imagine that some such device is used but we delete specific reference to it.

All the above is imagined to take place at some time t^n. What is required is the status of the system at a new time $t^{n+1} = t^n + \Delta t$, where Δt is a "time step." The method outlined here is called an "explicit" method because the prediction of the system's behavior at t^{n+1} depends only on knowledge of what is happening at t^n. More complicated "implicit" schemes employ an iteration between t^n and t^{n+1} quantities and equations such as was done in the previous section for spatial integrations. The latter schemes have decided advantages, but at the cost of increased complexity. In an explicit method physical quantities march forward in time in a simple way as follows.

Define a time-centered, or average, velocity between t^n and t^{n+1} as

$$\dot{r}_i^{n+1/2} = \dot{r}_i^{n-1/2} + \Delta t^{n+1/2}\, \ddot{r}_i^{n} \tag{7.76}$$

where $\dot{r}_i^{n-1/2}$ is presumed known from a previous time step and $\Delta t^{n+1/2}$

$$\Delta t^{n+1/2} = \frac{t^{n+1} - t^n}{2} . \tag{7.77}$$

The new radial position is then

[7] An excellent pedagogic introduction to shocks is Chapter 1 of *Physics of Shock Waves and High Temperature Hydrodynamic Phenomena* by Zel'dovich and Raizer (1966). That chapter is also available as a separate monograph published in 1968.

$$r_i^{n+1} = r_i^n + \Delta t \, \dot{r}_i^{n+1/2} . \tag{7.78}$$

Since the acceleration $\ddot{r}_i^n$ is known from (7.75), all the interface radii may be updated in this fashion. Having these, and assuming that mass is neither lost nor gained within a shell, we may similarly update the densities by applying the mass equation (7.73) using the appropriate new radii. Note that in applying the acceleration equation the pressure external to the last radius at the surface must be specified. The most simple procedure is to set it to zero as if a vacuum were present. (Fancier and more accurate choices are possible.) Thus if the model has K interfaces labeled $i = 1, 2, \cdots, K$ (with the stationary center at $r_0 = 0$), then $P_{K+1/2} = 0$.

The determination of what to use for Δt involves another technical point of some importance. We have not affixed a superscript to this quantity but it should be obvious that its choice depends on how fast the system is changing at the time in question. If the system is evolving rapidly then Δt should be small. A slowly evolving system need not imply that long steps may be taken, however. There is an upper bound on the length of a time step and this is determined by considerations of numerical stability. In an explicit scheme such as this, the time step must be some fraction less than the time it takes for a sound wave to traverse a shell, and all shells must be examined to find this upper bound. (This is known as the Courant condition.) This means that even though the system may be evolving *very* slowly (as in normal stellar evolution on a nuclear time scale), the computation must proceed at a pace comparable to sound travel times across shells—and this is the price paid for computational simplicity. An implicit scheme, which incorporates information about the state of the system at t^{n+1}, is computationally more difficult but is the one used in slow evolution analysis.

To continue further in time requires finding the new accelerations at t^{n+1}, and this requires computation of $P_{i+1/2}$ in (7.75) evaluated at time t^{n+1}, which we call $P_{1+1/2}^{n+1}$. Straight mechanics will not yield this pressure and the remainder of the stellar structure equations must enter. The first is the energy equation (6.2), which, in difference form at t^{n+1}, becomes

$$\begin{aligned} \frac{E_{i+1/2}^{n+1} - E_{i+1/2}^n}{\Delta t} = & - \frac{P_{i+1/2}^{n+1} + P_{i+1/2}^n}{2} \frac{V_{i+1/2}^{n+1} - V_{i+1/2}^n}{\Delta t} - \\ & - \frac{\left(\mathcal{L}_{i+1}^{n+1} - \mathcal{L}_i^{n+1}\right) + \left(\mathcal{L}_{i+1}^n - \mathcal{L}_i^n\right)}{2\mathcal{M}_{i+1/2}} + \\ & + \frac{\varepsilon_{i+1/2}^{n+1} + \varepsilon_{i+1/2}^n}{2} . \end{aligned} \tag{7.79}$$

Note that the luminosity is defined on interfaces whereas ε is defined within a shell (where densities, temperatures, etc., are defined). Also note that time averaging and differencing have been used in accordance with the methods outlined in §7.2.4.

The internal energy, pressure, and energy generation rate in (7.79) are functions of temperature, density, and composition (which, for simplicity, is assumed constant in time here). If only diffusive radiative transfer holds then $\mathcal{L}_r$ is given by (4.25) in Lagrangian form as

$$\mathcal{L}_r = -\frac{(4\pi r^2)^2 ac}{3\kappa}\frac{dT^4}{d\mathcal{M}_r} . \tag{7.80}$$

The mass derivative in (7.80) can be put into difference form in several ways depending on how terms are combined. The simplest, although not necessarily the best, way is to let $r \to r_i$, $\kappa \to \kappa_{i+1/2}$, and let the mass gradient of temperature (to the fourth) be modeled after the mass gradient of pressure in the acceleration equation. In any case, it is easy to see that $\mathcal{L}_i$ will contain temperatures computed on either side of interface i; that is, we need $T_{i\pm 1/2}$. It will also contain reference to $\rho_{i\pm 1/2}$ through the opacity but these are already known (as are the r_i) from the dynamics that got us to time t^{n+1}. The pressure, internal energy, and energy generation rate may also be computed at this time if the temperatures are known. The conclusion is that the temperatures in the K shells are the only unknowns at t^{n+1}. Note, however, that these temperatures are implicitly highly coupled in space in the energy equation because $\mathcal{L}_i$ and $\mathcal{L}_{i+1}$ both appear. Thus the energy equation at shell $i + 1/2$ contains $T_{i\pm 1/2}$ and $T_{i+3/2}$ so that three temperatures are associated with the shell at $i+1/2$ (and this will result in a tridiagonal system as indicated below).

To proceed further and find the K temperatures is relatively straightforward and a Newton–Raphson relaxation scheme is the method of choice. The energy equation in the $(i + 1/2)$th shell is linearized to yield corrections to guessed temperatures at t^{n+1}, which we denote as $\Delta T_{i+1/2}$. A close inspection of the procedure will reveal that a lot of thermodynamics is required. For example, finding out how the $P\,dV$ term in the energy equation responds to a change in temperature $T_{i+1/2} \to T_{i+1/2} + \Delta T_{i+1/2}$ involves computing the partial of pressure with respect to temperature at constant density. In the end, a set of linear equations in $\Delta T_{i+1/2}$ are obtained of the form

$$\mathbf{M} \cdot \mathbf{\Delta T} = -\mathbf{f} . \tag{7.81}$$

Here $\mathbf{f}_{i+1/2}$ is the differenced energy equation at $i + 1/2$ containing temperature guesses, etc., as cast in the form $f = 0$. If the unknown temperatures are ordered in the column vector $\mathbf{\Delta T}$ as

$$\left(\Delta T_{1/2}, \Delta T_{3/2}, \ldots, \Delta T_{K-3/2}, \Delta T_{K-1/2}\right)^{\mathrm{T}} \tag{7.82}$$

then the matrix $\mathbf{M}$ contains the relevant partials of $\mathbf{f}$ with respect to temperature and is tridiagonal in form (zeros everywhere except along the main diagonal and the two diagonals immediately on either side of it). This system is simple to solve (as in *Numerical Recipes*, §2.4). It would be worth your while to try to construct such a linear system of equations on your own if for no other reason than to see how difficult a bookkeeping job is involved.

The sequence in this explicit hydrodynamics scheme is then: compute the accelerations and velocities, update the radii and compute the new densities, and then use the energy equation to find the new temperatures. The transition to a calculation that does not do hydrodynamics but rather insists upon hydrostatic equilibrium is now not too difficult to imagine. The acceleration equation (7.75) is replaced by the hydrostatic equation that is the righthand side of (7.75) set to zero, but, to maintain numerical stability, all the quantities in that equation are time-averaged over t^n and t^{n+1} (as in the energy equation). The nasty part of the calculation is that the radii at the future time t^{n+1} must be solved for simultaneously along with all the other variables. This is what makes stellar evolution calculations so difficult. There are many variations on Henyey integrations of the time-dependent stellar problem. Some of these involve choosing clever combinations of variables or rephrasing the structure and evolution equations. For examples of some possibilities see Schwarzschild (1958), Kippenhahn, Weigert, and Hofmeister (1967), and Kippenhahn and Weigert (1990, §11).

If nuclear transformations are present, as they almost always are, changes in abundances must also be accommodated. Changes in mean molecular weight due to ionization usually take place so rapidly that they may be regarded as taking place instantaneously and are therefore incorporated directly into the equation of state. Abundance changes that are very slow compared to other time scales in the system are easy: update the composition after a time step is taken using as simple a difference scheme in time as possible. However, in the most complicated situations, where abundances change rapidly or particular nuclear species must be followed carefully in time for some purpose, then the rate equations for transmutations (or, possibly, an equilibrium version thereof) must be included among the stellar structure equations implicitly. Needless to say, such a full calculation takes its toll of time and patience—but that's the name of the game.

Returning to dynamic problems and, in particular, supernovae, explosive nuclear burning presents its own difficulties. A typical reaction network (such as shown in Fig. 6.12) consists of solving a coupled set of differential equations for the abundances of the reactants. The time scales associated with the creation and destruction of the reactants may be slow, fast, *really* fast, and everything in between. There may be situations where the creation and destruction of a pair of nuclear species connected by forward and backward reactions (such as capture of a proton on nucleus X to form Y followed by photodisintegration back to X) is such that the abundances hardly change although the rates of the reactions are rapid (as in Nuclear Statistical Equilibrium). The net result is that the differential equations may well become "stiff" and solutions can drift off from reality in a hurry. Ordinary methods (such as Runge–Kutta) fail miserably for stiff equations and special steps must be taken. For an excellent review of such problems (in the supernova context) see Arnett (1996, §4.4).

And now, after this long digression, back to polytropes.

7.2.7 The Eddington Standard Model

A simple example of the use of polytropes in making a stellar pseudo-model is the "Eddington standard model" in which the energy equation and the equation of diffusive radiative transfer are incorporated together in an approximate way.

Recall that the actual run of temperature versus pressure in situations where there is no convection is given by (see Eqs. 7.8–7.11)

$$\nabla = \frac{d\ln T}{d\ln P} = \frac{3}{16\pi ac}\frac{P\kappa}{T^4}\frac{\mathcal{L}_r}{G\mathcal{M}_r} \,. \tag{7.83}$$

We may also express ∇ in different terms by introducing the radiation pressure $P_{\rm rad} = aT^4/3$ and unwinding the derivatives in the definition of ∇ to obtain

$$\nabla = \frac{P}{T}\left(\frac{dT/dr}{dP/dr}\right) = \frac{1}{4}\frac{P}{P_{\rm rad}}\frac{dP_{\rm rad}}{dP} \,. \tag{7.84}$$

Solving for the pressure derivative and combining this with (7.83) yields

$$\frac{dP_{\rm rad}}{dP} = \frac{\mathcal{L}\kappa}{4\pi cG\mathcal{M}}\frac{\mathcal{L}_r/\mathcal{L}}{\mathcal{M}_r/\mathcal{M}} \tag{7.85}$$

where $\mathcal{L}$ and $\mathcal{M}$ are the total luminosity and mass.

The ratio $(\mathcal{L}_r/\mathcal{L})/(\mathcal{M}_r/\mathcal{M})$ in (7.85) is a normalized average energy generation rate, as may be seen from considering the energy equation in thermal balance. If $(d\mathcal{L}_r/d\mathcal{M}_r) = \varepsilon$ is integrated over $\mathcal{M}_r$, then define

$$\langle\varepsilon(r)\rangle = \frac{\int_0^r \varepsilon\, d\mathcal{M}_r}{\int_0^r d\mathcal{M}_r} = \frac{\mathcal{L}_r}{\mathcal{M}_r} \tag{7.86}$$

with $\langle\varepsilon(\mathcal{R})\rangle = \mathcal{L}/\mathcal{M}$. It is then traditional to introduce $\eta(r)$ as

$$\eta(r) = \frac{\langle\varepsilon(r)\rangle}{\langle\varepsilon(\mathcal{R})\rangle} = \frac{\mathcal{L}_r/\mathcal{L}}{\mathcal{M}_r/\mathcal{M}} \tag{7.87}$$

so that (7.85) becomes

$$\frac{dP_{\rm rad}}{dP} = \frac{\mathcal{L}}{4\pi cG\mathcal{M}}\,\kappa(r)\,\eta(r) \,. \tag{7.88}$$

Thus far we have made no other assumptions aside from thermal balance and pure diffusive radiative transfer.

We now formally integrate the last expression from the surface to an interior point r, assuming that the surface pressure is zero and find

$$P_{\rm rad}(r) = \frac{\mathcal{L}}{4\pi cG\mathcal{M}} \langle \kappa(r)\,\eta(r)\rangle\, P(r) \tag{7.89}$$

where the averaged expression is the combination

$$\langle \kappa(r)\,\eta(r)\rangle = \frac{1}{P(r)} \int_0^{P(r)} \kappa\eta\, dP \ . \tag{7.90}$$

This is now put in final form by recalling the definition of β, the ratio of ideal gas to total pressure of equation (3.106), so that

$$1 - \beta(r) = \frac{P_{\rm rad}(r)}{P(r)} \tag{7.91}$$

and, thus, after substituting in (7.89),

$$1 - \beta(r) = \frac{\mathcal{L}}{4\pi cG\mathcal{M}} \langle \kappa(r)\,\eta(r)\rangle \ . \tag{7.92}$$

To make further progress we now examine $\langle \kappa\eta \rangle$. The following will contain the key to the standard model as it was introduced by Eddington (1926) well before stellar processes were completely understood. Yet, as we shall see, this model is not only of historical interest but it also yields insights into how some kinds of stars work.

A reasonable opacity to insert in (7.92) is a combination of electron scattering and Kramers' (see Chap. 3). Thus let

$$\kappa = \kappa_{\rm e} + \kappa_0 \rho T^{-3.5}. \tag{7.93}$$

Except for the inclusion of an H^- opacity source, this is a good approximation for most main sequence stars. The important thing to note is that this opacity increases outward with radius if ρ does not decrease outward faster than $T^{3.5}$—which it will not in this model.

The quantity $\eta(r)$ is proportional to the average energy generation rate and, if we restrict ourselves to main-sequence-like objects, it should decrease outward fairly rapidly to unity because of the relatively high positive temperature exponent of ε. Thus the product $\kappa\eta$ should not vary as strongly with position as does either of its components. The crucial assumption in the standard model is that $\kappa\eta$ varies so weakly with position that it may be taken as a constant throughout the star. This means that the righthand side of (7.92) is constant and so is β.

The constancy of β may be translated into a temperature versus density relation as follows. If we assume that the pressure is made up of the sum of ideal gas plus radiation pressure only, then

$$P_{\text{rad}} = \frac{1-\beta}{\beta} P_{\text{gas}} = \frac{1-\beta}{\beta} \frac{N_{\text{A}}k}{\mu} \rho T = \frac{1}{3} a T^4 \tag{7.94}$$

from previous work. Solving for temperature then yields

$$T(r) = \left(\frac{3N_{\text{A}}k}{a\mu} \frac{1-\beta}{\beta} \right)^{1/3} \rho^{1/3}(r) \, . \tag{7.95}$$

Note that this relation is similar to that found using the virial theorem approach (of 1.36) but for the constant multiplying $\rho^{1/3}$. What is more important, however, is that this is not a relation between virial average quantities but rather gives the run of T versus ρ through the star.

To proceed, we now use

$$P = \frac{P_{\text{gas}}}{\beta} = \frac{N_{\text{A}}k}{\mu} \frac{\rho T}{\beta} \tag{7.96}$$

to find the pressure-density relation

$$P(r) = \left[\left(\frac{N_{\text{A}}k}{\mu} \right)^4 \frac{3}{a} \frac{1-\beta}{\beta^4} \right]^{1/3} \rho^{4/3}(r) \, . \tag{7.97}$$

If we restrict ourselves to situations where μ is constant (homogeneous, constant state of ionization, etc.), then the term within the brackets is a constant (because β is) and what we are left with is a polytropic equation of the form $P \sim \rho^{4/3}$. The exponent on the density immediately tells us that we are dealing with a polytrope of index $n = 3$ and that the coefficient K of (7.16) is

$$K = \left[\left(\frac{N_{\text{A}}k}{\mu} \right)^4 \frac{3}{a} \frac{1-\beta}{\beta^4} \right]^{1/3} . \tag{7.98}$$

On the other hand, K is also given by (7.40) which, for $n = 3$, is

$$K = \frac{(4\pi)^{1/3}}{4} \frac{G\mathcal{M}^{2/3}}{\left[\xi^4 \left(-\theta_3' \right)^2 \right]_{\xi_1}^{1/3}} \, . \tag{7.99}$$

We now equate the two expressions for K, substitute the relevant polytropic quantities from Table 7.1, and find

$$\frac{1-\beta}{\beta^4} = 0.002996 \, \mu^4 \left(\frac{\mathcal{M}}{\mathcal{M}_\odot} \right)^2 . \tag{7.100}$$

To find the temperature, combine (7.95) and (7.100):

$$T(r) = 4.62 \times 10^6 \beta\mu \left(\frac{\mathcal{M}}{\mathcal{M}_\odot} \right)^{2/3} \rho^{1/3}(r) \, . \tag{7.101}$$

Note that β and $\mathcal{M}$ in (7.101) are not independent but are connected through relation (7.100) so that things are not as simple as they may look. Table 7.2 summarizes this connection. The general trend is now clear and agrees with what we know about the zero-age main sequence: more massive stars have higher temperatures and radiation pressure plays a greater role (lower βs).

Table 7.2. Eddington Standard Model

$\mu^2\mathcal{M}/\mathcal{M}_\odot$	β
1.0	0.9970
2.0	0.9885
5.0	0.9412
10.0	0.8463
50.0	0.5066

What information does the standard model *not* provide? First of all, it does not yield absolute numbers for temperatures, densities, and pressures unless we specify both the mass and radius of the configuration. For example, (7.101) yields the run of temperature versus density but where is the normalization of density to be found? Equation 7.42 contains the central density but the *average* density is required and, hence, the mass and radius. The reason for this need is that, although we have shown the standard model to be a polytrope of index three using the presumed constancy of $\langle\kappa\eta\rangle$, we have not then gone back and really solved the energy and heat transport equations. In this sense, the analysis is incomplete. You may look into this further (q.v., Chandrasekhar, 1939, Chap. VI) but that will lead us far afield (and the standard model can only be pushed so far). However, we shall assume that both the mass and total radius of a model are specified and see how well the standard model does with that information.

Given the mass and radius, and thus average density, the central density for an $n = 3$ polytrope is $\rho_c = 54.18\langle\rho\rangle$ from Table 7.1. The run of T versus ρ then follows from (7.101). We can then compare this result to that from a ZAMS model with the same mass and total radius. Figure 7.4 shows the run of temperature versus density for a ZAMS solar model with $X = 0.743$, $Y = 0.237$, and $Z = 0.02$ (yielding $\mu \approx 0.6$ with complete ionization). The average density is $\langle\rho\rangle = 2.023$ g cm^{-3} and central density $\rho_c = 82.49$ g cm^{-3}. A standard model with this average density has $\rho_c = 109.6$ g cm^{-3}; we use this to compute $T(r)$ versus $\rho(r)$ as shown in the figure. The standard model does remarkably well compared to the ZAMS sun model in the inner regions where density and temperature are high but does not fare so well in the outer regions. These outer regions starting at $\rho \approx 0.1$ g cm^{-3} behave more like an $n = 3/2$ polytrope in this instance because of convection but this convective region constitutes only about 0.6% of the mass of the star. Most of the star, in mass units, has a structure close to that of an $n = 3$ polytrope. Note that if the interior pressure is controlled by an ideal gas, then $P \propto T^4$, as can be inferred from (7.97) and (7.101).

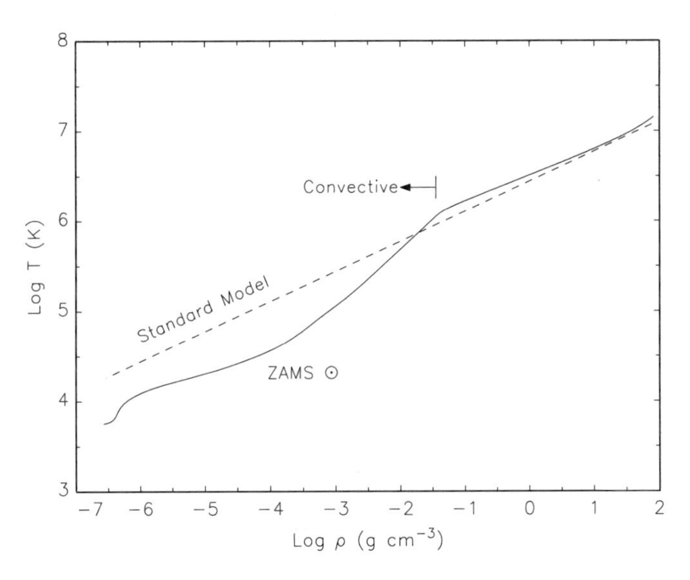

Fig. 7.4. The solid line is the temperature–density relation through a full-blown model of a ZAMS sun with total radius of 6.168×10^{10} cm or $0.886\ \mathcal{R}_\odot$. The dashed line shows the standard model result where the same total radius has been assumed. The centers of the models are at the right.

7.2.8 Applications to Zero-Temperature White Dwarfs

We are sure that it has occurred to you that polytropes also represent excellent approximations to zero-temperature white dwarfs in either the limits of completely nonrelativistic or relativistic degeneracy (that is how we started our discussion). Therefore, recall that the nonrelativistic equation of state is $P_e = 1.004 \times 10^{13} \, (\rho/\mu_e)^{5/3}$ dyne cm^{-2} from either (3.65) or (7.14). This is a polytropic equation of state and, hence, we know the regular solution is a polytrope of index $n = 3/2$. From (7.16) we may readily identify the equation of state coefficient K and equate it to the polytropic K given by (7.40), use Table 7.1, and find

$$K = \frac{1.004 \times 10^{13}}{\mu_e^{5/3}} = 2.477 \times 10^{14} \left(\frac{\mathcal{M}}{\mathcal{M}_\odot}\right)^{1/3} \left(\frac{\mathcal{R}}{\mathcal{R}_\odot}\right) . \tag{7.102}$$

This yields the mass–radius relation

$$\frac{\mathcal{M}}{\mathcal{M}_\odot} = 2.08 \times 10^{-6} \left(\frac{2}{\mu_\mathrm{e}}\right)^5 \left(\frac{\mathcal{R}}{\mathcal{R}_\odot}\right)^{-3} \tag{7.103}$$

which is close to what we found in the virial estimate of (3.63). It also reproduces the mass–radius relation (3.68–3.69) for small $\mathcal{M}$, as you can check by neglecting the mass term in the brackets of (3.68)—and it better do so!

The corresponding completely relativistic result is found using the equation of state (3.66), $P_\mathrm{e} = 1.243 \times 10^{15} \left(\rho/\mu_\mathrm{e}\right)^{4/3}$ dyne cm^{-3}, and by realizing that we are dealing with an $n = 3$ polytrope—namely,

$$K = \frac{1.243 \times 10^{15}}{\mu_\mathrm{e}^{4/3}} = 3.841 \times 10^{14} \left(\frac{\mathcal{M}}{\mathcal{M}_\odot}\right)^{2/3} \tag{7.104}$$

or

$$\frac{\mathcal{M}_\infty}{\mathcal{M}_\odot} = 1.456 \left(\frac{2}{\mu_\mathrm{e}}\right)^2 . \tag{7.105}$$

This is the Chandrasekhar limiting mass discussed in §3.52 (and see 3.67). Real white dwarfs will be discussed in Chapter 10.

7.3 The Approach to Real Models

This section will deal with some aspects of modeling real stars. We have already introduced sketches of various numerical modeling techniques and now we will discuss some special problems. We will start with what is to be done near model centers and then skip to the surface layers.

7.3.1 Central Expansions

As we saw in treating regular polytropes, the center of a stellar model presents some peculiar problems. The hydrostatic equation and the expressions for ∇_rad and ∇ (of 7.8 and 7.11) contain indeterminate ratios at $r = 0$, as may easily be verified by inspection. We require, however, that everything be regular at the origin. What is done in practice is to replace the stellar structure equations near the model center by simple expansions in r, much as was done for polytropes. The procedure is straightforward once regularity and symmetry are accounted for. Application of the boundary conditions $\mathcal{M}_r \to 0$ and $\mathcal{L}_r \to 0$ as $r \to 0$ and the requirement of zero spatial gradients of temperature, density, and pressure at the center yield

$$\mathcal{M}_r \longrightarrow \frac{4\pi}{3} \rho_\mathrm{c} r^3 \tag{7.106}$$

$$P(r) \longrightarrow P_\mathrm{c} - \frac{2\pi}{3} G \rho_\mathrm{c}^2 r^2 \tag{7.107}$$

$$\mathcal{L}_r \longrightarrow \frac{4\pi}{3} \rho_\mathrm{c} \varepsilon_\mathrm{c} r^3 \tag{7.108}$$

where, as usual, subscript "c" means the central value of a quantity. The behavior of temperature near the center depends on the mode of energy transport and is

$$T(r) \longrightarrow T_{\rm c} - \frac{1}{8ac}\frac{\kappa_{\rm c}\rho_{\rm c}^2\varepsilon_{\rm c}}{T_{\rm c}^3}\, r^2 \tag{7.109}$$

if there is no convection, and

$$T(r) \longrightarrow T_{\rm c} - \frac{2\pi}{3} G\,\nabla_{\rm ad,c}\,\frac{\rho_{\rm c}^2 T_{\rm c}}{P_{\rm c}}\, r^2 \tag{7.110}$$

if adiabatic (efficient) convection is present (so that $\nabla = \nabla_{\rm ad}$). An easy way to verify the temperature relations is to eliminate r^2 between $P(r)$ and $T(r)$ (either version) and construct a numerical logarithmic derivative. Note that (7.109) may be replaced with (7.110) if $\nabla_{\rm ad}$ is replaced by $\nabla_{\rm rad}$ (of Eq. 4.30) in the latter equation.

How to treat the surface is more difficult.

7.3.2 The Radiative Stellar Envelope

The term "envelope" is used in various ways in stellar physics and you will see it used many ways here also. Examples of its use are as follows. When describing the structure of red supergiants, which have a dense degenerate core surrounded by an immense diffuse region (both of which may contain substantial mass), the latter region is often referred to as the envelope. On the other hand, the "envelope" of a white dwarf is considered to be the very thin (in mass and depth) nondegenerate region overlying the massive degenerate core. For the purposes of this section, an envelope consists of the portion of a star that starts at the surface, continues inward, but contains negligible mass, has no thermonuclear or gravitational energy sources, and is in hydrostatic equilibrium. This implies, among other things, that the luminosity in the envelope is fixed at its surface value with $\mathcal{L}_r = \mathcal{L}$ and that $\mathcal{M}_r$ is essentially just the total mass $\mathcal{M}$. As we shall see, these requirements allow for considerable simplification in determining the structure of the envelope.

The Structure of the Envelope

Here we look first at the structure of envelopes in which convection is negligible or nonexistent. The primary reference is Cox (1968, Chap. 20), who gives many variations on the following.

If convection is neglected, then ∇ is equal to $\nabla_{\rm rad}$ with

$$\nabla = \frac{d\ln T}{d\ln P} = \frac{3}{16\pi acG}\frac{P\kappa}{T^4}\frac{\mathcal{L}}{\mathcal{M}} \tag{7.111}$$

where we have set $\mathcal{M}_r = \mathcal{M}$, $\mathcal{L}_r = \mathcal{L}$, and

$$\frac{3\mathcal{L}}{16\pi acG\mathcal{M}} = 7.59 \times 10^9 \left(\frac{\mathcal{M}}{\mathcal{M}_\odot}\right)^{-1} \left(\frac{\mathcal{L}}{\mathcal{L}_\odot}\right) . \tag{7.112}$$

Assume for convenience that the pressure is due solely to an ideal gas with no radiation pressure. Once the envelope solution is obtained, this assumption can be examined and corrections made. The opacity in (7.111) is then written in the interpolation form $\kappa = \kappa_0 \rho^n T^{-s}$, which, for an ideal gas, becomes

$$\kappa = \kappa_g P^n T^{-n-s} \tag{7.113}$$

where

$$\kappa_g = \kappa_0 \left(\frac{\mu}{N_A k}\right)^n . \tag{7.114}$$

With this substitution the differential equation (7.111) now contains only P and T as the active variables and may be rewritten

$$P^n\, dP = \frac{16\pi acG\mathcal{M}}{3\kappa_g \mathcal{L}} T^{n+s+3}\, dT . \tag{7.115}$$

If T_0 and P_0 refer to some upper reference level in the envelope (such as the photosphere) with $P(r) \geq P_0$ and $T(r) \geq T_0$, then an easy integration of (7.115) yields

$$P^{n+1} = \frac{n+1}{n+s+4} \frac{16\pi acG\mathcal{M}}{3\kappa_g \mathcal{L}} T^{n+s+4} \left[\frac{1-(T_0/T)^{n+s+4}}{1-(P_0/P)^{n+1}}\right] \tag{7.116}$$

for $n+s+4 \neq 0$. If this sum of exponents does equal zero through some unlikely numerical accident, then the solution differs from the above but is still easy to find.

We now examine some likely combination of exponents n and s to see what (7.116) implies for the run of pressure versus temperature in the envelope. Note first that if $n+s+4$ and $n+1$ are both positive then the terms $[T_0/T(r)]^{n+s+4}$ and $[P_0/P(r)]^{n+1}$ get small rapidly as we go to deep depths in the envelope. Thus for $(n+s+4)$, $(n+1) > 0$ and $T(r) >> T_0$, $P(r) >> P_0$

$$P^{n+1} \longrightarrow \frac{n+1}{n+s+4} \frac{16\pi acG\mathcal{M}}{3\kappa_g \mathcal{L}} T^{n+s+4} . \tag{7.117}$$

If T_0 and P_0 refer to photospheric values, then, in this instance, the solution for pressure versus temperature deep in the envelope below the photosphere converges rapidly to a common solution independent of photospheric conditions. Thus, as far as the interior structure is concerned, we could just as well have used zero boundary conditions for pressure and temperature. Examples include Kramers' opacity ($n = 1$, $s = 3.5$) and electron scattering ($n = s = 0$).

An important counterexample to the above is where the envelope opacity is due to H^- as in the estimate given by (4.65). Here $n = 1/2$, $s = -9$

with $n+s+4=-4.5$ and photospheric boundary conditions have a strong influence on the underlying layers. It is also true that in cool stars where H^- opacity is important, the underlying layers are convective and the above analysis does not apply. We shall return to this shortly.

If we assume (7.117) holds, then the logarithmic run of temperature versus pressure at depth is

$$\nabla(r) \longrightarrow \frac{n+1}{n+s+4} = \frac{1}{1+n_{\text{eff}}} \tag{7.118}$$

where $n_{\text{eff}} = (s+3)/(n+1)$ is the "effective polytropic index." The reason for this name is as follows. Equation (7.117) also yields

$$P = K' T^{1+n_{\text{eff}}} \tag{7.119}$$

where it is assumed that we may use zero boundary conditions on T and P, and where K' may readily be established from (7.117). If the gas is ideal and of constant composition, with $P \propto \rho T/\mu$, then $P \propto \rho^{1+1/n_{\text{eff}}}$ and the structure is polytropic (as in 7.22). Recalling our discussion of polytropes and ideal gases, the coefficient K' is the same as that identified in (7.27) and, from the present analysis, is equal to

$$K' = \left[\frac{1}{1+n_{\text{eff}}} \frac{16\pi acG\mathcal{M}}{3\kappa_0 \mathcal{L}} \left(\frac{N_{\text{A}}k}{\mu}\right)^n\right]^{1/(n+1)} \tag{7.120}$$

from which we may calculate the polytropic constant K by way of (7.28).[8] A practical difficulty in carrying out the analysis further in terms of polytrope language is that the solution (7.117) must eventually be joined to the rest of the star, and the present analysis cannot do that. Suffice it to say that you might well imagine how this might be done (with appropriate conditions of continuity, etc.). For now we remark that the polytropic-like solution in the envelope corresponding to $\theta_n(\xi)$ of the previous sections need not be of the complete E-solution variety and may be of F- or M-solution character.

The constancy of ∇ in (7.118) implies that the combination $P\kappa/T^4$ is a constant by virtue of (7.111). Specifically, if Kramers' opacity holds in the envelope, then (with $n = 1$, $s = 3.5$, and $n_{\text{eff}} = 3.25$) the solution for P versus T is

$$P(r) = \left[\frac{1}{4.25} \frac{16\pi acGM}{3\kappa_0 \mathcal{L}} \frac{N_{\text{A}}k}{\mu}\right]^{1/2} T^{4.25}(r) \tag{7.121}$$

and $\nabla = 0.2353$. If no ionization processes are taking place and $\Gamma_2 = 5/3$, then $\nabla_{\text{ad}} = 1 - 1/\Gamma_2 = 0.4$ and thus $\nabla < \nabla_{\text{ad}}$, which implies no convection; the envelope is radiative as originally assumed. For electron scattering with

[8] When we think "polytrope," there is always the danger of confusing the polytropic index n with the opacity exponent n in the present analysis. Don't fall into that trap!

$n = s = 0$, $n_{\text{eff}} = 3$ (an "$n = 3$" polytrope again!), $\nabla = 0.25$ and the same conclusion holds. Note, however, these results are only a rough guide and must be applied with caution; ionization processes and convection, albeit almost negligible, occur in the outer layers of nearly all stars and a complete and accurate integration including all effects is necessary in modeling real stars. The present analysis remains a rough guide but, on the other hand, and even in practice, constant luminosity and negligible envelope mass are often used to simplify envelope integrations.

The Radiative Temperature Structure

If the envelope is radiative with n_{eff} constant and certain restrictions given below apply, then temperature as a function of radius may be found. We assume that n and s are such that zero boundary conditions are adequate (as discussed above). This means that we know $\nabla = 1/(1 + n_{\text{eff}})$. Thus rewrite the equation of hydrostatic equilibrium in the form

$$\frac{dP}{dr} = \frac{P}{\nabla}\frac{1}{T}\frac{dT}{dr} = -\frac{G\mathcal{M}}{r^2}\rho \tag{7.122}$$

where $\mathcal{M}_r = \mathcal{M}$ is still assumed. If the gas is ideal and $P = \rho N_{\text{A}} kT/\mu$ is used to replace the pressure in the middle term of (7.122), we find

$$(n_{\text{eff}} + 1)\frac{dT}{dr} = -\frac{G\mathcal{M}\mu}{N_{\text{A}} k}\frac{1}{r^2} \,. \tag{7.123}$$

This is then integrated to yield T(r) as

$$\begin{aligned} T(r) &= \frac{1}{1 + n_{\text{eff}}}\frac{G\mathcal{M}\mu}{N_{\text{A}} k}\left(\frac{1}{r} - \frac{1}{\mathcal{R}}\right) \\ &= \frac{2.293 \times 10^7}{1 + n_{\text{eff}}}\mu\left(\frac{\mathcal{M}}{\mathcal{M}_\odot}\right)\left(\frac{\mathcal{R}}{\mathcal{R}_\odot}\right)^{-1}\left(\frac{1}{x} - 1\right)\ \text{K} \end{aligned} \tag{7.124}$$

where $x = r/\mathcal{R}$. Note that if Kramers' opacity could be used everywhere in the solar envelope (so that $n_{\text{eff}} = 3.25$) and the composition were ionized Pop I ($\mu \approx 0.6$), then the temperature at a level only 1% below the surface ($x = 0.99$) would be, say, for a pseudo-sun, about 33,000 K as compared to T_{eff} of 5,780 K. In other words, large positive values of the temperature and density exponents of opacity imply a rapid increase of temperature inward and the outer boundary conditions matter little.

We may also find the envelope mass using the above. If the gas is ideal $P = \rho N_{\text{A}} kT/\mu$ may be equated to $P = K' T^{1+n_{\text{eff}}}$ of (7.119) and the density is $\rho(r) = K'\mu T^{n_{\text{eff}}}(r)/N_{\text{A}} k$. But since $T(r)$ is given by (7.124), we then have $\rho(r)$. The latter is integrated (after weighting by $4\pi r^2$) from some envelope level r to $\mathcal{R}$ to give the mass, $\mathcal{M} - \mathcal{M}_r$, above that level. As an exercise for you, consider electron scattering opacity (κ_e, $n = s = 0$, $n_{\text{eff}} = 3$), use (7.120) to compute K', and show that

$$1 - \frac{\mathcal{M}_r}{\mathcal{M}} = \frac{\pi^2 ac}{12\kappa_e}\left(\frac{\mu G}{N_A k}\right)^4 \frac{\mathcal{M}^3}{\mathcal{L}} \times \times \left[\frac{x^3}{3} - \frac{3}{2}x^2 + 3x - \frac{11}{6} - \ln x\right] . \tag{7.125}$$

If x is very close to unity, then the term in the brackets is approximately $(1 - x)^4/4$. For a solar mass and luminosity, equation (7.125) implies, for example, that traversing 15% of the total radius inward from the surface uses up only a little less than 1% of the mass. This confirms our assumption that $\mathcal{M}_r \approx \mathcal{M}$ through the envelope.

7.3.3 Completely Convective Stars

We know from previous discussions that cool stars tend to have convective envelopes. In this section we shall carry this to the extreme and discuss some of the properties of *fully* convective stars and how such objects come to be. The analysis may become algebraically tedious in spots, but the result will bear directly on pre-main sequence evolution.

Consider a cool star whose surface layer is dominated by H^- opacity, which is, from (4.65),

$$\kappa_{H^-} \approx 2.5 \times 10^{-31}\left(\frac{Z}{0.02}\right)\rho^{1/2}T^9 \quad \text{cm}^2\ \text{g}^{-1} \tag{7.126}$$

for hydrogen mass fractions X around 0.70. We already know that the exponents for this opacity ($n = 1/2$, $s = -9$) spell trouble for a simple envelope analysis because the outer boundary conditions are felt deep down into the envelope. Thus, from now on, we use photospheric boundary conditions with temperature $T_p = T_{\text{eff}}$ and pressure $P_p = 2g_s/3\kappa_p$ as derived in §4.3. Now to find the structure of the envelope.

Consider the pressure–temperature relation (7.116) and transform it into an equation for ∇ as a function of temperature. You may easily verify that the result is

$$\nabla(r) = \frac{1}{1 + n_{\text{eff}}} + \left(\frac{T_{\text{eff}}}{T(r)}\right)^{n+s+4}\left[\nabla_p - \frac{1}{1 + n_{\text{eff}}}\right] \tag{7.127}$$

with $n_{\text{eff}} = (s+3)/(n+1)$ as given by (7.118). The "p" means photospheric, ∇_p is ∇ evaluated at the photosphere, and (see 7.111)

$$\nabla_p = \frac{3\kappa_0 \mathcal{L}}{16\pi acG\mathcal{M}}\left(\frac{\mu}{N_A k}\right)^n \frac{P_p^{n+1}}{T_{\text{eff}}^{n+s+4}} = \frac{3\mathcal{L}}{16\pi acG\mathcal{M}} P_p \kappa_p T_{\text{eff}}^4 . \tag{7.128}$$

At the photosphere $P_{\rm p} = 2g_s/3\kappa_{\rm p}$, $g_s = G\mathcal{M}/\mathcal{R}^2$, and, of course, $\mathcal{L} = 4\pi\mathcal{R}^2\sigma T_{\rm eff}^4$. Inserting this information into (7.128) yields $\nabla_{\rm p} = 1/8$. The run of ∇ below the photosphere is obtained from (7.127) which, for H^- opacity (with $n_{\rm eff} = -4$), is

$$\nabla(r) = -\frac{1}{3} + \frac{11}{24}\left[\frac{T_{\rm eff}}{T(r)}\right]^{-9/2} . \tag{7.129}$$

Note that since temperature increases with depth, so does ∇. The implication of this observation is that at some depth ∇ must eventually become larger than the thermodynamic derivative $\nabla_{\rm ad}$. If, as an approximation, we assume that $\nabla_{\rm ad}$ is given by its ideal gas value $\nabla_{\rm ad} = 0.4$ (with no ionization taking place), then we can estimate where (in temperature) ∇ is equal to, and thereafter would exceed, $\nabla_{\rm ad}$. Thus we can set $\nabla(r)$ of (7.129) equal to 0.4 and solve for $T(r)$ at this critical depth—and we shall do this in just a bit. For now, observe that if $\nabla > \nabla_{\rm ad}$, then the stellar material becomes convective and, for simplicity, we assume that the convection is adiabatic. Thus at depths deeper than the critical depth, $\nabla(r) = \nabla_{\rm ad} = 0.4$. As remarked upon at the end of §7.2.1, this behavior of $\nabla(r)$, along with the ideal gas assumption, implies a polytrope of index 3/2 and

$$P = K' T^{5/2} \tag{7.130}$$

gives the run of pressure with temperature (see 7.27). The picture is then that of a photosphere from which escapes the visible radiation, underlain by a radiative layer of depth to be determined (and probably a shallow layer at that), and, under that, convection. This represents the outer layers of the sun as we know it.

In writing down (7.130) we note the following. In the extreme case where convection continues down to the stellar center, the constant K' cannot be arbitrary because (7.130), as a complete polytrope, must have solutions corresponding to a complete model with appropriate central boundary conditions. In other words, given $\mathcal{M}$ and $\mathcal{R}$, K' must satisfy the combination of relations (7.40) and (7.28) for K' and K given earlier for ideal gas polytropes. One way to approach this is to recast pressure and temperature in the dimensionless variables discussed by Schwarzschild (1958, §13), where

$$p = \frac{4\pi}{G}\frac{\mathcal{R}^4}{\mathcal{M}^2}P \tag{7.131}$$

$$t = \frac{N_{\rm A}k}{G}\frac{\mathcal{R}}{\mu\mathcal{M}}T . \tag{7.132}$$

Equation (7.130) then becomes

$$p = E_0 t^{5/2} \tag{7.133}$$

with

$$E_0 = K'4\pi \left(\frac{\mu}{N_A k}\right)^{5/2} G^{3/2}\mathcal{M}^{1/2}\mathcal{R}^{3/2}. \tag{7.134}$$

But for an ideal gas, $n = 3/2$, E-solution polytrope, K' is given by (7.28) as

$$K'_{n=3/2} = \left(\frac{N_A k}{\mu}\right)^{5/2} K_{n=3/2}^{-3/2} \,. \tag{7.135}$$

After substituting for K of (7.40) this becomes

$$\begin{aligned} K'_{n=3/2} &= \frac{2.5^{3/2}}{4\pi}\left[\xi_{3/2}^{5/2}\left(-\theta'_{3/2}\right)^{1/2}\right]_{\xi_1} \times \\ &\times \left(\frac{N_A k}{\mu}\right)^{5/2} \frac{1}{G^{3/2}\mathcal{M}^{1/2}\mathcal{R}^{3/2}} \,. \end{aligned} \tag{7.136}$$

Putting this together we find the surprising result that E_0 does not depend on any of the physical parameters of the model (mass, radius, composition) but rather contains only the surface values of the polytropic variables and is the constant

$$E_0 = \left(\frac{-125}{8}\,\xi_{3/2}^5\theta'_{3/2}\right)_{\xi_1}^{1/2} = 45.48 \tag{7.137}$$

using the results from Table 7.1.

After these introductory remarks we now compute some of the parameters of a completely convective star. In particular, we seek a relation between mass, luminosity, effective temperature, and composition. This will take a few steps. We first need the temperature and density at that level in the star below the photosphere where $\nabla = \nabla_{\text{ad}} = 0.4$. This may be found from (7.127) using $\nabla_{\text{p}} = 1/8$ and the exponents $n = 1/2$ and $s = -9$ from the estimate for the H^- opacity. You may readily check that the temperature at that level, denoted by T_f, is given by $(T_f/T_{\text{eff}}) = (8/5)^{2/9} \approx 1.11$; that is, T_f is a mere 11% higher than T_{eff}. The implication is that convection starts just below the photosphere. The pressure, P_f, at the top of the convective interior is found by rewriting (7.116) in the form

$$\left(\frac{P}{P_{\text{p}}}\right)^{n+1} = 1 + \frac{1}{1+n_{\text{eff}}}\,\frac{1}{\nabla_{\text{p}}}\left[\left(\frac{T}{T_{\text{eff}}}\right)^{n+s+4} - 1\right] \tag{7.138}$$

which, for the case in question, yields $P_f = 2^{2/3}P_{\text{p}}$.

We now apply (7.130) in the form $P_f = K'T_f^{5/2}$. The polytropic parameter K' is obtained from the combination of equations (7.134) to (7.137) and is

$$K' = \frac{3.564 \times 10^{-4} E_0}{\mu^{2.5}}\left(\frac{\mathcal{M}}{\mathcal{M}_\odot}\right)^{-1/2}\left(\frac{\mathcal{R}}{\mathcal{R}_\odot}\right)^{-3/2} \tag{7.139}$$

or, in functional dependence, $K' = K'(\mathcal{M}, \mathcal{R}, \mu)$.[9] To express K' in terms of T_{eff} and $\mathcal{L}$, use $\mathcal{L} = 4\pi\sigma\mathcal{R}^2 T_{\text{eff}}^4$ and find $K' = K'(\mathcal{M}, T_{\text{eff}}, \mathcal{L}, \mu)$. Now for $P_f = 2^{2/3} P_{\text{p}}$. The photospheric pressure is $2g_s/3\kappa_{\text{p}}$, which is a function of $\mathcal{M}$ and $\mathcal{R}$ (from g_s) and T_{eff} and photospheric density (from $\kappa_{\text{p}} = \kappa_0 \rho_{\text{p}}^n T_{\text{eff}}^{-s}$). The density is eliminated using the ideal gas equation of state to yield

$$P_{\text{p}} = \left(\frac{2}{3}\frac{G\mathcal{M}}{\kappa_0 \mathcal{R}^2}\right)^{1/(n+1)} \left(\frac{\mu}{N_{\text{A}} k}\right)^{-n/(n+1)} T_{\text{eff}}^{(n+s)/(n+1)} \tag{7.140}$$

which has the dependence $P_{\text{p}} = P_{\text{p}}(\mathcal{M}, \mathcal{R}, \mu, T_{\text{eff}})$. Again get rid of $\mathcal{R}$ and find $P_{\text{p}} = P_{\text{p}}(\mathcal{M}, T_{\text{eff}}, \mathcal{L}, \mu)$. It should now be clear that the polytropic equation $P_f = K' T_f^{5/2}$ becomes a power law equation containing only the variables $\mathcal{M}$, T_{eff}, $\mathcal{L}$, κ_0, and μ (and, in the general case, n and s). For our estimate of H^- opacity, this relation becomes

$$T_{\text{eff}} \approx 2600 \mu^{13/51} \left(\frac{\mathcal{M}}{\mathcal{M}_\odot}\right)^{7/51} \left(\frac{\mathcal{L}}{\mathcal{L}_\odot}\right)^{1/102} \text{ K.} \tag{7.141}$$

The strange exponents appearing here are a good indication of how messy this calculation is. Also, as will be pointed out later, the temperature coefficient of 2,600 K is too low in this simple calculation. It should be more like 4,000 K. In any case, this relation shows up as a set of nearly vertical lines in the Hertzsprung–Russell diagram with $\mathcal{M}$ being the parameter labeling the lines and where T_{eff} is virtually independent of $\mathcal{L}$ for a given $\mathcal{M}$. Thus completely convective stars (with a radiative photosphere) of a given mass and (uniform) composition in hydrostatic equilibrium lie at nearly constant low effective temperature independent of luminosity. Or, phrased another way, the effective temperatures of such stars are nearly independent of how the energy is generated. The next section discusses to what kinds of stars these models correspond.

A Question of Entropy

To interpret the above, we return to a comment made in Chapter 5 (§5.1 and see Ex. 5.1) about the role of entropy in convective stars (and we shall follow the excellent historical review of Stahler, 1988, in much of what follows). For infinitesimal and reversible changes, the first and second laws state that

$$T\,dS = dE + P\,d\left(\frac{1}{\rho}\right) = dE - \frac{P}{\rho^2}\,d\rho \tag{7.142}$$

where E is in erg g^{-1} and the entropy, S, has the units erg g^{-1} K^{-1}. We wish to transform this to something more familiar in the stellar context. To do so,

[9] From now on, we shall write down little in the way of explicit formulas, but will wait until the very end to give the answer. You are advised to work out all the tedious numbers as we go along.

express E and ρ in terms of P and T; that is, let $E = E(P,T)$ and $\rho = \rho(P,T)$. The differentials are then expanded out into partials with respect to P and T. These partials are then transformed using standard thermodynamic rules (as in Landau and Lifshitz, 1958, §16; and Cox, 1968, §9.14) and the relation

$$c_P \nabla_{\text{ad}} = \frac{P}{\rho T} \frac{\chi_T}{\chi_\rho} \tag{7.143}$$

(from 3.99) is applied to finally arrive at

$$\frac{dS}{dr} = c_P \left(\nabla - \nabla_{\text{ad}} \right) \frac{d \ln P}{dr} . \tag{7.144}$$

The key here is the presence of $\nabla - \nabla_{\text{ad}}$. Since hydrostatic equilibrium requires that $d \ln P/dr \leq 0$ everywhere, then the following are true:

1. If the star is locally radiative with $\nabla < \nabla_{\text{ad}}$, then $dS/dr > 0$ and entropy increases outward at that location.
2. If $\nabla > \nabla_{\text{ad}}$ so that the star is convectively unstable, then $dS/dr < 0$ and entropy decreases outward. In the special case of very efficient adiabatic convection, ∇ exceeds ∇_{ad} by so little that $\nabla - \nabla_{\text{ad}} = 0^+$ and we may effectively set $\nabla = \nabla_{\text{ad}}$. If we restrict ourselves to this situation, then S is effectively constant through a convective region.

Combinations of these statements are shown in Fig. 7.5 for three stars; one is completely convective, the second entirely radiative, and the last has a radiative interior but convective envelope where, if you look closely, the convective stars have thin radiative photospheres.

Having settled on the above behavior for the run of entropy, we now determine some of the thermodynamics of entropy for an ideal gas. First rewrite (7.142) as

$$T\,dS = \left(\frac{\partial E}{\partial T} \right)_\rho dT + \left(\frac{\partial E}{\partial \rho} \right)_T d\rho - \frac{P}{\rho^2}\, d\rho \tag{7.145}$$

where for a constant-composition ideal monatomic gas (assumed not to be in the process of ionization) $E = 3N_{\text{A}}kT/2\mu$ erg g^{-1} and $P = \rho N_{\text{A}} kT/\mu$. Performing the indicated operations yields

$$\frac{dS}{dr} = \frac{N_{\text{A}}k}{\mu} \frac{d \ln \left(T^{3/2}/\rho \right)}{dr} \tag{7.146}$$

which, to within an additive constant after integration, becomes

$$S(r) = \frac{N_{\text{A}}k}{\mu} \ln \left[T^{3/2}(r)/\rho(r) \right] . \tag{7.147}$$

This may be recast into two other convenient forms by using the equation of state to eliminate either temperature or density. Thus

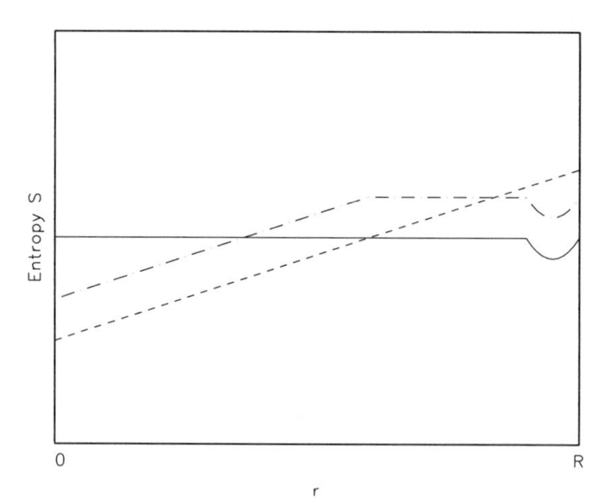

Fig. 7.5. The run of entropy with radius for a completely convective star with very efficient convection (solid line), a star that is radiative throughout (dashed line), and one with a radiative core and convective envelope (dash-dot). Note that the convective stars have thin radiative layers near their photospheres under which is a transition layer of inefficient convection.

$$S(r) = \frac{N_A k}{\mu} \ln\left[T^{5/2}(r)/P(r)\right] \quad (7.148)$$

$$S(r) = \frac{3}{2}\frac{N_A k}{\mu} \ln\left[P(r)/\rho^{5/3}(r)\right] . \quad (7.149)$$

Note that various constants have been absorbed into the constant terms implicit in the righthand sides of these expressions.

The first application of the last two equations is evident if $\nabla = \nabla_{\rm ad}$ in an ideal gas where $\Gamma_2 = 5/3$: namely, these conditions imply that $P \propto \rho^{5/3}$ and $P \propto T^{5/2}$ so that adiabatic convection means constancy of entropy as in Fig. 7.5. A second observation allows us to estimate the relation between entropy and, say, total radius in a star. If the star is in hydrostatic equilibrium, then, from dimensional considerations, pressure is approximately $P \approx G\mathcal{M}^2/\mathcal{R}^4$ and $\rho \approx \mathcal{M}/\mathcal{R}^3$. Equation (7.149) then yields the total entropy $S_{\rm tot} \approx \mathcal{M}(3N_A k/2\mu) \ln\left(G\mathcal{M}^{1/3}\mathcal{R}\right)$, or

$$\mathcal{R} \propto \exp\left(\frac{2\mu}{N_A k}\frac{S_{\rm tot}}{\mathcal{M}}\right) . \quad (7.150)$$

This states that the total radius is a sensitive increasing function of entropy for fixed mass and (uniform) composition. Thus if Fig. 7.5 represents the run of entropy for stars of fixed mass, composition, *and* radius, then the integrals over mass of the entropy must be very nearly the same. This is why, for example, the entropy in the inner regions of the radiative star is shown to be lower than for the convective star, and the reverse is true in the outer layers. The intermediate case in the figure shows what happens if just the outer layers are convective. As an afternote, the standard model is completely radiative according to the criteria given above (as it should be for consistency). Namely, $\rho \propto T^3$ for this model (from 7.101) so that $S(r)$ increases outward by application of (7.147).

We are now in a better position to understand the result given by (7.141). Equation (7.148) may be evaluated for the entropy at the photosphere by substituting $T_{\rm eff}$ and the photospheric pressure for the temperature and pressure in the logarithm. For H^- opacity, the entropy is easily shown to be

$$S_{\rm p} = \frac{N_{\rm A}k}{\mu} \ln \left(3T_{\rm eff}^{11.5} \rho_{\rm p}^{0.5} / 2\kappa_0 g_s \right) . \qquad (7.151)$$

Thus as $T_{\rm eff}$ is reduced, so is $S_{\rm p}$. However, $T_{\rm eff}$ cannot fall too low for two reasons. Too low an effective temperature means that the photosphere can become optically thin (as opacity decreases) and this violates our notion of a photosphere. In addition, the entropy cannot drop below the interior value in this simple picture because otherwise the very outer layers will have decreasing entropy implying convection—and this is a contradiction. There is then a minimum to $S_{\rm p}$ and a corresponding minimum for $T_{\rm eff}$.

Application to Pre-Main Sequence Evolution

A direct application of this discussion is to the evolution of pre-main sequence stars. If we suppose that they have no interior thermonuclear energy sources (although burning of deuterium, which may have been present in the protostar nebula, may play a role), then contraction from a protostellar cloud will eventually yield high luminosities at large radii, and large luminosities usually require convection. If accretion of matter onto the forming star may be neglected (and this is not really true), the object follows a path on the Hertzsprung–Russell diagram along a path of effective temperature qualitatively similar to that given by (7.141)—and see Fig. 2.2. At any given stage its effective temperature cannot fall below that value because of the arguments given above. Were this not true, then the star would enter into the "forbidden region" on the diagram at lower effective temperatures as shown long ago by Hayashi (1961). These paths are appropriately known as "Hayashi tracks" and are those taken by the T Tauri stars (see below). As the star continues to contract, however, its luminosity may decrease to the point where the deep interior ultimately becomes radiative and the foregoing discussion does not apply. To see what happens then, assume that the deep interior opacity is Kramers' and apply homology arguments. If the interior gas is ideal and there is still no thermonuclear energy generation, then a simple exercise yields the mass–luminosity effective temperature relation (show this because it's easy stuff from Chap. 1)

$$\mathcal{L} \propto \mathcal{M}^{22/5} T_{\text{eff}}^{4/5} . \tag{7.152}$$

The implication is that when the luminosity of the contracting star falls below a critical value, evolution proceeds along a track given by (7.152) to higher effective temperatures until, finally, interior temperatures reach the point of hydrogen ignition and the main sequence stage of evolution begins. Note that if the mass is *too* low, then the track given by (7.152) may lie below the intersection of the main sequence and the Hayashi track. If so, then hydrogen burning will commence at that intersection but the star will remain completely convective on the main sequence. If the protostar is even less massive than this, then hydrogen burning may never begin and the result is a brown dwarf. From such calculations, the minimum mass of a hydrogen-burning main sequence star is estimated to be close to 0.1 $\mathcal{M}_\odot$. Less massive objects result in brown dwarfs (see §2.2.2).

The essentials of the above results were summarized in Fig. 2.2 where Pre-main sequence evolutionary tracks were shown on the HR diagram for various mass stars. The direction of evolution is down the Hayashi track to lower luminosities until the equivalent of relation (7.152) is reached and then follows the march to the main sequence. Also shown in Fig. 2.2 were the locations of observed T Tauri stars, which are now believed to be stars in the process of contracting to the main sequence. The heavy line that defines an upper envelope to where these stars appear implies that we have not told the whole story—namely, why are these T Tauri stars not seen above this "birthline" (Stahler 1988)? A major reason is that we have neglected the actual hydrodynamical processes of star formation from interstellar clouds. Among these processes is accretion of gas onto the forming star. This provides a high luminosity at the accretion surface but this is obscured by dust and gas. It is only after the accretion ends that the star is fully revealed below the birthline.

7.4 Exercises

Exercise 7.1. This problem has to do with uniqueness of solutions to the stellar structure equations. Namely, suppose we wish to make a stellar model of mass $\mathcal{M}$ and, in this exercise, uniform composition X, Y, and Z. Such a model would correspond to a star on a main sequence. (Note that we say "a main sequence" and not necessarily a hydrogen main sequence.) The question is whether a unique solution exists or whether there may be multiple solutions with different radii, luminosities, etc. One way to look at this is to consider a radius-mass relation of the form (see 1.70)

$$R \propto \mathcal{M}^{\alpha_\mathcal{R}}$$

where

$$\alpha_\mathcal{R} = \alpha_\mathcal{R}(\chi_\rho, \chi_T, \lambda, \nu, s, n) .$$

Here the arguments of α_R are the power-law exponents defined in sections §1.5 and §1.6 of the text. In line with the text's discussion of dimensional analysis and homology of §1.6, we assume that these exponents are constant through a model but may differ from model to model. Now consider the possibility of a double-solution main sequence where, for some range of masses, two values of $\mathcal{R}$ are possible for a given mass. One example of this (and see below) is that the solutions with larger radii might correspond to a normal main sequence where as mass increases so does radius. A lower branch, with smaller radii, would correspond to pathological solutions that should not appear in real life. (For a discussion of this issue see, Hansen, C.J. 1978, ARA&A, 16, 15.) An example of a double-valued *helium* main sequence is discussed in

▷ Hansen, C.J., & Spangenberg, W.H., 1971, ApJ, 168, 71,

where there is a minimum mass at $\mathcal{M} \approx 0.3\,\mathcal{M}_\odot$. Models more massive than this are either electron nondegenerate (and "normal") or degenerate (and pathological). Models with mass less than $0.3\mathcal{M}_\odot$ do not exist; that is, there are no solutions to the stellar structure equations for masses less than $0.3\mathcal{M}_\odot$. Now for the problem.

It should be apparent that the minimum mass for a double-valued sequence occurs at that combination of exponents for which $\alpha_R \to \infty$. Derive that relation between the exponents. (This is really an easy problem. It just takes a long time to explain.) You should find that the "∞" condition is equivalent to requiring that

$$(4 - 3\chi_\rho)(s + 4 - \nu) = \chi_T(3n + 3\lambda + 4)$$

and it is well-satisfied near the center of the minimum mass model of Hansen and Spangenberg (1971). (Don't peek, but Eq. 1.75 gives a good clue.) The above condition is equivalent to the "Jeans' criterion" for secular stability (the subject of Hansen, 1978) where, in the case of double-valued main sequences, the pathological sequence can drift out of thermal balance while maintaining hydrostatic equilibrium. This is such an arcane subject.

Exercise 7.2. It is easy to say that a model looks like a polytrope of such-and-such index, but don't be fooled. Go back to Tables 2.5 and 2.6 that list the properties of ZAMS models and compute $\rho_c/\langle\rho\rangle$ for a large range of masses. See if you can make sense of what you find. And, by-the-way, note how the density ratio reaches a maximum just a few tenths of a solar mass after $1\mathcal{M}_\odot$. Something unusual is going on here (and is in line with our "mass cut" of about $1.5\,\mathcal{M}_\odot$ emphasized in Chap. 2.).

Exercise 7.3. Verify Equation 7.48 for the series expansion of $\theta_n(\xi)$ about the origin.

Exercise 7.4. Use some decent integrator to construct a polytrope with index n in the interval $2 \le n \le 4$. The simplest method is to shoot for a solution.

Exercise 7.5. Use the same integrator (as in Ex. 7.4) to construct a zero temperature white dwarf with $\mu_e = 2$ and $\mathcal{M} \approx \mathcal{M}_\infty/2$. You will have to use (3.53–3.55) for the thermodynamics. Warning: if x becomes very small or large watch out for cancellation of terms. Setting up the basic differential equation is the subject of many of the references.

Exercise 7.6. Derive Equations (7.103–7.105) for the limiting forms of the mass–radius relations for zero temperature white dwarfs.

Exercise 7.7. Derive the central expansions (7.106–7.110).

Exercise 7.8. Derive (7.141) for the T_{eff}, $\mathcal{M}$, $\mathcal{L}$ relation of a completely convective star with H^- opacity near the surface. (The algebra and arithmetic are so messy we hope we got it right.)

7.5 References and Suggested Readings

§7.1: The Equations of Stellar Structure

The subject of multiple solutions to the stellar structure equations has been reviewed by

▹ Hansen, C.J. 1978, ARA&A, 16, 15

in the context of secular stability.

§7.2: Polytropic Equations of State and Polytropes

The material on polytropes in this chapter is standard. We recommend

▹ Chandrasekhar, S. 1939, *Introduction to the Study of Stellar Structure* (New York: Dover)

for more details than you might ever want. We warn you that the mathematics can get a bit rough at times. Other references include §23.1 of

▹ Cox, J.P. 1968, *Principles of Stellar Structure* (New York: Gordon and Breach)

and Chapter 19 of

▹ Kippenhahn, R., & Weigert, A. 1990, *Stellar Structure and Evolution* (Berlin: Springer-Verlag).

Note that the symbols used for the polytropic variables in the last reference are not standard.

We strongly recommend that you have

▹ Press, W.H., Teukolsky, S.A., Flannery, W.T., Vetterling, W.T. 1992, *Numerical Recipes, The Art of Scientific Computing*, 2d ed. (Cambridge: Cambridge University Press)

on your bookshelf. A calculation poorly done is an abomination in this day of the high-speed computer (PCs and Macs included). Note that the earliest editions of *Numerical Recipes* have several mistakes in the `FORTRAN` computer programs accompanying the text. These have hopefully all been corrected. These programs are also available on floppy disks (as are "C" versions).

Again we recommend

▷ Hayashi, C., Hōshi, R., & Sugimoto, D. 1962, PTPhJS, No. 22.

Their use of the U-V plane is a virtuoso performance. Earlier examples are discussed in the text by

▷ Schwarzschild, M. 1958, *Structure and Evolution of the Stars* (Princeton: Princeton University Press).

The most complete discussion of model making is still to be found in

▷ Kippenhahn, R., Weigert, A., & Hofmeister, E. 1967, MethCompPhys, 7, 53.

The original "Henyey method" appears in

▷ Henyey, L.G., Forbes, J.E., & Gould, N.L. 1964, ApJ, 139, 306.

The original LANL (then called LASL) method for one-dimensional hydrodynamics is discussed in

▷ Cox, A.N., Brownlee, R.R., & Eilers, D.D. 1966, ApJ, 144, 1024.

There are newer techniques and some of these are reviewed in Press et al. (1992).

The most lucid introduction to shock phenomena we know of is the first chapter of

▷ Zel'dovich, Ya.B., & Raizer, Yu.P. 1966, *Physics of Shock Waves and High Temperature Hydrodynamic Phenomena*, in two volumes, eds. W.D. Hayes and R.F. Probstein (New York: Academic Press).

These two volumes contain much of interest for the astrophysicist. Chapter 1 has been reprinted separately in 1966 as *Elements of Gasdynamics and the Classical Theory of Shock Waves* (also from Academic Press). There are more recent texts containing numerical methods for dealing with shock waves but you cannot go wrong with

▷ Richtmyer, R.D., & Morton, K.W. 1967, *Difference Methods for Initial Value Problems* 2d ed. (New York: Wiley Interscience).

▷ Arnett, D. 1996, *Supernovae and Nucleosynthesis* (Princeton: Princeton University Press)

gives a short review (in §4.4) of numerical techniques necessary to solve systems of stiff differential equations for following abundances in supernovae.

Sir Arthur Eddington's name appears in many astronomical contexts during the earlier years of the 20th century, ranging from the gravitational bending of light to stellar interiors and variable stars. If you wish to find out how well science can be explained read through

▷ Eddington, A.S. 1926, *Internal Constitution of the Stars* (Cambridge: Cambridge University Press).

It is also available in a 1959 Dover edition.

The original work on constructing zero temperature white dwarfs appears in

▹ Chandrasekhar, S. 1939, *Introduction to the Study of Stellar Structure* (New York: Dover),

where the notion of polytropes is extended to include the combination of nonrelativistic and relativistic degenerate equations of state.

§7.3: The Approach to Real Models

Chapter 20 of

▹ Cox, J.P. 1968, *Principles of Stellar Structure* (New York: Gordon and Breach)

has a more complete discussion of envelope construction than we have attempted. Included are fairly realistic convective envelopes and the use of the dimensionless Schwarzschild variables.

As in Chapter 2 we recommend the review by

▹ Stahler, S.W. 1988, PASP, 100, 1474

for a discussion of pre-main sequence evolution. See also the original paper of

▹ Hayashi, C. 1961, PASJ, 13, 450

for Hayashi tracks. For some thermodynamic conversions see

▹ Landau, L.D., & Lifshitz, E.M. 1958, *Statistical Physics* (London: Pergamon).

8 Asteroseismology

"Shake, rattle, and roll."
Title and most of the lyrics of a popular song.
— Bill Haley & the Comets (1954)

With few exceptions we have treated stars as hydrostatic objects in which gravitational and pressure gradient forces are everywhere in balance. The exceptions have been supernovae, novae, and intrinsically variable stars. This chapter is about the last class. Variable stars may not be as spectacular as their violent cousins, but they have a charm of their own, and they have a special place in astronomy. Section 2.10 took us on a tour of the variable star zoo, but here we shall see what makes them tick. The final two chapters will go deeper into the sun as a variable star and discuss variable white dwarfs but, as here, a constant theme will be how observations of variable stars (supplemented by lots of theory) are used to peek into stellar interiors in ways that you can't do otherwise. In this sense, the title of this chapter is self-explanatory: we shall use observed surface luminosity, radius, and color variations to probe stellar interiors in much the same spirit as in terrestrial seismology. This is not a new subject but it has blossomed in recent years. A review of the list of variables we gave in Chapter 2 makes it clear that the subject spans all phases of stellar evolution.

The plan of the chapter is as follows. We first treat small-amplitude motions that are strictly periodic and radially symmetric. If the motions are strictly periodic, then the time-averaged energy content of the star remains constant, which is the same as saying that the oscillations are *adiabatic*. This is, of course, an approximation to a real situation where energy redistribution within the star takes place over time scales that are very long compared to a period of oscillation (or pulsation, or one of a few other terms used to describe the variability). We shall then introduce nonadiabatic effects to briefly explore the causes of variability. Finally, we shall see what happens when the motions are not radially symmetric. Since there are two excellent texts in the literature (Cox, 1980; Unno et al., 1989; and see the monograph by Christensen–Dalsgaard, 1997) our treatment of these topics will sometimes be quick and dirty. Examples of applications to real stars will be given where appropriate.

8.1 Adiabatic Radial Pulsations

We have frequently emphasized the point that a star is an object whose structure is primarily determined by mechanics. To understand this more clearly in the present context recall, from Chapter 1, that dynamic times ($t_{\rm dyn}$) are usually short compared to times characteristic of internal energy redistribution within a star (e.g., $t_{\rm KH}$). This is not strictly true for all stars, or even the outer portions of most stars, but it forms the basis of the "adiabatic approximation" for the study of stellar pulsation. In this approximation it is assumed that *all* heat exchange mechanisms may be ignored so that the system is purely mechanical. The problem then reduces to a rather complicated exercise equivalent to studying the normal modes of a coupled system of pendulums and springs or, more appropriately, the behavior of sound waves confined in a box. The adiabatic approximation is remarkably useful in variable star theory because not only does it greatly simplify the analysis but it also yields accurate models of dynamic response for most stars. The penalty paid is severe, however, because it cannot tell us what causes *real* stars to pulsate. In this section we shall restrict the discussion to radially symmetric motions. This means that the star is always radially symmetric and all effects due to rotation, magnetic fields, etc., may safely be ignored.

Since heat transfer is ignored in the adiabatic approximation, we can completely describe the mechanical structure with only the mass and force equations

$$\frac{\partial \mathcal{M}_r}{\partial r} = 4\pi r^2 \rho \tag{8.1}$$

$$\ddot{r} = -4\pi r^2 \left(\frac{\partial P}{\partial \mathcal{M}_r}\right) - \frac{G\mathcal{M}_r}{r^2} \tag{8.2}$$

where we explicitly introduce partial derivatives to make sure derivatives with respect to time appear only where appropriate. If the star were purely static, then $\ddot{r}$ would be zero everywhere. Imagine that this indeed is initially the case but, by some means, the star is forced to depart from this initial hydrostatic equilibrium state in a radially symmetric, but otherwise arbitrary, manner. Furthermore, and to make the problem tractable, suppose that any departures from the static state are small in the following sense. Let a zero subscript on radius (r_0) or density (ρ_0) denote the local values of these quantities in the static state at some given mass level $\mathcal{M}_r$. As the motion commences both radius and density will, in general, depart from their static values at that same mass level and be functions of time and the particular mass level in question. This constitutes a Lagrangian description of the motion because we follow a particular mass level on which, we can imagine, all particles are painted red for identification to distinguish them from particles at other mass levels. We now describe the motion by letting[1]

[1] The following discussion can be made more general, as it will for nonradial oscillations, but we think it best to go gently here.

$$r(t,\mathcal{M}_r) = r_0(\mathcal{M}_r)\left[1+\frac{\delta r(t,\mathcal{M}_r)}{r_0(\mathcal{M}_r)}\right] \tag{8.3}$$

$$\rho(t,\mathcal{M}_r) = \rho_0(\mathcal{M}_r)\left[1+\frac{\delta \rho(t,\mathcal{M}_r)}{\rho_0(\mathcal{M}_r)}\right] \tag{8.4}$$

where δr and $\delta\rho$ are the *Lagrangian perturbations* of density and radius. These last two quantities are used to describe the motion through time at a given mass level. The requirement that departures from the static state be small is $|\delta r/r_0| \ll 1$ and $|\delta\rho/\rho_0| \ll 1$.

We now "linearize" the mass and force equations by replacing the position (radius) and density at a mass level by the perturbed values of (8.3–8.4) and, in the result, keeping only those terms that are of first or lower order in $\delta r/r_0$ and $\delta\rho/\rho_0$. (Recall that we did the same sort of thing back in §1.1 in examining a variational principle.) To see how this goes, first consider the mass equation

$$\frac{\partial \mathcal{M}_r}{\partial\left[r_0(1+\delta r/r_0)\right]} = 4\pi\left[r_0(1+\delta r/r_0)\right]^2\left[\rho_0(1+\delta\rho/\rho_0)\right] . \tag{8.5}$$

Now carry through the derivative in the denominator of the lefthand side and expand out the products on the right. The first operation yields a new denominator $(1+\delta r/r_0)\,\partial r_0 + r_0\,\partial(\delta r/r_0)$. The derivative ∂r_0 is then factored out so that the overall lefthand side contains the factor $\partial\mathcal{M}_r/\partial r_0$. The small terms remaining in the denominator are then brought up using a binomial expansion to first order to yield

$$\frac{\partial\mathcal{M}_r}{\partial r_0}\left[1-\frac{\delta r}{r_0}-r_0\frac{\partial\left(\delta r/r_0\right)}{\partial r_0}\right] .$$

The righthand side of the mass equation is simpler because all we need do is expand out the factors to first order to obtain

$$4\pi r_0^2\rho_0\left(1+2\frac{\delta r}{r_0}+\frac{\delta\rho}{\rho_0}\right) .$$

When the two sides of the linearized mass equation are set equal we find that the result contains the zero-order equation

$$\frac{\partial\mathcal{M}_r}{\partial r_0} = 4\pi r_0^2\rho_0$$

which is the mass equation for the unperturbed configuration. Since this is automatically satisfied, we take advantage of the equality and subtract this from the linearized equation. (This is a typical result of linearization about an equilibrium state.) After some easy rearrangement we find that the following relation between the Lagrangian perturbations must be satisfied so that mass is conserved as the configuration evolves in time:

$$\frac{\delta\rho}{\rho_0} = -3\,\frac{\delta r}{r_0} - r_0\,\frac{\partial\,(\delta r/r_0)}{\partial r_0}\ . \tag{8.6}$$

Note that part of this equation is familiar because, if we ignore the derivative term, it is merely the logarithmic form of the homology relation between density and radius given by (1.65) of §1.6.

The force equation is relatively straightforward and we leave it as an exercise to show that its linearization yields

$$\rho_0 r_0\,\frac{d^2\,\delta r/r_0}{dt^2} = -\left(4\,\frac{\delta r}{r_0} + \frac{\delta P}{P_0}\right)\frac{\partial P_0}{\partial r_0} - P_0\frac{\partial\,(\delta P/P_0)}{\partial r_0}\ . \tag{8.7}$$

Implicit in the derivation of this equation are the conditions $\ddot{r}_0 = 0$ and $\dot{r}_0 = 0$, which must apply since the reference state is completely static.

At this point in the analysis we take a familiar path in perturbation theory and assume that *all* perturbations prefixed by δs may be decomposed into Fourier components with the time element represented by exponentials. Thus, for example, introduce the space component of relative fluid displacement, $\zeta(r_0)$, by

$$\frac{\delta r(t, r_0)}{r_0} = \frac{\delta r(r_0)}{r_0}\,e^{i\sigma t} = \zeta(r_0)\,e^{i\sigma t} \tag{8.8}$$

where the exponential takes over the duty of describing the time evolution of displacement and $\zeta(r_0)$ [or $\delta r(r_0)/r_0$], which depends only on r_0 (i.e., the mass level), can be considered to be the shape of the displacement at zero time. Note that both the frequency σ (in radians per second) and $\zeta(r_0)$ can be complex. The lefthand side of the force equation now becomes $-\rho_0 r_0 \sigma^2 \zeta(r_0)\,e^{i\sigma t}$.

It should be clear that we are now in trouble because the two linearized equations for force and mass contain the three variables $\zeta(r_0)$ and the space parts of the pressure and density perturbations. This comes about because we have neglected the energetics of the real system and so our description is incomplete. To make sure that this is a purely mechanical problem we now couple $\delta\rho$ and δP in the adiabatic approximation by recalling the Lagrangian relation between changes in pressure to changes in density given by (3.93)

$$\Gamma_1 = \left(\frac{\partial \ln P}{\partial \ln \rho}\right)_{\text{ad}}\ . \tag{8.9}$$

Since this is shorthand for $P \propto \rho^{\Gamma_1}$ and δ is a Lagrangian differential operator, we take logarithmic δ–derivatives to find

$$\frac{\delta P}{P_0} = \Gamma_1\,\frac{\delta\rho}{\rho_0}\ . \tag{8.10}$$

This relation takes the place of any energy and heat transfer equations that would normally appear and we now have as many variables as equations.

There are several paths we could take now but we choose the following: (1) make sure all perturbations are replaced by their spatial Fourier components with common factors of $e^{i\sigma t}$ cancelled; (2) replace all occurrences of $\delta\rho$ by δP using the adiabatic condition; (3) rearrange the two linearized equations so that space derivatives appear on the lefthand side; (4) replace partial derivatives by total space derivatives (with the understanding that all variables depend only on r_0); (5) delete all reference to zero subscripts because all that really appears are perturbations and quantities from the static configuration. The result is

$$\frac{d\zeta}{dr} = -\frac{1}{r}\left(3\zeta + \frac{1}{\Gamma_1}\frac{\delta P}{P}\right) \tag{8.11}$$

$$\frac{d\left(\delta P/P\right)}{dr} = -\frac{d\ln P}{dr}\left(4\zeta + \frac{\sigma^2 r^3}{G\mathcal{M}_r}\zeta + \frac{\delta P}{P}\right) \tag{8.12}$$

where the factor $r^3/G\mathcal{M}_r$ appears as a result of using the hydrostatic equation to get rid of some terms containing dP/dr (which you may wish to retain rather than introducing $\mathcal{M}_r$). Note that $\sigma^2 r^3/G\mathcal{M}_r$ looks suspiciously like a big piece of a period–mean density relation.

We now have a set of coupled first-order differential equations, but we need boundary conditions. The first of these is simple because we require that δr be zero at the center ($r = 0$). To see how this comes about consider a particle of infinitesimal extent at the very center of the equilibrium star. There is no place for this particle to move to ($\delta r \neq 0$) without violating the condition of radial symmetry. Physical regularity of the solutions also requires that both $\zeta = \delta r/r$ and $d\zeta/dr$ be finite at the center. The only way to arrange for all this to be true is to have the term in parenthesis on the righthand side of (8.11) vanish at stellar center. This yields the first boundary condition

$$3\zeta + \frac{1}{\Gamma_1}\frac{\delta P}{P} = 0 \quad \text{at} \quad r = 0\,. \tag{8.13}$$

The second boundary condition is applied at the surface. For our purposes it is adequate to assume zero boundary condition for the static model star (as in §7.1). Specifically, we assume $P \to 0$ as $r \to \mathcal{R}$. More complicated conditions are possible—such as for a photospheric surface—but they add nothing of real importance for our discussion. The first thing to realize is that the leading coefficient of the righthand side of the linearized force equation (8.12) is just $1/\lambda_{\rm P}$ where $\lambda_{\rm P}$ is the pressure scale height of (3.1). This latter quantity rapidly goes to zero as the surface is approached so that in order for the relative pressure perturbation, $\delta P/P$, to remain finite we must have

$$4\zeta + \frac{\sigma^2\mathcal{R}^3}{G\mathcal{M}}\zeta + \frac{\delta P}{P} = 0 \quad \text{at} \quad r = \mathcal{R}\,. \tag{8.14}$$

Though not immediately evident, this condition is equivalent to requiring that all interior disturbances be reflected at the surface (as it itself moves) back into the interior; that is, no pulsation energy is lost from the star because all is reflected back inward from the surface.

So far, so good. We have an equal number of differential equations and boundary conditions. But, all the equations derived thus far are linear and homogeneous in ζ and $\delta P/P$ so the question remains as to how these quantities are to be normalized. As it stands, any scaling is permitted for either perturbation at some unspecified point in the star and the overall solution may be as small or large as we like (but not zero, otherwise everything is zero!). To pin things down we must choose a nonzero normalization. This is completely arbitrary but certain choices are preferred (and differ among different investigators). We choose

$$\zeta = \frac{\delta r}{r} = 1 \quad \text{at} \quad r = \mathcal{R} \,. \tag{8.15}$$

We now realize that this places an additional constraint on the problem; and, in effect, we have exceeded the permissible number of boundary conditions. The way out of this apparent dilemma is to recognize that the (perhaps complex) frequency σ has not been specified. In fact, it can only take on a value (or values) such that all boundary conditions are satisfied including the normalization condition. (Note that σ cannot depend on the normalization condition because the latter just scales the solutions.) Thus σ or, more properly, σ^2—because only that quantity appears in our equations—is an *eigenvalue* and the corresponding perturbations are *eigenfunctions* for that particular σ^2. We now discuss the properties of the eigenvalues for this adiabatic problem and this will involve a little mathematics.

8.1.1 The Linear Adiabatic Wave Equation

First, we leave it to you to collapse the two first-order differential equations for ζ and $\delta P/P$ down into one second-order equation in ζ. (This involves differentiating 8.11 and then eliminating any reference to $\delta P/P$ or its derivative by using 8.11 and 8.12.) The result is

$$\mathbf{L}(\zeta) \equiv -\frac{1}{\rho r^4}\frac{d}{dr}\left(\Gamma_1 P r^4 \frac{d\zeta}{dr}\right) - \frac{1}{r\rho}\left\{\frac{d}{dr}\left[\left(3\Gamma_1 - 4\right)P\right]\right\}\zeta = \sigma^2\zeta \,. \tag{8.16}$$

Here $\mathbf{L}$ is a second-order differential operator that is shorthand for the middle part of the whole equation, where, in this case, ζ is the operand. We can write the above in simple form as $\mathbf{L}(\zeta) = \sigma^2\zeta$. It may not look like it at first sight but, with some hindsight, this equation is a wave equation and, in this context, is called the *linear adiabatic wave equation* or LAWE.[2]

This all may look pretty formidable but there are redeeming features. (You should consult any decent text on mathematical physics—such as Arfken and Weber, 1995—for what follows if you wish to do serious work with the theory

[2] Some investigators refer to more general versions of this equation as the LAWE. Cox (1980), for example, includes nonradial motions in his formulation.

of pulsating stars.) All the quantities in $\mathbf{L}$ are well-behaved and $\mathbf{L}$ is a Sturm–Liouville operator. Furthermore, we can symbolically integrate over the star and show (subject to our boundary conditions and other constraints—see Cox 1980, §8.8) that

$$\int_0^{\mathcal{M}} \zeta^* \left[\mathbf{L}(\zeta)\right] r^2 \, d\mathcal{M}_r = \int_0^{\mathcal{M}} \zeta \left[\mathbf{L}(\zeta)\right]^* r^2 \, d\mathcal{M}_r \qquad (8.17)$$

where ζ^* is the complex conjugate of ζ. Now we may as well be doing quantum mechanics because this equality means that the Sturm–Liouville operator $\mathbf{L}$ is self-adjoint (or Hermitian) and the following statements about σ^2 and its eigenfunctions are true (among other nice properties):

1. All eigenvalues σ^2 of the system are real, as are the corresponding eigenfunctions. There are then two possibilities. If $\sigma^2 > 0$ then σ is purely real and the complete eigenfunction $\zeta(r)\, e^{i\sigma t}$ is oscillatory in time by virtue of the temporal factor $e^{i\sigma t}$. (That is, we get sines and cosines of σt.) Otherwise, if $\sigma^2 < 0$, then σ is pure imaginary and the perturbations grow or decay exponentially with time. We shall only concern ourselves with the first possibility in practical situations. (But note that the $\sigma^2 < 0$ possibility implies a gross dynamic event such as collapse of the star.) Thus if $\sigma^2 > 0$, then σ is the angular frequency of the oscillation with corresponding period $\Pi = 2\pi/\sigma$.
2. There exists a minimum value for σ^2 which, were we doing quantum mechanics, would correspond to the ground state.
3. If ζ_j and ζ_k are two eigenfunction solutions for eigenvalues σ_j^2 and σ_k^2, then
$$\int_0^{\mathcal{M}} \zeta_j^* \, \zeta_k \, r^2 \, d\mathcal{M}_r = 0 \quad \text{if } j \neq k \, . \qquad (8.18)$$
The eigenfunctions are then said to be *orthogonal* (as in the scalar product of two perpendicular vectors).

What we then have for $\sigma^2 > 0$ are standing waves such that the star passes through the equilibrium state twice each period.

8.1.2 Some Examples

To get an idea of what is going on here first consider the (unrealistic) case where both ζ and Γ_1 are supposed constant throughout the star. (Were such a situation possible it would correspond to homologous motions.) The LAWE then reduces to[3]

[3] There aren't many easily solved analytic problems using the LAWE. To see what might be involved, do Ex. 8.3, where you are to find the pulsation eigenvalues and eigenfunctions for the constant density model. One result—among others—you will obtain is (8.20).

$$-\frac{1}{r\rho}\left(3\Gamma_1 - 4\right)\frac{dP}{dr}\,\zeta = \sigma^2\zeta\,. \tag{8.19}$$

In the simple case of the constant-density model $[\rho(r) = \langle\rho\rangle]$ we replace $-(1/\rho r)\,dP/dr$ by $G\mathcal{M}_r/r^3$, which becomes $4\pi G\langle\rho\rangle/3$. The result is

$$\left(3\Gamma_1 - 4\right)\frac{4\pi G}{3}\langle\rho\rangle = \sigma^2\,. \tag{8.20}$$

If $\Gamma_1 > 4/3$, then σ is real and the corresponding period is

$$\Pi = \frac{2\pi}{\sigma} = \frac{2\pi}{\sqrt{\left(3\Gamma_1 - 4\right)\langle\rho\rangle\, 4\pi G/3}}\,. \tag{8.21}$$

This is just the "period–mean density" relation derived in §1.3.5, but it is now clear how Γ_1 enters. If, on the other hand, $\Gamma_1 < 4/3$ we know enough to expect trouble. Here σ is imaginary and the e-folding time for either growth or decay of the motions is

$$\tau = \frac{1}{|\sigma|} = \frac{1}{\sqrt{\left|3\Gamma_1 - 4\right|\langle\rho\rangle\, 4\pi G/3}}\,. \tag{8.22}$$

This is the free-fall time (corrected for various factors), $t_{\rm dyn}$, discussed in §1.3.3.

More realistic examples are periodic oscillations in polytropes. Recall from §7.2.1 that the dependent variable for polytropes is θ_n, which is related to the pressure by $P = P_{\rm c}\theta_n^{1+n}$ where n is the polytropic index. The density is given by $\rho = \rho_{\rm c}\theta_n^n$ and the independent variable is ξ, which is proportional to radius. We introduce these variables into the two differential equations for adiabatic radial pulsations (8.11) and (8.12), use various relations derived in §7.2.1, and find

$$\frac{d\zeta}{d\xi} = -\frac{1}{\xi}\left[3\,\zeta + \frac{1}{\Gamma_1}\frac{\delta P}{P}\right] \tag{8.23}$$

$$\frac{d\left(\delta P/P\right)}{d\xi} = -(1+n)\frac{\theta_n'}{\theta_n}\left[4\,\zeta + \omega^2\frac{\xi/\theta_n'}{\left(\xi/\theta_n'\right)_1}\,\zeta + \frac{\delta P}{P}\right] \tag{8.24}$$

where ω^2 is the dimensionless frequency (squared)

$$\omega^2 = \frac{\mathcal{R}^3}{G\mathcal{M}}\,\sigma^2\,. \tag{8.25}$$

The prime on θ_n denotes $d\theta_n/d\xi$ and the subscript 1 in the middle term of (8.24) means that the term is to be evaluated at the surface of the polytrope where $\xi = \xi_1$. Both $\theta_n(\xi)$ and $\theta_n'(\xi)$ are known from the equilibrium polytrope solution. We could have phrased the above in terms of $\delta\xi/\xi$ and $\delta\theta_n/\theta_n$, but we prefer keeping the more physical variables $\zeta = \delta r/r$ and $\delta P/P$.

As an example, consider the oscillations of an $n = 2$ polytrope with Γ_1 of 5/3.[4] The insert in Fig. 8.1 shows $\theta_2(\xi)$ versus ξ/ξ_1 for that polytrope. The main curves show three solutions for ζ, each corresponding to a different eigenvalue ω^2. The curves are labeled "fundamental" for the smallest $\omega^2(= 4.001)$, "first overtone" for the next largest ($\omega^2 = 13.34$), and "second overtone" for the third largest ($\omega^2 = 26.58$). The nomenclature for these modes of oscillation agrees with that used in acoustics. We could also correctly have called the fundamental the "first harmonic," and the first overtone the "second harmonic," and so on. But beware, astronomers are not consistent (nor always correct) in their use of these terms and you will often see the first overtone given the name "first harmonic."

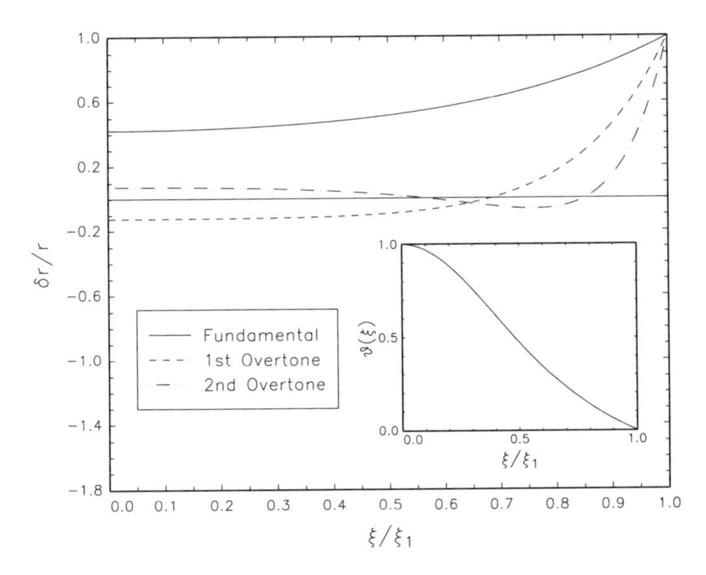

Fig. 8.1. Shown are the relative radial displacements $\zeta = \delta r/r$ for three modes in an $n = 2$ polytrope. The abscissa is the polytropic variable ξ given in units of its value at the surface (or, in other words, $r/\mathcal{R}$). Also shown in the insert is the run of $\theta_2(\xi)$ with ξ for the equilibrium polytrope.

You should note that the fundamental has no nodes (crossings of the $\zeta = 0$ axis) for $\delta r/r$, whereas each successive overtone has one more node than the preceding. Thus there is a one-to-one correspondence of complexity in eigenfunction with ordering of eigenvalue and we might as well be discussing vibrating strings, the hydrogen atom, or almost any other wave phenomenon. The fluid motions of the fundamental mode are not homologous because ζ is not a constant but they are still simple since the star expands and contracts

[4] With some care you can do this yourself as a computing exercise. See, for example, §38.3 in Kippenhahn and Weigert (1990), and Ex. 8.4.

almost uniformly and in phase.[5] (Some refer to this as a "breathing mode.") As a rough rule of thumb, the contrast between the central value of ζ to its surface value (here normalized to one) is $\zeta(\mathcal{R})/\zeta(r=0) \sim \rho_c/\langle\rho\rangle$. In our example, the degree of central concentration for the $n=2$ polytrope is $\rho_c/\langle\rho\rangle = 11.4$ and the corresponding ratio of the displacements is a little over three. The motion of the fundamental mode, and even more so the overtones, has been likened to a whip whose heavy handle moves only slightly in contrast to the wild excursions of the whip's end. An $n=2$ polytrope is only mildly centrally condensed but, on the other hand, a highly evolved star, such as a classical Cepheid variable, has a core of very high density and a comparatively low average density. For such variable stars the fluid displacements near the center are miniscule and, in practice, the central regions are often ignored in pulsation calculations.

Before we press on, we wish to emphasize that the eigenfunctions shown in Fig. 8.1 represent real (at least for polytropes) fluid motions. Each mode is shown at an instant of time when the surface is at maximum expansion. In the case of the fundamental, a quarter cycle later (at $t = \Pi/4$) the "star" has already undergone enough subsequent compression that it is just passing through the equilibrium state and the time-dependent eigenfunction is zero everywhere. In another quarter cycle maximum compression is reached and the eigenfunction is the mirror image of what is shown with $\delta r/r$ negative everywhere. Expansion then starts again and, a half cycle later, we are back from where we started.

8.1.3 Asymptotic Analysis

A listing of the eigenfrequencies of the $n=2$ polytrope suggests yet another property of radial pulsations. In order of increasing overtone we have $\omega = 2.00$, 3.65, and 5.16. The spacing between successive frequencies here is certainly not constant but it does hover around 1.6. Were we to compute successive overtones to high radial order (in terms of number of nodes) we would find that the spacing in frequency between modes would approach a constant. This is a common characteristic of mechanical systems and it follows from an "asymptotic analysis" of the wave equation. To show how this works we first convert the LAWE (8.16) into something that looks more like a wave equation by the substitution (see Tassoul and Tassoul, 1968)

$$w(r) = r^2 \left(\Gamma_1 P\right)^{1/2} \zeta(r) \,. \tag{8.26}$$

[5] The quantity Q, defined as $Q = \Pi\sqrt{\langle\rho\rangle/\langle\rho_\odot\rangle}$ is equal to 0.058 days for the fundamental mode of an $n=2$, $\Gamma_1 = 5/3$ oscillating polytrope. The corresponding value for Q in the standard model ($n=3$, see §7.2.7) is 0.038 days. This last value was used in the remarks made after equation (1.40) for the period–mean density relation. Q depends on the central concentration of the model with the largest value belonging to a constant-density ($n=0$) model.

A bit of manipulation then yields the new wave equation in $w(r)$

$$\frac{d^2w}{dr^2} + \left[\frac{\sigma^2\rho}{\Gamma_1 P} - \phi(r)\right] w = 0 \tag{8.27}$$

where $\phi(r)$ is the not very edifying function

$$\begin{aligned}\phi(r) &= \frac{2}{r^2} + \frac{2}{\Gamma_1 Pr}\frac{d\left(\Gamma_1 P\right)}{dr} - \left[\frac{1}{2\Gamma_1 P}\frac{d\left(\Gamma_1 P\right)}{dr}\right]^2 + \\ &+ \frac{1}{2\Gamma_1 P}\frac{d^2\left(\Gamma_1 P\right)}{dr^2} - \frac{1}{\Gamma_1 Pr}\frac{d}{dr}\left[\left(3\Gamma_1 - 4\right) P\right] .\end{aligned} \tag{8.28}$$

The first term in the brackets of the new wave equation is simpler than it appears because the combination $\Gamma_1 P/\rho$ is just the square of the local sound speed v_s^2 (as in Eq. 1.38). Thus

$$\frac{d^2w}{dr^2} + \left[\frac{\sigma^2}{v_s^2} - \phi\right] w = 0 . \tag{8.29}$$

Now assume that the wave function $w(r)$ may be represented as

$$w(r) \propto e^{ik_r r} \tag{8.30}$$

where k_r is a wave number which, in general, is a function of r. Because we shall eventually consider only large wave numbers (small wavelengths), the terms in the brackets on the lefthand side of (8.29) can be taken as roughly constant over a wavelength $\lambda_r = 2\pi/k_r$. Hence radial derivatives of k_r shall be assumed to be negligible; that is, k_r is *nearly* constant. The wave equation in disguised form then becomes

$$k_r^2 = \frac{\sigma^2}{v_s^2} - \phi \tag{8.31}$$

which is a *dispersion relation* for k_r. Thus if $\sigma^2 > v_s^2\phi$, then the argument of the exponential in w of (8.30) is purely imaginary and w is sinusoidal. This is characteristic of *propagating* or standing waves. However, if $\sigma^2 < v_s^2\phi$ then the solution for w contains exponentially decreasing or increasing components and the solution is said to be *evanescent*. The quantum mechanical analogue here is a particle in a potential well. If the particle's energy (which might be proportional to some σ^2) is greater than that of the bottom of the well, then the eigenfunction is sinusoidal within the well. The likelihood is small, however, of finding the particle in regions where its energy is less than the local potential because the solutions decay exponentially in those regions.

We now take the road into "asymptopia" and imagine that $\sigma^2/v_s^2 \gg \phi$ and $k_r r \gg 1$ within some interval of radius, say, $a \le r \le b$. This means that there are many wavelengths packed into that interval and the "particle" is high above the bottom of the potential well. We idealize the situation further by supposing that the potential well at a and b is so steep that the wave function is contained solely within those endpoints. Under these conditions the eigenfunction is trapped within the well and a mode, which consists of a standing wave for the mechanical system, must have an integer or half-integer number of wavelengths within $[a, b]$. To measure the number of wavelengths we must integrate k_r over the interval because k_r is still a function of r. (The simple expression $k_r[b-a]$ is not correct.) Thus our requirement for a true mode is, as you may verify with a little thought

$$\int_a^b k_r \, dr = (n+1)\pi \tag{8.32}$$

where n is the number of nodes the eigenfunction has within the interval not including the endpoints.[6] Thus $n = 0$ corresponds to a half-wave and the fundamental mode, and $n = 1$ is the first overtone with one complete wavelength packed into $[a, b]$, and so on. We are, of course, not in this low-overtone domain, but this is the general scheme.

Since ϕ is assumed to be small, then $k_r = \sigma/v_s$, which gives us

$$\sigma = (n+1)\pi \left[\int_a^b \frac{dr}{v_s} \right]^{-1} . \tag{8.33}$$

If we define the constant frequency $\sigma_0 = \pi / \int_a^b v_s^{-1} \, dr$, then the asymptotic behavior for σ is

$$\sigma = (n+1)\sigma_0 \tag{8.34}$$

so that the eigenfrequencies σ are equally spaced from one mode to the next with spacing σ_0. To date there is only one star for which we can actually observe high-order radial modes and this is the sun. Whether of high overtone or not, these modes are often called radial pressure modes (p-modes), or radial acoustic modes. The reason for this designation should be obvious from §1.3.5 where we considered the sound travel time across a star. Sound waves are propagating pressure disturbances and the preceding discussion containing ratios of dr/v_s, which measure travel times for sound waves, is the same thing.

Adiabatic radial pulsations in stars are relatively well understood and modern calculations do a good job in representing these purely mechanical motions. This is a desirable state of affairs because many, and perhaps

[6] We use n here to count the number of nodes and to denote mode order. This is the usual nomenclature used by those doing solar seismology. The white dwarf seismology community, on the other hand, tends to use k for the same purpose. Our apologies to our colleagues among the white dwarfers, but k is too easily confused with wave number. And please do not confuse this n with the opacity exponent or polytropic order. There are just so many letters available.

most, classes of intrinsically variable stars are radial pulsators (as discussed in §2.10). However, "intrinsically" implies that it is something inside the star that causes the pulsation. What we have done thus far cannot tell us what that something is because the purely adiabatic periodic motions we have considered have no beginning, end, or apparent cause—which means we have to relax the constraint of adiabaticity.

8.2 Nonadiabatic Radial Motions

By nonadiabatic we mean that heat may be exchanged between moving fluid elements of the star. The take-off point for the analysis is the energy equation given in time-dependent form. A good deal of the work was already done in Chapter 6, where gravitational energy sources were discussed. We now combine equations (6.4) and (6.5) from that chapter to obtain an expression for the mass gradient of luminosity that implicitly contains gravitational terms. Thus

$$\frac{\partial \mathcal{L}_r}{\partial \mathcal{M}_r} = \varepsilon - \frac{P}{\rho\,(\Gamma_3 - 1)} \left[\frac{\partial \ln P}{\partial t} - \Gamma_1 \frac{\partial \ln \rho}{\partial t} \right] \tag{8.35}$$

where ε is the thermonuclear energy generation rate. What we wish to do, for reasons to be made clear later, is eventually to replace the time derivative of pressure with one of temperature. This requires some thermodynamics, and the first step is to use (3.97) to convert the multiplier of the brackets to

$$\frac{P}{\rho\,(\Gamma_3 - 1)} = \frac{c_{V_\rho} T}{\chi_T} \,.$$

The energy equation then becomes, after minor rearrangement,

$$\frac{\partial \ln P}{\partial t} = \Gamma_1 \frac{\partial \ln \rho}{\partial t} + \frac{\chi_T}{c_{V_\rho} T} \left[\varepsilon - \frac{\partial \mathcal{L}_r}{\partial \mathcal{M}_r} \right] . \tag{8.36}$$

(Note that the adiabatic case is regained if the last term is always zero for thermal balance.) Now use the two middle terms of (3.100) to get rid of Γ_1. It is then relatively easy to manipulate χ_T and χ_ρ and ratios of partial derivatives to obtain the desired result,

$$\frac{\partial \ln T}{\partial t} = (\Gamma_3 - 1) \frac{\partial \ln \rho}{\partial t} + \frac{1}{c_{V_\rho} T} \left[\varepsilon - \frac{\partial \mathcal{L}_r}{\partial \mathcal{M}_r} \right] . \tag{8.37}$$

We now linearize the energy equation with the replacements

$$\begin{aligned} T &\longrightarrow T_0 + \delta T \\ \rho &\longrightarrow \rho_0 + \delta\rho \\ \varepsilon &\longrightarrow \varepsilon_0 + \delta\varepsilon \\ \mathcal{L}_r &\longrightarrow \mathcal{L}_{r,0} + \delta\mathcal{L}_r \end{aligned}$$

where, as usual, zero subscripts refer to the equilibrium state. (You may easily confirm later that we need not vary c_{V_ρ} or Γ_3 because the resulting variations will not appear in our final expressions.) If we insist that the equilibrium state is in both hydrostatic equilibrium and thermal balance (with $\varepsilon_0 = \partial\mathcal{L}_{r,0}/\partial\mathcal{M}_r$ and zero time derivatives of T_0 and ρ_0) then, after dropping zero subscripts as before,

$$\frac{\partial\,(\delta T/T)}{\partial t} = (\Gamma_3 - 1)\frac{\partial\,(\delta\rho/\rho)}{\partial t} + \frac{1}{c_{V_\rho} T}\left[\delta\varepsilon - \frac{\partial\,\delta\mathcal{L}_r}{\partial\mathcal{M}_r}\right]. \tag{8.38}$$

Finally let all perturbations vary as $e^{i\omega t}$ to find our final form of the linearized energy equation

$$\delta\varepsilon - \frac{\partial\,\delta\mathcal{L}_r}{\partial\mathcal{M}_r} = i\,\omega\,c_{V_\rho} T\left[\frac{\delta T}{T} - (\Gamma_3 - 1)\,\frac{\delta\rho}{\rho}\right]. \tag{8.39}$$

The δs again refer only to the space parts of the perturbations. (The perhaps complex frequency ω used here is not the dimensionless one introduced for polytropes.) Note, first off, that this equation contains the imaginary unit $i = \sqrt{-1}$ and we suspect, correctly, that the nonadiabatic problem results in eigenvalues and eigenfunctions that are complex. The solutions then automatically contain either exponentially growing ("driving" or "unstable") or decaying ("damping" or "stable") properties. An intrinsically variable star is one in which nonadiabatic effects drive the star to pulsational instability.

To go on and discuss the full nonadiabatic problem is beyond the scope of this text. You are invited to peruse Cox (1980) to appreciate the difficulty of the subject. We shall, however, give some relevant pointers to what is going on by first considering an analysis due to Sir Arthur S. Eddington (1926) that is reviewed in Cox (1980, §9.4) and Clayton (1968, §6–10). His approximation hinges on the assumption that the motions in a variable star are *almost* adiabatic in that the time scale for growth of an instability is long compared to the period of an oscillation. That is, over the bulk of a star the mechanics (operating on short time scales) dominate over the slower effects of heat exchange. This is paramount to saying that if we observe a variable star at the beginning of one cycle of oscillation (whenever that may be), then, if we wait for one mechanically determined period, the star will return *almost* precisely back to that beginning state. In fact, to the lowest order of approximation, we shall assume that initial and final states are thermodynamically identical. What we shall look for are higher-order effects that give us a sense of how big is the "almost" in the above.

Consider then a shell of mass $\Delta\mathcal{M}$ at some radius in a spherical star. If the physical properties in this shell vary over some portion of a pulsation cycle we can use the first law of thermodynamics to describe the relation between the work done (dW) by the shell on its surroundings, the internal energy (dE) gained by the shell, and the heat added to the shell (dQ):

$$dQ = dE + dW. \tag{8.40}$$

For a complete cycle this becomes

$$\oint dQ = \oint dE + \oint dW \tag{8.41}$$

where $\oint$ means that we compute the integral over only one cycle. If the shell of mass truly returns to its initial thermodynamic state over one cycle then the whole process is reversible and $\oint dE$ vanishes because E is a state variable. We are left with $W = \oint dQ$ as the work done by the shell on its surroundings in one cycle. To proceed further, the change in entropy for this reversible system is given by $dS = dQ/T$. Suppose we start the temperature at T_0 in the initial state. After some time has elapsed, the temperature of the shell will have changed to, say, $T = T_0 + \delta T$. To first-order the entropy change between the two states will be

$$dS = \frac{dQ}{T} \approx \frac{dQ}{T_0} - \frac{\delta T}{T_0^2}\, dQ \tag{8.42}$$

where dQ is the heat added over that time. Over one reversible cycle, however, $\oint dS = 0$ because S is also a state variable (and as such is a perfect differential). This leaves us with

$$\oint \frac{dQ}{T} = \oint \frac{dQ}{T_0} - \oint \frac{\delta T}{T_0^2}\, dQ = 0. \tag{8.43}$$

We may pull the constant T_0 from out of the first integral in the middle term (leaving just $\oint dQ$) and substitute into the expression for W to find, to first-order,

$$W = \oint \frac{\delta T}{T}\, dQ \tag{8.44}$$

where we have dropped the zero subscript on T_0 since it is no longer needed. This result looks peculiar because $\delta T/T$ now appears in the integrand for W. What has happened here is that W would be precisely zero over a cycle if the star really returned to its initial state. What we have picked up by our argument is that small piece of W that may differ from zero.

Finally, we consider all mass shells in the star by integrating over mass to find the total work done

$$W_{\text{tot}} = \int_{\mathcal{M}} \oint \frac{\delta T}{T}\, dQ\, d\mathcal{M}_r\,. \tag{8.45}$$

This is interpreted as follows. If $W_{\text{tot}} > 0$, then the pulsating star has done work on itself over one cycle (by means still to be explored) and any initial perturbation will increase. Note that some mass elements may contribute negatively to the whole but the overall effect is positive. In this case we say that the star is driving the pulsations and is unstable. If we take the integral apart, we see that *a positive contribution comes about if heat is absorbed*

($dQ > 0$) *when temperature is on the increase* ($\delta T > 0$) *but heat must be lost when temperatures are on the decline.* In this sense a variable star is a self-contained heat engine. An ordinary stable star cannot accomplish self-driven motion because were it to try, so to speak, perturbations would be damped out because the preponderance of perturbed mass elements would lose heat to their surroundings as temperatures increase and gain heat as they cool.

We now associate dQ with $(\delta\varepsilon - \partial\,\delta\mathcal{L}_r/\partial\mathcal{M}_r)\,dt$ (as in 8.38) because the latter is the heat added to (or subtracted from) the mass shell over a time dt. Thus $W_{\rm tot}$ becomes

$$W_{\rm tot} = \int_{\mathcal{M}} \oint \frac{\delta T}{T}\left[\delta\varepsilon - \frac{\partial\,\delta\mathcal{L}_r}{\partial\mathcal{M}_r}\right] dt\, d\mathcal{M}_r\ . \tag{8.46}$$

This tells us something about the reality of eigenvalues and eigenfunctions because the term in brackets sits at the lefthand side of equation (8.39). Suppose that ω, δT, and $\delta\rho$ are all real. This implies that $[\delta\varepsilon - \partial\delta\mathcal{L}_r/\partial\mathcal{M}_r]$ is pure imaginary and we are led to a contradiction as seen from the following. One of the integrals for $W_{\rm tot}$ is over time, so we must include the factor $e^{i\omega t}$ in all variations. Since the spatial part of $dQ\,dt$—which is the righthand side of (8.39) multiplied by dt—contains i it must be 90° out of phase with $\delta T\, e^{i\omega t}$. If you sketch the time behavior of the integrand of $W_{\rm tot}$ on the complex plane and then transfer the real part of the integrand to a real-time axis (to get physical results) you will immediately see that the result is a purely periodic curve over one period. That is, $W_{\rm tot}$ must be zero and the system is strictly conservative. Thus, in order for stars to be self-excited, the frequency ω and associated quantities must be complex so that the resultant phasing in $W_{\rm tot}$ yields a nonzero result.

8.2.1 The Quasi-Adiabatic Approximation

Another way to approach the question of instability is to derive a differential equation for $\delta r(t, \mathcal{M}_r)/r$ in nonadiabatic form. The result will be an expanded version of the LAWE which, when used in a useful approximation, will give estimates of the imaginary part of ω and hence the e–folding times for driving or damping.

We first differentiate the linearized force equation (8.7) with respect to time. The result will contain a term like $\partial\left[\delta P/P_0\right]/\partial t$, which can be eliminated using a linearized version of the energy equation (8.36). (This is where we depart from the adiabatic analysis.) Terms in $\partial\left[\delta\rho/\rho_0\right]/\partial t$ are dealt with by using the mass equation (8.6). A fair amount of algebraic simplification (see Cox 1980, §7.7) and introduction of $\delta r/r = \zeta\, e^{i\omega t}$ then yields

$$i\,\omega\,\mathbf{L}(\zeta) - i\omega^3\zeta = -\frac{1}{r\rho}\frac{d}{dr}\left[\rho\,(\Gamma_3 - 1)\left(\delta\varepsilon - \frac{\partial\,\delta\mathcal{L}_r}{\partial\mathcal{M}_r}\right)\right] \tag{8.47}$$

where $\mathbf{L}$ is the linear operator of the LAWE (8.16). We use this as follows. Suppose we separate ω into two parts so that

$$\omega = \sigma + i\,\kappa_{\rm qa} \tag{8.48}$$

where both σ and $\kappa_{\rm qa}$ are real. If our suspicions are correct, then σ (as the frequency of pulsation) should be determined primarily by adiabatic and mechanical processes and thus, as a presumably good approximation, we set it equal to the adiabatic frequency. The quantity $\kappa_{\rm qa}$ will then measure the rate of driving or damping because $e^{i\omega t} = e^{i\sigma t - \kappa_{\rm qa} t}$; that is, $-\kappa_{\rm qa}$ is an inverse e–folding time with $\kappa_{\rm qa} < 0$ (> 0) implying driving (damping). We expect $\sigma \gg |\kappa_{\rm qa}|$ because of the assumed rapidity of mechanical versus thermal effects.

The next step is essentially the equivalent of first-order perturbation theory in quantum mechanics. If we have a process that dominates the behavior of a system—such as in the mechanical oscillations of a star—but this is modified by a weaker process (the nonadiabatic effects), then the shift in the energy of the system (the shift in eigenvalue) may be estimated by using only the eigenfunctions of the mechanical system in an integral method without actually solving the complete problem. It goes like this.

Assume that any eigenfunctions required in the application of (8.47) are those obtained from an adiabatic calculation (which gives real eigenfunctions). Thus we replace any occurrence of ζ by $\zeta_{\rm ad}$ where the subscript "ad" implies adiabatic. Now multiply (8.47) on the left by $\zeta_{\rm ad}$ and integrate the result over $r^2\, d\mathcal{M}_r$ through the entire mass. This yields

$$\begin{aligned} i\,\omega \int_{\mathcal{M}} \zeta_{\rm ad}\, \mathbf{L}(\zeta_{\rm ad})\, r^2\, d\mathcal{M}_r - i\,\omega^3 \int_{\mathcal{M}} \zeta_{\rm ad}\zeta_{\rm ad}\, r^2\, d\mathcal{M}_r &= \\ = -\int_{\mathcal{M}} \frac{\zeta_{\rm ad}}{r\rho} \frac{d}{dr}\left[\rho\,(\Gamma_3 - 1)\left(\delta\varepsilon - \frac{\partial \delta\mathcal{L}_r}{\partial \mathcal{M}_r}\right)_{\rm ad}\right] r^2\, d\mathcal{M}_r \\ \equiv -C_{\rm qa} \end{aligned} \tag{8.49}$$

and defines $C_{\rm qa}$ as the integral on the second line. The "qa" subscript that appears here and in $\kappa_{\rm qa}$ means "quasi-adiabatic" because some of what we are doing is adiabatic, whereas other parts contain nonadiabatic elements.[7] Since σ is large compared to $|\kappa_{\rm qa}|$, the $i\,\omega^3$ in the second term becomes approximately equal to $\sigma^2\,(i\sigma - 3\kappa_{\rm qa})$ after small terms are dropped. The factor $i\omega$ is just $i\omega = i\sigma - \kappa_{\rm qa}$. We now recall from (8.17) that the integrated LAWE is, in our new notation for the adiabatic eigenfunction,

$$\int_{\mathcal{M}} \zeta_{\rm ad}\, \mathrm{L}(\zeta_{\rm ad})\, r^2\, d\mathcal{M}_r = \sigma^2 \int_{\mathcal{M}} \zeta_{\rm ad}\zeta_{\rm ad}\, r^2\, d\mathcal{M}_r \tag{8.50}$$

which allows us to get rid of the first integral in (8.49). We shall assume that the star is not dynamically unstable so that σ^2 is positive. Collecting all terms and solving for $\kappa_{\rm qa}$ gives

[7] The particular version of the quasi-adiabatic derivation given here is due to H.M. Van Horn from correspondence dating back to the middle 1970s.

$$\kappa_{\text{qa}} = -\frac{C_{\text{qa}}}{2\sigma^2 \int_{\mathcal{M}} \zeta_{\text{ad}}^2 \, r^2 \, d\mathcal{M}_r} = -\frac{C_{\text{qa}}}{2\sigma^2 J} \tag{8.51}$$

where J, the *oscillatory moment of inertia*, is

$$J = \int_{\mathcal{M}} \zeta_{\text{ad}}^2 \, r^2 \, d\mathcal{M}_r \, . \tag{8.52}$$

The quantity J often appears in pulsation theory. The "moment of inertia" part of its name comes about because $\int r^2 \, d\mathcal{M}_r$ is the moment of inertia for the star. Since both σ^2 and J are positive, the sign of the growth rate κ_{qa} depends only on the sign of C_{qa}. We now show how κ_{qa} and W_{tot} of (8.46) are related.

First change the integration variable in C_{qa} to r rather than $\mathcal{M}_r$ by using the mass equation $d\mathcal{M}_r = 4\pi r^2 \rho \, dr$. An integration by parts in (8.49) yields the constant term

$$4\pi r^3 \zeta_{\text{ad}} \rho \left(\Gamma_3 - 1\right) \left(\delta\varepsilon - \frac{\partial \delta \mathcal{L}_r}{\partial \mathcal{M}_r}\right)\Bigg|_0^{\mathcal{R}}$$

which vanishes at both limits if the density at the outer surface is taken to be zero. To deal with the remaining integral, note that

$$\frac{d\left(4\pi r^3 \zeta_{\text{ad}}\right)}{dr} = 4\pi r^2 \left(3\zeta_{\text{ad}} + r\frac{d\zeta_{\text{ad}}}{dr}\right) .$$

But, from the linearized mass equation (8.6),

$$\left(\frac{\delta\rho}{\rho}\right)_{\text{ad}} = -3\zeta_{\text{ad}} - r\frac{d\zeta_{\text{ad}}}{dr} \tag{8.53}$$

which we use to get rid of all reference to ζ_{ad} in favor of the adiabatic density variation. The next-to-last step is to realize that adiabatic density versus temperature perturbations are connected through Γ_3 by (3.95) which, in this context, reads

$$\left(\frac{\delta\rho}{\rho}\right)_{\text{ad}} \left(\Gamma_3 - 1\right) = \left(\frac{\delta T}{T}\right)_{\text{ad}} . \tag{8.54}$$

Finally, after converting back to a mass integration, we arrive at an expression for the quasi-adiabatic growth rate

$$\kappa_{\text{qa}} = -\frac{\int_{\mathcal{M}} \left(\delta T/T\right)_{\text{ad}} \left(\delta\varepsilon - \partial \, \delta\mathcal{L}_r/\partial\mathcal{M}_r\right)_{\text{ad}} \, d\mathcal{M}_r}{2\sigma^2 J} \, . \tag{8.55}$$

Elements of this result should be suspiciously familiar because the numerator looks like the mass integral part of W_{tot} of (8.46). Note that the two integrals in W_{tot} give the work done on the star by itself over one period. If we divide that by the total kinetic energy of oscillation, then the resulting ratio gives us a rate of change increase (or decrease) of pulsation energy per period. The kinetic energy of oscillation is derived from

$$\text{KE of oscillation} = \left| \int_{\mathcal{M}} \frac{1}{2} \left(\frac{\delta \dot{r}}{r} \right)^2 r^2 \, d\mathcal{M}_r \right| = \frac{\sigma^2 J}{2} \tag{8.56}$$

which is $mv^2/2$ in disguise. But a time derivative of δr yields $i\sigma\delta r$ or $i\sigma r\zeta_{\text{ad}}$ in the adiabatic case. Therefore the kinetic energy of oscillation is just $\sigma^2 J/2$, as indicated in (8.56). The last step is to recognize that κ_{qa} measures the e–folding time for a change in amplitude of a perturbation, whereas energies go as the square of amplitudes. To first-order this introduces a factor of 1/2 into $W_{\text{tot}}/\left(\sigma^2 J/2\right)$. The two formulations ($W_{\text{tot}}$ versus κ_{qa}) tell us the same thing if we accept the validity of the quasi-adiabatic approximation.

How may we use all this? Consider first the role of $\delta\varepsilon$ in equation (8.55) for κ_{qa} and ignore for now the presence of $(\partial\delta\mathcal{L}_r/\partial\mathcal{M}_r)$. As we have often done before, first write ε in the power law form $\varepsilon \propto \rho^\lambda T^\nu$. (See, e.g., 1.59 or 6.48.) Then, treating the perturbation operator as a differential, we find

$$\left(\frac{\delta\varepsilon}{\varepsilon} \right)_{\text{ad}} = \lambda \left(\frac{\delta\rho}{\rho} \right)_{\text{ad}} + \nu \left(\frac{\delta T}{T} \right)_{\text{ad}} \tag{8.57}$$

where the adiabatic subscript is appended to remind us—for the last time—that all the eigenfunctions are derived from an adiabatic calculation. We shall leave the subscript off from now on, but only for clarity. To get everything in terms of $\delta T/T$ we use the adiabatic relation (8.54), which introduces $(\Gamma_3 - 1)$ so that the integrand in κ_{qa} is proportional to $(\delta T/T)^2$. Except for some very unusual circumstances not often (if ever) met in stars, λ and ν (and certainly $[\Gamma_3 - 1]$) are all positive. In other words, the integrand of κ_{qa} is always positive and, hence, κ_{qa} is negative—implying instability. Thus the effect of thermonuclear reactions in stars is to push them toward instability by the "ε mechanism." This should be obvious because ε increases and adds heat to a compressing element experiencing a rise in temperature and this is our criterion for driving. The fact of the matter is, however, that no intrinsically variable star has been unambiguously shown to be unstable due to this effect (although very massive upper main sequence stars may feel the effects of this mechanism—and we still exclude novae and the like in this discussion). So, what physical mechanisms are present in stable stars that override thermonuclear destabilization, and what causes variable stars to pulsate? To answer this we have to look into the mass gradient term of the luminosity perturbation in (8.55).

8.2.2 The κ- and γ-Mechanisms

Thus far we have not considered how heat is transported. Since there are still outstanding problems associated with the interaction of convection and pulsation, we shall consider only heat transport by radiative diffusion and how it is modulated by pulsation.

We first linearize the diffusion equation, say, in the form given by (4.25). An easy exercise yields the relative variation of radiative luminosity,

$$\frac{\delta\mathcal{L}_r}{\mathcal{L}_r} = 4\zeta - \frac{\delta\kappa}{\kappa} + 4\frac{\delta T}{T} + \frac{1}{dT/dr}\frac{d}{dr}\left(\frac{\delta T}{T}\right) \tag{8.58}$$

where κ is the opacity. This expression is the same as given by Cox (1980, Eq. 7.11a). For our analysis we shall ignore the derivative term and, in the spirit of the quasi-adiabatic calculation, assume that the variations are adiabatic. Then, using: (1) the power law expression for the opacity, $\kappa \propto \rho^n T^{-s}$ (as in 1.62); (2) the linearized mass equation (8.6) without the derivative term; (3) the adiabatic relation between $\delta\rho/\rho$ and $\delta T/T$ (8.54), we find

$$\frac{\delta\mathcal{L}_r}{\mathcal{L}_r} \approx -\left(\frac{4/3+n}{\Gamma_3-1}\right)\frac{\delta T}{T} + (s+4)\frac{\delta T}{T}\,. \tag{8.59}$$

Now use this in the quasi-adiabatic expression (8.57) which, without the ε term, is

$$\kappa_{\text{qa}} = \frac{\int_{\mathcal{M}}(\delta T/T)\,(\partial\,\delta\mathcal{L}_r/\partial\mathcal{M}_r)\,d\mathcal{M}_r}{2\sigma^2 J}\,. \tag{8.60}$$

Recall that a positive value of κ_{qa} means that the star is stable. We can get that positive value if, through most of the star, the mass gradient of $\delta\mathcal{L}_r$ is positive upon compression (i.e., when $\delta T > 0$).

An example is a region within a star where Kramers' opacity operates with $n = 1$ and $s = 3.5$. From our discussion of opacities in §4.4 et seq., this opacity dominates in the righthand downward slopes of Fig. 4.2. There ionization is taking place but not as vigorously as it does near the half-ionization points (as the dashed line in Fig. 4.2). Thus set $\Gamma_3 = 5/3$ (or just a tad less) assuming a nonionizing ideal gas. Putting these numbers into (8.59) yields $\delta\mathcal{L}_r/\mathcal{L}_r = 4(\delta T/T)$. The mass derivative is then

$$\frac{\partial\,\delta\mathcal{L}_r}{\partial\mathcal{M}_r} \approx 4\mathcal{L}_r\,\frac{\partial(\delta T/T)}{\partial\mathcal{M}_r} \tag{8.61}$$

if the gradient of $\mathcal{L}_r$ can be neglected, as it can where there is no energy generation going on ($\partial\mathcal{L}_r/\partial\mathcal{M}_r = 0$).

In the simple case of a fundamental mode the mass gradient of δT is positive upon adiabatic compression because the absolute values of the variations in radius or density are small near the center and increase outward. Equation (8.61) then implies that the mass gradient of $\delta\mathcal{L}_r$ is also positive on compression in this case of Kramers' opacity. Hence, in the circumstance we describe, the quasi-adiabatic approximation states that the particular region (or entire star) is stable because κ_{qa} of (8.60) is positive. What is happening here is consistent with our earlier analysis. If $\partial\delta\mathcal{L}_r/\partial\mathcal{M}_r > 0$ upon compression, then more radiative power leaks out of a mass shell than is entering it (as is evident from 6.2). Thus the mass shell loses energy when compressed and this was our criteria for stability. You may turn the argument around by considering expansion and come to the same conclusion about the stabilizing nature of the process. The root cause of the stabilization is that Kramers' opacity decreases upon compression with its accompanying increase of temperature: the material becomes leakier to radiation.

If, on the other hand, we are in an active ionization zone, then κ either may not decrease nearly as rapidly as in the above case—which means s is considerably less than 3.5—or may actually increase with temperature—as in the positive slope portions of Fig. 4.2. Furthermore, ionization means that Γ_3 is less than 5/3. The net combination of these two circumstances can force the factors in front of δT in (8.59) to cause the mass gradient of $\delta \mathcal{L}_r$ to be *negative* when δT is positive. This situation is destabilizing: a mass element *gains* heat upon compression. In the extreme form of this argument, the destabilization due to opacity comes about because increases in opacity due to increases in temperature tend to dam up the normal flow of radiation from out of an ionizing mass element. The mass element, in effect, thus heats up relative to its surroundings. Destabilization in this instance is due to the "κ-mechanism."

It may be the case that the exponent s in the opacity law differs only slightly from a Kramers'–like law in an ionization zone but $\Gamma_3 - 1$ is still small. Such is the case for second helium ionization, which takes place at about 5×10^4 K (or a bit hotter and see Fig. 4.3), where the last electron of helium is being removed and recombined. (Fig. 4.2 shows the seemingly slight effect on opacity of second helium ionization for a typical Pop I mixture.) The effect on Γ_3 alone may be enough to cause instability when (8.59) is applied. What happens here is that the work of compression goes partially into ionization and temperatures do not rise as much if ionization is not taking place. Thus an ionization region tends to be somewhat cooler than the surrounding non-ionizing regions upon compression and heat tends to flow into the ionizing region. This part of the destabilization process is called the γ-mechanism. Note that in most instances the κ- and γ-mechanism go hand-in-hand.

To go further than this local treatment requires a lot of work. You may wish to investigate the various "one-zone models" reported in the literature. In these, assumptions are made that expand on just hand-waving but the analysis does go on a bit. The best summary of such models may be found in Cox (1980, §13.1–§13.4). The conclusions are substantially as we have described.

The Epstein Weight Function and Cepheids

In any case, stars generally have zones that are ionizing and others that are not. Which win out? Another related question is what regions are the most important in establishing the mechanical response of a star. This last may be answered by rearranging the LAWE in its integrated form (8.50) to read

$$\sigma^2 = \frac{\int_{\mathcal{M}} \zeta \, \mathbf{L}(\zeta) \, r^2 \, d\mathcal{M}_r}{\int_{\mathcal{M}} \zeta^2 \, r^2 \, d\mathcal{M}_r} \, . \tag{8.62}$$

All this says is that the eigenvalue is equal to a ratio of integrals containing its eigenfunctions. It is then reasonable to suppose that the integrand of the numerator on the righthand side of this equation is a "weight function" that tells us where it is in the star that most of the action occurs that determines σ^2 (see Cox, 1980, §8.13). This was investigated long ago by Epstein (1950). He specifically computed the weight functions for the fundamental modes of centrally condensed stellar models—such as those expected for evolved classical Cepheid variables—and found that the weight function reached a strong peak at $r/\mathcal{R} \approx 0.75$. If this radius does correspond to the region where we expect the pulsation properties of a centrally condensed model to be primarily determined, then what are the thermal properties of that region with respect to ionization?

We can easily compute the temperature of a star at $r/\mathcal{R} = 0.75$ if we assume that a simple envelope integration is justified. The temperature is then given as a function of radius by (7.124). If, in that expression, we use $\beta = 1$, $\mu = 0.6$, and $n_{\text{eff}} \approx 3.25$ (i.e., for Kramers' opacity), then

$$T(r/\mathcal{R} = 0.75) \approx 1.1 \times 10^6 \, \frac{\mathcal{M}/\mathcal{M}_\odot}{\mathcal{R}/\mathcal{R}_\odot} \; \text{K} \, . \tag{8.63}$$

A "typical" classical Cepheid variable has a mass of $7\,\mathcal{M}_\odot$ and radius $100\,\mathcal{R}_\odot$. This combination yields $T(r/\mathcal{R} = 0.75)$ of about 8×10^4 K, and $\rho \approx 7 \times 10^{-8}$ g cm^{-3} (from 7.121 with bound–free opacity from Fig. 4.6 coupled with 4.63), which corresponds closely to the second ionization zone of helium. (See Exs. 3.1 and 4.6, and the opacity plot for pure helium in Fig. 4.3.) Detailed nonadiabatic studies of classical Cepheid models have indeed confirmed that it is this ionization zone that is responsible for their variability. Other regions of the star may try to stabilize but this zone ultimately wins out.

8.2.3 Nonadiabaticity and the Cepheid Strip

Thus far we have relied on the quasi-adiabatic approximation as a diagnostic tool. This is not always wise because some regions in variable stars react very nonadiabatically to any perturbation. Our measure of how good we expected the quasi-adiabatic approximation to be was based on the ratio of thermal to dynamic time scales. If that ratio is large then motions are close to adiabatic. The dynamic time scale, t_{dyn}, for a variable star is determined by the period, Π, of oscillation for the whole star and, for the fundamental mode, it may be approximated by the period–mean density relation. The thermal time scale for a particular region in a star, on the other hand, is a measure of how long it takes heat to be transported through the region. We can estimate this time scale for a shell of mass content $\Delta\mathcal{M}$ by dividing its total heat content by the luminosity that must pass through it. (This is similar to what we did in deriving the Kelvin–Helmholtz time scale of §1.3.2.) As an approximation to the heat content, we use $c_{V_\rho} T\,\Delta\mathcal{M}$, where c_{V_ρ} and T are suitable averages within the shell. Thus the thermal time scale, $t_{\rm th}$, is

$$t_{\rm th} \approx \frac{c_{V_\rho} T\,\Delta\mathcal{M}}{\mathcal{L}} . \tag{8.64}$$

The ratio we use to estimate how adiabatically or nonadiabatically the shell responds to perturbations is $t_{\rm th}/t_{\rm dyn}$ with

$$\Phi(\Delta\mathcal{M}) \equiv \frac{t_{\rm th}}{t_{\rm dyn}} = \frac{c_{V_\rho} T\,\Delta\mathcal{M}}{\Pi\mathcal{L}} . \tag{8.65}$$

If Φ is large then the motion should be nearly adiabatic and the quasi-adiabatic approximation should be close to the truth. If, however, Φ happens to be small then the thermal time scale is short compared to the period and a full nonadiabatic treatment is warranted. We are *not* going to show you how to do a nonadiabatic analysis here but rather we shall extract from such analyses some properties of the classical Cepheid instability strip discussed in §2.10.

The place we expect Φ to be small is in the outer envelope of a star where temperatures are low and the heat content is correspondingly small. We therefore define the "transition temperature," $T_{\rm TR}$, as the temperature at that point in the envelope where Φ is unity. Interior to that point motions are quasi–adiabatic, whereas to the exterior and up to the surface all must be treated nonadiabatically. If $\Delta\mathcal{M}$ is now the mass of the envelope above this transition level then $T_{\rm TR}$ is given by

$$\Phi(\Delta\mathcal{M}) = \frac{c_{V_\rho} T_{\rm TR}\,\Delta\mathcal{M}}{\Pi\mathcal{L}} = 1 \tag{8.66}$$

where our "suitable average" for T is $T_{\rm TR}$ itself.

We shall not go into the details here but the region of maximum driving in a classical Cepheid variable coincides with the location of the transition temperature which, from our previous discussion, should then also be the temperature for second helium ionization. (Arguments for this and for what follows are discussed in Cox, 1968, §27.7; and Cox, 1980, §10.1.) Given this, we can then estimate the slope of the Cepheid strip in the $\mathcal{L}$–$T_{\rm eff}$ HR diagram.[8] To start, we set $\Phi(\Delta\mathcal{M})$ to unity as in the above, assume c_{V_ρ} is roughly constant, and delete reference to $T_{\rm TR}$ because it too is constant. What is left

[8] The following dimensional estimates give remarkably good answers, so much so that they must, as is said, "be almost too good to be true."

is a relation between $\Delta\mathcal{M}$, $\mathcal{L}$, and Π which, ignoring constants, is $\Delta\mathcal{M} \propto \Pi\mathcal{L}$. If the mass shell $\Delta\mathcal{M}$ is assumed to be thin, then $\Delta\mathcal{M}$ can be eliminated by using an estimate of the pressure at the transition point needed to support the overlying mass $\Delta\mathcal{M}$; that is, $P \approx G\mathcal{M}\,\Delta\mathcal{M}/\mathcal{R}^4$. Now equate this to the envelope solution for pressure (from 7.119) $P = K'T^{1+n_{\rm eff}}$, where K' is given by (7.120). The temperature T here is just a constant ($T_{\rm TR}$). Note that K' goes as $K' \sim \sqrt{\mathcal{M}/\mathcal{L}}$ for Kramers' opacity. Thus $\Delta\mathcal{M} \propto \mathcal{R}^4/(\mathcal{M}\mathcal{L})^{1/2}$. The period–mean density relation gives us the estimate $\Pi \propto \mathcal{M}^{-1/2}\mathcal{R}^{3/2}$ for the fundamental mode so that our condition for $\Phi = 1$ becomes, after some simple algebra,

$$\mathcal{L} \propto \mathcal{R}^{5/3}\,. \tag{8.67}$$

This is then equated to $\mathcal{L} \propto \mathcal{R}^2 T_{\rm eff}^4$ finally to reveal

$$\mathcal{L} \propto T_{\rm eff}^{-20} \tag{8.68}$$

which gives a very steep line on the $\mathcal{L}$–$T_{\rm eff}$ HR diagram tilting slightly to cooler $T_{\rm eff}$. The slope is very nearly correct and matches what was shown for the Cepheid Strip of Fig. 2.23.

We can squeeze more out of this by using the rough mass–luminosity relation (from evolutionary calculations) for helium core–burning stars crossing the Cepheid Strip derived from Iben (2000), which is

$$\mathcal{L} \propto \mathcal{M}^{9/2}\,. \tag{8.69}$$

Combining this with (8.67) and the period–mean density relation between Π, $\mathcal{M}$, and $\mathcal{R}$ gives

$$\Pi \propto \mathcal{R}^{1.31}\,. \tag{8.70}$$

Recent observations by Gieren et al. (1999) of Large and Small Magellanic Clouds, and galactic Cepheids gives

$$\Pi \propto \mathcal{R}^{(1.47\pm0.037)} \tag{8.71}$$

which is gratifyingly close to our back-of-the-envelope estimate.

Finally, Iben (2000) plots a theoretical Π–$\mathcal{L}$ relation that is reasonably represented by a power law. Comparing his to ours, which is easily derived from the above, we have

$$\Pi \propto \mathcal{L}^{0.84} \quad \text{(Iben)}; \quad \Pi \propto \mathcal{L}^{0.79} \quad \text{(our simple stuff)}\,. \tag{8.72}$$

Nonadiabatic calculations can nicely predict the blue (hot) edge of the instability strip but the red edge is another matter. On comparing models of Cepheids with the observed red edge, it is found that vigorous envelope convection begins near the low-temperature side of the strip. (See Fig. 5.7 for a sketch of where we expect efficient envelope convection on the HR diagram.) Thus, in one way or the other, it is thought that convection must inhibit instability. Exactly how it does so is still not clear despite some heroic attempts (as partially reviewed by Cox, 1980, §19.3, and see particularly Toomre, 1982). The trouble is that convection is difficult enough without having to couple it to pulsation.

Note that we have not even mentioned Cepheids that pulsate in modes other than the fundamental. There are first overtone pulsators and variables that operate in both modes. This means another handle on structure for the latter stars and masses may be determined for some of them.

Our discussion only hints at what has been accomplished over the last 45 years (or more) with the Classical Cepheids. But we must emphasize their importance. They are not only testbeds for advanced evolution studies but, since they are so luminous and their luminosity is correlated with period, they are among the key standard candles for cosmology.

A Footnote on Nonlinear Modeling

There have also been many successful studies of radially variable stars using one-dimensional hydrodynamics. The methods used are like those briefly discussed in §7.2.6. One particularly satisfying result that illustrates instability is described in Cox (1968, §27.8). A model classical Cepheid envelope is constructed in hydrostatic equilibrium with parameters (total mass, luminosity, etc.) in the range expected for a variable. The hydrodynamics are then turned on and the model is followed in time. By virtue of numerical noise in the initial model (no computer or calculation is perfect) the "Cepheid" quivers and this quivering is amplified as nonadiabatic effects begin to be felt. After an elapsed time corresponding to about 400 periods (in this particular calculation) maximum nonlinear amplitudes are reached and the model sitting in the computer acts like a real Cepheid you might observe in the sky. This type of work is very different from what we have described for linear theory. There are no imposed infinitesimal bounds on amplitudes and the dynamic model finds its own final pulsating state. The modeling of what processes (such as atmospheric shocks) are responsible for limiting the amplitude is still an active field of study.

Finally, we mention the work of J.R. Buchler and his collaborators. (For a review see Buchler, 1990, and the recent paper by Kolláth et al., 2002, for additional references.) Since we are dealing with very complex nonlinear systems we can expect some surprises. One of the most interesting fields of endeavor in the physical sciences in recent years is that of nonlinear dynamics. It has been shown that even some seemingly simple systems are prone to behavior that is almost counterintuitive. Under certain conditions, whether in the laboratory or the computer, these systems allow for regular limit cycles (what we have been discussing), irregular pulses and period doubling, and, in some instances, chaotic behavior. The minor adjustment of a laboratory or computer parameter can lead the system from one of these states to another with astonishing rapidity. So it may be with stars. Not all variable stars behave as nicely as we have led you to believe in this chapter. If you review the variables discussed in §2.10, some are indicated as being "irregular." It may well be that such variables, and others, are in this murky land of near-chaos. For a taste of what is being done in this field you may want to look through articles such as those by Kovács and Buchler (1988a,b), where theoretical nonlinear calculations are described for Type II Cepheids and RR Lyrae variables. They can do strange things on the computer.

8.3 An Introduction to Nonradial Oscillations

What we shall develop here are the tools necessary to describe how a star may oscillate in modes that do *not* preserve radial symmetry. These are called *nonradial* modes. The prime references here are Cox (1980), Unno et al. (1989), and Ledoux and Walraven (1958).

8.3.1 Linearization of the Hydrodynamic Equations

The path we shall take differs somewhat from the preceding discussion of radial pulsations. Instead of starting with the standard stellar structure equations, we shall delve into simple fluid mechanics although, for the sake of simplicity, adiabatic motions will still be assumed. In this vein we neglect all dissipative effects (such as viscosity) and assume that the stellar fluid cannot support shear stresses. (What happens when shear is introduced—as would be the case for the earth—will be briefly touched upon later.) The equations required to describe how the fluid behaves dynamically are Poisson's equation for the gravitational potential Φ (as introduced by Eq. 7.1), the equation of continuity, and the equation of motion. In that order, these equations are (cf. Landau and Lifshitz, 1959, §15)

$$\nabla^2 \Phi = 4\pi G \rho \tag{8.73}$$

$$\frac{\partial \rho}{\partial t} + \boldsymbol{\nabla} \bullet (\rho \mathbf{v}) = 0 \tag{8.74}$$

$$\rho \left(\frac{\partial}{\partial t} + \mathbf{v} \bullet \boldsymbol{\nabla} \right) \mathbf{v} = -\boldsymbol{\nabla} P - \rho \, \boldsymbol{\nabla} \Phi \, . \tag{8.75}$$

Here $\mathbf{v} = \mathbf{v}(\mathbf{r}, t)$ is the fluid velocity and Φ is the gravitational potential, which is related to the local (vector) gravity by $\mathbf{g} = -\boldsymbol{\nabla}\Phi$. (The scalar gravity g we have been using in this text is then the negative of the radial component of $\mathbf{g}$.) As phrased, these provide a *Eulerian* description of the motion wherein we place ourselves at a particular location, $\mathbf{r}$, in the star and watch what happens to $\mathbf{v}(\mathbf{r}, t)$, $\rho(\mathbf{r}, t)$, etc., as functions of time. In a nonrotating hydrostatic star, $\mathbf{v}$ is zero everywhere. (We ignore fluid motions associated with convection.)

Observe first that we know the values of all the above physical variables in the unperturbed spherical star as solely a function of $r = |\mathbf{r}|$. Now imagine that each fluid element in the star is displaced from its equilibrium position at $\mathbf{r}$ by an arbitrary, but infinitesimal, vector distance $\boldsymbol{\xi}(\mathbf{r}, t)$. (The radial component of this quantity previously was called δr.) Remember that this kind of displacement—which takes an identifiable parcel of fluid and moves it somewhere else—is a Lagrangian displacement. If $\mathbf{v} = 0$, then the Eulerian and Lagrangian perturbations of $\mathbf{v}$, denoted respectively, by $\mathbf{v}'$ and $\delta\mathbf{v}$, are equal and are given by

$$\mathbf{v}' = \delta\mathbf{v} = \frac{D\boldsymbol{\xi}}{Dt} \tag{8.76}$$

where D/Dt is the Stokes derivative

$$\frac{D}{Dt} = \frac{\partial}{\partial t} + \mathbf{v} \bullet \boldsymbol{\nabla} \, . \tag{8.77}$$

(For a complete derivation of these statements see Cox, 1980, §5.3. The Stokes derivative was also used in §5.1.2 when discussing convection.) We shall continue to use $'$ and δ to denote the Eulerian and Lagrangian perturbations of quantities.

As the fluid is displaced, all other physical variables are perturbed accordingly. Thus, for example, the pressure $P(\mathbf{r})$, which was originally associated with the fluid parcel at $\mathbf{r}$, becomes $P(\mathbf{r}) + \delta P(\mathbf{r}, t)$ when the parcel is moved to $\mathbf{r} + \boldsymbol{\xi}(\mathbf{r}, t)$. The same statement applies to the density and its perturbation $\delta\rho(\mathbf{r}, t)$. Note again that these are Lagrangian displacements and thus require the δ operator.

If the motion is adiabatic, then the relation between δP and $\delta\rho$ is the same as that used for radial oscillations:

$$\frac{\delta P}{P} = \Gamma_1 \frac{\delta\rho}{\rho} \, . \tag{8.78}$$

Note that we cannot use a like relation for the Eulerian perturbations $P'(\mathbf{r}, t)$ and $\rho'(\mathbf{r}, t)$ because these perturbations are used to find the new pressures and densities at a *given* point $\mathbf{r}$ without saying where the fluid came from. The Eulerian and Lagrangian variations are connected, however, by an easily derived relation found in any book on hydrodynamics (or see Cox, 1980, §5.3); namely, and using density as an example,

$$\delta\rho = \rho' + \boldsymbol{\xi} \bullet \boldsymbol{\nabla}\rho \, . \tag{8.79}$$

Such relations will be used extensively later.

The analysis proceeds by replacing P, ρ, Φ, and $\mathbf{v}$, by $P + P'$, $\rho + \rho'$, $\Phi + \Phi'$, and $\mathbf{v}'$ in equations (8.73–8.75), multiplying everything out, and keeping terms to only first order in the perturbations. We are again performing a linear analysis of the system.[9] Thus, for example, the force equation becomes

[9] Note that we use Eulerian perturbations here, but we could have used Lagrangian forms. It is a matter of taste and tradition, but see Pesnell (1990).

$$\rho \frac{\partial^2 \boldsymbol{\xi}}{\partial t^2} = -\boldsymbol{\nabla} P - \rho \boldsymbol{\nabla} \Phi - \boldsymbol{\nabla} P' - \rho \boldsymbol{\nabla} \Phi' - \rho' \boldsymbol{\nabla} \Phi \,. \tag{8.80}$$

The first two terms on the righthand side cancel because

$$-\boldsymbol{\nabla} P - \rho \, \boldsymbol{\nabla} \Phi = 0 \tag{8.81}$$

is the equation of hydrostatic equilibrium for the unperturbed star. What is left is an equation that contains only the perturbed quantities as first-order variables. Similarly, the continuity and Poisson equations become

$$\rho' + \boldsymbol{\nabla} \bullet (\rho \, \boldsymbol{\xi}) = 0 \tag{8.82}$$

$$\nabla^2 \Phi' = 4\pi G \rho' \,. \tag{8.83}$$

In setting down the linearized form of the continuity equation we have given what results after an integration over time and the removal of a constant of integration by insisting that $\rho' = 0$ when $\boldsymbol{\xi} = 0$.

Even though we have linearized the equations, the above set of partial differential equations is still daunting because the system is second-order in time and fourth-order in space. To reduce the system further requires a bit more work. The aim will be to convert what we have to *ordinary* differential equations. We first assume, as was done for radial oscillations, that all the variations may be Fourier analyzed with $\boldsymbol{\xi}$, P', ρ', and Φ' being proportional to $e^{i\sigma t}$ where σ is an angular frequency. Thus, for example,

$$\boldsymbol{\xi}(\mathbf{r}, t) = \boldsymbol{\xi}(\mathbf{r}) \, e^{i\sigma t} \,. \tag{8.84}$$

With this substitution the time variable is separated out and all variations become functions solely of the radius vector $\mathbf{r}$.

The second step we take is completely to ignore the variation in gravitational potential, Φ'. This step, called the "Cowling approximation" (Cowling 1941), is remarkably good, provided that little mass is thrown around during the motion of the fluid. We cannot justify it here (see the above references) but it introduces only minor errors in many cases of practical interest and it reduces our labors by nearly a factor of two. With this in mind, let's go on.

First observe that the continuity equation, (8.82), expands out to

$$\frac{\rho'}{\rho} = -\boldsymbol{\nabla} \bullet \boldsymbol{\xi} - \frac{\boldsymbol{\xi} \bullet \boldsymbol{\nabla} \rho}{\rho} \tag{8.85}$$

with minor rearrangement and where all perturbations are only functions of $\mathbf{r}$. The last term here, however, is just the last term of (8.79) divided by the density. Thus these two equations yield the following for the Lagrangian variation of density:

$$\frac{\delta \rho}{\rho} = -\boldsymbol{\nabla} \bullet \boldsymbol{\xi} \,. \tag{8.86}$$

(A little thought about the meaning of $\nabla \bullet \boldsymbol{\xi}$, which is sometimes called the dilatation, would have yielded this immediately.) This result also relates the relative Lagrangian variation of pressure to $\nabla \bullet \boldsymbol{\xi}$ through the adiabatic condition (8.78).

Now to decide on geometry. Since the unperturbed star is spherical, doesn't rotate, and has no magnetic fields, the natural coordinate system is, naturally, spherical coordinates. Thus we will talk about the radial (r), co-latitude angle (θ), and angular azimuthal (φ) components of displacement of $\boldsymbol{\xi}(r, \theta, \varphi)$. Call these $\xi_r(r, \theta, \varphi)$, $\xi_\theta(r, \theta, \varphi)$, and $\xi_\varphi(r, \theta, \varphi)$. Using these will enable us to make a separation of variables leading to a well-known angular function of mathematical physics that will make things simple (at last). If you choose to follow in detail the derivation coming up, you will need to review what gradients, etc., look like in our geometry.

We now expand (8.80) in vector components dealing with the radial component first, that is,

$$\sigma^2 \xi_r = \frac{1}{\rho} \frac{\partial P'}{\partial r} - \frac{\rho'}{\rho^2} \frac{dP}{dr} \tag{8.87}$$

where $\nabla \Phi$ has been replaced by $-\nabla P/\rho$ using the hydrostatic condition (8.81) for the unperturbed star. Note that P is a function of r only and thus we do not need partials for its gradient. Now, for reasons to be made apparent later, the term containing P' is manipulated so that the radial derivative acts on P'/ρ instead and, as you may easily check, (8.87) becomes

$$\sigma^2 \xi_r = \frac{P'}{\rho^2} \frac{d\rho}{dr} + \frac{\partial}{\partial r} \left(\frac{P'}{\rho} \right) - \frac{\rho'}{\rho^2} \frac{dP}{dr} . \tag{8.88}$$

The next devious steps are aimed at converting this into final form. We keep the second term on the right-hand side as it stands but manipulate the sum of the first and third terms so that P' and ρ' are replaced by their Lagrangian forms δP and $\delta \rho$. Use (8.79) to do this. Then use the adiabatic condition (8.78) to replace δP by $\delta \rho$. Finally, get rid of $\delta \rho$ in favor of $\nabla \bullet \boldsymbol{\xi}$ with the help of (8.86). After the smoke clears we find

$$\sigma^2 \xi_r = \frac{\partial}{\partial r} \left(\frac{P'}{\rho} \right) - A_{\rm s} \frac{\Gamma_1 P}{\rho} \nabla \bullet \boldsymbol{\xi} \tag{8.89}$$

for the radial equation where $A_{\rm s}(r)$ is the Schwarzschild discriminant of (5.31), which played an important role in convection (with $A_{\rm s} > 0$ implying convective instability). That is,

$$A_{\rm s}(r) = \frac{d \ln \rho}{dr} - \frac{1}{\Gamma_1} \frac{d \ln P}{dr} .$$

Note that $\Gamma_1 P/\rho$ in (8.89) is the square of the local sound speed v_s. The combined presence of v_s and $A_{\rm s}$ will mean that nonradial oscillations come in two distinct flavors and not just acoustic waves as was the case for radial pulsations.

The equations for the other two components of $\boldsymbol{\xi}$ are straightforward. Gradient terms are easy and, remembering that ρ depends only on r, we find

$$\sigma^2 \xi_\theta(r,\theta,\varphi) = \frac{\partial}{\partial\theta}\left[\frac{1}{r}\frac{P'(\mathbf{r})}{\rho}\right] \tag{8.90}$$

$$\sigma^2 \xi_\varphi(r,\theta,\varphi) = \frac{1}{\sin\theta}\frac{\partial}{\partial\varphi}\left[\frac{1}{r}\frac{P'(\mathbf{r})}{\rho}\right] . \tag{8.91}$$

The final stage for the separation of variables is now set.

8.3.2 Separation of the Pulsation Equations

The three equations (8.89–8.91) contain the four unknowns ξ_r, ξ_θ, ξ_φ, and $P'(\mathbf{r})/\rho$. We shall now show that there are really only *two* independent unknowns and that the remaining two degrees of freedom collapse down into a well-known function from mathematical physics. Our method for demonstrating this is to give the answer and then see if it works.

We propose the following solution for $\boldsymbol{\xi}(\mathbf{r})$ and $P'(\mathbf{r})/\rho$:

$$\begin{aligned} \boldsymbol{\xi}(r,\theta,\varphi) &= \xi_r(r,\theta,\varphi)\,\mathbf{e}_r + \xi_\theta(r,\theta,\varphi)\,\mathbf{e}_\theta + \xi_\varphi(r,\theta,\varphi)\,\mathbf{e}_\varphi \\ &= \left[\xi_r(r)\,\mathbf{e}_r + \xi_t(r)\,\mathbf{e}_\theta\frac{\partial}{\partial\theta} + \xi_t(r)\,\mathbf{e}_\varphi\frac{1}{\sin\theta}\frac{\partial}{\partial\varphi}\right] Y_{\ell m}(\theta,\varphi) \end{aligned} \tag{8.92}$$

and

$$\frac{P'(\mathbf{r})}{\rho} = \frac{P'(r)}{\rho} Y_{\ell m}(\theta,\varphi) . \tag{8.93}$$

Here the $\mathbf{e_i}$ are the dimensionless unit vectors in spherical coordinates, and $\xi_t(r)$ and $P'(r)/\rho$ are new functions of r only, which are related to each other by

$$\xi_t(r) = \frac{1}{\sigma^2}\frac{1}{r}\frac{P'(r)}{\rho} . \tag{8.94}$$

The function $\xi_t(r)$ effectively replaces P'/ρ and the θ and φ components of $\boldsymbol{\xi}$ and it will be referred to as the "tangential (or transverse) displacement." Finally, the angle-dependent function $Y_{\ell m}(\theta,\varphi)$ is the *spherical harmonic* (or surface harmonic) of combined indices ℓ and m. This function, which does the angular separation for us, arises frequently in physics (such as in the hydrogen atom and applications in electricity and magnetism) and is the regular solution of the second-order partial differential equation

$$\frac{1}{\sin\theta}\frac{\partial}{\partial\theta}\left(\sin\theta\,\frac{\partial Y_{\ell m}}{\partial\theta}\right) + \frac{1}{\sin^2\theta}\frac{\partial^2 Y_{\ell m}}{\partial\varphi^2} + \ell(\ell+1)Y_{\ell m} = 0 . \tag{8.95}$$

Here ℓ must be zero or a positive integer and m can only take on the integer values $-\ell$, $-\ell+1$, $\cdots$, 0, $\cdots$, $\ell-1$, ℓ. Thus for a given value of ℓ there are only $2\ell+1$ permitted values of m. The separation given above gives rise to "spheroidal modes." Another separation is possible, resulting in "toroidal modes," but we shall ignore them. Note that $\ell = 0$ is a special case because $Y_{00}(\theta, \varphi)$ is a constant and, more to the point, it does not depend on either θ or φ. In other words, solutions for $\ell = 0$ depend only on r and are radial modes. Since we have already discussed them we shall concentrate only on solutions for which $\ell > 0$. Some of the properties of the $Y_{\ell m}$ will be discussed in a bit but the important point for now is the following.

If the angular components of equations (8.92) and (8.94) are introduced into the angular components of the force equation (8.90) and (8.91), then the derivatives of $Y_{\ell m}$ cancel out of both sides of the resulting equations. For example, the lefthand side of (8.90) is (using 8.92)

$$\sigma^2 \xi_\theta(r, \theta, \varphi) = \sigma^2 \xi_t(r) \frac{\partial Y_{\ell m}}{\partial \theta}$$

while the righthand side becomes (using 8.93–8.94)

$$\frac{\partial}{\partial \theta}\left[\frac{1}{r}\frac{P'(\mathbf{r})}{\rho}\right] = \sigma^2 \frac{\partial\left[\xi_t(r) Y_{\ell m}\right]}{\partial \theta} = \sigma^2 \xi_\theta(r, \theta, \phi) .$$

Thus the θ component of the force equation is satisfied with our choice of ξ_t and separation of variables. You may easily show that the φ component is also consistent but aside from consistency it yields no further information. We need the radial component for this.

The difficult term in the radial equation (8.89) is the divergence of $\boldsymbol{\xi}(r, \theta, \varphi)$. Written out in full it is

$$\begin{aligned} \nabla \bullet \boldsymbol{\xi} &= \frac{1}{r^2}\frac{d}{dr}\left(r^2 \xi_r\right) + \\ &+ \frac{1}{r \sin\theta}\frac{\partial}{\partial \theta}\left(\xi_\theta \sin\theta\right) + \frac{1}{r \sin\theta}\frac{\partial}{\partial \varphi}\left(\xi_\varphi\right) . \end{aligned} \tag{8.96}$$

But if (8.92) is inserted here, then you may easily verify that you regain the first two terms of the differential equation for $Y_{\ell m}$ (of Eq. 8.95) so that

$$\nabla \bullet \boldsymbol{\xi} = \frac{1}{r^2}\frac{d}{dr}\left(r^2 \xi_r\right) Y_{\ell m} - \frac{\ell(\ell+1)}{r} \xi_t Y_{\ell m} \tag{8.97}$$

where ξ_r now depends only on r. On the other hand, $\nabla \bullet \boldsymbol{\xi}$ is, from (8.86), the same as $-\delta\rho/\rho$, which can be written (from parts of the derivation leading to 8.89 and 8.94) as

$$\frac{\delta \rho}{\rho} = \frac{\rho}{\Gamma_1 P}\left(\sigma^2 r\, \xi_t - g\, \xi_r\right) \tag{8.98}$$

where the equation of hydrostatic equilibrium has been used to replace the pressure derivative by $-\rho g$ and common factors of $Y_{\ell m}$ have been eliminated. The net result of these manipulations is that we obtain a first-order ordinary differential equation after equating the divergence and $-\delta\rho/\rho$. The final form of this will be given shortly after the following important frequencies are defined.

The first of these frequencies is the Brunt-Väisälä frequency N given by (5.31) and (5.33) as

$$N^2 = -A_s g = -g \left[\frac{d \ln \rho}{dr} - \frac{1}{\Gamma_1} \frac{d \ln P}{dr} \right] . \tag{8.99}$$

Recall that N, in the simplest interpretation, is the frequency of oscillation associated with a perturbed parcel of fluid in a convectively stable medium ($N^2 > 0$).

The second frequency is the *Lamb frequency* or *critical acoustic frequency*, S_ℓ, defined by

$$S_\ell^2 = \frac{\ell(\ell+1)}{r^2} \frac{\Gamma_1 P}{\rho} = \frac{\ell(\ell+1)}{r^2} v_s^2 . \tag{8.100}$$

In addition, we introduce the *transverse wave number*, k_t (with units cm^{-1}),

$$k_t^2 = \frac{\ell(\ell+1)}{r^2} = \frac{S_\ell^2}{v_s^2} . \tag{8.101}$$

If we relate a transverse wavelength $\lambda_t = 2\pi/k_t$ to k_t then S_ℓ^{-1} is the time it takes a sound wave to travel the distance $\lambda_t/2\pi$.

The differential equation resulting from equating $\boldsymbol{\nabla} \bullet \boldsymbol{\xi}$ and $-\delta\rho/\rho$ now becomes, after some easy algebra and using (8.97–8.98),

$$r \frac{d\xi_r}{dr} = \left[\frac{k_t^2 g r}{S_\ell^2} - 2 \right] \xi_r + r^2 k_t^2 \left[1 - \frac{\sigma^2}{S_\ell^2} \right] \xi_t . \tag{8.102}$$

A second differential equation is also gotten from the radial force equation (8.89), the definition of ξ_t from (8.94), and using $\delta\rho/\rho$ of (8.98) instead of $\boldsymbol{\nabla} \bullet \boldsymbol{\xi}$ in (8.97). You may verify that the result is

$$r \frac{d\xi_t}{dr} = \left[1 - \frac{N^2}{\sigma^2} \right] \xi_r + \left[\frac{r}{g} N^2 - 1 \right] \xi_t . \tag{8.103}$$

We can't apologize for the algebra you have had to go through to reach this point, but we're almost done.

Equations (8.102) and (8.103) taken together constitute a second-order ordinary differential equation. If we had kept in the potential field variations (Φ') the system would have ended up as fourth order with the additional variables Φ' and its radial derivative. Except for its use in precise applications, the second-order Cowling approximation captures the essence of the low-amplitude behavior in stars.

The boundary conditions on our set of two equations derive from the behavior of the oscillating star at the surface and the center. We refer you to Cox (1980, §17.6) and Unno et al. (1989, §14.1) for the details of how these are derived. The central boundary condition depends on how various hydrostatic quantities such as $A_{\rm s}$ and P vary with radius near the center and the insistence that ξ_r and ξ_t be finite there. These regular solutions go as $(\xi_r, \xi_t) \propto r^{\ell-1}$ for small r and the relation between them is

$$\xi_r(r) = \ell\, \xi_t(r) \quad \text{as} \quad r \to 0\,. \tag{8.104}$$

The condition at the surface depends on the atmospheric conditions of the static star. In the simplest instance of zero boundary conditions (as discussed in §4.3), the perturbations must be such that the surface pressure remains zero. This is the same as requiring that δP be zero at the surface. To express this as a relation between ξ_r and ξ_t consider (8.98), which can be rewritten as

$$\frac{\delta P}{P} = \frac{\rho}{P}\left[\sigma^2 r\, \xi_t - g\, \xi_r\right]\,.$$

Just under the surface the ratio $\delta P/P$ should be finite from physical considerations. As we approach the surface this should remain true even as P goes to zero. However, the factor ρ/P outside the expression in brackets in the above tends to infinity because, in the case of an ideal gas, for example, it is inversely proportional to temperature and T goes to zero at the surface (as in 4.124). Thus for the relative pressure perturbation to remain finite at the surface we require that

$$\xi_t(\mathcal{R}) = \frac{g_s \xi_r(\mathcal{R})}{\sigma^2 \mathcal{R}} \tag{8.105}$$

where g_s is the surface gravity. As in the case of radial oscillations this gives complete reflection and standing waves. We remark here that under some conditions perfect reflection is not possible for real atmospheres and pulsation energy may escape through the surface causing heating of circumstellar material and also pose a drain on the pulsations.

We now have two ordinary differential equations and two independent boundary conditions. But, as for radial oscillations, our equations and boundary conditions are linear and homogeneous and we thus have to fix a normalization. Again the choice of normalization is arbitrary and we choose $\xi_r(\mathcal{R})/\mathcal{R} = 1$. The system is now overdetermined and σ^2 is an eigenvalue and the perturbations $\xi_r(r)$ and $\xi_t(r)$ are the eigenfunctions.

At this juncture we note an important mathematical difference between radial and nonradial oscillations. Recall that the LAWE for radial oscillations was Sturm–Liouville and Hermitian, which led to a nice ordering of the eigenfrequencies and eigenfunctions with a definite lower bound for σ^2. Such is not the case here even in the Cowling approximation, although the system is still self-adjoint. This means that even though the eigenvalues, σ^2, are still real, there is no guarantee that nonradial modes are ordered in any simple way and, in particular, there may be no lower bound on σ^2. We still retain, however, orthogonality of eigenfunctions (see Unno, et al., 1989, §14.2, for a derivation of the self-adjointness).

We now return briefly to the properties of the $Y_{\ell m}(\theta, \varphi)$. These angular functions are given by

$$Y_{\ell m}(\theta, \varphi) = \sqrt{\frac{2\ell + 1}{4\pi} \frac{(\ell - m)!}{(\ell + m)!}} \, P_\ell^m(\cos\theta) \, e^{im\varphi} \tag{8.106}$$

where the $P_\ell^m(\cos\theta)$ are the associated Legendre polynomials generated by

$$P_\ell^m(x) = \frac{(-1)^m}{2^\ell \ell!} \left(1 - x^2\right)^{m/2} \frac{d^{\,\ell+m}}{dx^{\,\ell+m}} \left(x^2 - 1\right)^\ell . \tag{8.107}$$

Here x denotes $\cos\theta$.[10] As noted before, the restrictions on ℓ and m for these functions are $\ell = 0, 1, \ldots$ (an integer), and m is an integer with $|m| \le \ell$ for reasons of regularity and single-valuedness of solution. You may wish to play with these functions, but Fig. 8.2 shows what they look like on the surface of a sphere where light areas correspond to positive values of the real part of $Y_{\ell m}$ and dark areas to negative values. The symmetry axis defining $\theta = 0$ is almost vertical in the figure; it is actually tilted toward you by 10°. Modes with $m = 0$ are called zonal modes while those with $|m| = \ell$ resemble the segments of an orange and are called sectoral modes. Tesseral modes are those of mixed type.

Although the figure gives some idea of what is happening on spherical surfaces, the actual motion of the fluid is more complicated. The eigenfunctions have nodal lines on the surface of a sphere at any radius and instant of time but there may also be nodes at different radial positions within the star. This is very difficult to picture. In addition, the oscillatory time dependence means that fluid sloshes back and forth periodically. As a simple example, consider the sectoral mode $\ell = m = 1$. This has one angular node passing through the poles, but this node moves in a retrograde azimuthal direction (to smaller φ) because of the factor $e^{i(\sigma t + m\varphi)}$ coming from the time dependence and the exponential in $Y_{\ell m}$. That is, lines of constant phase are such that $d\varphi/dt = -\sigma/m$ and, for m positive, we have a running retrograde wave. The zonal ($m = 0$) case is easier to visualize because we can imagine the light and dark portions of a surface (as in the figure) alternating periodically in brightness with a period $\Pi = 2\pi/\sigma$ for σ^2 positive. For one view of nonradial motions see the vector displacement fields shown in Smeyers (1967) for massive upper main sequence stars.

8.3.3 Properties of the Solutions

A great deal can be learned about the solutions to the ordinary differential equations for ξ_r and ξ_t of (8.102) and (8.103) by performing a local analysis of

[10] Depending on the author and use, you may see other factors of $(-1)^m$ appearing in these formulas. These constitute different phase conventions but do not change the physics. We use the convention of Jackson (1999). Arfken and Weber (1995) and other texts use other choices of phase.

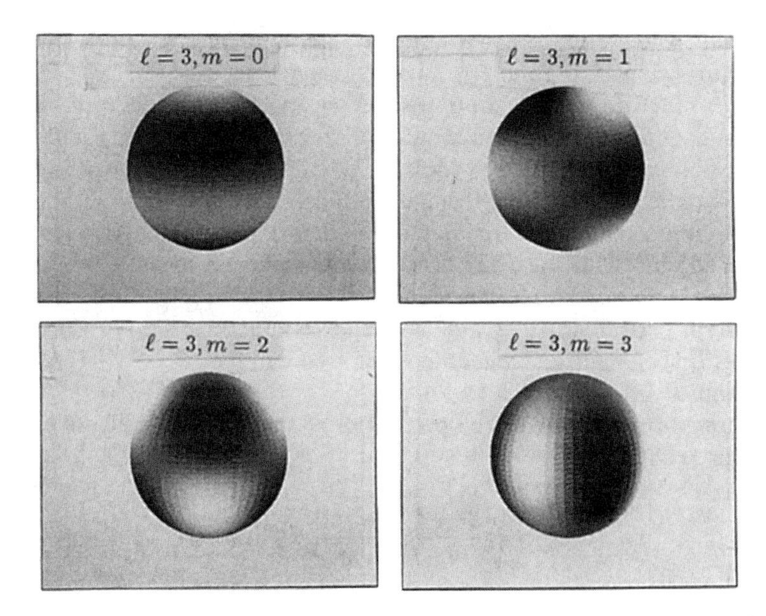

Fig. 8.2. Shown are the patterns of light and dark on the surface of a sphere corresponding to positive and negative values of the real part of the $Y_{\ell m}$. This figure gives illustrations for (clockwise from the upper left) Y_{30} (zonal), Y_{31} (tesseral), Y_{32} (tesseral), and Y_{33} (sectoral).

the system. We assume that ξ_r and ξ_t vary much more rapidly in space than do the other physical variables appearing in those equations—such as N^2—so that those variables can be considered constant over some limited range of radius. To quantify this, assume that both ξ_r and ξ_t vary spatially as $\exp(ik_r r)$, where the wave number k_r is very large compared to r. Thus both eigenfunctions have many wiggles over a short span of space. Inserting this complex exponential into the differential equations then yields a homogeneous set of algebraic equations in ξ_r and ξ_t whose coefficient determinant must be zero in order to obtain nontrivial solutions. Keeping terms dominant in k_r then yields the dispersion relation

$$k_r^2 = \frac{k_t^2}{\sigma^2 S_\ell^2}\left(\sigma^2 - N^2\right)\left(\sigma^2 - S_\ell^2\right) \tag{8.108}$$

where, as before, we assume σ^2 is positive. The implications of this are—

1. If σ^2 is greater or less than *both* of N^2 and S_ℓ^2, then $k_r^2 > 0$ and sinusoidal propagating solutions are present because $\exp(ik_r r)$ reduces to sines and cosines.
2. If σ^2 lies between N^2 and S_ℓ^2, then k_r^2 is negative and solutions show exponential, or evanescent, behavior.

Thus N^2 and S_ℓ^2 are critical frequencies for wave propagation.

You can use the dispersion relation (8.108) to solve for σ^2 in two limits for propagating waves. To facilitate this we define the total wave number, K, by $K^2 = k_r^2 + k_t^2$ (see Unno et al., 1979, §14). This gives more of the flavor of a wave that can travel in a combination of radial and transverse directions. The understanding is that K is large for a local analysis. Then, if σ^2 is much greater than both N^2 and S_ℓ^2, and $|N^2|$ is smaller than S_ℓ^2 (which is often the case), the "large" root of (8.108) is

$$\sigma_{\mathrm{p}}^2 \approx \frac{K^2}{k_t^2} S_\ell^2 = (k_r^2 + k_t^2)\, v_s^2 \qquad \text{for} \qquad \sigma^2 \gg N^2, S_\ell^2\,. \tag{8.109}$$

The subscript "p" has been appended to σ^2 to denote "pressure" because only the sound speed enters. These are *pressure* or *acoustic* modes but we shall often refer to them as "p-modes." You should note here the resemblance to radial modes where ℓ is zero. In that case k_t is zero and we regain (8.31, with $\phi = 0$) derived from our earlier asymptotic analysis.

The small root follows if σ^2 is much less than N^2 and S_ℓ^2 and is given by

$$\sigma_{\mathrm{g}}^2 \approx \frac{k_t^2}{k_r^2 + k_t^2} N^2 \qquad \text{for} \qquad \sigma^2 \ll N^2, S_\ell^2\,. \tag{8.110}$$

These are *gravity* or "g-modes," so-called because buoyancy in the gravitational field is the restoring force. Note that if N^2 is negative, implying convection, then σ_{g} is pure imaginary and the perturbation either grows or decays exponentially in time. These are called g^--modes, whereas those associated with $N^2 > 0$ are g^+-modes. We will only consider g^+-modes and refer to them just as g-modes. In any case, this is the pulsation analogue to our discussion of convective time scales in Chapter 5.

Thus, in summary, p-modes constitute the high-frequency end of the nonradial oscillation spectrum, whereas g-modes are of low frequency. (For very evolved and complicated models, modes may be of mixed character and this statement may not strictly hold true.)

If each mode in a spectrum is orthogonal with respect to the others, then the eigenfunctions corresponding to each eigenvalue σ^2 must differ in important respects. Following our local analysis as an approximation to what happens, k_r and ℓ must measure this difference. Since k_r is a wavenumber, the corresponding local wavelength is $\lambda_r = 2\pi/k_r$. The total number of nodes (denoted by n) in either eigenfunction is then $n \approx 2\int_0^{\mathcal{R}} dr/\lambda_r$ where the "2" counts the two nodes per wavelength. Thus $n \approx \int_0^{\mathcal{R}} k_r\, dr/\pi$. If (8.109) is integrated such that the integral of k_r appears by itself and if ℓ is small so that k_t^2 may be neglected (for simplicity), we again obtain the estimate

$$\sigma_{\mathrm{p},n} \approx n\pi \left[\int_0^{\mathcal{R}} \frac{dr}{v_s}\right]^{-1}\,. \tag{8.111}$$

Thus for large n the *p-mode frequencies are equispaced.* (For more exact treatments using JWKB methods see Unno et al., 1979, §15, or Tassoul, 1980. Our estimates hide many sins.) Note that the frequency spacing depends only on the run of the sound speed, which, for an ideal gas, depends primarily on temperature. Thus, in stars such as the sun, p-modes effectively sample the temperature structure, as will be discussed in the next chapter.

A corresponding estimate for the *periods* of g-modes is

$$\Pi_{g,n} = \frac{2\pi}{\sigma_g} \approx n \frac{2\pi^2}{[\ell(\ell+1)]^{1/2}} \left[\int_0^{\mathcal{R}} \frac{N}{r}\, dr \right]^{-1} = \frac{n\,\Pi_0}{[\ell(\ell+1)]^{1/2}} \tag{8.112}$$

which also defines Π_0 as a constant for the star (in our approximation). Here it is the *period* that is equally spaced in n (a fact to be taken advantage of when discussing variable white dwarfs in Chap. 10) and it depends sensitively on ℓ. Here $\Pi_0/\,[\ell(\ell+1)]^{1/2}$ is the period spacing between consecutive modes. Also, the frequencies (periods) decrease (increase) with n, in direct contrast to the p-modes. The reason the periods increase with increasing mode order is, if we may wave our hands around a bit, that while mass is being moved around, less massive elements are being moved as the distance between nodes decreases. Since bouyancy is the restoring force, less mass being moved means the restoring force is weaker and more sluggish—and sluggish implies longer periods. We shall give a numerical example of the relative ordering of nonradial modes in Fig. 8.3.

The same limits on σ^2 relative to N^2 and S_ℓ^2 also yield the following rough estimates for the ratio of radial to tangential eigenfunctions when used in the differential equations (8.102) and (8.103):

$$\left| \frac{\xi_r}{\xi_t} \right| \sim \begin{cases} rk_r & \text{p-modes} \\ \ell(\ell+1)/rk_r & \text{g-modes}\,. \end{cases} \tag{8.113}$$

For large radial wavenumber ($rk_r \gg 1$) the fluid motions for p-modes are primarily radial, whereas they are primarily transverse for g-modes.

Mode Classification

We have seen that the character of a particular mode depends on n, ℓ, and the relative amplitudes of the radial and tangential displacements. In addition, p-modes are of higher frequency than g-modes. The frequency of a given mode is denoted by $\sigma_{n\ell}$ where it is understood that two different modes (p- and g-) may exist for a given combination of n and ℓ. How does the azimuthal "quantum" number m enter this picture? If the unperturbed star is spherically symmetric, then the eigenvalue (the square of frequency) is independent of m even though the eigenfunction must depend on m through the appearance of $e^{im\varphi}$ in $Y_{\ell m}$. We know the former must be the case since the basic differential equations (8.102)–(8.103) and boundary conditions (8.104)–(8.105) make no mention of m. Another way to look at this is to realize that there is no preferred axis of symmetry in a spherically symmetric system. We may arbitrarily choose such an axis—and this establishes the pole for measuring the colatitude angle θ—but, since φ enters only as a phase factor in $e^{im\varphi}$ and we may choose any great polar circle to start measuring φ, it cannot enter in the final analysis for the eigenvalue. (The same is true for the isolated hydrogen atom where $Y_{\ell m}$ also appears in the eigenfunction: the energy eigenvalue does not depend on m.) If, on the other hand, there were effects that destroyed spherical symmetry—such as rotation or magnetic fields—then m would play a role and we would have to include that in the specification of σ. We will do this in a bit.

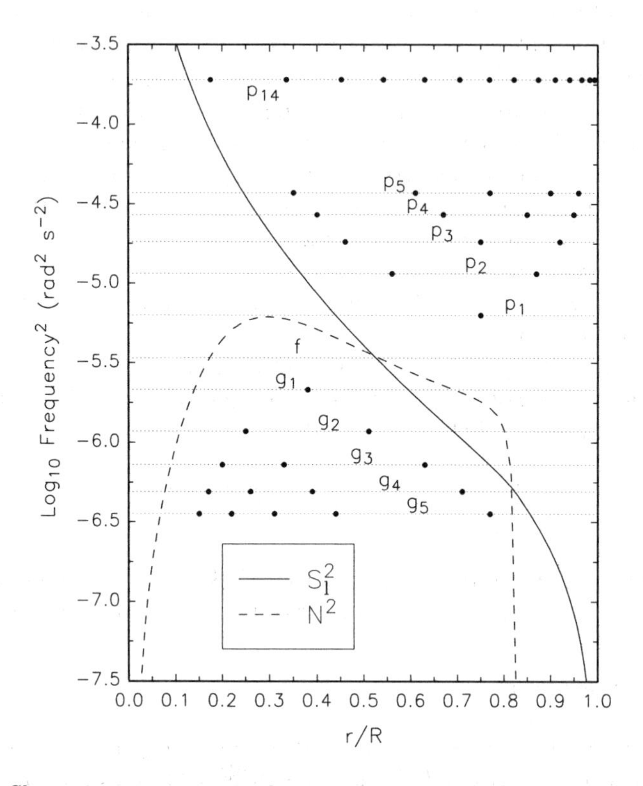

Fig. 8.3. Shown is a propagation diagram for a model of the ZAMS sun. See the text for details of how this figure was constructed and for its significance. The modes for this figure were computed using the ZAMS and PULS codes on the enclosed CD-ROM.

First let's look at a "typical" spectrum of nonradial modes. Figure 8.3 is the result of a series of calculations for $\ell = 2$ (quadrupole) modes in a ZAMS sun. The dashed line shows the run of $N^2 > 0$ versus $r/\mathcal{R}$ whereas the solid line is for S_ℓ^2. (The outer layers are convective with $N^2 < 0$.) The frequencies (i.e., σ^2 in radians s^{-2}) of twelve modes are indicated by the horizontal lines and they are labeled (starting from the top) p_{14} (and skip a few), p_5 down to p_1, f, and then g_1 down to g_5.

Before we go into more detail about the modes, first consider why the acoustic and Brunt–Väisälä frequencies look as they do. The behavior of S_ℓ^2 is understood as follows. As $r \to 0$, $S_\ell^2 = \ell(\ell+1)v_s^2/r^2$ approaches positive infinity because of the factor r^{-2}. Near the surface it approaches zero (negative infinity in the log) because the temperature of the ideal gas effectively goes to zero and thus does the sound speed. On the other hand, N^2 goes to zero at the center because it contains the factor g. It then increases to a maximum, tails off, and then drops to negative values starting at $r/\mathcal{R} \approx 0.83$. This last precipitous drop signals the onset of vigorous envelope convection.[11] You cannot see it in this figure but N^2 then rises to finite positive values as the photosphere is reached and convection turns off.

Leaving "f" and g_n aside for the moment, p_n is the nth p-mode with n nodes and the nodes (in $r/\mathcal{R}$) are indicted by the dots. As promised in (8.111), the highest order p-modes have the highest frequencies. (We left out p_{13} to p_6 for clarity.) Note that these modes only have nodes in the region to the right of the S_ℓ^2 curve so that this is where they "wiggle" and are propagating there, and this is consistent with what we discussed in the previous section. (We suppose, by the way, this is why a figure like this is called a "propagation diagram." You can find several more such diagrams in §15.4 of Unno et al., 1989. For highly evolved stars these diagrams can become quite complicated.) Outside that region the modes are evanescent. Thus the p-modes in this ZAMS sun model are primarily confined to the outer layers and, as mode order increases, more nodes are found nearer the surface. (And see Fig. 8.4 and discussion).

The g-modes are entirely different. As promised (by Eq. 8.112), their frequencies decrease (periods increase) as mode order increases. (Mathematically they approach a limit point at zero frequency as $n \to \infty$.) They are confined to the core in this model and only have nodes within the region bounded by the N^2 curve, which is again consistent with conclusions from (8.108). For these $\ell = 2$ modes, the shortest period is 71.6 minutes. Higher ℓ g-modes have shorter periods, since Π goes as $1/\ell(\ell+1)^{1/2}$ (as in 8.112).

Since this is a solar-like model, we ought to see solar oscillations somewhere. The latter have periods around five minutes and, if we look at our p_{14}, we find a calculated frequency of $\sigma = 1.386 \times 10^{-2}$ rad s^{-1} (2.2 mHz) or a period of $\Pi = 7.56$ minutes. We bet you that the sun is oscillating in this $\ell = 2$ p-mode even as we speak (and see Fig. 9.10).

[11] The base of the convection zone for the present-day sun is at $r/\mathcal{R} \approx 0.73$. Differences in composition, choice of mixing length, and evolutionary effects account for the ZAMS result.

Finally, Fig. 8.3 shows the location of the "f-mode," which we have not yet discussed. It has a frequency intermediate to those of the p- and g-modes, has no nodes, and is interpreted as a surface gravity wave that causes the whole structure to slosh in unison.

Just to confuse the issue, white dwarf oscillations are, in one important respect, entirely different. Their p-modes are confined to the core, whereas their g-modes are primarily envelope modes. As will be discussed in Chapter 10—and which should be obvious—N^2 and S_ℓ^2 look entirely different than they do in Fig. 8.3, and those frequencies are what determine where modes propagate.

The Eigenfunctions

Figure 8.4 shows the radial eigenfunctions for ($\ell = 2$) p_2 and g_2 modes of the ZAMS sun. The p_2 has two nodes (at the locations shown in Fig. 8.3) and the action is confined to the outer half of the model. On the other hand, the g_2 lives primarily in the deeper interior.[12]

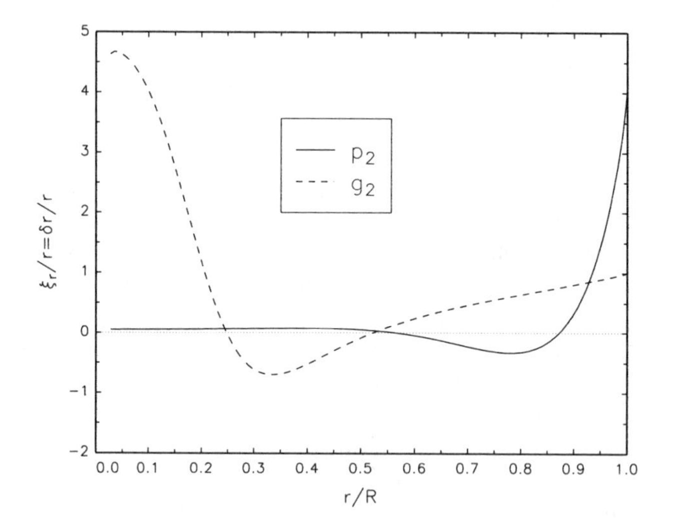

Fig. 8.4. Shown are the relative radial displacements, ξ_r/r, for $\ell = 2$ p_2 and g_2 modes as a function of radius for a ZAMS sun model. The p_2 mode eigenfunction has been normalized to $\xi_r/r = 4$ at the surface to make the nodes show up more clearly. These are from the same calculations used to make Fig. 8.3.

One way to interpret what is going on for the p_2 mode is to consider our short-wavelength result $\sigma_\mathrm{p}^2 \approx (k_r^2 + k_t^2)v_s^2$ of (8.109). Suppose a wave

[12] Solar g-modes have not been unambiguously observed yet and this is likely due to the deep character of these modes: they don't make it to the surface very well where they might be seen.

originates at the solar surface and travels inward. As it does so, it encounters an increase in sound speed because temperature increases inward. But since $\sigma_{\rm p}^2$ is a constant, this implies that $K^2 = k_r^2 + k_t^2$ must decrease. If we assume that k_r is the more rapidly varying term, then K decreases inward until $K = k_t = \sqrt{\ell(\ell+1)}/r_{\rm t}$ where $r_{\rm t}$ is the "turning radius" or "turning point." The reason for this terminology is that the vanishing of k_r means the wave can no longer progress radially inward but must start to move out again. The turning point is then established by the condition

$$\frac{v_s(r_{\rm t})}{r_{\rm t}} = \frac{2\pi f_{n\ell}}{[\ell(\ell+1)]^{1/2}} \tag{8.114}$$

where $f_{n\ell}$ is the frequency (in Hz) equal to $1/\Pi_{n\ell}$ for the nth mode of order ℓ. This is the same as the condition $\sigma_{\rm p}^2 = S_\ell^2(r_{\rm t})$. The picture is similar to that of refractive optics where ray paths are determined by the properties of the medium through which the light passes.

An approximate value for the turning radius for p-modes in our ZAMS sun (and for the real sun) that do not penetrate too deeply may be estimated with the following argument. The local sound speed for an ideal gas is $v_s = \sqrt{\Gamma_1 N_{\rm A} kT/\mu}$, where temperature as a function of radius is given by (7.124) for zero boundary conditions. (We use this despite the observation that the sun is convective just below the photosphere.) If Kramers' is the opacity source (so that $n_{\rm eff} = 3.25$), $\mu = 0.6$, and Γ_1 is 5/3, then, after little effort, (8.114) becomes

$$\frac{2.56 \times 10^8 f_{n\ell}^2}{\ell(\ell+1)} x_{\rm t}^3 + x_{\rm t} - 1 = 0 \tag{8.115}$$

where $x_{\rm t} = r_{\rm t}/\mathcal{R}$. For the $\rm p_2$, $\ell = 2$, mode of Fig. 8.4, the frequency is $f_{22} \approx 5.4 \times 10^{-4}$ Hz. Equation (8.115) then yields $r_{\rm t} \approx 0.37\mathcal{R}$. If you examine Fig. 8.3, this is the radius at which $\sigma^2 \approx S_\ell^2$ as should be expected: the wave becomes evanescent interior to that radius.

8.3.4 The Inverse Problem and Rotation

What we have discussed thus far for both radial and nonradial oscillations is usually referred to as the "forward problem." Namely, we calculate modes in a model and then compare them to observations. If the calculations and observations don't match, then there is a problem—and it is usually in the model. Theoretical opacities may be wrong or perhaps we have not quite got the evolutionary state correct. The trouble is that it is not always clear what has caused discrepancies or what to do about it. On the other hand, for the sun, and to a lesser extent some variable white dwarfs, there is a wealth of information contained in the oscillation spectrum with its many observed modes. These modes probe the interior to different depths depending on ℓ and n and whether they are p-modes or perhaps g-modes. The "inverse problem" consists of using these probes to either detect where things have gone astray in the model or to yield other information. We cannot treat this topic with anywhere near the detail it deserves and shall only give a hint to how it works and refer you to the next two chapters. For a fuller introduction from the textbook literature, see Unno et al. (1989, §41). The subject was originally developed by terrestrial seismologists, who have also contributed directly to the stellar analogue.

One way to look at the inverse problem is to consider the wave equation for adiabatic nonradial oscillations. We shall not derive it here (see, for example, Chandrasekhar, 1964; Lynden-Bell and Ostriker, 1967; Cox, 1980, §15.2), but it has the same form as (8.16) for adiabatic radial oscillations except we are dealing with vector displacements. In symbolic form it is really the force equation

$$\frac{\partial^2 \boldsymbol{\xi}}{\partial t^2} = -\mathcal{O}(\boldsymbol{\xi}) \tag{8.116}$$

where $\mathcal{O}$ is a linear operator acting on $\boldsymbol{\xi}(\mathbf{r}, t)$. Using the time factor $e^{i\sigma t}$, we find

$$\sigma^2 \boldsymbol{\xi} = \mathcal{O}(\boldsymbol{\xi}). \tag{8.117}$$

We now "dot" multiply each side of the above by $\boldsymbol{\xi}^*$ and integrate over mass and solid angle $[\Omega(\theta, \varphi)]$ to find

$$\sigma^2 = \frac{\int \boldsymbol{\xi}^* \bullet \mathcal{O}(\boldsymbol{\xi})\, d\mathcal{M}_r\, d\Omega}{\int \boldsymbol{\xi}^* \bullet \boldsymbol{\xi}\; d\mathcal{M}_r\, d\Omega} \,. \tag{8.118}$$

This is a variational expression for σ^2 analogous to (8.62). Thus if the eigenfunctions for the problem are known, then an appropriate integration over those eigenfunctions convolved with variables in the problem (in our case pressures, Lamb frequencies, etc.) yield back the eigenfrequencies. A specific formulation of this integral in terms of the variables we have used is given in Kawaler et al. (1985).

A useful property of this representation and its analogues in quantum mechanics is that the eigenfunctions may be changed by a considerable amount, while σ^2, as computed from (8.118), changes only slightly.[13] We can thus imagine the following thought experiment.

Suppose some quantity such as density is varied only slightly by an amount $\Delta\rho(r)$ through a model for which we already have an eigenfrequency, σ_0^2, and eigenfunction, $\boldsymbol{\xi}_0$. This variation in density will change the integrands in (8.118) by a small amount because the model has changed slightly. At first glance it would appear that to find the new eigenfrequency for the altered model we have to redo the complete pulsation problem. This is unnecessary, however, because of the stationary property of the integrals. All that needs

[13] The integral is "stationary" with respect to $\boldsymbol{\xi}$. Another way to phrase this is to realize that eigenvalues must be calculated very precisely in order to obtain accurate eigenfunctions.

to be done is to recast the integrand of the numerator of (8.118) in the form $K(\mathbf{X}_0, \boldsymbol{\xi}_0, r)\Delta\rho/\rho$ (following the nomenclature of Unno et al. 1989) where the function K is the first-order term in the linearization of the integrand with respect to $\Delta\rho$. The arguments of K are the untouched model parameters, $\mathbf{X}_0$, and the *original* eigenfunction $\boldsymbol{\xi}_0$. The relative change in the eigenvalue is then

$$\frac{\Delta\sigma^2}{\sigma_0^2} = \int_0^{\mathcal{M}} K(\mathbf{X}_0, \boldsymbol{\xi}_0, r) \, \frac{\Delta\rho}{\rho} \, d\mathcal{M}_r \, . \tag{8.119}$$

This procedure is essentially that followed in first-order perturbation theory in quantum mechanics.

If $\Delta\rho$ is specified in (8.119), then changes in the frequencies of various modes may be easily computed; that is, pick a model, calculate K, and find $\Delta\sigma^2$. But this is just the forward problem all over again. The inverse problem consists of computing σ_0^2 for many modes from a fiducial standard model (thus yielding K), calculating $\Delta\sigma^2 = \sigma_{\rm obs}^2 - \sigma_0^2$ as the difference between model and observed frequency $\sigma_{\rm obs}^2$, and, by some means, finding a best fit for what $\Delta\rho(r)$ must be in order to minimize $\Delta\sigma^2$ for many modes in, perhaps, a least squares sense. Here is where seismology comes in. If many modes are used and if they probe the star well, then we expect a good estimate for $\Delta\rho(r)$. The application of this inversion technique is both mathematically and computationally difficult but has yielded useful information about the run of the solar sound speed, for example.

Probing for Internal Rotation

Another application of inverse theory is deducing the internal rotation rate of stars. In the case of the sun, the equatorial rotation velocity of the solar surface is 2 km s^{-1}, which corresponds to a rotation period of about 25 days or an angular frequency of $\Omega = 2.9 \times 10^{-6}$ rad s^{-1}. The rotation is differential, however, in that the period increases as we move away from the equator to the poles. Before oscillations gave us a probe into the interior, the only evidence available for deducing the interior rotation properties came from observations of surface oblateness or small deviations in the advance of the perihelion of Mercury as calculated from Einstein's general theory of relativity. These are very indirect methods and depend sensitively on theoretical interpretation of the observations. Solar seismology is much more direct in this respect and relies on how pulsation modes are influenced by rotation.

The first thing to note is that rotation implies a preferred stellar axis (assuming that the axis implied by surface rotation is the same as that for the entire interior). This means that the frequency of an oscillation mode depends not only on the mode order, n, and ℓ, but also on m, which, in turn, means we must talk about $\sigma_{n\ell m}$. Note also that a typical angular frequency for a 5-minute mode in the sun is $\sigma \approx 0.02$ rad s^{-1}, a value much larger than Ω for the solar surface. The same probably holds true for variable white dwarfs, where g-mode periods are around 10 minutes, whereas rotation periods for white dwarfs with known periods are about an hour. We suspect, therefore, that the effect of rotation on pulsation is small and the $2\ell + 1$ frequencies $\sigma_{n\ell m}$ (with $m = -\ell, \cdots, 0, \cdots, \ell$) are close in value to those for no rotation ($\sigma_{n\ell \text{ any } m}$), which are degenerate with respect to m. This effect of removing degeneracy is often called "rotational splitting."

If rotation is "slow," as discussed above, then changes to the variational expression (8.118) are fairly straightforward and we will use it—eventually. Centrifugal forces can (probably) be neglected, which means that the star is still basically round. However, Coriolis forces, which go as $2\boldsymbol{\Omega}\times\mathbf{v}$ per unit mass, where $\mathbf{v}$ is the fluid velocity, cause deviations in the flow that were not accounted for in our earlier pulsation analysis. (The centrifugal force goes as the square of Ω, which is small in the first place.) We shall outline some of the steps to take for slow rotation but a more complete discussion may be found in Unno et al. (1989, §19.1 and §§31–34).

Suppose, as a simple case, the star were rotating uniformly. We then place ourselves in a noninertial reference frame corotating with the star. In that frame, the force equation is amended to take account of Coriolis forces by appending the term $-2\boldsymbol{\Omega}\times\mathbf{v}$ to the righthand side of the force equation (8.75). The velocity here is that caused by pulsation and, after linearization, becomes $d\boldsymbol{\xi}/dt$. This, in turn, causes terms of order $|\Omega\,\sigma\,\boldsymbol{\xi}|$ to appear in the righthand sides of the pulsation equations (8.89–8.91) after time has been separated out using $e^{i\sigma t}$. The resulting equations are very difficult to solve because there is no longer any guarantee that spherical harmonics will do the trick. But we assume the star rotates slowly in the sense that $\Omega << \sigma$. Thus the Coriolis term containing $\Omega\sigma$ is much smaller than the acceleration term containing σ^2 on the lefthand sides of (8.89–8.91) and the Coriolis force is only a perturbation to the solution for the nonrotating sun. This sounds like what we discussed for the inverse problem—namely, perturb the Hermitian operator (8.117) to account for small effects due to Coriolis forces (rather than small perturbations in density), use the eigenfunctions for the unperturbed problem, and then evaluate a few integrals. For uniform rotation this is straightforward (see Unno et al., 1979, §18, or Hansen, Cox, and Van Horn, 1977) and yields the following solution.

Let $\sigma_{n\ell m}$ be the eigenfrequency of a mode in the fictitious uniformly rotating star and $\sigma_{n\ell,0}$ be the eigenfrequency for the same mode not influenced by rotation (as computed, independent of m, by the methods outlined previously). The two are connected by the small perturbation in frequency $\Delta\sigma_{n\ell m}$ by

$$\sigma_{n\ell m} = \sigma_{n\ell,0} + \Delta\sigma_{n\ell m} \tag{8.120}$$

where, if we now measure eigenfrequencies in the external inertial frame of an observer,

$$\Delta\sigma_{n\ell m} = -m\Omega \left\{ 1 - \frac{\int_0^{\mathcal{M}} \left[2\xi_r\xi_t + \xi_t^2\right] d\mathcal{M}_r}{\int_0^{\mathcal{M}} \left[\xi_r^2 + \ell(\ell+1)\,\xi_t^2\right] d\mathcal{M}_r} \right\} . \tag{8.121}$$

Here the eigenfunctions $\xi_r(r)$ and $\xi_t(r)$ are those obtained from the nonrotating model (and depend only on n and ℓ) and thus the ratio of integrals, denoted by $C_{n\ell}$, depends only on n and ℓ. This solution has two parts: $-m\Omega$ and $m\Omega C_{n\ell}$. The first part comes about only because we are viewing the rotating star from an inertial frame where, for $m \neq 0$, we detect the additional red or blue Doppler shift due to the running wave in the azimuthal direction (Cox, 1984). That is, if we were to neglect Coriolis forces then the frequency of a mode as viewed in the rotating frame of the star would be $\sigma_{n\ell,0}$. But moving to the inertial frame means we see running azimuthal wave crests (from $e^{i\sigma t + im\varphi}$) either moving faster or slower with respect to us, depending on whether the waves are prograde ($m < 0$ and moving in the same sense as the rotation) or retrograde ($m > 0$ and moving in the opposite sense). The second term contains the effect of the Coriolis force.

In this example of uniform rotation it is easy to see that the modes are equally split in frequency with $\sigma_{n\ell\ m+1} - \sigma_{n\ell\ m}$ equal to $\sigma_{n\ell\ m} - \sigma_{n\ell\ m-1}$ and the degree of splitting (as given by these differences) is proportional to the rotation frequency Ω. We shall see that such a "picket fence" structure in frequency occurs in the oscillation spectra of some variable white dwarfs but the sun is more complicated. We know from the outset that it rotates differentially and not uniformly. Therefore the frequency splitting of solar p-modes is not necessarily uniform and thus should yield information on the rotation of the interior. This is again an inversion problem, where information is used from observations of many modes to, in effect, probe the Coriolis forces within the sun. Unfortunately the rotation frequency is actually a function of all three space variables and many degrees of freedom are implied (unlike inversion problems discussed earlier that assumed spherical symmetry). Yet, by assuming some reasonable constraints, considerable progress has been made.

Solid Stars?!

Finally, as a slightly off-beat topic, we mention what happens when astrophysicists get involved with terrestrial seismology as applied to stars. During the last stages of cooling of white dwarfs, their cores are thought to become crystalline. The surface layers of cooler neutron stars are also solid. Thus shear stresses and strains have to also be considered in a pulsation analysis. This may be standard fare for seismologists, but it is not easy. (Of course seismologists don't have to do nonadiabatic analyses, so we shouldn't feel too badly.) The equations for a nonradial adiabatic analysis in its full glory (still for round stars) is fourth-order in space. Put in a solid and you have to deal with a sixth-order system with some very nasty properties. For a summary of what kinds of modes may pop up in neutron stars (besides p- and g-modes) see, for example, McDermott et al. (1988). They find a total of seven different flavors of modes, all of which are found, in one form or the other, in the earth. Montgomery and Winget (1999) have looked into the effects of crystallization in the cores of white dwarfs using a simplified version of the full pulsation equations.

8.4 Exercises

Exercise 8.1. Derive Eq. 8.7. You may wish to consult Cox (1980, §5.3) for a discussion about how δs and time derivatives work with one another.

Exercise 8.2. Derive the Linear Adiabatic Wave Equation (8.16) using the hints given in the text.

Exercise 8.3. You may be beginning to think that we have something against the constant density model of §1.4 since we refer to it so often. Perhaps you're correct because it will now be forced to undergo adiabatic radial pulsations. This exercise will test the mathematical skills of some of you, but what we are about to embark on is standard fare for mathematicians and theoretical physicists. The aim is to find the first few eigenvalues and eigenfunctions for the oscillations. You may wish to consult §3.2 of

▹ Rosseland, S. 1964, *The Pulsation Theory of Variable Stars* (New York: Dover Publications)

which is a reprint of the original 1949 edition. And, yes, it is the same Svein Rosseland of the Rosseland opacity.

1. If Γ_1 is a constant, show that the following is equivalent to the LAWE of (8.16) with $\zeta(r) = \delta r / r$ in that equation:

$$\frac{d}{dr}\left[\frac{\Gamma_1 P}{r^2}\frac{d\left(r^2 \delta r\right)}{dr}\right] + \left(\sigma^2 + \frac{4g}{r}\right)\rho\,\delta r = 0\,.$$

2. Now introduce the potential Φ by setting

$$\delta r = \frac{d\Phi}{dr}\,.$$

Integrate over r (remembering that ρ is constant), let $x = r/\mathcal{R}$, put in the pressure $P(x)$ (from 1.41) and $g(r) = G\mathcal{M}_r/r^2$ for the constant density model, and show that the new wave equation is

$$\frac{1-x^2}{x^2}\frac{d}{dx}\left(x^2\frac{d\Phi}{dx}\right) + \frac{2}{\Gamma_1}(\omega^2 + 4)\phi = 0$$

where the dimensionless frequency is

$$\omega^2 = \sigma^2 \frac{\mathcal{R}^3}{G\mathcal{M}} \, .$$

To save space, let

$$\tilde{\omega}^2 = \frac{2}{\Gamma_1}(\omega^2 + 4) \, .$$

3. And now for something different. As with many second-order equations, including wave equations, it is advantageous to pose a series solution of the form

$$\Phi(x) = \sum_{\lambda=0}^{\infty} a_\lambda \, x^{k+\lambda}$$

where we assume that $a_0 \neq 0$. For a review of series solutions (in this case often called the method of Frobenius) see, for example, §8.5 of
   ▹ Arfken, G.B., & Weber, H.J. 1995, *Mathematical Methods for Physicists*, 4th ed. (New York: Academic Press).
4. Put this series into the new wave equation and show that

$$\sum_\lambda a_\lambda \, \lambda(\lambda+1) \, x^{\lambda-1} - \sum_\lambda a_\lambda \left[\lambda(\lambda+1) - \tilde{\omega}^2\right] x^{\lambda+1} = 0 \, .$$

5. Let, for example, $n = \lambda - 2$ in the first series and $n = \lambda$ in the second, and find the *recursion relation*

$$a_{n+2} = a_n \, \frac{n(n+1) - \tilde{\omega}^2}{(n+2)(n+3)} \, .$$

6. Show that the series diverges for large n; i.e., that $\lim_{n\to\infty} a_{n+2}/a_n \to 1$, which implies that the series diverges at the surface $x = r/\mathcal{R} = 1$. Thus conclude that the series must be terminated at some index m (or call it whatever) by the condition

$$m(m+1) = \tilde{\omega}^2$$

which means that, for a given m, only certain values of $\tilde{\omega}^2$ are allowed. Since we want solutions to be regular (finite) at the origin, argue that only even m are allowed.
7. Verify that the first three eigenvalues and eigenfunctions are those given in Table 8.1, where δr has been normalized so that $\delta r(\mathcal{R}) = 1$. You will have to show that only $k = 0$ in the original series is allowed. (Note that the solution for $n = 2$ is the fundamental and was given earlier by 8.20. This should come as no surprise.)
8. Finally, show that

$$\int_0^1 \frac{\delta r_2}{r} \frac{\delta r_4}{r} \, r^2 \, d\mathcal{M}_r = 0$$

as a part of showing the eigenfunctions are orthogonal (as in 8.18).

Table 8.1. Radial Eigenvalues and Eigenfunctions of the Constant Density Model

n	σ_n^2	$\delta r_n(x)$
2	$(G\mathcal{M}/\mathcal{R}^3)\,(3\Gamma_1 - 4)$	x
4	$(G\mathcal{M}/\mathcal{R}^3)2\,(5\Gamma_1 - 2)$	$x\left(-5 + 7x^2\right)/2$
6	$(G\mathcal{M}/\mathcal{R}^3)\,(21\Gamma_1 - 4)$	$x\left(35 - 126x^2 + 99x^4\right)/8$

Exercise 8.4. Try and reproduce the adiabatic radial eigenfunctions for the $n = 2$ (or an n of your choice) polytrope (i.e., Fig. 8.1). The discussion of numerics in §7.2 may be of some help. Can you write down the period–mean density relations from your calculations? By the way, one method of attacking this eigenvalue calculation is to treat it as an initial value problem. That is, make a guess at the eigenvalue, pick a normalization for ζ at the center (thus also picking $\delta P/P$ by way of 8.13), integrate out to the surface and test whether the boundary condition (8.14) at ξ_1 is satisfied. If not, change the eigenvalue by a little bit and do everything over again. You can set up an simple procedure to zero in on the eigenvalue this way.

Exercise 8.5. In an interesting short paper
▹ Fernie, J.D. 1995, AJ, 110, 2361
gives period–surface gravity ($g = G\mathcal{M}/\mathcal{R}^2$) relations that seem to hold up fairly well for *all* stars pulsating in the radial fundamental mode. As he points out, there is no real theoretical justification for his relations but they may prove useful in some circumstances. The one we chose (his Eq. 2) contains a term involving metallicity, which we will ignore because we assume our sample of stars all have the same Z. He finds that $g \propto \Pi^{(1.186\pm0.15)}$. See how well this holds up for Cepheid variables using the discussion leading to (8.72). If you went through that analysis, you will have all you need (and will find that Fernie's relation is not that surprising).

Exercise 8.6. Verify (8.115) for the turning point radius of p-modes in a solar-like star and compute $r_{\rm t}$ for the $\rm p_1$ $\ell = 2$ mode of Fig. 8.3. Check to see if $S_\ell^2 \approx \sigma^2$ at that radius.

Exercise 8.7. We have not talked that much about what nonradial eigenfunctions look like. So consider the following. How do radial and tangential displacements compare on the surface of a nonradially pulsating star? As a crude estimate use (8.105) for the surface boundary condition

$$\frac{\xi_t}{\xi_r} = \frac{g_s}{\sigma^2 \mathcal{R}}$$

to find ξ_t/ξ_r for: (1) a variable white dwarf g-mode of period 10 minutes; (2) a 5 minute p-mode for the sun. Assume the white dwarf has a typical mass of 0.6 $\mathcal{M}_\odot$ and you can use (3.68) to get $\mathcal{R}$. You will find that the displacement ratios are very different. You should discuss your results; i.e, why are the results so different? (Note that what you really want in order to compare to observation is the ratio averaged over the spherical surface and an oscillation cycle. This introduces a factor of $\sqrt{\ell(\ell+1)}$, as discussed in §4.3.2 of Christensen–Dalsgaard, 1997.)

Exercise 8.8. Just so we see nonradial g-modes in another, and more familiar, form, consider water waves. Suppose we have a wave traveling in the x-direction along the surface over a bottomless sea with z, the vertical coordinate, starting at zero at the water's surface. To a very good approximation, water is imcompressible, so assume that density is constant with depth and time.

1. If we have already decomposed displacements by $\exp(i\sigma t)$, then show that (8.82) implies $\nabla \bullet \boldsymbol{\xi}(x, y) = 0$, where $\boldsymbol{\xi}$ has components ξ_x and ξ_z. Since we take gravity, g, constant (and $\Phi' = 0$), take the divergence of (8.80) and show that $\nabla^2 P' = 0$.
2. Now try the solution $P' = w(z)\cos(kx)$ where $w(z)$ is to be found. Coupled with $\exp(i\sigma t)$ this describes a running wave. From what you have thus far, you ought to be able to show that $d^2w/dz^2 = k^2 w$ with solution $w(z) \propto \exp(-kz)$ if you exclude the solution that blows up with depth.
3. At the free surface the pressure perturbation, δP, should vanish, so use (8.79) with ρ replaced by P, work through the algebra, and show that $\sigma^2 = gk$. (Realize that equilibrium means $dP/dz = g\rho$.) OK, what this means is that short-wavelength waves (with $\lambda = 1/k$) have high frequencies, whereas long wavelengths are leisurely in how they heave up and down. (You knew this already!)
4. As an example, tsunamis (seismic sea waves, and don't call them "tidal waves") have wavelengths around 500 km (or more). What is their period of oscillation (in minutes)? (This is a bit of a fraud because tsunami wavelengths are so long that ocean depth has an important effect.) We call these g-modes because gravity is the restoring force. What is the velocity of the wave? Sketch out the waveform on the surface as a function of time.

8.5 References and Suggested Readings

Introductory Remarks

The most useful textbook references to the theory of variable stars are

▷ Cox, J.P. 1980, *Theory of Stellar Pulsation* (Princeton: Princeton University Press)

▷ Unno, W., Osaki, Y., Ando, H., & Shibahashi, H. 1979, *Nonradial Oscillations of Stars* (Tokyo: University of Tokyo Press)

and

▹ Unno, W., Osaki, Y., Ando, H., Saio, H., & Shibahashi, H. 1989, *Nonradial Oscillations of Stars*, 2d ed. (Tokyo: University of Tokyo Press).

The last chapter of

▹ Cox, J.P. 1968, *Principles of Stellar Structure*, in two volumes (New York: Gordon and Breach)

may also be consulted. A much more difficult, but classic, reference is

▹ Ledoux, P., & Walraven, Th. 1958, in *Handbuch der Physik*, ed. S. Flügge (Berlin: Springer-Verlag) Vol. 51, 353.

Be advised, however, because this work reverses Lagrangian (δs) and Eulerian ('s) operators compared to what is now the usual notation. The combination monograph and lecture note work by

▹ Christensen–Dalsgaard, J. 1997, *Stellar Oscillations*, 4th ed. (or later)

from Aarhus, Denmark, is very useful, if you can find it. He has a lot of material on solar oscillations. It might still be available at

`www.obs.aau.dk/~jcd/oscilnotes/`.

He also has an evolution program (for UNIX platforms) available on that site.

Review articles include

▹ Gautschy, A., & Saio, H. 1995, ARA&A, 33, 75

▹ Gautschy, A., & Saio, H. 1996, ARA&A, 34, 551

and

▹ Brown, T.M., & Gilliland, R.L. 1994, ARA&A, 32, 37.

§**10.1**: Adiabatic Radial Pulsations

Stellar pulsation theory can get very mathematical at times. A decent text, among a few of its kind, that will answer many of the questions we raise in this chapter is

▹ Arfken, G.B., & Weber, H.J. 1995, *Mathematical Methods for Physicists*, 4th ed. (New York: Academic Press).

Part VII of

▹ Kippenhahn, R., & Weigert, A. 1990, *Stellar Structure and Evolution* (Berlin: Springer-Verlag)

discusses some of the material in our chapter but does not go into any detail about helio- or white dwarf seismology. Their §38.3 gives examples of polytropic oscillations.

▹ Tassoul, M., & Tassoul, J.L. 1968, ApJ, 153, 127

discuss reduction of the LAWE to an equation that looks more like a wave equation. Their asymptotic results are much more complete than ours.

§10.2: Nonadiabatic Radial Pulsations

Although Eddington did not figure out exactly what physical processes were responsible for variable stars he did understand what the thermodynamics had to do. See §134 of

▷ Eddington, A.S. 1926, *The Internal Constitution of the Stars* (Cambridge: Cambridge University Press).

The analysis we quote is also discussed in §5–10 of

▷ Clayton D.D. 1968, *Principles of Stellar Evolution and Nucleosynthesis* (New York: McGraw-Hill).

Numerical investigation of weight functions for radial pulsations was originally reported by

▷ Epstein, I. 1950, ApJ, 112, 6.

▷ Iben, I., Jr. 2000, in *Variable stars as Essential Astrophysical Tools*, C. Ibanoğlu ed. (Dordrecht: Kluwer Academic Publishers), p. 437

gives many handy items about evolutionary and pulsation studies of Cepheids. The result we quote for the Cepheid period–radius relation is from

▷ Gieren, W.P., Moffett, T.J., & Barnes, T.G. III 1999, ApJ, 512, 553.

The tough and important problem of coupling pulsation to convection has not really improved since the review by

▷ Toomre, J. 1982, in *Pulsations in Classical and Cataclysmic Variable Stars*, eds. J.P. Cox & C.J. Hansen (JILA publication; Boulder, CO), p. 170.

One-dimensional hydrodynamic calculations of radially variable stars have been done since the early 1960s. Recently, however, these have been extended to look into questions of chaos, etc. Reviews and examples may be found in

▷ Buchler, J.R. 1990, *Nonlinear Astrophysical Fluid Dynamics*, Vol. 117 of *Annals of the New York Academy of Sciences*, p. 17

▷ Kovács, G., & Buchler, J.R. 1988, ApJ, 334, 971

▷ Ibid. 1988, ApJ, 324, 1026

▷ Kolláth, Z., Buchler, J.R., Szabó, R., & Csubry, Z. 2002, A&A, 385, 932

▷ Goupil, M.-J., & Buchler, J.R. 1994, A&A, 291, 481.

§10.3: An Introduction to Nonradial Oscillations

There are several good texts on fluid dynamics. One of our favorites is

▷ Landau, L.D., & Lifshitz, E.M. 1959, *Fluid Mechanics* (London: Pergamon).

Our main references for nonradial oscillations use Eulerian perturbations.

▷ Pesnell, W.D. 1990, ApJ, 363, 227

discusses the use of a Lagrangian formalism which, under some circumstances, is numerically superior when nonadiabatic calculations are being done.

The Cowling approximation is more than a pedagogical tool. It reduces computational labor without introducing gross errors in many circumstances. See

▷ Cowling, T.G. 1941, MNRAS, 101, 367.

The choice of phase for the spherical harmonics is liable to lead to confusion. We choose that of

▷ Jackson, J.D. 1999, *Classical Electrodynamics*, 3rd ed. (New York: John wiley & Sons).

You might be able to get a better idea of what nonradial motions look like by consulting

▷ Smeyers, P. 1967, BullSocRoySci Liège, 36, 35.

Asymptotic methods for nonradial modes can involve some difficult mathematics. Besides Unno et al., you might also try

▷ Tassoul, M. 1980, ApJS, 43, 469.

▷ Dziembowski, W. 1971, AcA, 21, 289

is one of a series of pioneering papers by Dziembowski. Many of us still use the variables he introduced. See, for example,

▷ Osaki, Y., & Hansen, C.J. 1973, ApJ, 185, 277.

Derivations of the nonradial wave equation, among other important items, may be found in

▷ Chandrasekhar, S. 1964, ApJ, 139, 644

▷ Lynden-Bell, D., & Ostriker, J.P. 1967, MNRAS, 136, 293.

One version of an integral formulation of the equation is given by

▷ Kawaler, S.D., Hansen, C.J., & Winget, D.E. 1985, ApJ, 295, 547.

The effects of both uniform and cylindrically symmetric slow rotation on nonradial frequencies is discussed in

▷ Hansen, C.J., Cox, J.P., & Van Horn, H.M. 1977, ApJ, 217, 151

and, for an easy introduction to rotational splitting, see

▷ Cox, J.P. 1984, PASP, 96, 577.

▷ Hansen, C.J., & Van Horn, H.M. 1979, ApJ, 233, 253

▷ McDermott, P.N., Van Horn, H.M., & Hansen, C.J. 1988, ApJ, 325, 725

▷ Montgomery, M.H., & Winget, D.E. 1999, ApJ, 526, 976

discuss the effects of solid material on nonradial oscillations in white dwarfs and neutron stars.

9 Structure and Evolution of the Sun

"Here comes the sun.
It's alright.
Sun, sun, sun, here it comes."
— George Harrison, Here Comes the Sun (1968)

On the other hand,

"I hate the beach. I hate the sun.
I'm pale and I'm red-headed.
I don't tan—I stroke."
— Woody Allen, Play It Again Sam (1972)

"These particles are so elusive that you do not notice
the hundred billion solar neutrinos that pass
through your thumbnail every second."
— John N. Bahcall (2001)

Perhaps the first astronomical object that we become aware of as children is our sun. Indeed, the sun is *the* prototype star, and before we can claim to understand the stars, we must claim some mastery of current ideas about our own sun's origin, its internal structure, and how it has evolved to this state. In this chapter, we shall emphasize not only what strides have been made towards understanding this star, but also what uncertainties remain.

Our proximity to the sun and the level of detail visible from our viewpoint on earth means that we see far more than we completely understand at present. Observations over the past century have revealed a rich variety of surface phenomena associated with magnetic fields and their almost cyclical behavior; magnetic fields, like convection, are difficult to include in stellar models. The general pattern of surface rotation has the equator rotating faster than higher latitude regions; interactions between this differential rotation, modulations in the magnetic field, and subsurface convection, remain particularly thorny subjects and active areas of research. Neutrinos have been observed emanating from the deep interior but at a rate less than that predicted from standard models (although, as we shall see, the models may be fine but our ideas about neutrinos might have to be modified). The sun is a variable star—albeit variable on only a low-amplitude scale—and we have to see what this can tell us.

So while the sun is our best observed star, the uncertainties and phenomena listed above, which we do not fully understand, are multiplied many-fold when extrapolated to other far more distant objects. In light of this we shall discuss some of these issues here.

9.1 Vital Statistics of the Sun

The mass of the sun is determined from measurements of the dynamics of planets and natural or artificial satellites in the solar system and it is known far more accurately than for any other stellar object, with the possible exception of some binary pulsars. A currently accepted figure is $\mathcal{M}_\odot = (1.9891 \pm 0.0004) \times 10^{33}$ g. The solar luminosity is known to almost the same precision although variations of up to 0.5% have been reported. These may be due to variations during its magnetic cycle or other short-term effects such as solar flares and the like. A more probable variation due to the solar magnetic cycle is only about 0.07% around a mean value for the solar flux of $(1.368 \pm 0.001) \times 10^6$ erg cm^{-2} s^{-1} at one Astronomical Unit (Willson et al., 1986, and see Newkirk, 1983). This yields $\mathcal{L}_\odot = (3.847 \pm 0.003) \times 10^{33}$ erg s^{-1}. Detection of a secular change in luminosity due to evolution is not possible at this time. The solar radius appears to be as stable in size as the luminosity and a value of 6.96×10^{10} cm is currently accepted for the radius at the optical photosphere.

The composition of the sun is not directly observable except at the photosphere, and even there it requires theoretical interpretation of spectral features or solar wind abundances with attendant uncertainties. If no mixing has occurred to change the surface composition of the sun during its evolution, then it is the same as that of the material from which the sun was formed in the first place (with the possible exception of some very reactive nuclei such as deuterium and lithium). It appears that the current ratio of the mass fraction of heavy elements (Z, in the nomenclature of Chap. 1) to hydrogen (X) is $Z/X = 0.0245$–0.0277 to quote from Grevesse and Noels (1993) (the lower figure) and Grevesse (1984) (the upper figure).

Not so well determined are the individual values of either X or Y (the helium mass fraction) but it is known that $Y \approx 0.25$. One way to establish the value of Y relies on constructing evolutionary sequences for the sun and then matching the present-day luminosity and radius to an estimated solar age (see below). Given Z/X, the value of Y that gives the best match is then the adopted mass fraction of helium. Note, however, that the validity of this procedure is no better than the input physics characterizing the opacity, nuclear reaction rates, and other processes such as convection. Another important complication is diffusion. As we will see in the extreme case of white dwarfs in the next chapter, under the influence of "strong" gravity, heavier elements in a stable stellar environment can sink, and lighter elements float upward. Over the 4.6-billion- year lifetime of the sun, a small but important fraction of helium has drained out of the upper layers of the sun. When taking this diffusion of helium into account, the initial helium abundance in the ZAMS sun was a few percent higher than the current abundances.

Granting these caveats, Bahcall et al. (2001), for example, find a "primordial" value for the helium composition of the sun of $Y = 0.2656$ using $Z/X = 0.0245$ in models that do not include diffusion (and $Y = 0.2735$ with an initial value of $Z/X = 0.0266$ in models that do). Since $X + Y + Z = 1$, this yields $X = 0.7169$ and $Z = 0.01757$ without diffusion ($X = 0.7077$ and $Z = 0.0188$ with diffusion), where the number of significant digits quoted will most probably turn out to be illusory.

The individual "mix" of heavy elements is also important because of its effect on opacities and, to some extent, the operation of the CNO cycles (see Chap. 6). Given in Table 9.1 are the Grevesse and Noels (1993) fractional abundances of the most abundant heavy elements, where Z_i is the nuclear charge for an element and n_i is the number density (in cm^{-3}) normalized to a total of unity ($\sum_i n_i = 1$). (Note that these are not isotopic abundances.) We see that carbon and oxygen are, by far, the most abundant elements at the sun's surface after hydrogen and helium. This table formed the basis for our earlier Fig. 1.2.

Table 9.1. Solar Heavy Element Abundances

	Z_i	n_i		Z_i	n_i		Z_i	n_i
C	6	0.24552	Al	13	0.00204	Ca	20	0.00159
N	7	0.06458	Si	14	0.02455	Ti	22	0.00008
O	8	0.51297	P	15	0.00020	Cr	24	0.00033
Ne	10	0.08321	S	16	0.01122	Mn	25	0.00017
Na	11	0.00148	Cl	17	0.00022	Fe	26	0.02188
Mg	12	0.02631	Ar	18	0.00229	Ni	28	0.00129

The age of the sun can be found in a way that is independent of the theory of stellar evolution via radiometric dating of terrestrial rocks (plus some geological estimates of melting and cooling times), lunar material, and meteorites. The time of condensation of solar matter is placed at somewhat less than five billion years ago. If the sun was formed on the main sequence before the planets and other material, then a best estimate for the present age of the sun is $(4.57 \pm 0.05) \times 10^9$ years. Stricter error bars are quoted by some—as in Guenther (1989) and Bahcall and Pinsonneault (1995)—but, at least at present, a possible error of several times 10^7 years is not at all serious for stellar evolution studies.

The significance of the sun for stellar structure and evolution is clear: it is the primary proving ground because we know its mass, luminosity, radius, effective temperature, initial composition, and age. If model-building procedures fail to reproduce these solar properties then all other studies using these procedures are in grave danger of being just plain wrong.

9.2 From the ZAMS to the Present

This section will review the structure and evolution of the sun from the zero-age main sequence (ZAMS) to the present in light of material developed in earlier chapters. After this is completed we shall go on to discuss how well our models compare to the real sun when it is looked at closely.

9.2.1 The Sun on the ZAMS

Constructing a consistent homogeneous ZAMS model for the sun is like doing stellar evolution in reverse; the model must be such that when evolved forward in time it yields the present-day sun at its present age. The "givens" for our ZAMS sun are its mass and composition (see §§2.2 and 7.1). Rotation and magnetic fields are assumed to have no effect on the evolution. Furthermore, mass loss is almost universally neglected for the main sequence stage of solar evolution because the present mass loss rate of $\dot{\mathcal{M}} \approx 10^{-14}\ \mathcal{M}_\odot\ \mathrm{yr}^{-1}$ (Cassinelli and MacGregor, 1986) amounts to a small fraction of the total mass even when integrated over the main sequence lifetime of 10^{10} years. If these effects are ignored, then current practice recognizes the following parameters which, aside from some practical limits, are varied in solar ZAMS modeling until 4.6×10^9 years of evolution yields a model matching the radius and luminosity of the present–day sun. These are—

- The present (and initial) helium content of the surface layers is not precisely determined by observation but it must be around $Y \approx 0.25$. The exact value (as reflecting the original solar content) affects the structure through the mean molecular weight of the mixture, the opacity, and the behavior of convection zones. If, as discussed previously, the metal content relative to hydrogen is fixed by observation, then Y is varied as a parameter affecting the overall composition.
- The method of treating convective transport must be decided at the outset. As we discussed in Chapter 5, the majority of model builders choose some version of the mixing length theory for this purpose because of computational simplicity. The results to be reported here do this also. There are, however, undetermined parameters in any version of the MLT. Foremost among these is the mixing length, ℓ. If the pressure scale height, λ_P, is taken as a measure of the mixing length, then the mixing length parameter, $\alpha = \ell/\lambda_P$, is the usual free parameter. From experience with both general and solar modeling, α should be close to unity. Note, however, that a single value of α is used everywhere and for all time. There is no guarantee that this is reasonable even in the context of the MLT.

The procedure is then to choose various combinations of Y (or Y_{init}) and α in different ZAMS models, evolve them to the solar age, and settle on the combination that gives the correct radius and luminosity.

The results quoted here are either abstracted from the Yale models of Guenther et al. (1989, 1992) or from calculations using similar codes and input physics. Of course, models that include the latest OPAL opacities, more sophisticated equations of state, and diffusion of helium (e.g.,. Bahcall et al., 2001) produce more accurate oscillation frequencies—but for the purposes of illustrating the general evolution of internal quantities, the simpler Yale models are more than adequate. These models are representative of "standard models," which use just the sort of physics and techniques outlined in this text and do not include the gravitational settling of helium. For example, the Yale models use the Anders and Grevesse (1989) mix of metals with Los Alamos opacities and auxiliary tables (§4.6) and an Eddington gray atmosphere (§4.3). The latter is computationally efficient and perfectly adequate for general studies although, for some purposes (e.g., solar oscillations), a real atmosphere should be used. To reproduce the present-day sun, a helium mass fraction of $Y = 0.288$ and mixing length to pressure scale height of $\alpha = 1.2$ are required under these adopted input physics values. The version of the MLT used in these calculations is that reviewed in our §5.1 (but with radiative leakage, of course) and, more fully, in Cox (1968, Chap. 14). An elapsed age of 4.5×10^9 years for the present-day sun is assumed.

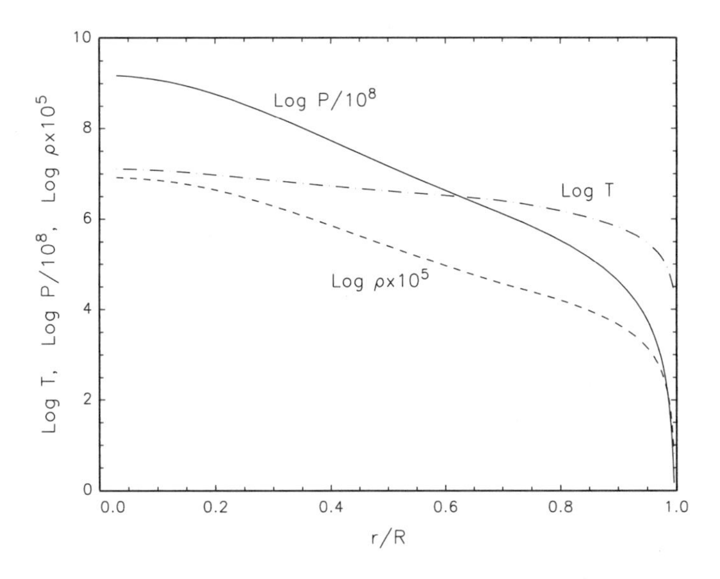

Fig. 9.1. Shown are the runs of pressure, temperature, and density for a model of the zero-age sun. Note that the pressure has been multiplied by 10^{-8} and the density by 10^5. The abscissa is the relative radius $r/\mathcal{R}$ where $\mathcal{R} = 0.886\,\mathcal{R}_\odot$.

Figures 9.1 and 9.2 show some results for the ZAMS sun. The runs of pressure, temperature, and density versus radius in Fig. 9.1 are smooth and show the rapid decrease in these variables to the surface. We have already remarked (in §7.2.7) that the Eddington standard model reproduces the behavior of these variables through most of the model remarkably well given, say, the average density.

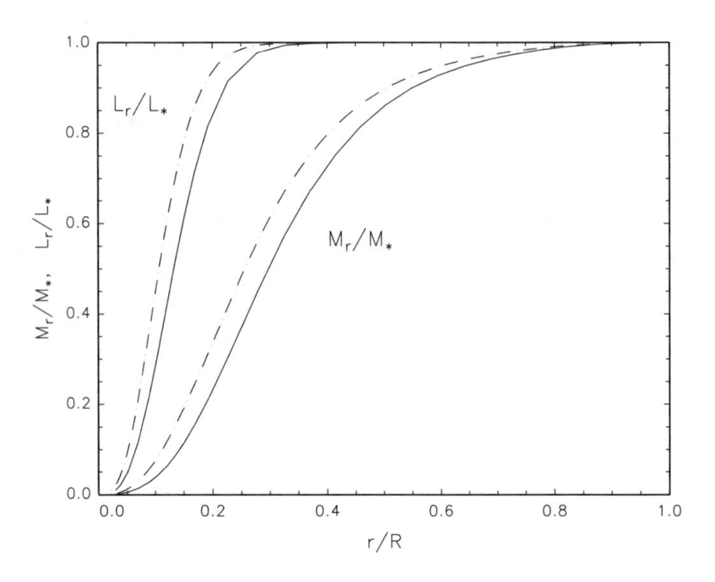

Fig. 9.2. Relative luminosity, $\mathcal{L}_r/\mathcal{L}_\star$, and mass, $\mathcal{M}_r/\mathcal{M}_\star$, are plotted versus relative radius for the ZAMS (solid lines) and present-day sun (dashed lines). Total radius and mass for the ZAMS model are $\mathcal{R} = 0.886\,\mathcal{R}_\odot$ and $\mathcal{L} = 0.725\,\mathcal{L}_\odot$.

The solid lines in Fig. 9.2 illustrate the run of relative luminosity $\mathcal{L}_r/\mathcal{L}$ (with $\mathcal{L} = 0.725\,\mathcal{L}_\odot$) and $\mathcal{M}_r/\mathcal{M}_\odot$ versus relative radius for the ZAMS sun. (The dashed lines refer to the present-day sun.) The total radius is $\mathcal{R} = 0.886\,\mathcal{R}_\odot$. We note immediately that the sun on the ZAMS, at 4.5×10^9 years in the past, was some 12% smaller and 25% less luminous than it is now. The luminosity shows the characteristic rapid rise at small radius and this is associated with the temperature sensitivity of the nuclear reaction rate. The proton-proton chains dominate over the CNO cycles and, as you may easily verify from our estimates for the reaction rates in (6.76) and (6.77), $\varepsilon_{\text{CNO}}/\varepsilon$ is less than 1% at model center. Radiative diffusion transports all the energy flux out to a relative radius of $r/\mathcal{R} \approx 0.73$. Thereafter, efficient convection takes over to levels just below the visible surface (as will be illustrated more fully for the present-day sun). Although this means that convection is most important for some 30% of the total radius, the corresponding figure in mass is only about 3% because of the greatly lower densities in the outer layers. Note also that $\mathcal{M}_r$ and $\mathcal{L}_r$ rise more rapidly in the present-day sun compared to the ZAMS. This is because the deep interior of the sun has contracted as evolution has gone on. And, as emphasized in Chapter 2, this core contraction has been balanced, so to speak, by expansion of the outer layers.

How does this structure change as nuclear transmutations take place and the sun evolves from the zero-age main sequence?

9.2.2 Evolution From the ZAMS

The most obvious changes in the sun as it evolves from the ZAMS to its present state are increases in both radius and luminosity. This is a characteristic of all standard calculations. To estimate the magnitude of the increase in luminosity, for example, we shall use simple dimensional arguments plus the fact that hydrogen burning to helium means an increase in mean molecular weight. In doing so, we follow the discussion of Endal (1981). Almost all of what we need can be found in the first chapter.

Recall the virial theorem analysis (§1.3.4), which yielded the relation

$$T \propto \mu \mathcal{M}^{2/3} \rho^{1/3} \tag{9.1}$$

for an ideal gas star in hydrostatic equilibrium. (Radiation pressure contributes much less than 1% to the solar pressure even at the center and may be safely ignored.) Here, as usual, μ is the mean molecular weight. If, furthermore, we assume that radiative diffusion controls the energy flow, then

$$\mathcal{L} \propto \frac{\mathcal{R}T^4}{\kappa \rho} \tag{9.2}$$

using (1.60). If Kramers' is the dominant opacity with $\kappa = \kappa_0 \rho T^{-3.5}$, then elimination of T and application of the mass equation $\mathcal{R} \propto (\mathcal{M}/\rho)^{1/3}$ yields

$$\mathcal{L} \propto \frac{\mathcal{M}^{5.33} \rho^{0.117} \mu^{7.5}}{\kappa_0} . \tag{9.3}$$

This expression can be used to calculate an estimate for the change in luminosity with time as composition changes.

Because of the small exponent of density (and, from knowing beforehand how relatively little the radius will change) we neglect the term $\rho^{0.117}$. From estimates of κ_0 (as in §4.4) we know that κ_0 does not vary strongly with either X or Y and so we neglect it also. And, of course, we keep mass fixed so it will not appear. Equation (9.3) is then rewritten in time-dependent form relating changes in $\mathcal{L}$ and μ from time $t = 0$ to some arbitrary time t—

$$\frac{\mathcal{L}(t)}{\mathcal{L}(0)} = \left[\frac{\mu(t)}{\mu(0)} \right]^{7.5} . \tag{9.4}$$

The task is now to find how μ varies with time.

If the bulk of the stellar interior is assumed to be completely ionized and the metal content is small compared to hydrogen and helium, then (1.55) is appropriate with

$$\mu(t) = \frac{4}{3 + 5X(t)} \tag{9.5}$$

where we explicitly indicate the time dependence in $X(t)$. But we know something about how $X(t)$ changes with time. Because hydrogen burning releases approximately $\mathrm{Q} = 6 \times 10^{18}$ ergs for every gram of hydrogen converted to helium, the instantaneous rate of change of a spatially averaged X is

$$\frac{dX(t)}{dt} = -\frac{\mathcal{L}(t)}{\mathcal{M}\mathrm{Q}} . \tag{9.6}$$

Taking the time derivative of μ then yields

$$\frac{d\mu(t)}{dt} = -\frac{5}{4}\mu^2(t)\frac{dX}{dt} = \frac{5}{4}\mu^2(t)\frac{\mathcal{L}(t)}{\mathcal{M}\mathrm{Q}} . \tag{9.7}$$

We now differentiate (9.4), substitute (9.7), get rid of $\mu(t)$ using (9.4), and find

$$\frac{d\mathcal{L}(t)}{dt} = \frac{75}{8}\frac{\mu(0)}{\mathcal{M}\mathrm{Q}}\frac{\mathcal{L}^{1+17/15}(t)}{\mathcal{L}^{-1+17/15}(0)} \tag{9.8}$$

with solution

$$\mathcal{L}(t) = \mathcal{L}(0)\left[1 - \frac{85}{8}\frac{\mu(0)\mathcal{L}(0)}{\mathcal{M}\mathrm{Q}}\,t\right]^{-15/17} . \tag{9.9}$$

Putting in numbers by expressing luminosities in units of $\mathcal{L}_\odot$, introducing the present solar age $t_\odot$ of 4.6×10^9 years, and letting $\mu(0) \approx 0.6$, then gives

$$\frac{\mathcal{L}(t)}{\mathcal{L}_\odot} \approx \frac{\mathcal{L}(0)}{\mathcal{L}_\odot}\left[1 - 0.3\,\frac{\mathcal{L}(0)}{\mathcal{L}_\odot}\frac{t}{t_\odot}\right]^{-15/17} . \tag{9.10}$$

If $t = t_\odot$ (i.e., $\mathcal{L}(t) = \mathcal{L}_\odot$), then the luminosity on the ZAMS must have been $\mathcal{L}(0) \approx 0.79\,\mathcal{L}_\odot$ from the solution of (9.10). The models we have quoted give an agreeably close value of 0.73.

The above result is interesting from not only our stellar evolution perspective, but it bears on how life must have evolved on earth. The earliest microorganisms appear in the fossil record about 3.5×10^9 years ago (or a little later). At that time, by application of the above with an adjustment given by the evolutionary models, the sun was nearly 25% less luminous than now. Because descendants of some of those same microorganisms are alive today in essentially unchanged form, there must be some explanation for how the earlier life forms could have survived and propagated with a significantly lower solar constant. The answer probably lies in the evolution of the earth's atmosphere which, as fascinating as that topic may be, we shall have to pass by.

Another way to look at the above is to consider what happens as the mean molecular weight increases with time in the hydrogen-burning core of a hydrostatic star. If $P \propto \rho T/\mu$, then an increase in μ without a corresponding increase in the product ρT would be accompanied by a decrease in pressure. But since the core must still support the unchanged mass layers above it, this situation would lead to an imbalance of forces and hydrostatic equilibrium would be impossible to maintain. The result would then be a compression of the core with a corresponding increase in density. This process would take place very rapidly compared to nuclear time scales because we know that the dynamic readjustment time of §1.3.3, t_{dyn}, is only about an hour for the sun. The conclusion is that ρT must increase. In particular, ρ must increase but so should T by virtue of the virial result (9.1). An increase in T then implies an increase in the energy generation rate (to the fourth power of T for the proton–proton reaction) and thus the overall rate of power output increases also.

9.2.3 The Present-Day Sun

Figure 9.3 shows the evolutionary track of a model of the sun on an HR diagram from the ZAMS to the point where it is clearly a red giant. Elapsed time is indicated by the labeled circles in Gyr (10^9 years). The present-day sun, even after some 4.6 Gyr of evolution, is still very close in $\mathcal{L}$, $\mathcal{R}$, and T_{eff} to its original ZAMS position. The inner core, however, has changed substantially.

Figure 9.2 showed the current run of relative luminosity and mass as a function of relative radius (as dotted lines). The figure shows that $\mathcal{L}_r/\mathcal{L}_\odot$ and $\mathcal{M}_r/\mathcal{M}_\odot$ for the present-day sun rise more steeply than for the ZAMS sun. In the case of the luminosity, the reason for this is that increased central temperatures have intensified the energy production and, from the energy equation (1.57), the luminosity gradient must steepen accordingly. (The contribution from gravitational energy sources due to contraction is always very small during the initial stages of evolution off the main sequence.) Contraction and the implied increase in density account for the steeper gradient in mass.

The run of pressure, temperature, and density versus relative radius at an elapsed age of 4.6 Gyr is very similar to that of Fig. 9.1 for the ZAMS, except that the central values are now increased to $T_c = 1.53\times10^7$ K, $P_c = 2.26\times10^{17}$ dyne cm^{-2}, and $\rho_c = 146$ g cm^{-3} as a consequence of the contraction of the inner regions. The burning of hydrogen to helium in the still radiative core of the sun has depleted the former, and this is shown in Fig. 9.4. No longer are the mass fraction profiles, $X(r)$ and $Y(r)$, flat—as was initially assumed for the ZAMS—and the central helium mass fraction Y_c has increased by about a factor of two, which is made up for by a corresponding decrease in X_c. Also shown in the figure is the ratio of the energy generation rate $\varepsilon(r)$ to its central value $\varepsilon_c = 16.2$ erg g^{-1} s^{-1} (an increase of 17% from the ZAMS).

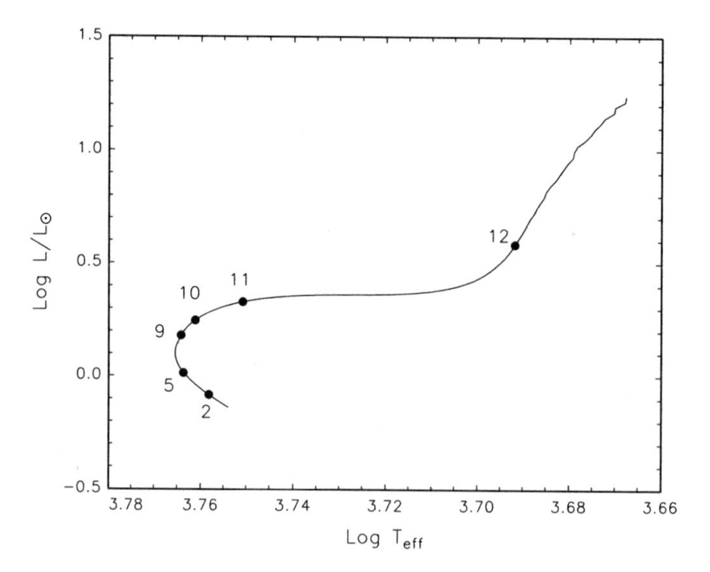

Fig. 9.3. A solar model evolution track in the Hertzsprung–Russell diagram from the ZAMS to the red giant stage. Elapsed evolutionary time from the ZAMS is indicated by the filled circles where the units are Gyr.

The contribution from the CNO cycles is now 7% and will continue to rise as central temperatures get hotter because of the high temperature sensitivity of the CNO cycles.

The convection zone of the sun is moderately extensive (but not compared to lower-mass main sequence stars) and occupies the outer 30% of the radius (but only 2% of the mass). At these high levels in the star the total luminosity is constant because little or no energy is being generated there. Thus Fig. 9.5, which shows the run of $\mathcal{L}_r/\mathcal{L}$ and convective to total luminosity ($\mathcal{L}_{r,\text{conv}}/\mathcal{L}_{\text{tot}}$), indicates that a major fraction of the solar energy flux is carried by convection before the luminous power is finally radiated to space at the photosphere. The detailed model results show that the convection is nearly adiabatic through almost all of the convection zone except for the very bottom and top of the zone. Phrased another way, the gradients ∇ and ∇_{ad} of Chapter 5 are very nearly equal (and see Fig. 5.2).

Another way to show the extent of the convection zone is to examine the square of the Brunt-Väisälä frequency

$$N^2 = g\left[\frac{1}{\Gamma_1 P}\frac{dP}{dr} - \frac{1}{\rho}\frac{d\rho}{dr}\right] = -\frac{\chi_T}{\chi_\rho}\left(\nabla - \nabla_{\text{ad}}\right)\frac{g}{\lambda_{\text{P}}} \tag{9.11}$$

which was originally given as (5.29). Remember that in a radiative zone N is the frequency of oscillation for a fluid blob that has been displaced from its equilibrium position when buoyancy is the restoring force (§5.1). In convectively unstable regions, where $\nabla > \nabla_{\text{ad}}$, N^2 is negative and $1/|N|$ measures

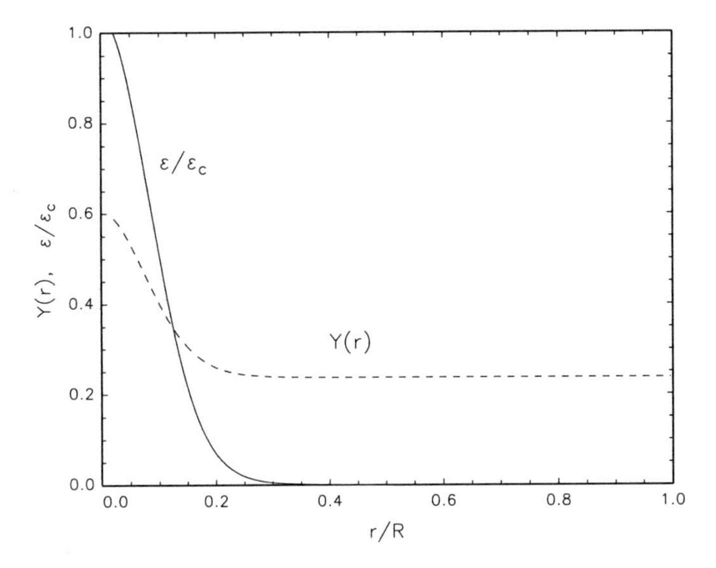

Fig. 9.4. Because of hydrogen burning in the radiative solar core, the mass fraction, Y, of helium in the present-day sun has increased while the mixture becomes less hydrogen-rich. Also shown is the energy generation rate $\varepsilon(r)$ as compared to its central value ε_c.

the e-folding time for increases in velocity and temperature perturbations. All the information necessary to construct N^2 as a function of radius may be found from the solar model and this is shown in Fig. 9.6. Here, the abscissa is $\log(1 - r/\mathcal{R}_\odot)$, which is a scale that heavily emphasizes the outer regions. Note that the stellar center is now at the right end of the figure. The ordinate is in the units of frequency^2 (Hz^2) to accommodate the other variable in the figure (the "Lamb frequency"), which will be used later when discussing solar oscillations, as will N^2. With this abscissa the bottom of the solar convection zone is at $\log(1 - r/\mathcal{R}_\odot) \approx -0.55$, while the top is at a value of approximately -3.7. The latter corresponds to a depth of only a couple of hundred kilometers below the photosphere—the exact value will depend on just how the atmosphere is constructed—and is located at a height where temperatures are sufficiently low that hydrogen recombination has finally been completed (as discussed in §3.4).

From all external appearances, the standard model reproduces the sun as we see it. It has been constructed so that the age, surface composition, luminosity, radius, and effective temperature match the object in our daytime sky. This is a significant achievement but, aside from the apparent consistency of thinking that we have the inside right because the outside looks right, are there other observations that probe beneath the visible surface that can reinforce our optimism or poke holes in it? In the next three sections, we explore some details of the solar interior that can be probed because it so

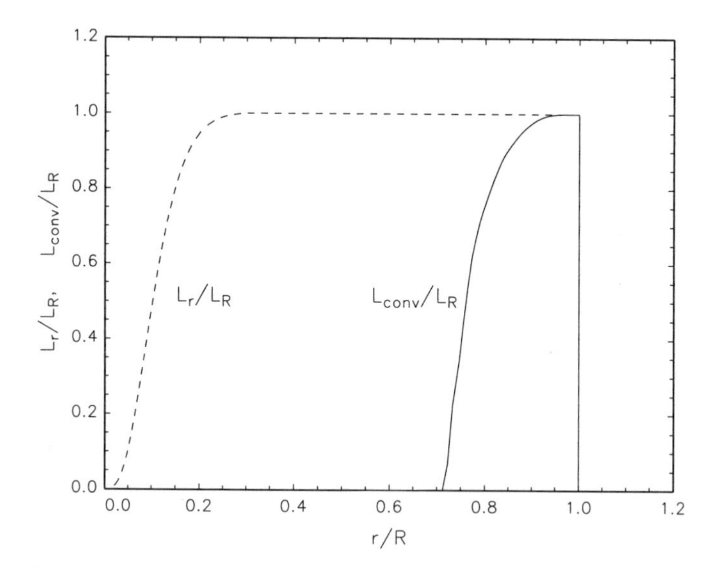

Fig. 9.5. The ratios $\mathcal{L}_{\rm tot}/\mathcal{L}_R$ (dashed line) and $\mathcal{L}_{\rm conv}/\mathcal{L}_R$ (solid line) are shown as a function of relative radius in the present-day sun. $\mathcal{L}_R$ is equal to the luminosity at the surface. The convective luminosity drops to zero just as the photosphere is reached at $r = R$.

close to us. With lots of photons (and neutrinos!) coming from so close at hand, we can look inside the sun as we can no other star. These up-close insights reveal how much we really do understand, but also how much we still need to learn.

The proximity of the sun to the earth means that we have sufficient photons, and a sufficiently detailed view of the solar surface, that we can take advantage of its very small oscillations (nonradial pulsations) and use asteroseismology to probe the internal structure and rotation of the sun. But first, we can use the fact that the sun is close to try to see, directly, one of the by-products of the nuclear reactions at the solar core.

9.3 The Solar Neutrino "Problem"

Since 1968, in the Homestake Mine, Kellogg, South Dakota, at a depth of nearly 1,500 meters, Raymond Davis Jr. and his collaborators have been detecting electron neutrinos emitted from deep within the sun. Using some 600 tons of the cleaning fluid compound tetrachloroethylene (C_2Cl_4)—and a great deal of ingenuity—they count the rate at which the radioactive isotope ^{37}Ar (half-life of 35 days) is produced by the reaction

$$\nu_{\rm e} + {}^{37}{\rm Cl} \longrightarrow {\rm e}^- + {}^{37}{\rm Ar}\,. \tag{9.12}$$

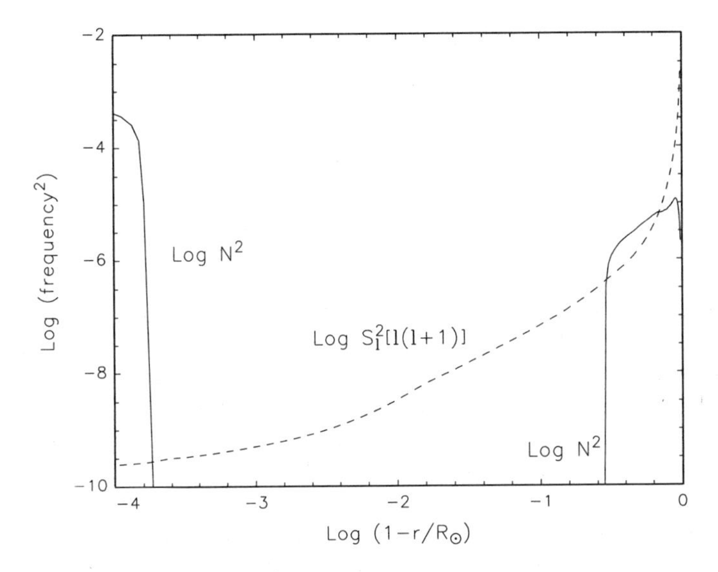

Fig. 9.6. Shown are N^2 and $S_\ell^2/[\ell(\ell+1)]$ in a standard solar model. The abscissa is $\log(1 - r/\mathcal{R}_\odot)$, which places the solar center at the righthand edge of the figure. S_ℓ^2 is the square of the Lamb frequency that we discussed in Chapter 8.

This requires an incoming neutrino with an energy exceeding 0.81 MeV for the reaction to proceed. (For a review of the history and results of this ^{37}Cl experiment see Rowley et al., 1985, and references therein.) Besides a small amount accounted for from extra-solar sources, these incoming neutrinos are produced from hydrogen-burning reactions in the sun. Ordinary material is remarkably transparent to neutrinos, whose typical absorption cross sections are in the 10^{-44}-cm^2 range for energies of order MeV. Thus, using this figure, a typical mean free path for neutrino absorption within the sun is $\lambda = 1/\sigma\langle n\rangle \sim 10^9\,\mathcal{R}_\odot$ (give or take a couple orders of magnitude) where $\langle n\rangle$ is the average number density of particles. Therefore, any neutrinos produced in the sun escape easily.[1]

The particular reactions responsible for producing neutrinos in the solar interior may be inferred (aside from two reactions to be given shortly) from Tables 6.1 and 6.2 listing the pp-chains and CNO cycles. These are

[1] An excellent overview of the neutrino problem is to be found in Bahcall (1989). We strongly suggest you look through that material if you want a complete picture up to 1989.

$$\begin{aligned}
{}^{1}\mathrm{H} + {}^{1}\mathrm{H} &\longrightarrow {}^{2}\mathrm{H} + \mathrm{e}^{+} + \nu_{\mathrm{e}} \\
{}^{7}\mathrm{Be} + \mathrm{e}^{-} &\longrightarrow {}^{7}\mathrm{Li} + \nu_{\mathrm{e}}(+\gamma) \\
{}^{8}\mathrm{B} &\longrightarrow {}^{8}\mathrm{Be} + \mathrm{e}^{+} + \nu_{\mathrm{e}} \\
&\cdots \\
{}^{13}\mathrm{N} &\longrightarrow {}^{13}\mathrm{C} + \mathrm{e}^{+} + \nu_{\mathrm{e}} \\
{}^{15}\mathrm{O} &\longrightarrow {}^{15}\mathrm{N} + \mathrm{e}^{+} + \nu_{\mathrm{e}} \\
{}^{17}\mathrm{F} &\longrightarrow {}^{17}\mathrm{O} + \mathrm{e}^{+} + \nu_{\mathrm{e}} \\
{}^{18}\mathrm{F} + \mathrm{e}^{-} &\longrightarrow {}^{18}\mathrm{O} + \nu_{\mathrm{e}}.
\end{aligned}$$

The first reaction is the pp-reaction from the PP–I chain and the next two are from the PP–II and PP–III chains. The last four are from the CNO(F) cycles. Neutrinos from the pp-reaction are emitted in a continuum up to an endpoint energy of 0.42 MeV. These cannot be detected by the Davis experiment because the endpoint energy is less than the threshold energy for the neutrino capture on ^{37}Cl. The decay of ^{8}B also yields a continuum of neutrinos with energies up to 15 MeV and these are accessible to the experiment. The ^{7}Be reaction is actually two reactions and the final neutrino energy depends on the final nuclear state of ^{7}Li. The result is a monoenergetic neutrino at an energy of either 0.862 MeV or 0.384 MeV. The first decay is more probable and occurs 90% of the time. The three CNO reactions yield continuum neutrinos with endpoint energies of 1.20, 1.73, and 1.74 MeV, respectively. Two additional pp-chain reactions—which we have not discussed—also yield neutrinos. These are the three-body reactions "pep" [$^{1}\mathrm{H}(\mathrm{p\,e}^{-},\nu_{\mathrm{e}})^{2}\mathrm{H}$] and "hep" [$^{3}\mathrm{He}(\mathrm{p\,e}^{-},\nu_{\mathrm{e}})^{4}\mathrm{He}$]. These rare reactions contribute almost nothing to energy generation in the sun but pep does emit line neutrinos at 1.44 MeV, while hep emits in a continuum up to 18.8 MeV.

We now examine what are the predicted fluxes of these neutrinos from a standard solar model and what should be the rate of detection by the chlorine experiment. It should be clear that the rate of neutrino emission by the above individual reactions must depend sensitively on temperature because temperature primarily determines the relative competition between PP–I through PP–III chains and the CNO cycles. Thus neutrino-detecting experiments potentially offer a unique probe into the solar interior and that is why so much effort has been expended in designing such experiments.

Figure 9.7 shows the predicted electron neutrino spectrum from a standard solar model (from Bahcall, 1989, §1.4; see also Bahcall and Ulrich, 1988; Bahcall et al., 1988). The units of the flux are cm^{-2} s^{-1} MeV^{-1} for continuum sources and cm^{-2} s^{-1} for line sources. All fluxes are calibrated so that they are those that should be seen at one astronomical unit from the sun. The flux from the more energetic ^{7}Be neutrino, which ^{37}Cl can see, is about 10^{10} neutrinos cm^{-2} s^{-1}. To find how many absorption reactions would take place per second per target on targets with absorption cross section $\sigma \sim 10^{-44}$ cm^{2}, we multiply σ by the flux to find 10^{-35} captures s^{-1} per target. This estimate is not entirely correct but the final result is not too far from what is observed by the Davis experiment. The convenient unit used in this business is the "solar neutrino unit," or SNU (pronounced "snoo"), and it is defined as 1 SNU $= 10^{-36}$ captures s^{-1} target^{-1}. We shall use these SNU units from now on.

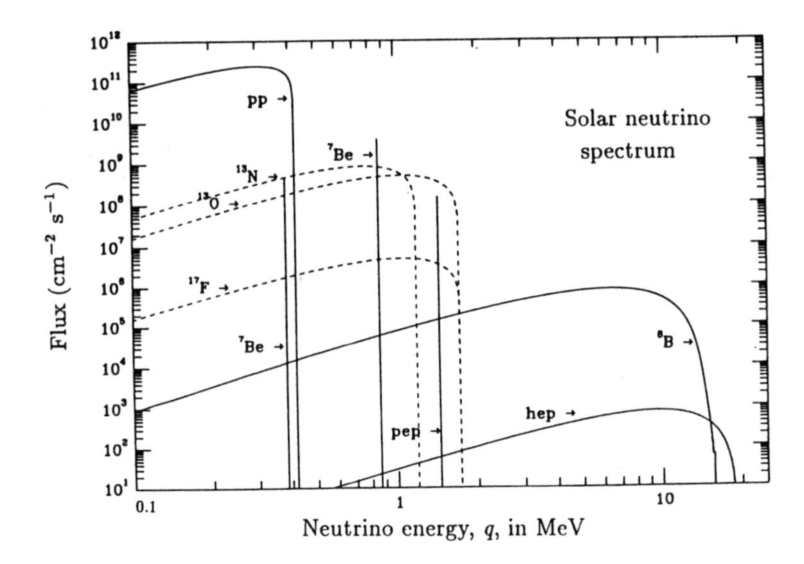

Fig. 9.7. The energy spectrum of electron neutrinos predicted from a standard solar model. The solid (dashed) lines are pp-chain (CNO) reaction neutrinos. The units for continuum neutrinos are cm^{-2} s^{-1} MeV^{-1} and, for line neutrinos, cm^{-2} s^{-1}. The fluxes are what should be observed at 1 AU. Reproduced with permission from Bahcall (1989).

The solar neutrino "problem" became apparent when the neutrino flux observed at the Homestake Mine was established as 2.07 ± 0.3 (1σ error) SNU over the period 1970–1988 (see the above references). The final value for the Homestake experiment is not much different at 2.56 ± 0.23 SNU over the 30 year lifetime of the experiment. The flux predicted for this experiment from standard models is 7.6 ± 1.2 SNU. Of the 7.6 SNU, 5.8 SNU is contributed by the ^{8}B decay, 1.2 SNU by the ^{7}Be electron capture, and less than 0.5 SNU from CNO neutrinos.

Thus the Homestake experiment reveals an effective 3σ deficit of neutrinos, where errors cover estimates of independent uncertainties in nuclear reaction rates, opacities, model-building techniques, idiosyncrasies of various researchers, and practically all the other items discussed thus far in this text (see Bahcall, 2001a). The theoretical prediction is well outside the experimental error bars.

A total of six neutrino experiments have run (or are running). Three use radiochemical techniques. In addition to the Davis experiment, the GALLEX and SAGE experiments use gallium, via the reaction $^{71}Ga(\nu_e, e^-)^{71}Ge$, with a threshold of only 0.23 MeV, which is low enough to detect the higher energy range of pp-reaction neutrinos. The gallium experiments are in principle sensitive to a large fraction of the predicted solar neutrino energy spectrum. The others rely on the Cerenkov radiation produced by electrons that participate in neutrino interactions in H_2O (Kamiokande and Super–Kamiokande), with a threshold of 6.5 to 7.4 MeV. Electron neutrino-induced deuterium dissociation in 1,000 tonnes heavy water (D_2O) is explored at the Sudbury Neutrino Observatory (SNO), which is over 2,000 meters below ground level in Ontario.[2] The water-based experiments are directional and can produce "neutrino images" of the sun despite being deep underground. *All* of these experiments detect neutrinos but at rates less than are predicted for the particular reactions involved using standard neutrino physics.[3]

The factor-of-three deficit in the chlorine experiment is different than the deficit seen in the water detectors: as Fig. 9.8 shows, the Kamiokande and Super–Kamiokande neutrino fluxes are 50% of the predicted rate. Since they are sensitive only to the ^8B rate, that means that their results are not consistent with the chlorine experiment, and cutting the predicted ^8B rate by 50% still results in a deficit of neutrinos in the chlorine experiment! Thus, as articulated by John Bahcall, there are really several solar neutrino "problems" to consider.

The pioneering work by Ray Davis and his colleagues, verified by the results from Kamiokande, has presented physics and astronomy with a profound dilemma that in turn has broad consequences—one of the reasons why Davis and Koshiba (the Kamiokande director) shared the 2002 Nobel Prize in physics for their work on neutrino astronomy.

The mystery deepens when we consider the gallium experiments. These are sensitive to all of the neutrino sources in the core of the sun. The predicted rates show that approximately half of the neutrinos seen in the gallium experiment come from the pp-reaction, and that the observed rates are consistent with those being the only neutrinos coming from the sun. Allowing for a deficit of 50% in the ^8B neutrinos, this in turn leaves no room for the ^7Be neutrinos—thus providing yet a third solar neutrino problem!

On one hand, we seem to have a good grasp on what makes the sun work, but, when looked at closely using one probe that senses the deep interior, something appears greatly amiss. There is insufficient space here to discuss at any length "nonstandard" solar models that try to address the neutrino

[2] For a semi-popular review of this experiment, see McDonald, Klein, and Wark (April, 2003), SciAm, 288 #4, 40.

[3] See, for example, the following articles: *Physics Today*, 1990, Vol. 43, No. 10, p. 17; *Science News*, 1992, Vol. 141, p. 388; *Physics Today*, 1992, Vol. 45, No. 8, p. 17, and *Nature*, July 2001, p. 29.

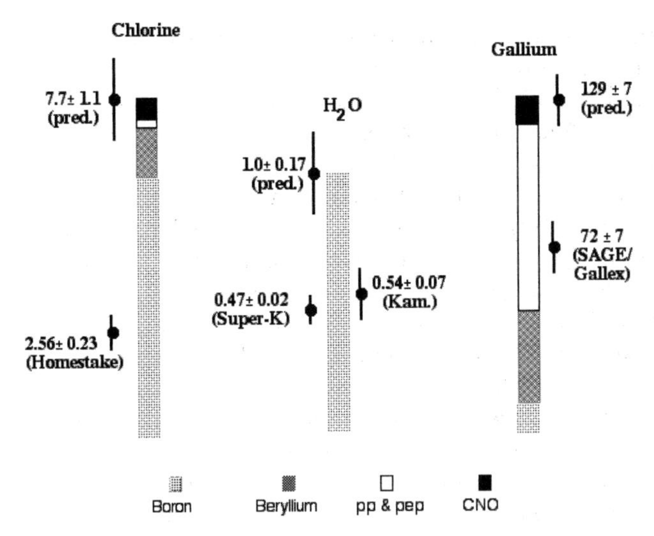

Fig. 9.8. Observed and predicted solar neutrino rates. The central columns of this bar graph shows the standard solar model predictions for neutrino fluxes in the chlorine (left), H_2O (center), and gallium (right) experiments. The error bars show the predicted ("pred") and observed fluxes (by name of experiment). Adapted from Bahcall (2001a).

problems. These include, for example, models that have inhomogeneous outer layers. If there were some way to cause elements to diffuse and separate in the sun during some earlier stage of evolution, then we may be fooled by the presently observed solar surface composition into thinking we know what the interior composition was when the sun formed. There are several mechanisms capable of causing elemental segregation, and numerical experiments using these can "solve" all of the neutrino problems. There are three difficulties with this and other solutions, however: the prescriptions tend to be ad hoc; they are underconstrained, leaving no observational way to test them; and typically one that solves one of the neutrino problems (say, the deficit of ^{8}B neutrinos) makes another problem worse. Finally, all nonstandard models do run afoul of the observed solar oscillation spectrum, which we shall discuss later in this chapter. No reasonable tinkering with standard models seems capable of removing the neutrino discrepancy.

Elementary particle physics offers a solution that leaves the standard model intact. In some grand unified theories ("GUTs"), the electron neutrino is not massless—and this is not ruled out by experiment—but the mass must be small. If so, it is possible that an electron neutrino may be converted into a muon or tau neutrino under the proper circumstances (or the other way around). Muon neutrinos, on the other hand, are not detectable by the chlorine and other experiments (except Kamiokande and Super–Kamiokande, but they can't distinguish between different kinds of neutrinos) and, hence, even though the sun produced electron neutrinos at exactly the rate predicted by the standard model, some fraction of them would not be detected at the earth. This process of "neutrino oscillation" may be enhanced when neutrinos pass through a material medium (the "MSW" effect—see Bahcall et al., 1988; Bahcall and Bethe, 1990).

It is believed by many that this is the most promising line of inquiry to follow and may yet turn out to do the trick. Here is how it may work. Since all neutrinos produced in the sun are electron neutrinos, if they change "flavor" among the three species of neutrinos on their way to the earth, then detectors that see only electron neutrinos should see a deficit. The chlorine experiment sees exactly that deficit, but the Kamiokande experiments see more than one-third. The chlorine experiment detects only electron neutrinos. The Super–Kamiokande detector, though mostly sensitive to electron neutrinos, can also detect muon and tau neutrinos. Therefore, the fact that it shows a significantly smaller deficit works in the right direction for the neutrino oscillation theory. However, Super–Kamiokande cannot distinguish individual detections as either electron, muon, or tau neutrino events. SNO, on the other hand, can identify electron neutrino events because of the extra neutron in the heavy–water molecules. Neutrinos from the sun detected by SNO can then be tagged as electron neutrinos and compared with the flavor-blind results from Super–Kamiokande in the same neutrino energy range. Results announced by the SNO collaboration in 2002 indicate that the prediction of the standard solar model is nearly exactly the observed electron neutrino rate when neutrino oscillations are taken into account. Thus it may be that the solar neutrino problem tells us more about physical nature at its most fundamental level than it does about stellar astrophysics.

9.4 The Role of Rotation in Evolution

It is fortunate for makers of solar models that the sun rotates slowly. Were it to rotate rapidly, at speeds close to breaking up by centrifugal forces, the solar test bed for stellar evolution would almost be uncomputable using present technology. On the other hand, the sun does rotate and the effects are observable.

Rapidly rotating stars are not unusual but, on the main sequence at least, almost all of these are massive and bright. Less massive main sequence stars tend to be slow rotators. The observational evidence for this is presented in Fig. 9.9 (from Kawaler, 1987), which shows the average equatorial rotation velocities versus mass of a large sample of main sequence stars compiled by Fukuda (1982). The crosses denote complete samples, whereas data indicated by circles exclude Am stars, which tend to rotate anomalously slowly, and Be stars, which rotate rapidly but show peculiar emission features. (The Am stars are of spectral class A but with peculiar abundances and they are almost exclusively members of binary systems.) The nearly horizontal line labeled $\langle v \rangle = v_{\rm crit}$ is the locus where stars of equatorial velocity v are at breakup; that is, where surface gravitational and centrifugal forces are equal with $v_{\rm crit} = \sqrt{G\mathcal{M}/\mathcal{R}}$. It is evident that the more massive stars are rapid rotators with equatorial velocities only about a factor of two less than breakup.

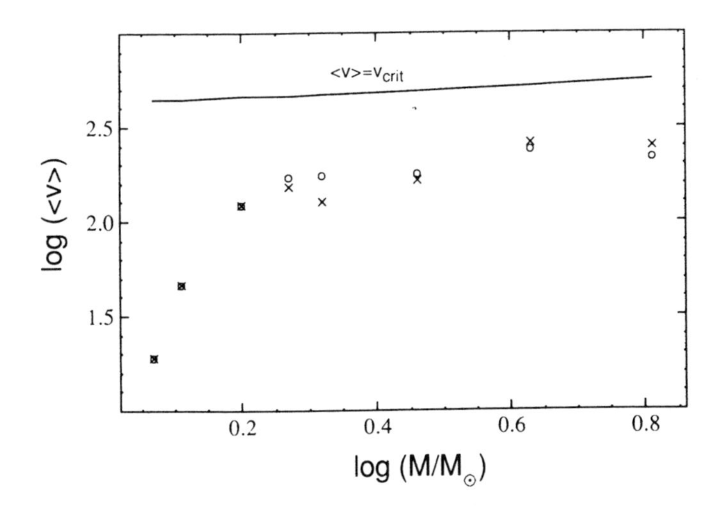

Fig. 9.9. The observed average equatorial rotation velocity $\langle v \rangle$ is plotted as a function of mass for main sequence stars. Note the clear change in the behavior of $\langle v \rangle$ at $\log(\mathcal{M}/\mathcal{M}_\odot) \approx 0.2$ ($\mathcal{M} \approx 1.5\,\mathcal{M}_\odot$). Reproduced, with permission, from Kawaler (1987).

Once masses drop below about $1.5\,\mathcal{M}_\odot$ ($\log \mathcal{M}/\mathcal{M}_\odot \approx 0.2$), however, average rotation velocities decrease precipitously. This mass corresponds to around spectral type F0 ($\mathcal{L} \approx 6\,\mathcal{L}_\odot$) on the main sequence and to the mass (and $T_{\rm eff}$) below which stars have significant envelope convection (see Fig. 5.7). Thus it appears that convective envelopes and slow rotation are connected in some way.

Rotational angular momentum can be carried away from a star by a stellar wind, but how much angular momentum is lost depends on the distance at which the wind decouples from the stellar interior in terms of rotation. For a simple wind, this decoupling occurs near the photosphere. However, if the wind couples to the stellar magnetic field, it can be forced to corotate with the star well beyond the photosphere. The magnetic field is rooted in the stellar interior and as the corotating wind moves beyond the photosphere it gains angular momentum at the expense of the interior. This gain of angular momentum by the wind is proportional to the square of the distance above the photosphere. Thus magnetic fields can greatly amplify the angular momentum loss a star experiences when coupled with a wind.

Since stars later (cooler) than F0 have significant surface convection zones that can drive stellar winds and, through dynamo action, produce and modulate internal magnetic fields, these stars will experience much more angular momentum loss than their higher mass cousins. Support for this idea comes from the observation that the more massive main sequence stars generally have feeble surface convection zones and weak magnetic fields. Even though some have strong winds, most rotate near the average value of $1/3\ v_{\rm crit}$. For more on the theory of magnetic braking, see the reviews by Mestel (1984), Collier Cameron et al. (1991), and Kawaler (1990). Observations of young stars that lend support to the theory are reviewed in Stauffer and Hartmann (1986) and Stauffer (1991).

The other side of this picture implies that the sun initially formed rotating much more rapidly than it does today. Could earlier rapid rotation have influenced evolution? Before we discuss this, it is worthwhile pointing out some of the subtle implications of rotation for stellar structure even when the rotation is slow.

9.4.1 von Zeipel's Paradox

Stellar rotation, of even the simplest kind, introduces complexities in the construction of realistic stellar models that are beyond our present capabilities. This is because rotation is inherently three dimensional. We shall only touch on one aspect of the subject. For an excellent introduction we recommend the monograph by Tassoul (1978) and, in particular for what follows, his Chapters 7 and 8.

Suppose we attempt to construct a chemically homogeneous stellar model in hydrostatic and thermal balance where heat is carried solely by diffusive radiative transfer. To complicate matters, however, let us require that the model rotate as a rigid body. A rigidly rotating body is naturally described in cylindrical coordinates $\mathbf{r} = \mathbf{r}(\varpi, \varphi, z)$ where the z-axis coincides with the rotation axis and the radial coordinate ϖ is measured from the z-axis. The equation of hydrostatic equilibrium must now include the effects of centrifugal forces which, in the rotating frame of the star, are given by the term $\rho\Omega^2\varpi\, \boldsymbol{e_\varpi}$ where $\boldsymbol{e_\varpi}$ is the radial unit vector. The potential corresponding to this force is $\Phi_{\rm cent} = -\Omega^2\varpi^2/2$. The total potential, centrifugal plus gravitational is then

$$\Phi_{\rm eff}(\varpi, z) = \Phi_{\rm grav}(\varpi, z) - \frac{\Omega^2\varpi^2}{2} \,. \tag{9.13}$$

Note that we have assumed azimuthal symmetry (no φ dependence) as seems reasonable. The equation of hydrostatic equilibrium then becomes

$$\frac{1}{\rho}\boldsymbol{\nabla} P = -\boldsymbol{\nabla}\Phi_{\rm eff} = -\boldsymbol{\nabla}\Phi_{\rm grav} + \Omega^2\varpi\, \boldsymbol{e_\varpi} = \mathbf{g}_{\rm eff} \,. \tag{9.14}$$

The acceleration $\mathbf{g}_{\rm eff}$ is the local effective gravity, which now includes centrifugal effects. In addition to the above we also need Poisson's equation, which is still given by

$$\nabla^2 \Phi_{\text{grav}} = 4\pi G \rho \tag{9.15}$$

(as in §7.2.1). Note that only Φ_{grav} appears here.

The first application of these equations comes about from considering "level" surfaces on which Φ_{eff} is constant. The gradient $\boldsymbol{\nabla}\Phi_{\text{eff}}$ evaluated on such a level surface is, of course, perpendicular to that surface. If $d\mathbf{r}$ is an infinitesimal unit of length lying tangent to the surface then, by usual arguments in the vector calculus, $d\mathbf{r} \bullet \boldsymbol{\nabla}\Phi_{\text{eff}} = d\Phi_{\text{eff}} = 0$ (as it must). But this implies, by dotting $\mathbf{r}$ into (9.14), that $dP = 0$ on a level surface or, equivalently, pressure is a constant on such a surface. Thus level surfaces are also "isobaric" surfaces with $P = P(\Phi_{\text{eff}})$. This may be reversed to read $\Phi_{\text{eff}} = \Phi_{\text{eff}}(P)$ from which follows (from 9.14) $\rho^{-1} = d\Phi_{\text{eff}}/dP$. Density is then also constant on a level surface (making it an "isopycnic" surface).

If the equation of state of our chemically homogeneous model is that of an ideal gas with $P = \rho N_{\text{A}} kT/\mu$, then $T = T(\Phi_{\text{eff}})$. Thus far, the only quantity in sight that is *not* a constant on a level surface is $\mathbf{g}_{\text{eff}}$. It is normal to the surface but, because the level surfaces need not be the same distance apart in (ϖ, φ, z)-space, it may vary in magnitude over the surface. (Note that if Ω is zero, then the level surfaces are concentric spheres and we regain the constancy of gravity for fixed radius.)

The second constraint we have placed on the model is that of thermal balance. That is, we require $d\mathcal{L}_r/dr = 4\pi r^2 \rho\varepsilon$ of (1.57). A more general way of putting this is in terms of the vector flux $\boldsymbol{\mathcal{F}}$. Thus write

$$\boldsymbol{\nabla} \bullet \boldsymbol{\mathcal{F}} = \rho\varepsilon \,. \tag{9.16}$$

The requirement of radiative diffusive transfer specifies how the flux is transported through a level surface so that, with a slight rearrangement of gradients,

$$\boldsymbol{\mathcal{F}} = -\frac{4ac}{3}\frac{T^3}{\kappa\rho}\frac{dT}{d\Phi_{\text{eff}}}\boldsymbol{\nabla}\Phi_{\text{eff}} \,. \tag{9.17}$$

Note that T is regarded as a function of Φ_{eff} as above. Furthermore, ε and κ are also functions of Φ_{eff} because they contain only $\rho(\Phi_{\text{eff}})$ and temperature. Thus, and still following Tassoul (1978), we write

$$\boldsymbol{\mathcal{F}} = f(\Phi_{\text{eff}})\,\boldsymbol{\nabla}\Phi_{\text{eff}} = -f(\Phi_{\text{eff}})\mathbf{g}_{\text{eff}} \tag{9.18}$$

where the function $f(\Phi_{\text{eff}})$ takes care of the Φ_{eff}-dependent terms in (9.17). Now take the divergence of this expression and use the fact that $\boldsymbol{\nabla} f$ and $\boldsymbol{\nabla}\Phi_{\text{eff}}$ are perpendicular to the level surface to find

$$\boldsymbol{\nabla} \bullet \boldsymbol{\mathcal{F}} = \frac{df}{d\Phi_{\text{eff}}}\left(\frac{d\Phi_{\text{eff}}}{dn}\right)^2 + f\nabla^2\Phi_{\text{eff}} = \rho\varepsilon. \tag{9.19}$$

Here $\mathbf{n}$ is an outward unit normal to the level surface.

A nearly final result is obtained by realizing that $d\Phi_{\text{eff}}/dn$ is the same as $|\mathbf{g}_{\text{eff}}|$ so that, using Poisson's equation and doing some rearranging, you should find

$$\frac{df}{d\Phi_{\text{eff}}}\,\mathbf{g}_{\text{eff}}^2 + f(\Phi_{\text{eff}})\left[4\pi G\rho - 2\Omega^2\right] = \rho\varepsilon\ . \tag{9.20}$$

In order to satisfy all the constraints set on the model, this equation must be satisfied everywhere. At first glance, nothing seems peculiar because it is just the divergence of the radiative flux set equal to the energy generation rate (per unit volume) for thermal balance. However, if the structure is such that any two level surfaces are not spaced everywhere equidistantly apart (as they are if there were no rotation), then there is trouble. To see this, note that $\rho\varepsilon$ and the terms within the brackets depend only on Φ_{eff} *but* $\mathbf{g}_{\text{eff}}$ is *not* constant on arbitrary level surfaces. Therefore the coefficient $(df/d\Phi_{\text{eff}})$ must be zero because everything else is a constant. Thus drop this term from (9.20) and what remains is a relation between variables on a level surface and, in particular, a requirement on ε. Solving for ε yields

$$\varepsilon \propto \left[1 - \frac{\Omega^2}{2\pi G\rho}\right]\ . \tag{9.21}$$

But how can this be? The energy-generation rate cannot be directly determined by how the star rotates!

The difficulty is that we have overly constrained the problem. This is von Zeipel's (1924) "paradox": namely, a uniformly rotating star (and the situation is more general than this) cannot be in steady-state radiative thermal equilibrium. Something must give.

The solution lies in relaxing the constraints. We refer you to Tassoul (1978, Chap. 8), where this is discussed. Briefly, it appears that either the angular rotation frequency Ω must depend on both ϖ and z, or fluid motions (e.g., "meridional" circulation) must take part in the transfer of heat. Consideration of these topics would take us into fascinating, very difficult, and not fully resolved territory. Our stance here is to back off and assume that rotation is sufficiently slow that—as a good approximation—many such effects can safely be ignored. But we do so at our peril with the realization that our description of the interiors of many stars is incomplete.

9.4.2 Rotational Mixing of Stellar Interiors

From the discussion above, we see that rotation should have little direct effect overall on the evolutionary changes in temperature, density, and pressure within the sun during the course of its evolution. However, some observed properties of the sun—and, in particular, the elemental abundance of rare species in the photosphere—may be telling us a significant story about the internal rotation of the sun and how it has changed with time. For example, the element ^{7}Li has an abundance in the solar photosphere that is a factor of 200 smaller than the abundance of ^{7}Li found in meteorites, terrestrial rocks, and younger stars. (See §6.3.) This would be an easy thing to explain if this ^{7}Li was destroyed by nuclear reactions within the solar convection zone. If so, then convective mixing would dilute the lithium abundance of the entire convection zone and lead to a depleted surface value compared to the primordial abundance. However, standard solar models indicate that the base of the solar convection zone never gets quite hot enough to burn ^{7}Li very much; the standard solar model depletes lithium to about one-third of its initial value. If the solar convection zone was deeper by about half a pressure scale height, then the observed level of ^{7}Li destruction could occur, but to do this would result in a solar model that is at odds with observations of other younger stars. Rotation provides us with a way out of this dilemma.

Recall that in our discussion of convection, convective material undergoes presumably turbulent mixing. Therefore, if such a convection zone is rotating, then convective mixing should result in angular momentum exchange, leaving the zone rotating essentially as a solid body. Radiatively stable material, on the other hand, can support a gradient in angular velocity with depth. Now consider the young sun, which started its life as a fully convective pre-main sequence star. As such, it initially rotated as a solid body, with a much larger angular momentum than it has today. As it settled onto the main sequence, the interior became radiative while the envelope remained convective (see the section on the ZAMS sun). During contraction, the early sun became more centrally concentrated. If specific angular momentum was conserved (that is, each mass shell had an angular momentum that did not change with time by transfer to other mass shells), then this central concentration resulted in a spin-up of the solar core, with decreasing angular velocity $\Omega(r)$ from the center outward.

Recall also that for stars with surface convection zones, angular momentum can be lost from the surface by a magnetized stellar wind. The surface convection zone, which should have continued rotating as a solid body, thus experienced a continual loss of angular momentum to space. Therefore, in addition to a smooth gradient in $\Omega(r)$ in the radiative interior, the surface spun down quickly and *somewhat* independently of the interior. At the base of the convection zone, then, a discontinuity in $\Omega(r)$ may have developed. Numerous analytic and laboratory studies of rotating fluids have shown that steep gradients of Ω can be hydrodynamically unstable (see, e.g., Zahn 1987). Thus these instabilities will trigger mixing of stellar material and its angular momentum to reduce the gradients in $\Omega(r)$ to a state of (at least) marginal stability. Such redistribution of angular momentum can occur on either short time scales (i.e., the free-fall time) or longer time scales (i.e., the thermal time scale), but it is hard to see how angular momentum redistribution can be completely avoided. Because the principal generator of shear is the braking of the convective envelope, the base of the envelope will be where the angular momentum redistribution and resulting mixing of material will be most noticeable.

Rotation therefore could result in mixing of material near the base of the solar convection zone, and the solar lithium depletion may be a signature of angular momentum redistribution. It should be clear that a discussion of computations of solar evolution that includes such mixing is beyond the scope of this text; it involves simultaneous solutions of the usual equations of stellar evolution and, in addition, treatment of hydrodynamic instabilities on many different time scales (and in three dimensions!). Some computational work in this area, with simplifying assumptions about distribution of angular momentum in latitude and longitude, has addressed the problem of rotational mixing. Comparison of these model results with observations of the lithium abundance seen in young stars and very old stars, as well as the sun, has met with remarkable success. Additional work in this area stems from the very important paper (but not one for the faint of heart) by Jean-Paul Zahn (1994) that considers the general subject of rotationally driven instabilities within stars and uses stars in binary systems to test the results.

We leave this very active subject at this point, and note that rotational effects are probably even more important to consider in the evolution of stars more massive than the sun. For results relevant to the sun, we refer you to papers by Mark Pinsonneault and colleagues, among others, for many more details. See, for example, Pinsonneault (1997) and the review by Sofia et al. (1991) and references therein.

9.5 Helioseismology

One of the early "solutions" to the solar neutrino problem invoked the fact that the rate of production of solar neutrinos is proportional to a very high power of the central temperature, so that anything that might lead to a suppression of the central temperature of the sun would reduce the predicted value. However, it was recognized quite early that despite variations in some of the input physics, construction of a standard solar model (one that matched the solar radius and luminosity at the solar age) produced a very consistent value for the central temperature. Some ad hoc solutions (such as an altered metallicity in the center of the sun, or rapid rotation of the solar core) could be ruled out by examining the nonradial pulsation frequencies, as introduced in the previous chapter.

In this section, we introduce helioseismology—the application of nonradial pulsation theory to the study of solar oscillations, and the determination of the structure of the solar interior through examination of its oscillation frequencies. We begin with a brief description of the observations, and then show how the observations compare with theoretical predictions. Using inversion theory, we then show how the standard solar model has been tested and revised and how the internal rotation profile has been measured. Though we have been able to produce some astounding results for the sun over the past three decades, the hope is that in the future we can make similarly detailed tests of stellar models. But for now, the results from helioseismology will have to stay our curiosity.

9.5.1 Observed and Predicted Pulsation Frequencies

The sun, and perhaps all stars at some level, is variable but you need keen instruments to detect its variability. Leighton, Noyes, and Simon (1962) first observed a five-minute correlation in velocity–induced Doppler shifts of absorption lines formed at the solar surface. These were (and are) interpreted as vertical oscillations of large patches of fluid having velocities of 1 km s^{-1} or less with a coherence time of around five minutes. It was not until nearly ten years later that Ulrich (1970) and Leibacher and Stein (1971) independently suggested that what was being observed was the result of global oscillations with periods of around five minutes wherein the sun acted as a resonant cavity for acoustic waves propagating through its interior. It is now well-established that the majority of these waves are nonradial acoustic modes.

We shall not discuss here how these waves are excited, but it is not by the same mechanism that is responsible for the variability of other overtly variable stars. The best current model involves the convection zone of the sun wherein the noise generated by turbulence effectively causes the whole sun to quiver in response.

Nonradial modes have now been observed by several methods and it is known that they consist of the incoherent superposition of millions of acoustic p-modes. Over the past decade, the literature on the subject, both observational and theoretical, has expanded at a tremendous rate as observations from ground–based networks, and dedicated satellites, has flooded astronomers. For very thorough discussions of the background theory and early observations—and to get an excellent sense for the enthusiasm with which astronomers embraced this field—we recommend the classic review articles by Christensen–Dalsgaard et al. (1985), Leibacher et al. (1985), Libbrecht and Woodard (1991), Toomre (1986), and Libbrecht (1988). More recent accounts of the results of "modern" observational and theoretical efforts can be found in Gough et al. (1996) (and subsequent articles in the same issue of *Science*), along with the review article by Christensen–Dalsgaard (2002). For an excellent pedagogical introduction see Gough (2003). You should also peruse the conference proceedings literature for reports of the annual meetings of the "Global Oscillation Network Group" (GONG). This consortium of astronomers operates a global network of stations to observe oscillations of the sun.

To give a representative picture of the observational and theoretical results, we first present Fig. 9.10 (from Libbrecht, 1988), which shows measured p-mode frequencies, $f_{n\ell}$ (in mHz with $2\pi f_{n\ell} = \sigma_{n\ell}$), versus ℓ. The individual points are observed modes (from Duvall et al., 1988) where n, the mode order,

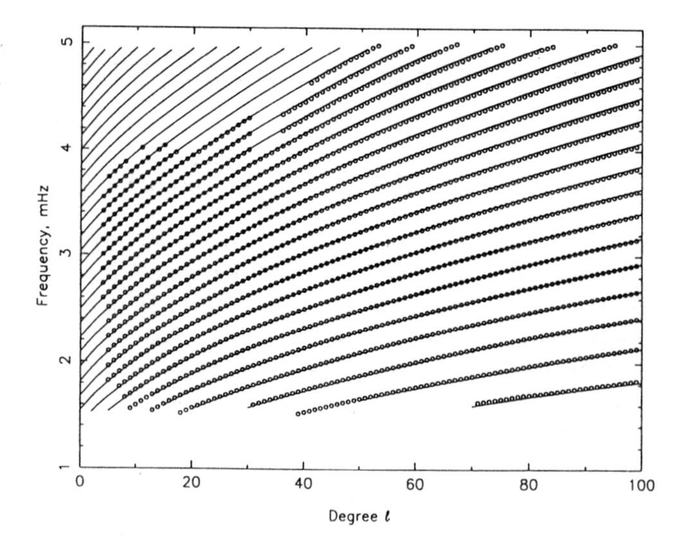

Fig. 9.10. Measured solar p-mode frequencies, $f_{n\ell}$, for $\ell < 100$ are shown by the circles. Theoretical results are indicated by the solid lines where individually computed frequencies for fixed n have been made continuous. The lowest "ridge line" has $n = 2$ and n increases by one for each higher ridge. Note that the frequency scale is in mHz so that a typical frequency of $f = 3$ mHz corresponds to a period of $\Pi = 1/f \approx 5$ minutes. From Libbrecht (1988) based on data from Duvall et al. (1988) . Reprinted by permission of Kluwer Academic Publishers.

is as indicated in the figure caption. Such measurements are extraordinarily difficult and involve two-dimensional Doppler imaging of the solar surface and other techniques. The solid lines are from theoretical nonradial pulsation calculations for an early standard model of the sun by Christensen–Dalsgaard et al. (1985).

Recall that for the discussion of mode spectra for nonradial modes in §8.3.3, we used a ZAMS sun model. From a seismic point of view, the present sun isn't all that different. The run of the Brünt–Väisälä frequency and Lamb frequencies with radius are similar, and therefore the modal structure will be very similar. The precise frequencies, however, are very sensitive to the total radius of the sun, the position of the base of the surface convection zone, and the position of composition gradients in the solar core.

9.5.2 Helioseismology and the Solar Interior

It would appear from Fig. 9.10 that the theoretical results using the early standard solar model of Christensen–Dalsgaard et al. (1985) for p-mode frequencies are very good indeed because the results match the observations as well as the eye can detect. This tells us that that model for the sun is "accurate" down to appreciable depths into the interior. However, there are discrepancies that cannot be ignored. In fact, results using a "modern" standard solar model (from Guenther et al., 1996) are shown in the left panel of Fig. 9.11, which shows differences between observed and calculated frequencies for ℓs ranging from $\ell = 0$ to 100. Note that the ordinate is in μHz. Though the results look quite good—and in fact are a factor of 10 better than results from a decade earlier—the differences between the model and the observations is many times larger than the uncertainties in the frequencies themselves.

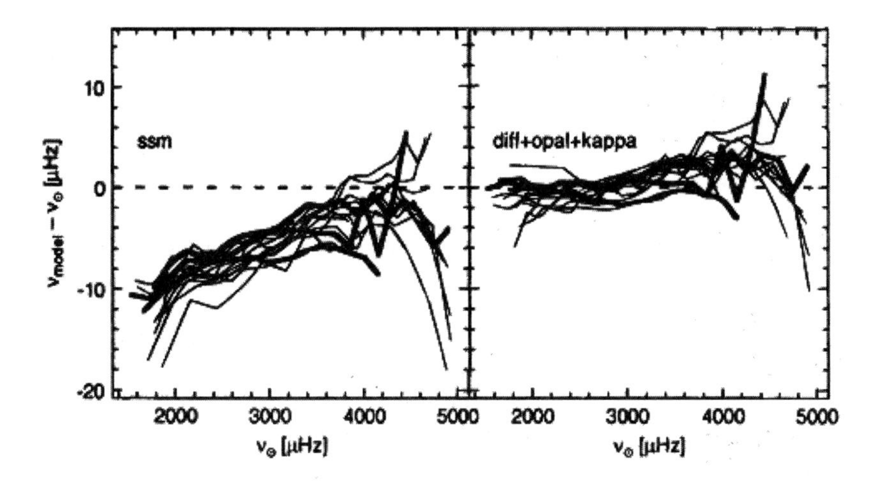

Fig. 9.11. The ordinate shows the difference, $\Delta\nu = \nu_{\rm model} - \nu_{\odot}$, with ν as frequency, between observed solar p-mode frequencies ($\nu_{\odot}$) and theoretical frequencies ($\nu_{\rm model}$). Thick lines refer to modes with ℓ of 10, 20, and 30. Taken, with permission, from Guenther et al. (1996) and the AAS. The left panel represents results using a basic standard solar model, while the right panel shows the improvements possible by including better equation of state and opacity data along with including helium diffusion.

The most obvious thing of note in the right panel of Fig. 9.11 is the overall improvement for all values of ℓ and all frequencies when the standard solar model is modified. In this case, Guenther et al. (1996) incorporated a more sophisticated and improved equation of state, modified opacities, and the effects of diffusion of helium and heavy elements. The overall result was the removal of overestimates of pulsation frequencies for low frequencies, and an improved trend at higher frequencies.

Structural Inversions

Figure 9.11 shows the general agreement between theoretical and observed oscillation frequencies as a "forward problem" wherein stellar models are made, oscillation frequencies computed and compared with observed frequencies, differences noted, and model adjustments made (and the process repeated). In Section 8.3.4, we discussed a different way to derive properties from the oscillation frequencies. Through inversion, assuming sufficient observed modes over a range of ℓ as well as frequency, (8.119, or a similar expression for the physical quantities of interest) can be used to calculate corrections to model structures based on the observed frequencies.

One property of significance is the local sound speed v_s, which, in the adiabatic limit, is proportional to $\sqrt{\Gamma_1 T}$. Sound speed inversions can reveal composition discontinuities and equation of state errors within the deep interior, and minimize effects of uncertainties in the structure of the (strongly nonadiabatic and very dynamic) outermost layers. In general, modern solar models depart from the inversion results by no more than about 0.5 km s^{-1} through most of the interior (where the sound speed ranges from over 500 km s^{-1} near the center down to about 80 km s^{-1} near the surface). Figure 9.12 shows the difference between the model sound speed and the sound speed obtained by inversion of various helioseismic data sets for the standard solar model from Bahcall et al. (2001). The agreement between data and the model is within 0.2% over nearly the entire interior—an extraordinary degree of precision for astrophysics!

Low-frequency p-modes probe deeply into the sun's interior. This is especially true for modes with low ℓ. The most deeply penetrating are the radial modes with $\ell = 0$ that pass right through the center; investigation of these modes bears directly on the solar neutrino problem. Much effort has thus gone into searching for possible errors and effects that make both theoretical oscillation and neutrino calculations match the observations. Various versions of nonstandard models have been tried with conflicting results. Thus, for example, models with lower amounts of metals in the deep interior ("low-Z models") help the neutrino problem but worsen the match for low ℓ modes. One promising avenue has been to postulate that the primordial helium abundance in the deep interior of the sun was higher than in the surface layers. This tends to bring low ℓ frequencies in line and decreases the calculated neutrino fluxes—but not by enough. Bahcall et al. (2001) and Basu et al. (2000) review such attempts but, sad to say, the results are that no solar model, standard or even imaginative, can match both the neutrino results and the oscillation data. Bahcall et al. (2001) and Bahcall et al. (1997) (and referenced works therein) conclude that the seismological constraints on the sound speed within the sun, as well as the general frequency matching, convincingly eliminate these models. The solar neutrino "problem" first elucidated by the work of Ray Davis has indeed exposed an ignorance of fundamental particle physics, and not of more traditional astrophysics.

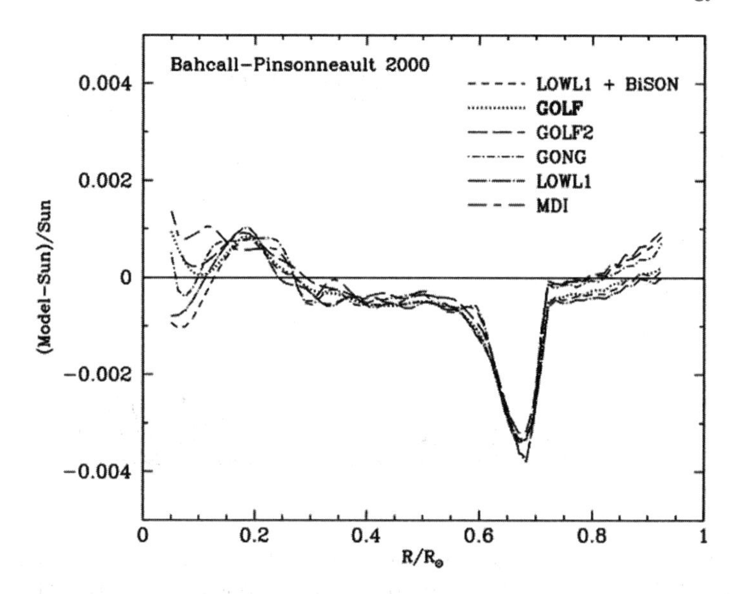

Fig. 9.12. Results of sound speed inversions of low degree helioseismic data, taken (with permission) from Bahcall et al. (2001) and the AAS. The difference between the model and the inversion result (denoted by "sun") is plotted as a function of fractional radius. Different lines correspond to inversions of the labeled helioseismic frequency sets. Note that despite the small differences between these sets, the spread in results is comparable to the overall difference between the model and the average of the data.

Rotational Inversions

Another interesting quantity that we have noted for the sun is its rotation rate. The surface of the sun displays differential rotation with latitude, with the equator rotating more quickly than the poles. Given the strong feedback between rotation and magnetic fields, and our desire to understand the origin of both in the sun (and their behavior with depth) one of the goals that helioseismology has is to explore the rotation rate as a function of latitude and depth within the sun.

Determining $\Omega(r,\theta)$ is an inversion problem, as discussed in Chapter 8. Again, by using modes over a large range of ℓ and frequency, one can invert the frequency splittings—or, more precisely, appropriate linear combinations of splittings that have associated kernels that are localized with respect to r—to obtain the rotation profile within the sun. Representative of these attempts are those shown in Fig. 9.13, where $\Omega(r,\theta,\varphi)$ is assumed to have the same θ dependence as at the photosphere. In this investigation the rotation down to $r/\mathcal{R}_\odot \approx 0.7$ is similar to that of the surface but for $r/\mathcal{R}_\odot \lesssim 0.6$ more closely resembles solid body rotation. The important questions these results pose is

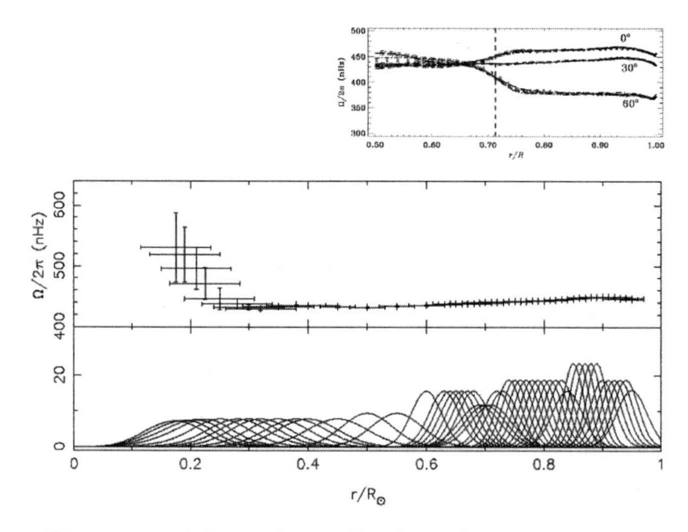

Fig. 9.13. The top panel shows the profile of angular rotation frequency for three latitudes of the sun as inferred from rotational splitting of the SOHO–MDI (satellite) data using inversion theory. The ordinate is in the units of nHz. The data for deeply penetrating p-modes is inadequate to resolve Ω for $r \lesssim 0.4\mathcal{R}_\odot$ clearly. Adapted, with permission, from Schou et al. (1998). The bottom panel shows the rotational inversion nearer the stellar core, using additional data for low ℓ modes from SOHO's GOLF experiment. The bottom portion of the lower panel represents the many localization kernels from the solar model used in the inversion. This portion is from Corbard et al. (1998).

how this rotation comes about and what consequences it has for the entire evolution of the sun.

Results quoted in §9.4.2 can now be directly tested with the observed rotation profile with depth. The most obvious general conclusion to draw from Fig. 9.13 is the nearly flat rotation rate at depth within the sun. Non-magnetic rotational instabilities alone cannot account for such flatness, suggesting that magnetic fields indeed play in important role in angular momentum transport in deep solar interior.

9.6 References and Suggested Readings

§9.1: The Vital Statistics of the Sun

Our sources for variations in solar luminosity are

▹ Newkirk, G., Jr. 1983, ARA&A, 21, 429

▹ Willson, R.C., Hudson, H.S., Frohlich, C., & Brusa, R.W. 1986, *Science*, 234, 1114.

The details of quoted solar surface abundances are bound to change through the years. We quote from the work of

▹ Grevesse, N. 1984, PhysScr, T8, 49

▹ Grevesse, N., & Noels, A. 1993, in *Origin and Evolution of the Elements*, ed. N. Prantzos, E. Vangioni–Flam, & M. Casse (Cambridge: Cambridge University Press), p. 15.

and

▹ Grevesse, N., & Sauval, A.J. 1998, SpSciRev, 85, 161.

The "age of the sun" is probably not known to within better then several times 10^7 years because of uncertainties of phasing between solar, planetary, and minor body formation within the solar system. For comments see

▹ Guenther, D.B. 1989, ApJ, 339, 1156

and an update using newer geochemical studies by G. Wasserburg in an appendix to the paper by

▹ Bahcall, J., & Pinsonneault, M. 1995, RevModPhys, 67, 781.

§9.2: From the ZAMS to the Present

Unlike luminous stars, the sun does not have much of a wind. The mass loss rate we quote is from

▹ Cassinelli, J.P., & MacGregor, K.B. 1986, in *Physics of the Sun* Vol III, ed. P.A. Sturrock (Boston: Reidel), p. 47.

"Standard" models have become more standard through the years as model makers have tended to agree upon the best strategies and physics for making models. (This does not say that everyone gets the same results, however!) Earlier papers describing the process in some detail are

▹ Guenther, D.B., Jaffe, A., & Demarque, P. 1989, ApJ, 345, 1022

▹ Guenther, D.B, Demarque, P., Kim, Y.–C., & Pinsonneault, M.H. 1992, ApJ, 387, 372.

The 1989 Yale models use the mix of

▹ Anders, E., & Grevesse, N. 1989, GeoCosmo, 53, 197.

More recent work has included the effects of diffusion on the solar abundances and revised estimates of the helium abundance. See

▹ Bahcall, J., Pinsonneault, M. & Basu, S. 2001, ApJ, 555, 990

who used the Grevesse and Sauval (1998) abundance determinations; other model details are outlined in

▹ Bahcall, J., & Pinsonneault, M. 1995, RevModPhys, 67, 781.

As usual, we recommend

▹ Cox, J.P. 1968, *Principles of Stellar Structure*, in two volumes (New York: Gordon and Breach)

for some in-depth perspectives on stellar structure.

It is always gratifying to use a simple calculation to help make sense out of all the numbers pouring out of a computer. The virial estimate for luminosity change of the evolving sun is from

▷ Endal, A.S. 1981, in *Variations of the Solar Constant*, NASA Conf. Pub. 2191, 175.

§9.3: The Solar Neutrino "Problem"

The observational results of the Homestake Mine experiment are reviewed in

▷ Rowley, J.K., Cleveland, B.T., & Davis, R. Jr. 1985, in *Solar Neutrinos and Neutrino Astronomy*, eds. M.L. Cherry, W.A. Fowler, and K. Lande (AIP: New York), pp. 1–21

while the SNO experiment is described in

▷ McDonald, A., et al. 2000, NuclPhysProcSupp, 91, 21.

A good overall review of neutrinos, solar and otherwise, may be found in

▷ Bahcall, J.N. 1989, *Neutrino Astrophysics* (Cambridge: Cambridge University Press).

For more on solar neutrinos see also

▷ Bahcall, J.N., Huebner, W.F., Lubow, S.H., Parker, P.D., & Ulrich, R.K. 1982, RevModPhys, 54, 767

▷ Bahcall, J.N., & Ulrich, R.K. 1988, RevModPhys, 60, 297

with more recent updates to the results of solar models in

▷ Bahcall, J.N., & Pinsonneault, M. 1995, RevModPhys, 67, 781

▷ Bahcall, J., Pinsonneault, M., & Basu, S. 2001, ApJ, 555, 990.

Solutions of the solar neutrino problem involving fundamental particle physics are discussed in

▷ Bahcall, J.N., Davis, R. Jr., & Wolfenstein, L. 1988, Nature, 334, 487

▷ Bahcall, J.N., & Bethe, H. 1990, PhysRevL, 65, 2233.

A concise but very readable summary of the solar neutrino story is

▷ Bahcall, J.N. 2001a, "Solar Interior: Neutrinos," in *Encyclopedia of Astronomy and Astrophysics*, (Basingstoke: Nature)

with an update in

▷ Bahcall, J.N. 2001b, Nature, 412, 29.

§9.4: The Role of Rotation in Evolution

The best textbook on the subject of stellar rotation is due to

▷ Tassoul, J.-L. 1978, *Theory of Rotating Stars* (Princeton: Princeton University Press)

and we recommend it for your general bookshelf.

Figure 9.9 is taken from

▷ Kawaler, S.D. 1987, PASP, 99, 1322

using material from

▷ Fukuda, I. 1982, PASP, 94, 271.

Papers concerning the role of magnetic braking and rotation include

▷ Mestel, L. 1984, in *3d Cambridge Workshop on Cool Stars, Stellar Systems, and the Sun*, eds. S. Baliunas and L. Hartmann (New York: Springer), p. 49

▹ Stauffer, J.R., & Hartmann, L. 1986, PASP, 98, 1233
▹ Kawaler, S.D. 1990, in *Angular Momentum and Mass Loss for Hot Stars*, eds. L.A. Willson and R. Stallio (Dordrecht: Kluwer), p. 55
▹ Collier Cameron, A., Li, J., & Mestel, L. 1991, in *Angular Momentum Evolution of Young Stars*, eds. S. Catalano and J.R. Stauffer (Dordrecht: Kluwer), p. 297
▹ Stauffer, J.R. 1991, in *Angular Momentum Evolution of Young Stars*, eds. S. Catalano and J.R. Stauffer (Kluwer: Dordrecht), p. 117.

The original von Zeipel's "paradox" are discussed in

▹ von Zeipel, H. 1924, MNRAS (London), 84, 665, 684.

General questions of instabilities induced by steep rotation gradients are discussed by

▹ Zahn, J.-P. 1987, in *The Internal Solar Angular Velocity*, eds. B. Durney and S. Sofia (Dordrecht: Reidel).

and

▹ Zahn, J.-P. 1994, A&A, 288, 829.

Some papers concerning the sun, rotation, and the lithium problem include

▹ Pinsonneault, M. 1988, *Evolutionary Models of the Rotating Sun and Implications for Other Low Mass Stars*, Ph.D. Dissertation, Yale University
▹ Sofia, S., Kawaler, S., Larson, R., & Pinsonneault, M. 1991, in *The Solar Interior and Atmosphere*, eds. A.N. Cox, W.C. Livingston, and M.S. Matthews (Tucson: University of Arizona), p. 140
▹ Sofia, S., Pinsonneault, M., & Deliyannis, C. 1991, in *Angular Momentum Evolution of Young Stars*, eds. S. Catalano and J.R. Stauffer (Dordrecht: Kluwer), p. 333,

and

▹ Pinsonneault, M., Steigman, G., Walker, T., & Narayanan, V. 2002, ApJ, 574, 398

along with the paper by Zahn cited above. A general review of results of computations of realistic stellar models that include chemical mixing (by rotation and other processes) is

▹ Pinsonneault, M. 1997, ARA&A, 35, 557.

§9.5: Helioseismology

Some seminal papers in helioseismology include—

▹ Leighton, R.B., Noyes, R.W., & Simon, G.W. 1962, ApJ, 135, 474
▹ Ulrich, R.K. 1970, ApJ, 162, 993
▹ Leibacher, J.W., & Stein, R.F. 1972, ApJ, L7,191.

We recommend the reviews by

▹ Christensen–Dalsgaard, J., Gough, D.O., & Toomre, J. 1985, Science, 229, 923

▷ Leibacher, J., Noyes, R.W., Toomre, J., & Ulrich, R.K. 1985, SciAm, 253, 48

▷ Libbrecht, K.G., & Woodard, M.F. 1991, Science, 253, 152

▷ Toomre, J. 1986, in *Seismology of the Sun*, ed. D.O. Gough (Boston: Reidel), p.1

▷ Libbrecht, K.G. 1988, SpSciRev, 47, 275.

Following the GONG results, newer reviews indeed show that helioseismology has lived up to many of its promises. See, for example, the reviews by

▷ Harvey, J. 1995 (October), Physics Today, 32,

and a more recent and comprehensive review of helioseismology can be found in

▷ Christensen–Dalsgaard, J. 2002, RevModPhys, 74, 1073.

Discussions of the results of analysis of GONG data on solar oscillations can be found in a special issue of the journal *Science*, (Vol. 272), from May 31, 1996. Despite a cover graphic that may represent the ultimate in inscrutable representations of massive data sets, many terrific review articles appear within. In particular, the overview paper,

▷ Gough, D., Leibacher, J., Scherer, P., & Toomre, J. 1996, Science, 272, 1281

sets the stage for reviews of the GONG instrumentation in

▷ Harvey, J.W., et al., 1996, Science, 272, 1284

and the observed acoustic–mode spectrum in

▷ Hill, F., et al., 1996, Science, 272, 1292.

Seismic inversions and the deduced structure of the sun are then described in another article by Gough and company in the same issue:

▷ Gough, D., et al., 1996, Science, 272, 1296.

Many more details in articles that accompany this review in the same issue of *Science*.

▷ Gough, D.O., 2003, Ap&SS, 284, 165

gives a nice introduction using analytic models.

Figure 9.10 is from

▷ Libbrecht, K.G. 1988, SpSciRev, 47, 275

who uses data from

▷ Duvall, T.L., Harvey, J.W., Libbrecht, K.G., Popp, B.D., & Pomerantz, M.A. 1988, ApJ, 324, 1158.

See also

▷ Libbrecht, K.G., Woodard, M.F., & Kaufman, J.M. 1990, ApJS, 74, 1129.

There have been several tabulations of solar p-mode frequencies from the pre–GONG era such as Duvall et al. (1988). For lower-order modes see

▷ Duvall, T.L. Jr., Harvey, J.W., Libbrecht, K.G., Popp, B.D., & Pomerantz, M.A. 1988, ApJ, 324, 1158.

Frequencies obtained by the GONG network, and space–based missions such as SOHO can be found online through several sites, More recent tabulations of frequencies from the GONG experiment can be glimpsed starting at the GONG website at `http://gong.nso.edu` which also has movies, and graphics that illustrate the frequency spectrum and solar physics results by the members of the GONG teams.

Figure 9.11 is an edited version of Figure 5 from

▷ Guenther, D., Kim, Y.–C., & Demarque, P. 1996, ApJ, 463, 382.

Exploration of how well non-standard solar models fare when dealing with the solar neutrino issues and helioseismic constraints are described in the paper by Bahcall, Pinsonneault and Basu (2001) mentioned above, with further details in

▷ Basu, S., Pinsonneault, M., & Bahcall, J.N. 2000, ApJ, 529, 1084.

The representative rotational inversions shown in Fig. 9.13 are described in more detail in

▷ Schou, J. et al. 1998, ApJ, 505, 390

and

▷ Corbard, T., et al. (1998), in *Structure and Dynamics of the Interior of the Sun and Sun-like Stars:* SOHO6/GONG98 *Workshop*, ed. S.G. Korzennik & A. Wilson, (ESA: Noordwijk) p. 741

but see also

▷ Schou, J., et al. 2002, ApJ, 567, 1234.

10 Structure and Evolution of White Dwarfs

"Any fool can make a white dwarf."
— *Icko Iben Jr. (1985)*

We have already discussed the evolutionary stages leading to the white dwarfs (§2.6) and described their internal structure as being determined by the combination of high gravities and an electron degenerate equation of state. This chapter will elaborate on their structure, evolution, and importance as the endpoint of evolution for most stars. The variable white dwarfs, discussed in the final section, will be shown to play an important role in this program. There is no one text that deals solely with these objects but, for further reading, we suggest Liebert (1980), Shapiro and Teukolsky (1983), Iben & Tutukov (1984), Tassoul et al. (1990), D'Antona & Mazzitelli (1990), Weidemann (1990), Trimble (1991, 1992), Fontaine et al. (2001), Koester (2002), and Hansen and Leibert (2003).

10.1 Observed Properties of White Dwarfs

In most respects white dwarfs form a remarkably homogeneous class of star. Figure 2.15 showed a color-magnitude diagram where a large sample of these stars resident in the galactic disk were plotted on an HR diagram, and we suggest you look at that figure again. They form a well-defined sequence to luminosities down to around $3 \times 10^{-5}\ \mathcal{L}_\odot$, below which, as far as can be determined, we do not find cooler objects (if we exclude those lurking in the halo of our galaxy, as discussed in §2.6.1). The tight correlation of luminosity with effective temperature (i.e., M_V with $B{-}V$) immediately demonstrates that their radii are all very nearly the same with $\mathcal{R} \approx 0.01\,\mathcal{R}_\odot \approx 7 \times 10^8$ cm. Spectroscopic observations coupled with theoretical stellar atmosphere calculations have determined that their surface gravities are near $\log g \approx 8$ ($g \approx 10^8$ cm s^{-2}), which, considering the radii, yields masses of $\mathcal{M} \sim 0.6\,\mathcal{M}_\odot$. Spectroscopic results for individual single stars of the most common types (DA and DB, as discussed later) firm this up further and indicate that an average mass is $0.6\ \mathcal{M}_\odot$ with a surprisingly low dispersion of only around $0.1\,\mathcal{M}_\odot$ about this figure.[1]

[1] The realization that most single white dwarfs have nearly the same mass is a relatively recent development. Since they have evolved from stars with different initial masses this uniformity must be telling us a lot about how mass is lost

White dwarfs in binary systems have a wider range of masses determined from reliable binary orbit solutions; for example, the mass of Sirius B (α CMa B) is $1.053 \pm 0.028\ \mathcal{M}_\odot$, while 40 Eri B ($\sigma^2$ Eri B) is below the single-star mean with $0.43 \pm 0.02\ \mathcal{M}_\odot$ (as reviewed in Liebert, 1980).

Spectroscopic observations also reveal that the atmospheric composition of white dwarfs may differ wildly from one to the next. Most common are those whose surfaces consist almost entirely of hydrogen with contamination by other elements exceeding, in some instances, no more than one part in a million by number of atoms. These are the DA white dwarfs and they make up some 80% of all white dwarfs, although the exact percentage does depend on effective temperature class. Next most common are the DB white dwarfs with helium atmospheres, which make up almost 20% of the total. The remainder consists of stars with hybrid atmospheres or those with peculiar abundances. The most commonly used spectroscopic classification scheme is summarized in Table 10.1 adapted from McCook and Sion (1999). Note that there is evidence that the surface abundance, and therefore spectral classification, for a given white dwarf may change as it evolves. As new data appear, we expect the classification scheme to evolve with time also. Apparently strange hybrids are possible, so that, for example, McCook and Sion list one star as having the classification DAZQO, which may, however, be an indicator of uncertainties in observing the spectrum.[2]

Effective temperatures for white dwarfs range from well over 100,000 K to lower than 4,000 K. The majority of known white dwarfs have temperatures higher than the sun and hence the "white" in their name. As in the MKK system (§4.7), numbers are attached to the DA, DB, etc., classification to indicate T_{eff}. Thus, for example, we find DA.25 for a DA with $T_{\text{eff}} \approx 200,000$ K and DA13 for $T_{\text{eff}} \approx 3,600$ K. As will soon be apparent, we can best explain the sequence in the HR diagram of Fig. 2.15 by cooling where, as time progresses, hot white dwarfs gradually evolve to lower temperatures along the sequence and, with a small number of important exceptions, become redder as they cool. The exceptions occur at the very lowest temperatures. At $T_{\text{eff}} \approx 3,600$ K the white dwarf surface is cool enough that the H_2 molecule can survive. Infrared "collision induced absorption" by this molecule is so efficient that the emergent spectrum looks bluer than for white dwarfs with somewhat higher effective temperatures. This is very new stuff and we suggest you peruse, for example, Saumon and Jacobson (1999) and Oppenheimer et al. (2001).

---

in the AGB stage. However, there are a small number of single objects whose masses lie in the high-mass tail of the distribution.

[2] McCook and Sion also use "n" to denote WDs with very sharp, narrow lines and "d" for those with very diffuse, broad lines. The significance of the line width is that sharp lines may be associated with exceptionally low surface gravities ($\log g \lesssim 7$) and "d" with exceptionally high gravities ($\log g \lesssim 9$) giving rise to pressure broadening. These designations are to be used only when there are good reasons to believe the gravity determinations.

Table 10.1. White Dwarf Spectroscopic Classification Scheme

Spectral type	Characteristics
DA	Balmer Lines only; no He I or metals present
DB	He I lines; no H or metals present
DC	Continuous spectrum with no readily apparent lines
DO	He II strong; He I or H present
DZ	Metal lines only; no H or He lines
DQ	Carbon features of any kind
P (suffix)	Magnetic WDs with detectable polarization
H (suffix)	Magnetic WDs without detectable polarization
X (suffix)	Peculiar or unclassifiable spectrum
E (suffix)	Emission lines are present
? (suffix)	Uncertain classification (: may be used)
V (suffix)	Variable white dwarf

White dwarfs are observed to rotate but with periods usually longer than a few hours (Greenstein and Peterson, 1973; Pilachowski and Milkey, 1987; and Koester and Herrero, 1988). This is a remarkable observation in itself because if, for example, we were to let the sun evolve to the white dwarf stage without losing either mass (an unlikely assumption) or angular momentum (equally unlikely), then the resultant carbon–oxygen object, with a radius of 5.6×10^8 cm (see 3.68), would have a solid body rotation period of only about 2.5 minutes. Angular momentum loss must therefore be a common feature of stellar evolution. We will explore rotation of variable white dwarfs in §10.4.

Many white dwarfs are variable stars (and probably the most common overtly variable stars in the universe) and a small number also have the strongest magnetic fields known for "normal" stars (perhaps exceeding 10^9 G and we exclude pulsars here for which there is only indirect evidence for even stronger fields). The magnetic white dwarfs will be the subject of §10.3.

10.2 White Dwarf Evolution

We have yet to establish that evolutionary models of white dwarfs actually do reproduce the observed objects, but, if our earlier ideas are correct, then the interior should be largely electron degenerate. On the other hand, the very surface cannot be degenerate because white dwarfs are observed to have high effective temperatures. The surface layers should therefore be nondegenerate, and this means very different equations of state and opacity sources. We shall see that energy is transported rapidly through the degenerate interior but has to diffuse gradually through the nondegenerate envelope. Thus the cooling of white dwarfs involves the properties of matter under a wider range of conditions than most other problems in physics or astrophysics. However, this degenerate core–nondegenerate envelope picture results in an elegant simplification that permits us to construct a very simple model for how white dwarfs evolve.

10.2.1 Cooling of White Dwarfs

The white dwarf model we shall now consider has the following elements. Imagine that the core of the star, which comprises nearly all of the mass and radius, is degenerate. Overlying the core is a thin envelope of nondegenerate material and the transition between degeneracy and nondegeneracy is assumed to take place abruptly at a radius $r_{\rm tr}$. If the electrons are nonrelativistic at $r_{\rm tr}$, then the relation between density and temperature there is given by (3.70), $\rho_{\rm tr} \approx 6 \times 10^{-9} \mu_{\rm e} T_{\rm tr}^{3/2}$. In the electron-degenerate core interior to $r_{\rm tr}$, electron conduction is very efficient at transporting heat (according to the arguments of §4.5) and only a mild temperature gradient is required to drive the flux. Thus, for simplicity, assume that the core is isothermal with temperature $T_{\rm core} = T_{\rm tr}$.[3]

To determine $r_{\rm tr}$ we need to be more specific about the model. If the envelope does not support convection (and this is *not* true for many white dwarfs), then the envelope approximations of §7.3.2 should describe the run of pressure versus temperature and density. For zero boundary conditions (7.119) states that $P = K' T^{1+n_{\rm eff}}$ where K' is given by (7.120) and $n_{\rm eff} = (s+3)/(n+1)$. The exponents n and s are those in $\kappa = \kappa_0 \rho^n T^{-s}$ of (1.62). We now use this information to establish a relation between $T_{\rm tr}$ (and hence $T_{\rm core}$), luminosity, and mass.

The pressure must be continuous across $r_{\rm tr}$. Above $r_{\rm tr}$ the gas is nondegenerate and we use the ideal gas law $P_{\rm tr} = \rho_{\rm tr} N_{\rm A} k T_{\rm tr} / \mu$, so that

$$P_{\rm tr} = K' T_{\rm tr}^{1+n_{\rm eff}} = \rho_{\rm tr} \frac{N_{\rm A} k}{\mu} T_{\rm tr} = 6 \times 10^{-9} \mu_{\rm e} \frac{N_{\rm A} k}{\mu} T_{\rm tr}^{1+3/2} \tag{10.1}$$

where the transition relation between density and temperature (3.70) has been used to eliminate density in the third term. The coefficient K' contains $\mathcal{L}$, $\mathcal{M}$, μ, and the opacity. At this point we have to decide what is the dominant opacity source, which also means specifying the composition. To make matters as simple as possible, suppose the white dwarf is composed entirely of elements with atomic masses heavier than ^{4}He. In fact, we expect the cores of most white dwarfs to be composed of some combination of ionized carbon and oxygen. This is because ^{12}C and ^{16}O, both with $\mu_{\rm e} = 2$, are the products of helium burning and, for single white dwarfs of average mass 0.6 $\mathcal{M}_\odot$, this is as far as core evolution has gone. Choosing the same composition for the surface layers is not so good but, as we shall see, the final result we obtain

[3] This assumption of isothermality also requires that there be no strong sources or sinks of energy in the core such as might be associated with nuclear burning, gravitational contraction, or neutrinos. It really only applies to WDs that are well past the PNN stage.

shall be quite reasonable. For the opacity we choose bound–free Kramers' which, from (4.63), is approximated by $\kappa_{\rm bf} \approx 4 \times 10^{25} \rho T^{-3.5}$ cm^2 g^{-1}. This analytic opacity, while crude, still gives the flavor of what goes on, so we use it here with no further apologies. There is no way to get what we want without a little judicious fudging (and see Ex. 10.1).

The model we are setting up is that of a highly conductive core surrounded by a thin insulating blanket. Heat flows easily out of the core but must work its way out through the envelope. When we discuss cooling, it is the envelope that controls the rate of cooling whereas the core supplies the heat.

Applying (7.120) for K' yields

$$K' \approx 8.1 \times 10^{-15} \mu^{-1/2} \left[\frac{\mathcal{M}/\mathcal{M}_\odot}{\mathcal{L}/\mathcal{L}_\odot} \right]^{1/2} \tag{10.2}$$

where we have used $\kappa_{\rm bf}$ and $n_{\rm eff} = 3.25$. Setting $\mu_{\rm e} = 2$ and solving the combination of (10.1) and (10.2) for luminosity gives us the relation

$$\frac{\mathcal{L}}{\mathcal{L}_\odot} \approx 6.6 \times 10^{-29} \mu \frac{\mathcal{M}}{\mathcal{M}_\odot} T_{\rm tr}^{7/2} . \tag{10.3}$$

Note that for $\mu = 12$ (^{12}C) and $\mathcal{M} = 0.6\,\mathcal{M}_\odot$, $\mathcal{L}/\mathcal{L}_\odot = 100$ (10^{-4}), $T_{\rm tr} = T_{\rm core}$ is about 2.4×10^8 K (4.6×10^6 K).

We can now estimate the thickness of the surface layer by using (7.124), which gives temperature as a function of $r/\mathcal{R}$ for a thin radiative envelope. You may easily check that for $\mathcal{L}/\mathcal{L}_\odot = 10^{-4}$, $r_{\rm tr}/\mathcal{R} \approx 0.99$, which means that the nondegenerate envelope can indeed be thin.

The next step is to find how the white dwarf cools. We still assume that there are no internal energy sources such as nuclear burning or gravitational contraction. The last means that the total radius remains roughly constant with time. This approximation becomes better and better as the star cools and the only energy source is the internal heat of the star.[4] These are the essentials of the now-classic Mestel (1952) cooling theory for white dwarfs which, except for some refinements, has stood the test of time.

To apply the above results first recall that the specific heat of a nonzero temperature degenerate gas is controlled by the ions. From (3.118), $c_{V_\rho} = 1.247 \times 10^8/\mu_{\rm I}$ erg g^{-1} K^{-1}. Since temperature is constant in the core and the core takes up essentially all the stellar mass, then the rate at which the ions release heat on cooling is

[4] Most stars heat up when they lose energy (see the virial theorem results of §1.3.2), which makes for a self-regulating mechanism even though it sounds peculiar—and it is because it means they have a negative specific heat overall. Thus it is that white dwarfs are odd because they follow what seems like the more reasonable path and cool when they lose energy. The reason has partially to do with the very low specific heats of degenerate electrons versus ions and the strong influence of density on the internal energy of the electrons (§3.7.1). See the discussion in Cox (1968, §25.3b).

$$\mathcal{L} = -\frac{dE_{\text{ions}}}{dt} = -c_{V_\rho}\mathcal{M}\frac{dT_{\text{tr}}}{dt} \, . \tag{10.4}$$

To find how luminosity changes with time, differentiate (10.3) with respect to time, use (10.3) to get rid of T_{tr}, and then use (10.4) to eliminate the temperature derivative. The resulting differential equation, which should now contain only the dependent variable $\mathcal{L}(t)$, is then integrated to obtain a "cooling time"

$$t_{\text{cool}} = 6.3 \times 10^6 \left(\frac{A}{12}\right)^{-1} \left(\frac{\mathcal{M}}{\mathcal{M}_\odot}\right)^{\frac{5}{7}} \left(\frac{\mu}{2}\right)^{-\frac{2}{7}} \left[\left(\frac{\mathcal{L}}{\mathcal{L}_\odot}\right)^{-\frac{5}{7}} - \left(\frac{\mathcal{L}_0}{\mathcal{L}_\odot}\right)^{-\frac{5}{7}}\right] \tag{10.5}$$

where t_{cool} is in years. Here A is the mean atomic weight of the nuclei in amu and $\mathcal{L}_0$ is the luminosity at the start of cooling. After long elapsed times the second term in the brackets becomes negligible compared to the first so we drop it for simplicity and, at the same time, change the leading coefficient to match more accurate results for cooling white dwarfs calculated by Iben and Tutukov (1984, and see Iben and Laughlin, 1989). The final result is

$$t_{\text{cool}} = 8.8 \times 10^6 \left(\frac{A}{12}\right)^{-1} \left(\frac{\mathcal{M}}{\mathcal{M}_\odot}\right)^{5/7} \left(\frac{\mu}{2}\right)^{-2/7} \left(\frac{\mathcal{L}}{\mathcal{L}_\odot}\right)^{-5/7} \text{yr} \, . \tag{10.6}$$

Figure 10.1 shows cooling curves for pure carbon white dwarfs derived from evolutionary calculations. Were these results pure Mestel (1952) cooling, we would only have straight lines in this log–log plot as shown by the dotted line in the figure for $0.6\mathcal{M}_\odot$ as derived from (10.6). The very close, but perhaps partly fortuitous, match indicates that we have captured the essentials of the physics of white dwarf cooling. It is evident that the hottest, and therefore most luminous white dwarfs, cool the fastest. At the cool end, we see by plugging $\mathcal{L} = 10^{-4.5}\mathcal{L}_\odot$ into (10.6) that the cooling time for the coolest white dwarfs is approximately 10^{10} years. We have already made this point in §2.6.1 where, in Fig. 2.16, we plotted the "drop-off" in the number of white dwarfs at low luminosities. For those of you who attempted Ex. 2.9 (and you should have) you will already have applied (10.6) to find Galactic disk and halo ages. And now you know where it came from.

10.2.2 Realistic Evolutionary Calculations

The ideal calculation of evolving white dwarfs has yet to be realized. In addition to physical processes that are difficult to model in the white dwarfs themselves, the starting conditions require knowledge of how white dwarfs are formed. This means that the origin of white dwarfs as planetary nebula nuclei must be understood to provide realistic starting models. Unfortunately, the process of planetary nebula formation is still somewhat of a mystery. White dwarfs can play a critical role in helping explain that stage because whatever PNN models are made must eventually reproduce the observed statistics of those stars.

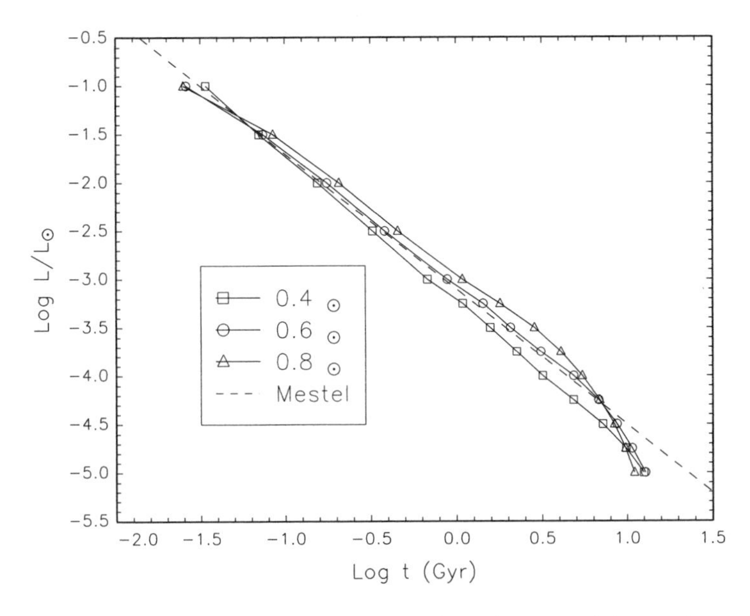

Fig. 10.1. Cooling curves for pure carbon white dwarf models as adapted from Winget et al. (1987). The dotted line is a Mestel (1952) cooling curve for a $0.6\mathcal{M}_\odot$ carbon white dwarf using (10.6).

What is usually done in white dwarf evolutionary studies is to either start off with a completely ab initio model of a very hot white dwarf where the initial structure (run of composition, $T(r)$, etc.) is specified beforehand or, with the hope of realism, to start a model from the main sequence and continue on to the white dwarf stage. The latter requires evolving the model into the asymptotic giant branch phase and then lifting off the outer layers in some reasonable manner to expose the underlying pre–white dwarf object. Both methods have their champions.

It is generally assumed that the cores of most white dwarfs are composed of a carbon–oxygen mixture. Exceptions to this may occur for special objects such as the white dwarfs in nova systems where there is not only evidence that these are more massive than average (about $1\,\mathcal{M}_\odot$); but, from observations of the ejecta, some may be rich in heavier elements such as those in the nuclear mass range Na to Al (as briefly discussed in §2.11.1). We shall ignore such anomalous objects.[5]

The surface layers of most white dwarfs consist initially of some combination of hydrogen and helium (but not necessarily both) contaminated by

[5] There may be a selection effect operating here because the more massive objects in close binary systems are expected to erupt more often and more brightly than their less massive counterparts. A paucity of ejecta for the less massive objects could also hide their true composition.

traces of heavier elements. The exact details depend on how much mass was lost in the AGB phase, how much nuclear processing has occurred, whether mixing has taken place, and whether stellar winds are active in the PNN phase. Determination of the surface composition of the just-formed and very hot object is difficult because temperatures exceeding 100,000 K put interesting details of the spectrum in the hard UV and soft x–ray. However, based on the observation that cooler white dwarfs show either nearly pure hydrogen (DA) or helium (DB) in their spectra, two different classes of model have been examined in detail. The first assumes that all hydrogen has been lost. These evolve into DB white dwarfs. The second has a hydrogen surface layer above a layer of helium. Whether the DB-like object has trace amounts of hydrogen (which can later "float" to the surface to convert the star to a DA) is a matter of controversy at the present time. Similarly, the thickness of the hydrogen layer in the DA objects is also not known very well (although one cool variable DA star most likely has a thin layer; see §10.4). If thin enough, convection at later stages could convert the DA to a spectroscopic DB. We shall avoid such unresolved issues here for the moment because, for the most part, they are refinements. However, the total mass of hydrogen cannot much exceed $10^{-4}\mathcal{M}_\odot$ because, if it did, nuclear burning would occur. Similarly, the helium layer mass should not exceed about $10^{-2}\,\mathcal{M}_\odot$.

Whether DA or DB, the newly formed objects have some common characteristics. The surface layers are still hot enough that shell CNO hydrogen or helium burning may still be taking place. These energy sources can, for the initial stages of evolution, provide the dominant energy source. There are also energy losses due to neutrino emission. Because of the extreme conditions in the deep interior, various processes come into play that produce neutrinos that easily pass through the star and carry away energy. We shall not discuss these processes here (see §6.8) but in very hot white dwarfs they are efficient in cooling the interior and may cause a central temperature inversion. Under the latter circumstance heat flows inward and some is eventually given to the neutrinos. The luminosity loss due to neutrinos can rival or exceed optical luminosity of the entire star (see below).

Several new pieces of physics enter as the white dwarf cools. We have already mentioned solid-state effects in §3.6 and these can radically alter the equation of state. Because these are phase changes (e.g., crystallization), the thermal properties of the medium are also effected. Thus, for example, if the stellar material crystallizes, then latent heat is liberated, which can slow the cooling of the star. All these effects are incorporated into modern models (as in Iben and Tutukov, 1984; Lamb and Van Horn, 1975; Koester and Schönberner, 1986; Tassoul et al., 1990, and see especially §3 in the review of Fontaine et al., 2001).

An additional subtlety comes about with crystallization in a C/O core. The two elements do not crystallize at quite the same rate and the tendency is for oxygen to crystallize first, separate out from the carbon, and then sink deeper into the interior. Not only does this affect the composition profile but the sinking of the heavier oxygen relative to the carbon liberates some energy (as, in effect, a form of gravitational contraction). This too affects the rate of cooling. Without us getting involved in this complicated (and still somewhat contentious) issue, we suggest you peruse Isern et al. (2000, and references therein).

In the simple radiative model of the preceding section we ignored convection completely. This is a serious omission. As the white dwarf cools, surface convection zones grow and die out, thus affecting the rate of cooling in these important outer layers (see, e.g., Fontaine and Van Horn, 1976; Tassoul et al., 1990; and Fontaine et al., 2001). As part of the modeling of these convection zones we also require adequate and consistent equations of state and opacities for the envelope. These can be very difficult to compute because of nonideal effects for the multicomponent gases involved.

Mixing of elements by convection is also possible, which can change the photospheric abundances and confuse the issue of spectral classification. We again remind you of the theoretical difficulties in describing stellar convection. Thus far, the MLT (see Chap. 5) is used for these evolutionary studies with its attendant problems.

Now how do we account for the near purity of elements in the atmospheres of DA and DB white dwarfs? This is not only an important observational issue but it also impacts on the evolution: hydrogen and helium are different. Even at the trace element level it has important consequences because, for example, opacities are strongly effected by even trace amounts of heavy elements. The prime cause of this purity is "gravitational settling." This term, although it is frequently used in the literature, is somewhat of a misnomer, although it is convenient to picture light elements floating and heavy elements sinking. It is true that the gravitational field is ultimately responsible for separation of heavy from light, but the immediate cause is the presence of pressure gradients and the resulting imbalance of forces on ions. The derivation of the rate of separation is beyond what we shall do here but it contains some very interesting physics. A classical derivation is contained in Chapman and Cowling (1960).

Countering the effects of gravitational settling is the normal process of diffusion whereby gradients in composition force elements to diffuse and thus reduce the gradients. In addition, "radiative levitation" can cause elements to rise by means of radiative forces acting on specific trace atoms through bound–bound and bound–free transitions. The net effect of these diffusive processes is very complex, but the bottom line is that evolutionary time scales in high-gravity white dwarfs are amply long for separation to become complete (although some elements in the deep core may resist). Other consequences of separation and diffusion for white dwarfs are discussed in Fontaine and Michaud (1979), Michaud and Fontaine (1984), and Iben and MacDonald (1985). Some combination of these processes must at least be partially responsible for what seems to be the changing faces of DB to DA and back to DB again as evolution goes on (as briefly discussed in §2.6.1). We also remark that these processes are of considerable interest for other stars as in the peculiar abundance spectral class A stars (Michaud, 1970).

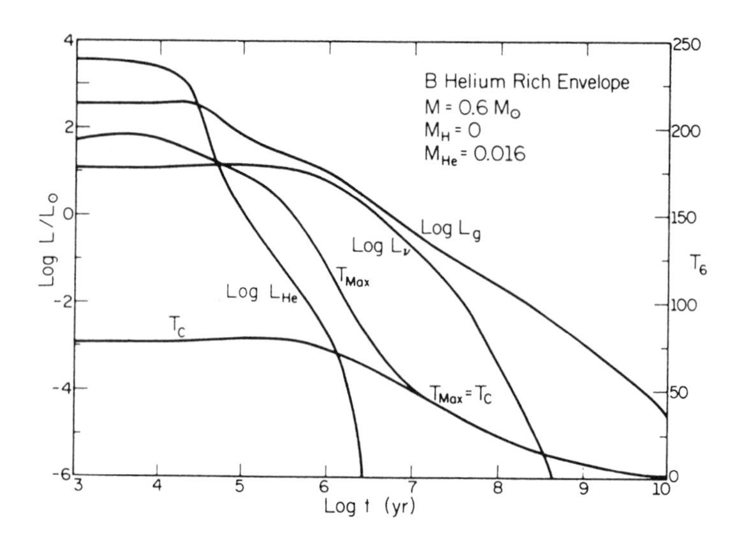

Fig. 10.2. Shown as a function of time are luminosities during the evolution of a $0.6\,\mathcal{M}_\odot$ helium atmosphere white dwarf. Maximum and central temperatures are also indicated. Reproduced, with permission, from Iben and Tutukov (1984).

Finally, we show some evolutionary results from the work of Iben and Tutukov (1984, who give lots of graphical summaries). Their methods parallel those of other authors and are representative of modern efforts. The model is that of a $0.6\,\mathcal{M}_\odot$ DB helium atmosphere white dwarf and the mass of the helium layer is $0.016\,\mathcal{M}_\odot$. Figure 10.2 shows the time evolution of various components of luminosity and temperature. The quantity $\mathcal{L}_g$ is the total of the luminosity released from internal thermal and gravitational potential energies. The luminosity generated by helium shell burning is denoted by $\mathcal{L}_{\rm He}$ and neutrino losses are represented by $\mathcal{L}_\nu$. Also shown is the maximum temperature in the model ($T_{\rm max}$) which, because of neutrino losses, is not necessarily located at the center of the model where the temperature is T_c.

The total photon luminosity for the same DB model is shown in Fig. 10.3 as "Model B" (as are the results for a DA white dwarf sequence). Also shown are the late evolutionary effects of liquefaction and crystallization. The legends involving $\dot{\mathcal{M}}_{\rm acc}$ indicate the luminosity released by gravitational potential energy if mass is accreted onto the stellar surface. All these effects are only noticeable after the total luminosity falls below $10^{-2}\,\mathcal{L}_\odot$. You may wish to compare this figure to the dotted line in Fig. 10.1, which shows the simple analytic cooling curve.

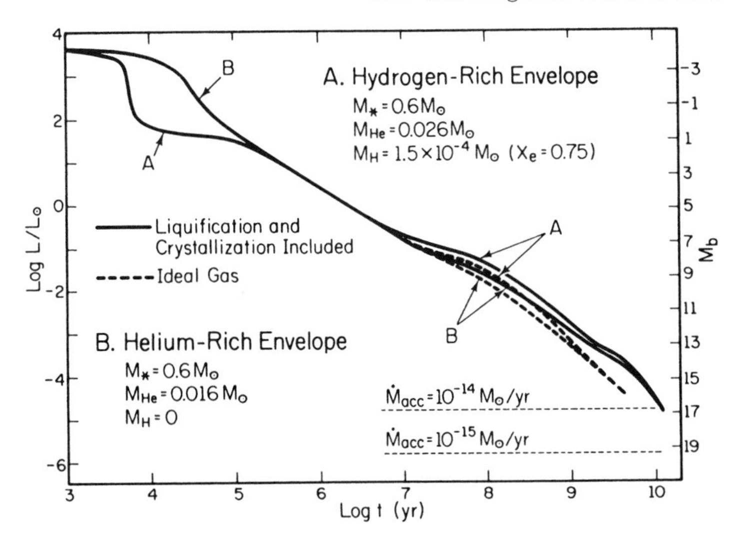

Fig. 10.3. Luminosity versus time for two evolutionary sequences. Model "B" is the same as in Fig. 10.3. Model "A" is a DA star with parameters given in the figure. Reproduced, with permission, from Iben and Tutukov (1984).

10.3 The Magnetic White Dwarfs

There are two primary ways of detecting magnetic fields in white dwarfs. The first, and most sensitive, is by measuring linear and quadratic Zeeman effects in spectral lines if those lines can be recognized in the strong field objects. This technique requires strong lines, but these are not always to be found, and, even if they are present, the inherent dimness of white dwarfs often defeats the observer. (A check on some results may be made by observing gravitational redshifts of the lines.) The second method depends on the detection of continuum circular polarization and is especially useful when magnetic field strengths are high. At the present time, the lower limit for detectable fields is about 10^4 G, except under unusually favorable circumstances, and that's 100,000 times bigger than the sun's average magnetic field!

The compilation of magnetic field strengths by Angel et al. (1981) lists measurements for over 100 white dwarfs. The results fall into three categories: (1) upper limits of a few thousand Gauss (these are the relatively rare cases where the observing conditions are favorable); (2) possible detections at around 10^5 G (where, in most cases, the errors bars on the measurements preclude a firm determination at that level); and (3) fields clearly in excess of 10^6 G (=1 MG). Thus it appears that white dwarfs either have "weak" fields or very strong fields. The number of these strong-field white dwarfs is, however, very small and they are now referred to as the "magnetic white dwarfs." Schmidt (1988) lists 24 known magnetic white dwarfs. The inferred strengths of the polar field (assuming dipole geometry) range from about 2 MG up to the strongest at perhaps over 500 MG for the star PG1031+234.

The polarization in this last object is modulated with a period of three and a half hours due to rotation and this allows the surface of star to be "scanned" as a function of time. From this, the geometry of the magnetic field may be inferred, and this is shown in Fig. 10.4 (from Schmidt et al., 1986). There appears to be a global field that is dipolar in nature but the axis of the field is inclined away from the rotation axis (an "oblique" rotator). In addition there is a magnetic "spot" on the surface whose central field may approach 1,000 MG (Latter et al., 1987)!

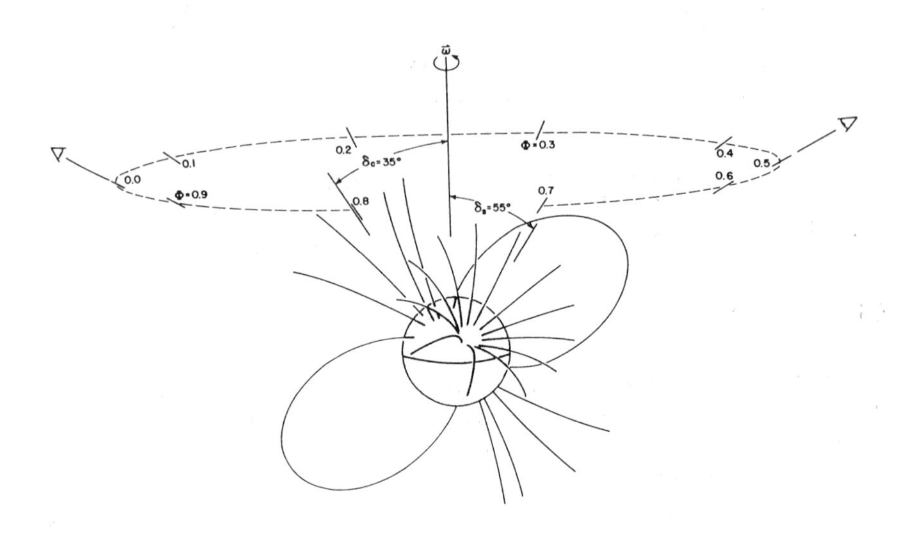

Fig. 10.4. The inferred field geometry of the magnetic white dwarf PG1031+234. A dipole-like field that is not aligned with the rotation axis is accompanied by a magnetic "spot." Fields on the surface of this star may approach 10^9 Gauss. Reproduced, with permission, from Schmidt et al. (1986).

A more recent survey is that of Schmidt and Smith (1996) who report on Zeeman spectropolarimetric observations of 48 DA white dwarfs. The mean error in the longitudinal magnetic field measurement is 8,600 G. An intensive "mini-survey" of a selected few white dwarfs brings the mean error for that sample down to a mere 2,000 G. They note that a null result (within the errors) may still not reveal the presence of a magnetic field because, as may well be the case, a tangled field, although strong locally, may escape detection. (Schmidt and Grauer, 1997, report the apparent absence of fields on three variable white dwarfs, but that is a story for later.)

10.3.1 Magnetic Field Decay

Are strong (or weak) fields in white dwarfs a surprise or not? The origin and evolution of magnetic fields in stars is still not at all well understood. Some fields may be the remains of interstellar medium magnetic fields that were trapped during the process of star formation. It is more likely, however, that fields are produced in situ by dynamo action caused by the interaction of convection and rotation (see, for example, the review by Parker, 1977). Important as it is, we shall not attempt to develop this topic. We can, on the other hand, do some simple calculations to see if we should be surprised at the strength of these fields.

The average surface magnetic field of the sun (as a typical white dwarf progenitor star) is of the order of one Gauss. Suppose this is representative of the interior down through the convection zone (although one Gauss is probably too low). If we were to now suddenly turn off the mechanism for producing this field, how long would the field persist? The time development of a magnetic field is given by (see Jackson, 1999)

$$\frac{\partial \mathbf{B}}{\partial t} = \nabla \times (\mathbf{v} \times \mathbf{B}) + \frac{c^2}{4\pi\sigma} \nabla^2 \mathbf{B} \,. \tag{10.7}$$

The units used here are Gaussian (see the Appendix of Jackson, 1999) and there is no clear agreement within the astronomical community on a common set of units. The quantity σ is the conductivity and has the units s^{-1}. If we ignore the first term on the righthand side (in the spirit that no currents are flowing), then the remainder is a diffusion equation, meaning that the magnetic field will decay away unless replenished. Taking a dimensional approach, a characteristic diffusion time is

$$\tau = \frac{4\pi\sigma L^2}{c^2} \tag{10.8}$$

where the length L is a measure of the spatial variation of the field. For the remainder of the discussion this will be set equal to the stellar radius $\mathcal{R}$ (or you can use a reasonable fraction thereof).

An estimate for the conductivity of the ideal ionized gas is given by Spitzer (1962, §5.4) which is, after changing to our units,

$$\sigma = \gamma_\mathrm{e} \frac{2(2kT)^{3/2}}{\pi^{3/2} m_\mathrm{e} Z e^2 \ln\Lambda} \,. \tag{10.9}$$

The quantity γ_e depends on the ionic nuclear charge, Z (for complete ionization), and ranges from unity for very large Z down to 0.582 for hydrogen. If we assume that the sun is pure hydrogen, then its conductivity is

$$\sigma(\mathrm{H}) \approx 1.4 \times 10^8 \, \frac{T^{3/2}}{\ln\Lambda} \tag{10.10}$$

where an estimate for Λ is (see Spitzer 1962)

$$\Lambda \approx 10^4 \frac{T^{3/2}}{n_e^{1/2}} .$$

For a typical virial estimate temperature of $T \approx 2 \times 10^6$ K from (1.36) and average density of $\langle \rho \rangle = 1.4$ g cm^{-3}, the solar magnetic field decay time is $\tau \sim 7 \times 10^{18}$ s or 2×10^{11} years. This is an overestimate (the time should be more like $\sim 10^{10}$ years), but the end result is that the field should persist at reasonable strength through the main sequence stage. If this field is not lost in later stages (and planetary nebula ejection might do just that), then the final white dwarf should retain the remnants of the field.

To estimate what the field strength in the just-formed white dwarf might be, we return to the original equation for the magnetic field evolution. Since the conductivity seems to be large (and, as we shall show, will remain so), we neglect the diffusion term; that is, we go to the infinite conductivity limit. In this limit the field lines are frozen into the plasma and the magnetic flux, Φ_{mag}, is conserved (in a Lagrangian sense—see Jackson, 1999). This may be put crudely as "$\Phi_{\text{mag}} \sim B\mathcal{R}^2$ remains constant as radius changes." Thus if the sun has a roughly 1-G field, then the just-formed white dwarf (with a radius of a little over $0.01\,\mathcal{R}_\odot$) should have a field of the order 10^4 G. Pushing all other uncertainties aside, this means that typical white dwarfs should have weak fields. Where do the magnetic objects come from?

A best guess is that the progenitors of the magnetic white dwarfs are the Ap stars. (See, for example, the review by Angel, 1978, and, for the Ap stars, the monograph by Jaschek and Jaschek, 1987.) These stars have anomalously intense magnetic fields that may be as high as 4,000 G and their population statistics are consistent with the number of magnetic white dwarfs versus weak field objects. And, not so incidentally, there is a subclass called the "rapidly oscillating Ap stars" (roAp) which are nonradial variables in which the magnetic field is aligned obliquely to the rotation axis. (For reviews see Kurtz, 1986; §9 of Unno et al., 1989; and our §2.10.)

Once having been formed with strong magnetic fields, these objects can retain their fields for long times. To substantiate this we now allow for diffusion and consider the finite conductivities of a nonrelativistically degenerate pure carbon plasma given by Wendell et al. (1987):

$$\sigma = 10^9 \frac{T^2}{\rho \kappa_{\text{cond}}} \quad \text{s}^{-1} . \tag{10.11}$$

Here κ_{cond} is the conductive opacity for which we gave an estimate in (4.72). Inserting that estimate for pure carbon we find that

$$\sigma \approx 2 \times 10^{15} \rho \quad \text{s}^{-1} . \tag{10.12}$$

For an average white dwarf density of 5×10^5 g cm^{-3} this yields a decay time of 2×10^{11} years. This is an overestimate (as in the case of the sun), but the detailed calculations of Wendell et al. (1987) indicate that the decay times for simple dipole fields are always longer than the evolutionary time. More complicated fields with higher multipole moments tend to decay faster, implying that after long times the geometry of the fields should simplify.

Finally, are strong magnetic fields a factor in evolution? A rough way to gauge their importance is to compare the magnetic field pressure $P_{\rm mag} = B^2/8\pi$ to the gas pressure. Wendell et al. (1987) point out that the central values of the fields in their models are some ten times the surface values. For a surface field of $B = 10^9$ G, as rough upper limit thus far, this implies a central magnetic pressure of $P_{\rm mag} \approx 4 \times 10^{18}$ dyne cm^{-2}. But this is far smaller than the hydrostatic estimate $P \sim G\mathcal{M}^2/\mathcal{R}^4 \sim 4 \times 10^{23}$ dyne cm^{-2} required for equilibrium in a typical white dwarf. They also suggest that $P_{\rm mag} \ll P$ holds for all times because flux conservation ($B\mathcal{R}^2$ a constant) implies $P_{\rm mag} \sim \mathcal{R}^{-4}$, which has the same dependence on $\mathcal{R}$ as the hydrostatic pressure. The conclusion is that we may safely neglect the effects of magnetic fields as far as deep interior evolutionary calculations are concerned. The same may not be true for regions of the star near the surface because the pressures are relatively low there. At the very least, opacities are affected by fields because of their effects on atomic energy levels, and this, after all, is how the fields may be detected in the first place.

10.4 The Variable White Dwarfs

It was once thought that white dwarfs were extremely stable in their light output, so much so that they could be used as luminosity standards for faint variable stars. Acting on this assumption, A.U. Landolt observed the white dwarf HL Tau 76 with the intent of using it as a standard star. To his surprise, and as reported in Landolt (1968), he found instead that this star was variable with a period of 12 minutes and with luminosity variations of over a tenth of a magnitude. Thirty-six these variable white dwarfs have been discovered and, from considering the statistics of the total white dwarf population, McGraw (1977, and see Cox, 1982) concludes that this class of variable star is the most common in the universe among those that are obviously variable. There are also 15 variables that we have not included at this point. These should perhaps more properly be called pre-white dwarfs because they are either PNNs or PNNs that have but recently lost their surrounding nebulosities. (We suppose the question is, "When is a white dwarf a white dwarf?" As you will see, we will often hedge on the question.) For an accessible compilation of most of the presently known variables see Bradley (1999). In any case, the following discussion will include the true white dwarfs and their immediate progenitors.

The study of variable white dwarfs is still in the stages of active observational and theoretical development and we can only touch on the high points here. Reviews include Winget (1988), Kawaler and Hansen (1989), and Fontaine et al. (2001).

10.4.1 The Observed Variables

As briefly discussed in §2.10, there are two major classes of variable white dwarfs. All are multiperiodic, with periods ranging from roughly 100 to 1,000 s. The coolest are the hydrogen-surfaced DAVs or "ZZ Ceti" variables. They lie in the effective temperature range $12,500 \gtrsim T_{\rm eff} \gtrsim 11,300$ K (Bergeron et al., 1995). The hot (cool) end of this range is called the "blue (red) edge" because of color. At just a bit over 1,000 K, the interval in temperature (and color) is narrow and well defined and is called the "instability strip."[6] (The same nomenclature applies to the variable stars of the Cepheid strip discussed in chaps. 2 and 8.) Twenty-eight of these stars have been discovered, and HL Tau 76 is among them.

The second class are the DBVs with helium surfaces lying in the temperature range $28,000 \gtrsim T_{\rm eff} \gtrsim 22,000$ (Beauchamp et al., 1999). There are eight known.

An additional class, as pre-white dwarfs, are not as well-defined as the above two. These are very hot ($8 \times 10^4 \lesssim T_{\rm eff} \lesssim 1.7 \times 10^5$ K) and are either DO (pre-)white dwarfs or nuclei of planetary nebulae. The prototype DOV is PG1159-035 discovered by McGraw et al. (1979) and is otherwise known as GW Vir. The coolest of these could very well be called very hot white dwarfs. The prototype PNN variable (PNNV) is K1–16 that is embedded in a planetary nebulosity, which makes it difficult to observe as a variable star (Grauer and Bond, 1984).

These are the immediate essentials of the observational characteristics of the white dwarf variables. More detailed observational material will be discussed below in considering what the variables tell us about structure and evolution.

10.4.2 White Dwarf Seismology

Because typical periods for well-studied variable white dwarfs are around a few hundred seconds, the observed oscillations cannot be acoustic modes. The estimate for the frequencies of p-modes given by (8.111) sets an upper limit for periods of $\Pi_{\rm p} \lesssim \pi \int v_s^{-1}\, dr$. But this is essentially the period–mean density relationship discussed in §1.3.5. A typical average density for white dwarfs is $\langle \rho \rangle \approx 10^6$ g cm^{-3}, so that using (1.40) with a coefficient of 0.04 day yields $\Pi_{\rm p} \lesssim 4$ s. This is too short by at least a factor of 20 to match the observations. What is left are gravity modes. The following outlines the

[6] There is some evidence that not all DAs in the instability strip pulsate. If not, then the name is a misnomer. See Kepler and Nelan (1993).

arguments originally developed by Chanmugam (1972), Warner and Robinson (1972), and put on a firmer numerical footing by Osaki and Hansen (1973).

The periods of gravity modes depend on the run of the Brunt–Väisälä frequency, N^2, as shown by (8.112). There is really no way we can estimate that quantity easily but it does have certain distinctive qualitative features in white dwarfs. For example, it is *very* small in the electron-degenerate interior. This may be seen by comparing (5.29), which gives a definition of N^2, and (3.116), which shows how χ_T varies with temperature for a degenerate gas. The point is that for the (relatively) low temperatures deep inside white dwarfs, N^2 is small. This is not necessarily the case in the envelope and typical values of a few thousand s^{-2} may be encountered. (It will, of course, be negative in convection zones.) On the other hand, the Lamb frequency S_ℓ is large in the interior but becomes very small in the envelope as was the case for the sun.

If we now recall our discussion of the conditions for wave propagation (see 8.108 and following), it becomes clear that g-modes propagate in the envelope regions, whereas p-modes (which do not seem to exist in these variables) tend to do so in the deeper interior. Note that this is the opposite from the sun. Thus we have the picture of gravity modes actively waving around in the surface regions but being excluded from the core because of very small values of N^2 deep inside. Detailed numerical calculations, as reviewed in the primary references, yield periods of the length observed.

The cause of the instability has been determined to be the same as that which drives more classical variable stars: it is associated with some combination of the ionization zones of hydrogen or helium and perhaps carbon in the hottest objects (Dziemboski and Koester, 1981; Dolez and Vauclair, 1981; Winget, 1981; Starrfield et al., 1982; Winget et al., 1982a; and see O'Brien, 2000).[7] Part of the great (and relatively recent) success of this program was the search for, and discovery of, the DB variables (by Winget et al., 1982b). The existence of these variables was predicted by theory. This is the first class of variable star not to have been found by accident.

The calculations that test for stability for g-modes have been remarkably successful for the DAV and DBV stars and the results agree reasonably well with the observed location of their respective blue edges. Although there are some differences of opinion on the details of precisely how much mass is tied up in surface hydrogen or helium layers, the cause of instability is now understood, and our knowledge of the overall structure of these stars is secure.

The situation for the very hot DOV and PNNV variables is not as rosy. Theoretical periods derived from adiabatic pulsation studies have no trouble

[7] Another destabilizing mechanism for the DAVs has been proposed by
▹ Goldreich, P., & Wu, Y. 1999, ApJ, 523, 805
that involves efficient surface convection where ionization is not explicitly responsible for the driving.

matching the observed periods for these stars but the exact cause of the instability is still somewhat uncertain. The difficulty is that the evidence from spectroscopy is not clear enough to determine the precise composition of their photospheres. If this were known, then more reasonable guesses could be made to model the interior layers close to the surface. It is known, however, that helium, carbon, and oxygen are present in the photospheres and it is very likely that ionization of some of these elements is sufficient to drive the star to instability. To confuse the issue further, non-variables coexist along with variables—and you can hardly tell some of them apart by their spectra (Werner, 1995).

Another problem that arises with the hot variables is that theoretical studies suggest that these stars should be driven unstable due to the ε-mechanism (§8.2.1) operating in hydrogen or helium burning shells left over from the previous evolutionary stages and that oscillations with periods of from 50 to 200 seconds should be seen (Kawaler, 1988; Kawaler et al., 1986). These calculations are based on standard evolutionary models in which active burning shells are present. The problem is that the hot variables show no evidence for such short periods (e.g., Hine and Nather, 1988). This is very disturbing and implies that either some adjustments have to be made in our evolutionary calculations or the pulsation work is somehow incorrect. If it is with the models, then our ideas about how white dwarfs are formed from AGB stars may be flawed. Sounding out such things is one of the roles of asteroseismology.

Another task is detecting evolution in action. This has been reported by Costa et al. (1999, and see Winget et al., 1991). Using data spanning 10 years (but, of course, not continuously), they were able to detect a secular change of period in one otherwise very stable oscillation period (at 516 s) in the hot and (presumably) rapidly evolving DOV star PG1159–035. The latest update (Costa et al., 1999) on the rate of period change for this 516-s g-mode is $\dot{\Pi} = (+13.07 \pm 3) \times 10^{-11}$ s s^{-1}, which corresponds to an e-folding time ($\Pi/\dot{\Pi}$) for an increase in period of about 10^5 years.[8] This time scale is a bit shorter than the e-folding time for luminosity decrease derived from evolutionary models of PG1159–035-type stars, and in fair agreement with pulsation calculations of these same models. By "fair" we mean within a factor of ten but, considering the complexity of the problem, this is really rather good. We shall return to PG1159–035 in a bit.

Detecting secular period changes in the cooler white dwarfs is much more difficult because they cool and evolve so slowly. Kepler et al. (2000) have reported a rate of period change for a $\Pi = 215$-s mode of $\dot{\Pi} = (2.3 \pm 1.4) \times 10^{-15}$ s s^{-1} in the DAV star G117–B15A. This corresponds to an e-folding time of 3×10^9 years. If this $\dot{\Pi}$ is due solely to processes intrinsic to the star

[8] Costa et al. (1999) quote some smaller error bars for this $\dot{\Pi}$, but, being conservative, we choose the largest they obtain. In any case, their $\dot{\Pi}$ implies that the 516-s period has increased by only about 40 ms over the 10 years of observation. See their paper to find out how such magic is performed.

(i.e., with no significant contributions from proper motion or orbital effects from a distant companion), then it is consistent with DA models for G117–B15A. So, for example, using models of Bradley (1998), this star should be contracting (by cooling) at a rate of 1 cm year^{-1} and the thickness of the hydrogen layer is a mere 1.5×10^{-4} of the total stellar mass. Remarkable.

White Dwarfs and the Whole Earth Telescope

Perhaps the best way to summarize the successes of white dwarf seismology is to review what we know about the DOV star PG1159–035 discussed above and how that information was obtained. This is a case history in asteroseismology.

One of the prime difficulties met in extracting information from a variable star are the constraints placed on observation by the rotation of the earth and the seasonal aspect of the constellations. If we observe from a single telescope, then information is lost during the daylight hours and roughly half a year is lost each year since the star is not in the nighttime sky (to say nothing of weather). This is a serious problem for variable white dwarfs because what is needed is *resolution* of the multiperiodic oscillation structure. As an example, consider a hypothetical variable that is pulsating in two modes whose frequencies are spaced a mere 4 μHz apart. If we were to observe this star for eight hours over only one night then there is no way that we could tell there were actually two periods present. This is because of the relation between length of observation and resolution in frequency, which we can show using the properties of Fourier transforms.

If we observe a sinusoidally periodic signal of frequency f_0 over a finite time span T, then the amplitude of the Fourier transform of that signal is *not* a delta function at f_0. Instead, we find a relatively broad peak around f_0 with a width in frequency of approximately 1/T with "sidelobes" of lower amplitude extending out on either side. This means that we may see a peak at f_0 but the uncertainty in the exact location of the frequency of that peak is perhaps as large as 1/T. (Heisenberg would be amused. This is just another version of the uncertainty principle between time and energy as applied to astronomical observations.) Thus if we observe for eight hours, the uncertainty in frequency is 1/(8 hours) or 3.5×10^{-5} Hz, or 35 μHz, which is far larger than the 4-μHz spacing between our two hypothetical modes. The net result is that we would not be able to resolve the two peaks in the Fourier transform and thus would not be able to tell there were two different signals present. The way around this is, obviously, to observe the star long and continuously. And this gets us back to PG1159–035.

Winget et al. (1991) reported observations of PG1159–035 taken nearly continuously around the 24-hour clock for a period of two weeks using the "Whole Earth Telescope" (WET). This remarkable instrument consists of up to (depending on the circumstances) 13 or so individual telescopes, with cooperating observers in attendance, spaced around the world in longitude whose duty it is to observe, during their individual nighttimes, a single white dwarf or other kind of variable as the earth turns. You might say that even though the earth turns, the telescope doesn't. The data gathered from the high-speed photometers is relayed by electronic mail to a single control site (presently in Ames, Iowa) where the information is analyzed in almost real time. The operation of the WET is reported in Nather et al. (1990).

The WET was used for two weeks of "dark time" (no moon) in 1989 to observe PG1159–035 almost continuously. A small sample—only about seven hours out of a total of 264 hours of data—of the light curve from that star is shown in the insert of Fig. 10.5 (earlier shown as Fig. 2.24), where the ordinate is the relative intensity, in visible light, around the mean. We have pictured this as a continuous curve but it really consists of about 2,400 individual points spaced 10 s apart, which is the integration time for the photometers. The full light curve is displayed in Winget et al. (1991).

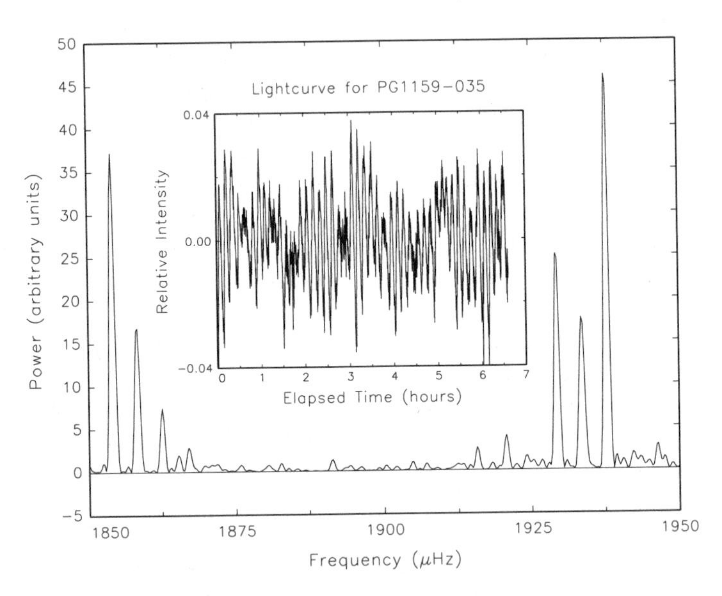

Fig. 10.5. Shown in the insert is a small segment of the light curve of PG1159–035 from a 1989 WET run. The main frequency peaks, around periods of 500 s, are shown in the main figure.

If you were to take a ruler and measure off a rough spacing between individual swings of data in the insert you would find that, on the whole, they represent a sinusoid-like signal, or signals, with a period of about 500 s. This is verified by the main curve in the figure, which shows the Fourier transform of the entire 264-hour data set for a very small interval in frequency. The largest peak (in power or amplitude squared) at $f \approx 1,937$ μHz represents a single g-mode oscillation whose period is 516.04 s (and is the same one whose rate of change of period has been measured). The peak to the left of this one is lower in frequency by 4.3 μHz. These peaks are well resolved because the two-week duration of WET implies a frequency uncertainty of only 1/(2 weeks) or about 0.8 μHz. You will note that there are three well-defined and equispaced peaks in the righthand portion of the figure. (The much smaller peaks around them are due to inevitable noise in the observations and sidelobes from gaps in the data.) This triplet is due to the effects of rotation on an $\ell = 1$ g-mode showing the $2\ell + 1$ m modes (see §8.3). If the rotation were uniform, then (8.121) gives a rotation period of 1.4 days, which is typical of white dwarfs. There may be no way in the foreseeable future to detect this rotation by spectroscopic means for this star.

Note that there is another triplet down at about 1,850 μHz (periods around 538 s) and this is another $\ell = 1$ rotationally split g-mode. From model calculations of DO stars, this mode and its neighbor at 516 s are of harmonic order $n \approx 20$ and the two modes differ by one in n. This complex of strong peaks in the vicinity of 500 s is the cause of the curious structure of the light curve, which shows a modulation in intensity with wide swings and nulling superimposed on the main ups and downs of 500 s. The situation is analogous to the musical interference beats heard from an orchestra whose members are playing *nearly* the same notes. Here we *see* the beats.

Space prevents us from presenting the entire Fourier spectrum for PG1159, but Winget et al. (1991) identified 101 modes in this star, including many more with $\ell = 1$ and a number of rotationally split quintuplets for which $\ell = 2$. Because even WET observations cannot resolve the disk of the star—unlike the sun—it is unlikely that we can detect modes with $\ell \gtrsim 3$, were they to be present, because of the effects of light cancellation over the disk of the complicated patterns of high ℓ spherical harmonics. Even so, from this wealth of data, it is possible to get a period spacing between modes of given ℓ. Using the $m = 0$ (the central component of the triplets in Fig. 10.5) as the mean for a multiplet, the average spacings between consecutive ns is $\Delta P_{\ell=1} = 21.6$ s and $\Delta P_{\ell=2} = 12.5$ s. Note that the ratio of these two spacings is consistent with (8.112, reproduced below as Eq. 10.13), which contains the factor $\sqrt{\ell(\ell+1)}$; that is, $21.6/12.5 = 1.73$ is the same as (within the error bars not quoted here) $\sqrt{2(2+1)/1(1+1)} = \sqrt{3} = 1.732$.

The significance of this result is the following. Recall that $\Pi_0/\sqrt{\ell(\ell+1)}$ is the period spacing in

$$\Pi_{\mathrm{g},n} = \frac{2\pi}{\sigma_{\mathrm{g}}} \approx n \frac{2\pi^2}{\left[\ell(\ell+1)\right]^{1/2}} \left[\int_0^{\mathcal{R}} \frac{N}{r}\, dr\right]^{-1} = \frac{n\,\Pi_0}{\left[\ell(\ell+1)\right]^{1/2}} \tag{10.13}$$

so that $\Pi_0 = 21.6\sqrt{2} = 30.5$ s, as will be used shortly.

PG1159 has an effective temperature of $140,000 \pm 5,000$ K and a rather uncertain $\log g = 7.0 \pm 0.5$ from spectroscopic observations. The luminosity is estimated to be $\log \mathcal{L}/\mathcal{L}_\odot = 2.7 \pm .5$ (Werner et al., 1991).[9] Kawaler and Bradley (1994), using a grid of evolutionary models and non-radial adiabatic pulsation calculations, have come up with the following fit between mass, luminosity and q_y (the fraction by mass of surface helium) versus Π_0 of (10.13):

$$\Pi_0 = 15.5 \left(\frac{\mathcal{M}}{\mathcal{M}_\odot}\right)^{-1.3} \left(\frac{\mathcal{L}}{100\,\mathcal{L}_\odot}\right)^{-0.035} \left(\frac{q_y}{10^{-3}}\right)^{-0.00012} . \tag{10.14}$$

Noting the very weak dependence on q_y implied by its very small exponent, we set that term to unity (but will make amends shortly). Then, using the values of Π_0 and $\log \mathcal{L}/\mathcal{L}_\odot$ given above, find that (10.14) yields $\mathcal{M} = 0.58$ $\mathcal{M}_\odot$ for PG1159. Recall that this figure is the typical mass for single white dwarfs derived from spectroscopy. The WET data and analysis yield the mass using remote seismology.

There are, however, small and systematic deviations from the average spacing between the $\ell = 1$ modes. This is due to the presence of the composition discontinuity between the helium/carbon/oxygen surface layer and the carbon/oxygen core. Since nature (and eigenfunctions) abhors discontinuities in physical properties, it is not surprising that the periods of some modes are affected in subtle—but *predictable* ways. It is similar to problems in quantum mechanics when considering a potential well with sharply varying depths. A particle passing over these depths feels their effect and the eigenfunction *and* eigenvalue (the frequency or period in our case) may behave in strange ways. For the stellar case, a composition layer may partially "trap" the eigenfunction when a node coincides, or nearly coincides, with the discontinuity in composition. In effect, the mode is tuned to the thickness of the layer and the period of the mode can differ from that predicted by simple asymptotic theory (i.e., 10.13). Modes that are "out of tune" more closely follow the asymptotic relation. The effects of trapping are fully discussed by Brassard et al. (1992) and depend on the thickness (in both radius and mass) of the surface layer and the severity of the discontinuity (e.g., by how much mean molecular weight changes). In any case, we need no longer expect that period spacings be exactly equal.

As an example, Kawaler and Bradley (1994, and see O'Brien, 2000) have explored the effects on PG1159 model period spacings by varying surface composition and layer thickness and have compared their results to those observed. They find that a composition of 27% by mass of helium with a 20%–60% mix of carbon to oxygen (consistent with spectroscopic results) in a layer approximately $q_y = 0.004$ of the stellar mass does very well (and now we know what q_y to put in 10.14). Our version of their Fig. 10 is shown

[9] The distance to PG1159 is not well determined and thus neither is its luminosity. The numerical result we obtain for the mass of PG1159, however, is within about ± 0.01 $\mathcal{M}_\odot$ using the quoted error bars on the luminosity. The important factor is the period spacing.

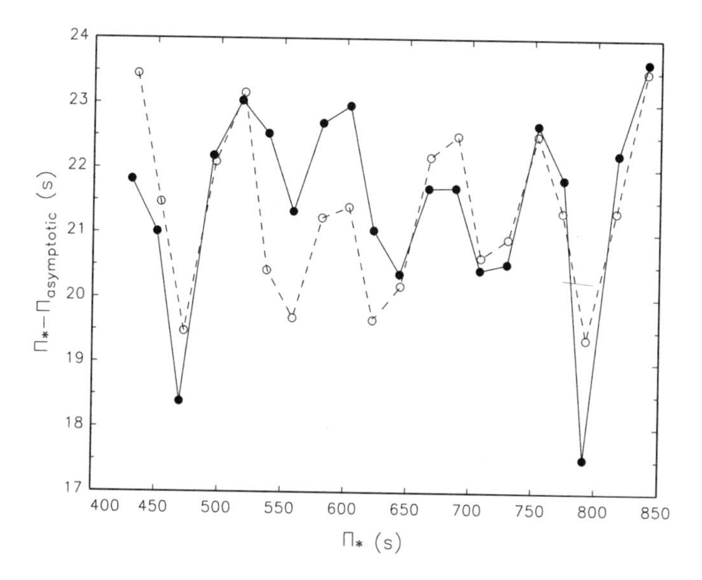

Fig. 10.6. Shown is the comparison of differential period spacings for 20 modes in a model of PG1159 (dashed curve with open circles) versus those seen in PG1159 itself (solid curve with filled circles) as a function mode period. "Differential period spacing" here is the difference between a period of either the star or model ($\Pi_\star$) and the period the mode *would* have if the set of periods were to follow the asymptotic relation of (10.13) as determined by a linear fit to the sequence of mode periods. The consecutive sequence starts with mode order $n = 18$ of period $\Pi = 430$ s.

in Fig. 10.6, and it compares the variation in period spacings found from their model to those observed in PG1159.[10] (Needless to say, the periods themselves are very close: the maximum discrepancy between model and star is only 1% and the average is 0.2% over the twenty modes shown.) The game, so to speak, is to match the ups and downs of deviations from equal period spacing. In the figure the maximum discrepancy between star and model is one and a half seconds (and remember that periods are in the 500-s range). A very satisfying result.

What about the effects of magnetic fields on the pulsations of white dwarfs? Jones et al. (1989) looked into this using the techniques outlined in §8.4 (and see 8.119). For simple field configurations they found that normal modes can be split in ways that, in principle, could be distinguished from splitting due to rotation. Winget et al. (1994) suggested that some strange splittings in GD358 (the first DBV discovered, otherwise known as V777 Her) could be due to magnetic fields with strength 1,300 G, or so, based on the

[10] Our version of this kind of figure differs from what you will find in the literature. Our aim is to compare the effects of mode trapping on what would otherwise be a smooth asymptotic sequence of periods. If the modes were equispaced, the curves would be horizontal lines.

Jones et al. (1989) analysis. Intensive spectropolarimetric observations by Schmidt and Grauer (1997), however, failed to turn up fields of that strength (at least within their error bars). To complicate matters, Vuille et al. (2000) and Kepler et al. (2003), using WET, found that GD358 shows strong evidence for a pulsation spectrum that is replete with signals composed of sums and differences of frequencies, among other things. These are thought (probably correctly) to be the result of nonlinear interactions between modes. The pulsation spectrum of that star is sufficiently complex that a magnetic field could be hiding there but we may never pick it out from all else that is going on.[11] So, thus far, we have struck out as far as magnetic fields are concerned, but there just might be a variable white dwarf out there that will surprise us. For the moment, on the other hand, as for seeing evolution in action, watching stars rotate, and dipping into the interior structure, we must say the program has been a gratifying success for asteroseismology—especially considering how tiny and dim these stars are.

10.5 Exercises

Exercise 10.1. In deriving the cooling time of (10.5, 10.6) we used Kramers' bound–free opacity as the opacity source in the surface layer of the white dwarf. Suppose, however, we had used electron scattering instead; that is, find $t_{\rm cool}$ for that much leakier opacity. Compare your result to Fig. 10.4. (Qualitatively you know what's going to happen.)

Exercise 10.2. Try to reproduce our Fig. 10.6 using the numbers you will find in Table 3 of Kawaler and Bradley (1994). Remember that you will have to fit the model and PG1159 periods to straight lines to obtain asymptotic sequences. You might also wish to simply perform numerical differences of the listed periods to get another version of our figure (as is usually done).

Exercise 10.3. Use what you can find in this text to estimate cooling times for PG1159 and G117–B15A and compare your results to the $\dot{\Pi}$s we quote for those stars. (The effective temperature of G117–B15A is 12,400 K and $\mathcal{R} \approx 9.6 \times 10^8$ cm.) Don't expect miracles. Be happy with a factor of ten, or even more, agreement, and be aware that cooling times need not, in any case, track period evolution exactly.

10.6 References and Suggested Readings

§10.1: Observed Properties of White Dwarfs

[11] Nonlinear interactions can be a nightmare and only the simplest situations have been explored for nonradial pulsators. For a taste of what is involved, see Buchler, J.R., Goupil, M.-J., & Hansen, C.J. 1997, A&A, 321, 159. Simple linear analyses, such as described here, can just be plain wrong.

There are several texts, papers, and reviews worth reading. Among these are chapters 3 and 4 of

▹ Shapiro, S.L., & Teukolsky, S.A. 1983, *Black Holes, White Dwarfs, and Neutron Stars* (New York: Wiley & Sons);

Chapter 25 of

▹ Cox, J.P. 1968, *Principles of Stellar Structure*, in two volumes (New York: Gordon and Breach),

and Chapter 35 of

▹ Kippenhahn, R., & Weigert, A. 1990, *Stellar Structure and Evolution* (Berlin: Springer–Verlag).

Important papers and reviews include

▹ Liebert, J. 1980, ARA&A, 18, 363

▹ Iben, I., Jr., & Tutukov, A.V. 1984, ApJ, 282, 615

▹ Tassoul, M., Fontaine, G., & Winget, D.E. 1990, ApJS, 72, 335

▹ Weidemann, V. 1990, ARA&A, 28, 103

▹ D'Antona, F., & Mazzitelli I. 1990, ARA&A, 28, 139

▹ Trimble, V. 1991, PASP, 104, 1, §10.5

▹ Trimble, V. 1992, PASP, 105, 1, §13.6

▹ Fontaine, G., Brassard, P., & Bergeron, P. 2001, PASP, 113, 409

▹ Koester, D. 2002, A&ARev, 11, 33

▹ Hansen, B.M., & Liebert, J. 2003, ARA&A, 41, 465.

Masses of single white dwarfs are discussed in

▹ Weidemann, V., & Koester, D. 1984, A&A, 132, 195

▹ Oke, J.B., Weidemann, V., & Koester, D. 1984, ApJ, 281, 276

▹ Weidemann, V. 1990, ARA&A, 28, 103

▹ Bergeron, P., Saffer, R.A., & Liebert, J. 1992, ApJ, 394, 228.

The classification scheme in Table 10.1 is relatively new and a form of it was first proposed by a group of astronomers who have had much to do with establishing the properties of white dwarfs. See

▹ Sion, E.M., Greenstein, J.L., Landstreet, J.D., Liebert, J., Shipman, H.L., & Wegner, G. 1983, ApJ, 269, 253.

Our table is adapted from Table 1 (plus discussion) of

▹ McCook, G.P., & Sion, E.M. 1999, ApJS, 121, 1.

This latter work is periodically updated by McCook and Sion, and it is an invaluable resource. It lists all known white dwarfs along with their colors, positions, etc.

Very cool white dwarfs are a rarity but they will turn out to be important for dating stellar populations. Two sample papers are

▹ Saumon, D., & Jacobson, S.B. 1999, ApJ, 511, L107,

▹ Oppenheimer, B.R., et al. 2001, ApJ, 550, 448.

Detecting rotation in white dwarfs is a difficult enterprise as you may see by reading

- ▹ Greenstein, J.L., & Peterson, D.M. 1973, A&A, 25, 29
- ▹ Pilachowski, C.A., & Milkey, R.W. 1987, PASP, 99, 836
- ▹ Koester, D., & Herrero, A. 1988, ApJ, 332, 910.

§10.2: White Dwarf Evolution

The essentials of white dwarf cooling were spelled out fifty years ago in the classic paper by

- ▹ Mestel, L. 1952, MNRAS, 112, 583.

The cooling time of (10.6) is derived from

- ▹ Iben, I. Jr., & Tutukov, A.V. 1984, ApJ, 282, 615

and see

- ▹ Iben, I. Jr., & Laughlin, G. 1989, ApJ, 341, 312

for more material. The cooling curves of Fig. 10.2, on the other hand, are from

- ▹ Winget, D.E., Hansen, C.J., Liebert, J., Van Horn, H.M., Fontaine, G., Nather, R.E., Kepler, S.O., & Lamb, D.Q. 1987, ApJ, 315, L77.

A selection of evolutionary calculations for white dwarfs includes

- ▹ Lamb, D.Q., & Van Horn, H.M. 1975, ApJ, 200, 306
- ▹ Iben, I. Jr., & Tutukov, A.V. 1984, ApJ, 282, 615
- ▹ Koester, D., & Schönberner, D. 1986, A&A, 154, 125
- ▹ Tassoul, M., Fontaine, G., & Winget, D.E. 1990, ApJS, 72, 335.

Section 3 of

- ▹ Fontaine, G., Brassard, P., & Bergeron, P. 2001, PASP, 113, 409

contains an excellent historical survey.

We defer any detailed discussion of crystallization and its effect on cooling to Fontaine et al. (2001), and see

- ▹ Hansen, B.M. 1999, ApJ, 520, 680

with further comments by

- ▹ Isern, J., Garcia-Berro, E., Hernanz, M., & Chabrier, G. 2000, ApJ, 528, 397.

An older, but still useful, study of the effects of convection is the work of

- ▹ Fontaine, G., & Van Horn, H.M. 1976, ApJS, 31, 467.

Nonideal effects in multicomponent mixtures is discussed in

- ▹ Fontaine, G., Graboske, H.C. Jr., & Van Horn, H.M. 1977, ApJS, 35, 293.

The diffusion of heavy versus light elements is discussed in

- ▹ Chapman, S., & Cowling, T.G. 1960, *The Mathematical Theory of Non-Uniform Gases* (Cambridge: Cambridge University Press).

Astrophysical applications, for both white dwarfs and other stars, may be found in

▹ Michaud, G. 1970, ApJ, 160, 641
▹ Fontaine, G., & Michaud, G. 1979, ApJ, 231, 826
▹ Michaud, G., & Fontaine, G. 1984, ApJ, 283, 787
▹ Iben, I. Jr., & MacDonald, J. 1985, ApJ, 296, 540.

§10.3: The Magnetic White Dwarfs

A series of paper and reviews concerning the magnetic white dwarfs includes

▹ Angel, J.R.P., Borra, E.F., & Landstreet, J.D. 1981, ApJS, 45, 457
▹ Schmidt, G.D., West, S.C., Liebert, J., Green, R.F., & Stockman, H.S. 1986, ApJ, 309, 218
▹ Schmidt, G.D. 1988, in IAU Colloquium No. 95, *Second Conference on Faint Blue Stars*, eds. A.G.D. Philip, P.S. Hayes, & J. Liebert (Schenectady: L. Davis Press), p. 377
▹ Latter, W.B., Schmidt, G.D., & Green, R.F. 1987, ApJ, 320, 308.
▹ Schmidt, G.D, & Smith, P.S. 1995, ApJ, 448, 305

list the presence (or not, depending on the error bars) of magnetic fields in 42 DA white dwarfs.

▹ Schmidt, G.D., & Grauer, A.D. 1997, ApJ, 488, 827

have looked at three variable white dwarfs and find no magnetic field to the limits of their measurements.

Generation of magnetic fields in stars is discussed in

▹ Parker, E.N. 1977, ARA&A, 15, 45.

Other references pertinent to our discussion include

▹ Wendell, C.E., Van Horn, H.M., & Sargent, D. 1987, ApJ, 313, 284
▹ Jackson, J.D. 1999, *Classical Electrodynamics*, 3nd ed. (New York: Wiley & Sons)
▹ Spitzer, L. 1962, *Physics of Fully Ionized Gases* (New York: Interscience)
▹ Angel, J.R.P. 1978, ARA&A, 16, 487
▹ Kurtz, D.W. 1990, ARA&A, 28, 607
▹ Unno, W., Osaki, Y., Ando, H., Saio, H., & Shibahashi, H. 1989, *Nonradial Oscillations of Stars* (Tokyo: University of Tokyo Press)
▹ Jaschek, C. , & Jaschek, M. 1987, *The Classification of Stars* (Cambridge: Cambridge University Press).

§10.4: The Variable White Dwarfs

The initial discovery of the variable white dwarfs was reported in

▹ Landolt, A.U. 1968, ApJ, 153, 151.

That these variables are the most common in (at least) our galaxy has been discussed by

▹ McGraw, J.T. 1977, *The ZZ Ceti Stars: A New Class of Pulsating White Dwarfs*, Ph.D. Dissertation, University of Texas, p. 228

▷ Cox, J.P. 1982, Nature, 299, 402.

The dissertation by McGraw reports the first comprehensive study of these stars.

▷ Bradley, P. 1999, in *Allen's Astrophysical Quantities*, §16.3, ed. A.N. Cox (New York: Springer-Verlag)

lists the known white dwarf and pre-white dwarf variables as of 1999.

Useful reviews include—

▷ Winget, D.E. 1988, in *Advances in Helio– and Asteroseismology*, eds. J. Christensen-Dalsgaard and S. Frandsen (Reidel: Dordrecht), p. 305

▷ Kawaler, S.D., & Hansen, C.J. 1989, in IAU Colloquium 114, *White Dwarfs*, ed. G. Wegner (Berlin: Springer–Verlag), p. 97

▷ Fontaine, G., Brassard, P., & Bergeron, P. 2001, PASP, 113, 409

and see

▷ Gautschy, A., & Saio, H. 1995, ARA&A, 33, 75

▷ Gautschy, A., & Saio, H. 1996, ARA&A, 34, 551.

The width of the ZZ Ceti instability strip is discussed in

▷ Bergeron, P., et al. 1995, ApJ, 449, 258.

Temperatures for the DBVs were established by

▷ Koester, D., et al. 1985, A&A, 149, 423

▷ Liebert, J., et al. 1986, ApJ, 309, 241.

The results we quote are from

▷ Beauchamp, A., et al. 1999, ApJ, 516, 887.

But perhaps not all DAs in the strip pulsate. See

▷ Kepler, S.O., & Nelan, E.P. 1993, AJ, 105, 608.

The prototype of the DOVs was discovered by

▷ McGraw, J.T., Starrfield, S.G., Liebert, J., & Green, R.F. 1979, in IAU Coll. 53, *White Dwarfs and Variable Degenerate Stars*, eds. H.M. Van Horn and V. Weidemann (Rochester: University of Rochester), p. 377

and, for K1-16, see

▷ Grauer, A.D., & Bond, H.E. 1984, ApJ, 277, 211.

The essentials of the arguments for the g-mode character of the white dwarf variables was laid down by

▷ Chanmugam, G. 1972, Nature PhysSci, 236, 83

▷ Warner, B., & Robinson, E.L. 1972, Nature PhysSci, 234, 2.

Shortly after these key papers, the first numerical experiments were performed by

▷ Osaki, Y., & Hansen, C.J. 1973, ApJ, 185, 277.

The cause of variability of the DAV and DBV variables was established by

▷ Dziemboski, W., & Koester, D. 1981, A&A, 97, 16

▷ Dolez, N., & Vauclair, G. 1981, A&A, 102, 375

▷ Winget, D.E. 1981, *Gravity Mode Instabilities in DA White Dwarfs*, Ph.D. Dissertation, University of Rochester, Rochester, N.Y.

▷ Starrfield, S.G., Cox, A.N., Hodson, S., & Pesnell, W.D. 1982, in *Pulsations in Classical and Cataclysmic Variable Stars*, eds. J.P. Cox & C.J. Hansen (Boulder: Joint Institute for Laboratory Astrophysics) p. 46

▷ Winget, D.E., Van Horn, H.M., Tassoul, M., Hansen, C.J., Fontaine, G., & Carroll, B.W. 1982a, ApJ, 252, L65.

The discovery of the first DBV is reported in

▷ Winget, D.E., Robinson, E.L., Nather, R.E., & Fontaine, G. 1982b, ApJ, 262, L11.

Causes of instability in the DOV variables is discussed in

▷ Starrfield, S., Cox, A., Kidman, R., & Pesnell, W.D. 1985, ApJ, 293, L23

and see

▷ O'Brien, M.S. 2000, ApJ, 532, 1078.

Most of the theoretical papers cited above use some version of the nonadiabatic methods of

▷ Saio, H., & Cox, J.P. 1980, ApJ, 236, 549

to search for instability. For non-variable DOVs in the "strip" see

▷ Werner, K 1995, Baltic Astron., 4, 340.

Possible shell-burning instabilities in the DOV's are reviewed in

▷ Kawaler, S.D., Winget, D.E., Hansen, C.J., & Iben, I. Jr. 1986, ApJ, 306, L41

▷ Kawaler, S.D. 1988, ApJ, 334, 220

but, for a null search for variables with the requisite periods, see

▷ Hine, B.P., & Nather, R.E. 1988, in IAU Colloquium 95, *The Second Conference on Faint Blue Stars*, eds. A.G.D. Philip, D.S. Hayes, & J. Liebert (Schenectady: L. Davis Press), p 627.

The very time-consuming measurements of the secular drift of period in one of the modes of PG1159 were first reported in

▷ Winget, D.E., Kepler, S.O., Robinson, E.L., Nather, R.E., & O'Donoghue, D. 1985, ApJ, 292, 606

▷ Winget et al. 1991, ApJ, 378. 326.

Corresponding measurements for G117-B15A are given by

▷ Kepler, S.O., et al. 1991, ApJ, 378, L45

and constraints on core composition for this star based on evolutionary models have been computed by

▷ Bradley, P.A., Winget, D.E., & Wood, M.A. 1992, ApJ, 391, L33.

An updated period change for G117–B15A is reported by

▷ Kepler, S.O., et al. 2000, ApJ, 534, L185.

They discuss, among other results, the model calculations of

▷ Bradley, P.A. 1998, ApJS, 116, 307

as applied to this star.

The operation of the WET is described in

▷ Nather, R.E., Winget, D.E., Clemens, J.C., Hansen, C.J., & Hine, B.P. 1990, ApJ, 361, 309.

We report further observations and analysis of PG1159 given in

▷ Costa, J.E.S., Kepler, S.O., & Winget, D.E. 1999, ApJ, 522, 973

that refine (and correct) the results of the earlier papers.

Figure 10.5 is derived from the original 1989 WET data for PG1159.

▷ Kawaler, S.D., & Bradley, P.A. 1994, ApJ, 427, 415

describe how period spacing is related to stellar mass. Spectroscopic determinations of gravity and T_{eff} plus the estimate for the luminosity of PG1159 are from

▷ Werner, K., Heber, U., & Hunger, K. 1991, A&A, 244, 437.

More recent results for PG1159 stars (but not PG1159 itself) are given in

▷ Werner, K. 1995, Baltic Astron., 4, 340

▷ Dreizler, S., & Heber, U. 1998, A&A, 334, 618.

Layering of elements and the resulting deviations in period spacings for DAVs is discussed in

▷ Brassard, P., Fontaine, G., Wesemael, F., & Hansen, C.J. 1992, ApJ, 80, 369.

▷ Jones, P.W., Pesnell, W.D., Hansen, C.J., & Kawaler, S.D. 1989, ApJ, 336, 403

reported on the effects of simple magnetic fields on the g-mode pulsations in white dwarfs.

▷ Winget, D.E., et al. 1994, ApJ, 430, 839

suggested that some peculiar mode splitting observed in GD358 could be due to magnetic fields of roughly 1300 G in the outer stellar layers. Intensive Zeeman measurements by

▷ Schmidt, G.D., & Grauer, A.D. 1997, ApJ, 488, 827

failed to find any overt fields but their measurement sensitivity was on the ragged edge of that suggested for GD358. The WET observations of

▷ Vuille, F., et al. 2000, MNRAS, 314, 689

and

▷ Kepler, S.O., et al. 2003, A&A, in press

show that GD358 is a very complicated pulsator.

A Mini Stellar Glossary

This short glossary of elementary astronomical terms associated with stars is not intended to be complete or very detailed. It is meant mostly for those of you who have no earlier experience in the subject. For the most part, we only list terms not specifically treated in the main text. An excellent overall reference to this material is

▷ Mihalas, D., & Binney, J. 1981, *Galactic Astronomy*, 3d ed. (San Francisco: Freeman & Co.)

and, on a more elementary level,

▷ Böhm–Vitense, E. 1989, *Introduction to Stellar Astrophysics*, Vol. 2 (*Stellar Atmospheres*) (Cambridge: Cambridge University Press).

1. **Stellar populations**: These are useful shorthand designations for stars sharing common properties of kinematics, location in a galaxy, and composition.
   a) **Population I** stars have a small scale height (confined to the disk of a spiral galaxy, if that's where you're looking), rotate with the disk, generally have a surface composition not too different from the sun's, and have a large range of masses since the young ones are still on the main sequence. Also look for them in any galaxy having active star-forming regions.
   b) **Population II** stars usually have a very large scale height (mostly found in the halo of a spiral galaxy), high space velocities, are poor in metals, and are of low mass since the more massive stars have already evolved. Hence Pop II stars are old stars.
   c) **Population III:** These are stars that must have been around at the time when the first stars were forming. They should have contained no metals because none were produced in the Big Bang. Thus far none have been observed either because they destroyed themselves early on or, more likely, the remaining ones accreted metal-rich material and are hence in disguise.
2. **Star clusters**: These are useful since they are generally composed of many stars of roughly the same composition and age. The turn-off point from the main sequence provides an age when compared to models or relative ages when one cluster is compared to another.

a) **"O–B" associations** consist of a loose cluster dominated in light by bright stars that are still associated with the interstellar gas that begat them. The cluster is not gravitationally bound and the stars are associated only because of their tender youth.

b) **Open clusters** or **galactic clusters** are galactic disk (young Pop I or somewhat older Pop I) stars or stars in regions of star formation in other types of galaxies. The clusters are bound together by gravitation. They contain both massive and low-mass stars. The Milky Way (our galaxy) contains at least 1,200 clusters. Many others must be present but we cannot see them because of intervening dust and gas in the disk (where we reside). The Large and Small Magellanic Clouds (galaxies relatively nearby to us) contain a total of more than 6,000. They are conspicuous because of the bright massive stars. Compared to globular clusters they have fewer stars in total number. Stellar membership ranges (at least) between 10 and 200. The Pleiades is a conspicuous open cluster in the northern night sky (and see Fig 2.6).

c) **Globular clusters** are gravitationally tightly bound and contain many low-mass Pop II stars. Their stars have surface metal contents between 1/2 and 1/200th that of the sun. They are associated with the halo of a galaxy and were formed early in the history of the galaxy. Our galaxy contains about 150 clusters with memberships of roughly 2000–10^6 stars. Figure 2.7 shows a HR diagram for one of them (M3).

3. **Observation of stars—Photometry**: Photometry refers to observing stars over one or more wavelength bands where details of the spectrum are not necessarily important.

   a) The **magnitude scale** is an astronomical scale for brightness constructed along the same lines as the decibel scale for sound. It is logarithmic, as are all such scales, so that the mind can handle the broad dynamic range of real external stimuli. There are two main magnitude scales. (Note that in what immediately follows we assume that the stars in question are observed over the same band of wavelengths or colors.)

      i. **Apparent magnitude**, m, is the brightness of a star as observed from earth and is thus a function of the intrinsic brightness of the star, its distance, and what is between us and the star. Unlike the decibel scale it runs backwards: the larger the magnitude, the apparently dimmer the star. An arithmetic *difference* of 5 in magnitude means a *multiplicative* factor of 100 in brightness (in, say, apparent luminosity). If b_i is the apparent brightness in physical units of star i, then the rule is

$$\frac{b_1}{b_2} = 2.512^{m_2 - m_1}.$$

A rough guide to apparent magnitudes is that a star with magnitude 6 is just visible to the naked eye while stars of magnitude 0 are among the brightest in the sky.

ii. The **absolute magnitude**, M, of a star is the apparent magnitude the star would have if we placed the star at a standard distance of 10 parsecs (1 pc equals 3.086×10^{18} cm or 3.2616 light years). If absorption of light by intervening gas or dust may be ignored, then the relation between apparent and absolute magnitude is

$$M = m + 5 - 5 \log d$$

where d is the actual distance to the star in parsecs. The difference $m - M$ is called the distance modulus.

iii. The **bolometric magnitude** is the magnitude integrated over all wavelengths (i.e., the entire electromagnetic spectrum). Since absolute magnitudes are standardized by the common distance 10 parsecs there must be a relation between luminosity and absolute bolometric magnitude, M_{bol}. Using the sun as a normalization this relation for a star ($\star$) is

$$\log(\mathcal{L}_\star/\mathcal{L}_\odot) = [M_{\text{bol}}(\odot) - M_{\text{bol}}(\star)]/2.5$$

where $M_{\text{bol}}(\odot)$ is +4.75.

b) **Colors** of stars are a reflection of the relative dominance of various wavelengths in their spectra and hence their effective temperatures. Observational magnitudes are always quoted for some range of wavelengths, which is standardized. An example of a standardized system is the Johnson–Morgan UBV system. The U filter lets in a band of light centered around 3,650Å in the ultraviolet. The B (blue) and V (visual) filters are centered around 4,400Å and 5,500Å , respectively. The magnitudes in these wavelength ranges are usually denoted by their letters. Thus the absolute magnitude of the sun in the V range of the UBV system is $V = -26.7$. The "color" of a star can be described by the difference in brightness in two filters. Thus, for example, and keeping in mind the backward scale for magnitudes, $(B–V)$ is negative for blue (hot) stars and positive for red (cool) stars.

c) **Bolometric correction**: To obtain the bolometric magnitude you must assume an energy distribution as a function of wavelength to find the spectrum at unobserved wavelengths (as, e.g., in the far ultraviolet for surface-based telescopes). Since the color, as well as the energy distribution, is related to the temperature, given the color you can estimate the correction to a magnitude to give the bolometric magnitude. Specifically, the bolometric correction, $B.C.$, is defined as

$$M_{\text{bol}} = M_V + B.C.$$

where M_V is the absolute visual magnitude (usually V).

d) **Color–magnitude diagram**: This is the observer's version of the Hertzsprung–Russell diagram (as in Figs. 2.6, 2.7, and more in the text). The abscissa is color (B–V, for example) and the ordinate is some magnitude (V, M_V, etc.). Color is the effective temperature surrogate and the single magnitude plays the role of luminosity. Needless to say, there is a lot of witchcraft involved in passing back and forth between these variables.

e) **Time-resolved photometry** is what the name implies. Snapshots are taken of the star to get magnitudes or intensities over intervals of time. The resulting times series is then used to infer dynamic properties such as those seen in variable stars.

4. **Observations of stars—Spectroscopy**: Here the details are important and one or more spectral absorption or emission lines are used for diagnostic purposes.

a) **Spectral types or classes**: This refers to the Henry Draper classification scheme, which is based on the appearance of spectral lines of hydrogen, helium, and various metals. An "O" star shows strong HeII (second ionization) lines of helium. Since the ionization potential is high, such stars are the hottest in terms of effective temperature. Spectral type "B" stars show strong HeI plus some HI lines. The Balmer lines reach their peak in the "A" stars and begin to disappear in "F." Calcium H&K lines strengthen in "G" and the "K" stars show other metallic lines. The "M" stars have strong molecular bands (particularly TiO). The effective temperatures decrease in the order O, B, A, F, G, K, M. The historical reasons for the lettering of this sequence are worth looking up in the literature. Subdivisions such as G0, G1, G2, $\cdots$ are in order of decreasing temperature. The sun is a G2 spectral class star. Our Figure 4.8 shows spectra for the normal range of classes for main sequence stars (and see accompanying discussion). Our §4.7 briefly discusses the new classes L and T for dwarfs (luminosity class V below).

b) **Luminosity classes** were introduced because stars of the same spectral type may have very different luminosities. The Morgan–Keenan (MK) classification uses spectral class and luminosity class as a two-dimensional scheme to phrase the HR diagram another way. The implementation of the scheme depends on details of spectral lines (such as width and depth) and reflects the density and gravity of the photosphere. Luminosity class I stars are the most intrinsically luminous. Because they are usually large (especially the cooler ones) they are also called **supergiants**. Luminosity class II (**giants**) are intrinsically bright, but not as bright as class I. The numbers increase until luminosity class V is reached. These are the **dwarfs** and primarily refer to main sequence stars. The sun is a G2V star. An M3I star is huge, intrinsically very luminous, and very cool. There are some intermediate classes such as Ia and Ib. White dwarfs are not included in this luminosity scheme.

c) **Abundances** are derived from spectral lines and are usually compared to the sun or the abundance of hydrogen in the star—or both. Thus the number density of iron, $n(\text{Fe})$, in a star ($\star$) as compared to the sun might be expressed as

$$[\text{Fe}/\text{H}] = \log\left[n(\text{Fe})/n(\text{H})\right]_\star - \log\left[n(\text{Fe})/n(,H)\right]_\odot .$$

As an example, the metal-poor and old globular cluster M3 (of Fig. 2.7) has [Fe/H] = −1.57. Since the surface hydrogen abundance of the sun is probably not too different from stars in M3, the conclusion is that the sun's surface abundance of iron is roughly 40 times greater.

B Table of Symbols and Physical Constants

Values for most of the fundamental constants and conversions in this list are from

▹ Mohr, P.J., & Taylor, B.N. 1999, JPhysChemRefData, 28, 1713.

Note that many symbols have multiple meanings, and those used only once or so may not be listed. In some cases we refer to the number of the equation that defines or contains the symbol as it appears early in the text.

a	Radiation constant, 7.56577×10^{-15} erg cm^{-3} K^{-4}
Age of $\odot$	$(4.57 \pm 0.05) \times 10^9$ years
amu	Atomic mass unit, 1 amu $= 1.6605402 \times 10^{-24}$ g, 941.494 MeV in energy units
A	Coefficient of energy for completely degenerate gas (Eq. 3.54)
A_i	Mass of nuclear species i, atomic mass units
$A_{\rm s}$	Schwarzschild discriminant (Eq. 5.32)
AU	Astronomical unit, 1.496×10^{13} cm
b	A reaction rate factor (Eq. 6.38)
B	Coefficient of pressure for completely degenerate gas (Eq. 3.52)
$\mathbf{B}$	Magnetic (induction) field (Eq. 10.7)
$B(T)$	Planck function, erg cm^{-2} s^{-1} (Eq. 3.22)
$B_\nu(T)$	Frequency-dependent Planck function, erg cm^{-2} (Eq. 4.7)
$B_{\rm C}$	Coulomb barrier height, usually in MeV (Eq. 6.19)
$B_{\mathcal{E}}$	Binding energy, usually in MeV (Eq. 6.16)
$B_{\mathcal{E}}/A$	Binding energy per nucleon
c	Speed of light in vacuum, $2.99792458 \times 10^{10}$ cm s^{-1} (exact)
c_P	Specific heat at constant pressure, erg g^{-1} K^{-1} (Eq. 3.83)
c_V	Specific heat at constant volume, erg cm^{-3} K^{-1} (Eq. 3.83)

c_{V_ρ}	Specific heat at constant specific volume, erg g^{-1} K^{-1} (Eq. 3.85)
$\mathcal{D}$	Diffusion constant
$d\Omega$	Differential solid angle
e	Elementary charge, $4.8032068 \times 10^{-10}$ esu
e^-	electron
e^+	Positron
eV	Electron volt, $1\ \mathrm{eV} = 1.60217733 \times 10^{-12}$ erg
E	Thermodynamic internal energy, erg g^{-1} or erg cm^{-3}, depending on context
E_{rad}	Energy density of radiation field, erg cm^{-3} (Eq. 3.18)
$\mathcal{E}$	Kinetic energy of individual particle
$\mathcal{E}_0$	Peak of the Gamow peak (Eq. 6.40)
$\mathcal{E}_F$	Fermi energy (Eq. 3.48)
f	Frequency in Hz (for example); oscillator strength (Eq. 4.73); "shape factor" (Eq. 6.30)
$f_{n\ell}$	Frequency of orders n and ℓ (Eq. 8.114)
$\mathcal{F}$	Energy flux, erg cm^{-2} s^{-1} (Eq. 4.3)
$\boldsymbol{\mathcal{F}}$	Vector energy flux, erg cm^{-2} s^{-1}
$\mathcal{F}_{\mathrm{conv}}$	Convective flux (Eq. 5.36)
$\mathcal{F}_{\mathrm{rad}}$	Radiative flux (Eq. 4.3)
$\mathcal{F}_{\mathrm{tot}}$	Total flux (Eq. 5.41)
g	Local gravitational acceleration (Eq. 1.1); gravity mode (g–mode)
$\mathbf{g}_{\mathrm{eff}}$	Effective gravity, $\mathbf{g}_{\mathrm{eff}} = -\boldsymbol{\nabla}\Phi_{\mathrm{eff}}$ (Eq. 9.14)
$g_\oplus$	Standard surface gravity of earth, 980.665 cm s^{-2} (exact)
$\mathbf{g}$	Vector local gravitational acceleration
G	Gravitational constant, $G = 6.6726 \times 10^{-8}$ g^{-1} cm^3 s^{-2}
h	Planck's constant, $6.6260688 \times 10^{-27}$ erg s; step size in an integrator (Eq. 7.46)
$\hbar$	$h/2\pi$
$H(a,u)$	Voigt function (Eq. 4.81)
H^-	H-minus ion

$I_\nu(\vartheta)$	Specific intensity (Eq. 4.1)
$\Im$	Imaginary part of a number
J	Ocillatory moment of inertia (Eq. 8.52)
$j_\nu(\vartheta)$	Mass emission coefficient, erg s g^{-1} (before Eq. 4.8)
k	Boltzmann's constant, 1.380650×10^{-16} erg K^{-1}, 8.617386×10^{-5} eV K^{-1}; wave number
$\mathbf{k}$	Vector wave number
k_r	Radial wave number (Eq. 8.30)
k_t	Transverse wave number, cm^{-1} (Eq. 8.101)
K	Kinetic energy (Eq. 1.19); polytropic constant (Eq. 7.16)
K'	Polytropic constant for ideal gas (Eq. 7.28)
$\mathcal{K}$	Diffusion constant
$\mathbf{L}$	Linear operator, as in the LAWE (Eq. 8.16); vector angular momentum
LAWE	Linear Adiabatic Wave Equation (Eq. 8.16)
ℓ	Mixing length; latitudinal index of $Y_{\ell m}(\theta,\varphi)$; angular momentum quantum number
$\mathcal{L}$	Luminosity, erg s^{-1} (Eq. 1.57)
$\mathcal{L}_{\mathrm{conv}}$	Convective luminosity
$\mathcal{L}_{\mathrm{rad}}$	Radiative luminosity
$\mathcal{L}_{\mathrm{tot}}$	Total luminosity
$\mathcal{L}_\odot$	Solar luminosity, $(3.847 \pm 0.003) \times 10^{33}$ erg s^{-1}
m	Mass, grams; azimuthal order in $Y_{\ell m}(\theta,\varphi)$
m_e	Electron rest mass, 9.1093898×10^{-28} g, 5.4858×10^{-4} amu, 0.5109991 MeV c^{-2}
m_n	Neutron rest mass, $1.6749286 \times 10^{-24}$ g, 1.0086649 amu, 939.56563 MeV c^{-2}
m_p	Proton rest mass, $1.6726231 \times 10^{-24}$ g, 1.00727647 amu, 938.27231 MeV c^{-2}
$\mathcal{M}$	Stellar mass, usually in units of grams
$\mathcal{M}_\odot$	Solar mass, $(1.9891 \pm 0.0004) \times 10^{33}$ g
$\mathcal{M}_r$	Mass contained within a sphere of radius r (Eq. 1.3)
$\mathcal{M}_\infty$	Chandrasekhar limiting mass (Eq. 3.67)

Symbol	Meaning
n	Density exponent of opacity (Eq. 1.62); number density (Eq. 1.51); polytropic index (Eq. 7.16); mode order (Eq. 8.34)
$n_{\rm e}$	Electron number density, $\rm cm^{-3}$ (Eq. 1.48)
$n_{\rm I}$	Ion number density, $\rm cm^{-3}$ (Eqs. 1.44–1.45)
$n_{\rm eff}$	Effective polytropic index (Eq. 7.118)
n_γ	Number density of radiation field, $\rm cm^{-3}$ (Eq. 3.15)
N	Brunt–Väisälä frequency (Eq. 5.28); number density per gram
$N_{\rm A}$	Avogadro's number, 6.022142×10^{23} $\rm mole^{-1}$
p	Linear momentum, g cm $\rm s^{-1}$
$\mathbf{p}$	Vector linear momentum
p-	Pressure (p-) mode
p_F	Fermi momentum (before Eq. 3.48)
pc	Parsec, 3.086×10^{18} cm
P	Pressure, usually in dyne $\rm cm^{-2}$; orbital period
$P_{\rm c}$	Central pressure
$P_{\rm e}$	Electron pressure
$P_{\rm g}$	Ideal gas pressure (Eq. 3.104)
$P_{\rm I}$	Ion pressure
$P_{\rm rad}$	Radiation pressure (Eq. 3.17)
P_ℓ	Barrier penetration factor (Eq. 6.20)
$P_\ell(x)$	Legendre polynomial of degree ℓ with argument $x = \cos\theta$ (Eq. 8.107)
$P_\ell^m(x)$	Associated Legendre polynomial of degree ℓ and order m with argument $x = \cos\theta$ (Eq. 8.107)
Pr	Prandtl number (Ex. 5.3)
Q	Specific heat content, erg $\rm g^{-1}$ (Eq. 1.11); pulsation Q
Q	Q-value of nuclear reaction
$-\mathcal{Q}$	Coefficient of thermal expansion
$\rm Q_{eff}$	Effective Q-value for pp-chains (Eq. 6.75)
r	Radius, cm
$\mathbf{r}$	Radius vector
r_n	Polytropic radial scale (Eq. 7.24)

r_{t}	Turning point or radius (Eq. 8.114)
$r_{\alpha\beta}$	Sample nuclear reaction rate, cm^{-3} (Eq. 6.22)
R	Nuclear radius (Eq. 6.17)
$\mathcal{R}$	Total radius; gas constant, $N_{\mathrm{A}}k = 8.314511 \times 10^7$ erg K^{-1} mole^{-1}
$\mathcal{R}_\odot$	Solar radius, $\mathcal{R}_\odot = 6.96 \times 10^{10}$ cm
$R_\oplus$	Radius of earth, 6.38×10^8 cm
Ra	Rayleigh number (Ex. 5.4)
$\Re$	Real part of a number
s	The negative of the temperature exponent of opacity (Eq. 1.62); impact parameter
S	Entropy, erg gm K^{-1} or erg cm^{-3} K^{-1} (Eq. 1.11)
$S(\mathcal{E})$	Nuclear "S" factor (Eq. 6.36)
S_ν	Source function (before Eq. 4.8, Ex. 4.1)
S_ℓ	Lamb frequency (Eq. 8.100)
t	Time
t_{cool}	Cooling time of white dwarfs (Eq. 10.5)
t_{dyn}	Dynamic time scale (Eq. 1.33)
t_{KH}	Kelvin–Helmholtz time scale (Eq. 1.31)
t_{ML}	Mass-loss time scale
t_{th}	Thermal time scale (Eq. 8.64)
t_{nuc}	Nuclear time scale (Eq. 1.89)
T	Temperature (K)
T_{c}	Central temperature
T_n	Temperature in units of 10^n (K)
T_{eff}	Effective temperature, K (Eq. 1.92)
$T_{\mathrm{eff}}(\odot)$	Solar effective temperature, 5780 K
T_{TR}, T_{tr}	Transition temperature in pulsation (Eq. 8.66) or white dwarfs (Eq. 10.1)
u	Energy density of radiation field (Eq. 3.19 et seq.)
u_ν	Energy density of radiation field per unit frequency (Eq. 3.19)
u_λ	Energy density of radiation field per unit wavelength (Eq. 3.20)

U	Variable in the U–V plane (Eq. 7.52)
U	Total internal energy (Eq. 1.10)
v	Velocity
$\mathbf{v}$	Velocity vector
v_s	Sound speed (Eq. 1.38)
V	Variable in the U–V plane (Eq. 7.52)
V	Volume
V_ρ	Specific volume, $V_\rho = 1/\rho$ (Eq. 3.12)
W	Total energy, erg (Eq. 1.9)
W_{tot}	Work done over pulsation cycle (Eq. 8.44)
x_F or x	Dimensionaless Fermi momentum (Eq. 3.48); $x = \cos\theta$
X	Hydrogen mass fraction (Eq. 1.52)
X_i	Mass fraction of nuclear species i
y_i	Ionization fraction of species i (Eq. 1.47)
Y	Helium mass fraction (Eq. 1.52)
$Y_{\ell m}(\theta, \varphi)$	Spherical harmonic of degree ℓ and azimuthal order m (Eqs. 8.95, 8.106)
Z	Mass fraction of metals (Eq. 1.52)
Z_i	Integer nuclear charge of species i
α	Mixing length parameter, dimensionless
$\alpha_{\mathcal{R}}$	Radius homology exponent, $\mathcal{M}^{\alpha_{\mathcal{R}}}$ (Eq. 1.70)
α_ρ	Density homology exponent, $\mathcal{M}^{\alpha_\rho}$ (Eq. 1.71)
$\alpha_{\mathcal{L}}$	Luminosity homology exponent, $\mathcal{M}^{\alpha_{\mathcal{L}}}$ (Eq. 1.73)
α_T	Temperature homology exponent, $\mathcal{M}^{\alpha_T}$ (Eq. 1.72)
β	Ratio of gas to total pressure, $\beta = P_{\mathrm{g}}/P$ (Eq. 3.106); $kT/m_{\mathrm{e}}c^2$; $-dT/dz$ (Eq. 5.2)
γ	As in a "γ-law equation of state" (Eq. 1.24); ratio of specific heats (Eq. 3.92); damping constant (Eq. 4.73)
γ^2	Reduced width of a nuclear energy state (Eq. 6.20)
Γ	Width of a nuclear energy state (usually MeV) (Eq. 6.12)
Γ_C	Coulomb interaction factor (dimensionless) (Eq. 3.78)
Γ_1	Adiabatic exponent $\Gamma_1 = (\partial \ln P/\partial \ln \rho)_{\mathrm{ad}}$ (Eqs. 1.39, 3.93)

Γ_2	Adiabatic exponent $\Gamma_2/(\Gamma_2-1)=(\partial\ln P/\partial\ln T)_{\rm ad}$ (Eq. 3.94)
Γ_3	Adiabatic exponent $\Gamma_3-1=(\partial\ln T/\partial\ln\rho)_{\rm ad}$ (Eqs. 1.81, 3.95)
δ	δ refers to the Lagrangian perturbation of a quantity (e.g., Eq. 8.3); infinitesimal operator
Δ	Width of Gamow peak (Eq. 6.41)
ΔT	Temperature excess of a convective parcel, $\Delta T = T' - T$ (Eq. 5.20)
$\Delta\nu_D$	Doppler width (Eq. 4.78)
∇	Logarithmic gradient $d\ln T/d\ln P$ (Eq. 4.28)
$\boldsymbol{\nabla}$	Gradient or divergence operator
$\nabla_{\rm ad}$	Adiabatic gradient $(\partial\ln T/\partial\ln P)_{\rm ad}$ (Eq. 3.94)
$\nabla_{\rm rad}$	Radiative gradient (Eq. 4.30)
ε	Energy generation rate, erg g^{-1} s^{-1} (Eq. 1.57)
$\varepsilon_{\alpha\beta}$	Sample energy generation rate (Eq. 6.47)
$\varepsilon_{3\alpha}$	Triple-α energy generation rate (Eq. 6.80)
ε_{ν}	Neutrino energy loss rate
$\varepsilon_{\rm CNO}$	Simple energy generation rate for the CNO cycle (Eq. 6.77)
$\varepsilon_{\rm eff}$	Energy generation rate for pp–chains (Eq. 6.76)
$\varepsilon_{\rm grav}$	Gravitational energy generation rate (Eq. 6.3)
ζ	Relative radius variation $\delta r/r$ (Eq. 8.8)
η	Degeneracy parameter; Sommerfeld factor (Eq. 6.35)
θ or ϑ	Colatitude angle
θ_n	Dependent polytropic variable (Eq. 7.20)
κ	Opacity (mass absorption coefficent), cm^2 g^{-1} (Eq. 1.62)
$\kappa_{\rm qa}$	Quasi-adiabatic growth rate, s^{-1} (Eq. 8.48)
$\kappa_{\rm bb}$	Bound–bound opacity
$\kappa_{\rm bf}$	Bound–free opacity (Eq. 4.63)
$\kappa_{\rm e}$	Electron scattering opacity (Eq. 4.60)
$\kappa_{\rm ff}$	Free–free opacity (Eq. 4.61)
$\kappa_{\rm H^-}$	H$^-$ opacity (Eq. 4.65)
$\kappa_{\rm cond}$	Conductive "opacity" (Eq. 4.67)

$\kappa_{\rm tot}$	Total opacity cm^2 g^{-1} (Eq. 4.69)
λ	Density exponent of the energy generation rate (Eq. 1.59); wavelength; decay constant
$\lambdabar$	DeBroglie wavelength (Eq. 6.29)
λ_P	Pressure scale height, $\lambda_P = P/g\rho$ (Eq. 3.1)
$\lambda_{\rm phot}$	Photon mean free path
μ	Mean molecular weight (Eqs. 1.50, 1.55); chemical potential (Eq. 3.3)
$\mu_{\rm e}$	Mean molecular weight of electrons (Eqs. 1.49, 1.53)
$\mu_{\rm I}$	Mean molecular weight of ions (Eqs. 1.46, 1.54)
ν	Temperature exponent of the energy generation rate (Eq. 1.59); photon frequency; kinematic viscosity (Ex. 5.3)
$\nu_{\rm e}$	Electron neutrino
$\overline{\nu}_{\rm e}$	Electron antineutrino
ν_T	Thermal diffusivity (or conductivity), cm^2 s^{-1} (Eq. 5.16)
ξ	Independent polytropic radius, dimensionless (Eq. 7.24); nonradial displacement
$\boldsymbol{\xi}$	Vector nonradial displacement (Eq. 8.76)
ξ_1	Location of first zero of polytropic $\theta_n(\xi)$ (Eq. 7.31)
ξ_r	Radial displacement (Eq. 8.87)
ξ_t	Tangential displacement (Eq. 8.92)
Π	Period of oscillation or variability (Eq. 1.37); orbital period; parity
$\Pi_{\rm p}$	P-mode period
$\Pi_{\rm g}$	G-mode period
$\dot{\Pi}$	Rate of period change, s s^{-1}
Π_0	Period spacing times $\sqrt{\ell(\ell+1)}$ (Eq. 8.112)
ϖ	Radial distance from axis in cylindrical coordinates
ρ	Density, g cm^{-3}
$\rho_{\rm c}$	Central density
$\langle\rho\rangle$	Average density, g cm^{-3}
$\langle\rho_\odot\rangle$	Average density of the sun, 1.41 g cm^{-3}
σ	Stefan–Boltzmann constant, 5.67040×10^{-5} erg cm^{-2} K^{-4} s^{-1}; angular frequency ($2\pi f = \sigma$)

σ_{p}	P-mode frequency (Eq. 8.109)
σ_{g}	G-mode freqency (Eq. 8.110)
$\sigma_{\alpha\beta}$	Sample nuclear cross section, cm^2 or barn (Eq. 6.21)
$\langle\sigma v\rangle_{\alpha\beta}$	Average nuclear cross section times velocity cm^3 s^{-1} (Eq. 6.25)
τ	Mean- or half-life (Eq. 6.12); reaction rate parameter (Eq. 6.43)
τ_ν	Optical depth (Eq. 4.8)
ϕ or φ	Azimuthal angle; phase
Φ	Gravitational potential (Eq. 7.18)
Φ_{eff}	Effective potential, gravitational plus centrifugal (Eq. 9.13)
χ_ρ	Density exponent of pressure (Eq. 1.67)
χ_T	Temperature exponent of pressure (Eq. 1.67)
χ_{H}	Ionization potential of hydrogen (Eq. 3.31)
$\Psi(\mathcal{E})$	Maxwell–Boltzmann distribution (Eq. 6.24)
ω	Angular frequency; dimensionless frequency (Eq. 8.25)
$(\omega\gamma)_r$	Fowler's factor for nuclear reactions (Eq. 6.54)
Ω	Total gravitational energy (erg) (Eq. 1.7); solid angle; rotation frequency;
$\odot$	Sun symbol
'	' refers to the Eulerian perturbation of a quantity
*	Denotes complex conjugate
$(\ )_{\mathrm{ad}}$	Indicates an adiabatic process

C List of Journal Abbreviations

The journal abbreviation and reference style used in this text is modeled after the vanilla-flavored ones favored by *The Astrophysical Journal* and the WWW site of the NASA Astrophysics Data System (ADS) archives, even though some of these abbreviations tend to be obscure. Some popular journals we name in full, such as *Nature*, or as they are pronounced by everyone, such as PhysRev or ApJ. Some periodicals mentioned in the text are not listed here (as obscure examples) and, in those rare cases, we spell out the title in the text.

A typical reference appears as:

▹ Writer, I'ma 2001, ApJ, 706, 19

where the author's name comes first, then the year of publication, abbreviated journal name, volume number, and, lastly, page number. (The more useful older scheme had the volume number in bold print to keep the various numbers straight.) In this example ApJ means "The Astrophysical Journal."

AAS	*American Astronomical Society*
A&A	*Astronomy and Astrophysics*
A&AS	*Astronomy and Astrophysics Supplement*
A&ARev	*Astronomy and Astrophysics Reviews*
AcA	*Acta Astronomica*
ADNDT	*Atomic Data & Nuclear Data Tables*
AJ	*Astronomical Journal*
AmJPh	*American Journal of Physics*
ApJ	*Astrophysical Journal*. Note that the "Letters" issue of this journal is signaled by an "L" preceding the page number.
ApJS	*Astrophysical Journal Supplement*
Ap&SS	*Astrophysics and Space Science*
ARA&A	*Annual Review of Astronomy and Astrophysics*
ARNS	*Annual Review of Nuclear Science*
ARN&PS	*Annual Review of Nuclear & Particle Science*
ASP	*Astronomical Society of the Pacific*
AstNachr	*Astronomische Nachrichten*
AustJPhys	*Australian Journel of Physics*
BAAS	*Bulletin of the American Astronomical Society*

BAN	*Bulletin of the Astronomical Institute of the Netherlands*
CanJPhys	*Canadian Journal of Physics*
ComAp	*Comments on Astrophysics*
CoPhC	*Computational Physics Communications*
GeoCosmo	*Geochimica et Cosmochimica*
IAU	*International Astronomical Union*
JFlMech	*Journal of Fluid Mechanics*
JPhB	*Journal of Physics, Part B*
MNRAS	*Monthly Notices of the Royal Astronomical Society*
MethCompPhys	*Methods of Computational Physics*
Nature	*Nature*
Obs	*The Observatory*
PASJ	*Publications of the Astronomical Society of Japan*
PASP	*Publications of the Astronomical Society of the Pacific*
PhysRep	*Physics Reports*
PhysRev	*Physical Review* (with final letters A-E depending on subject)
PhysRevL	*Physical Review Letters*
PhysScr	*Physica Scripta*
PRSocL	*Proceedings of the Royal Society of London*
PTPhJS	*Progress of Theoretical Physics, Japan, Supplement*
QJRAS	*Quarterly Journal of the Royal Astronomical Scoiety*
RepProgPhys	*Reports of Progree in Physics*
RevModPhys	*Reviews of Modern Physics*
RusAJ	*Russian Astronomical Journal*
Science	*Science*
SciAm	*Scientific American*
Sky&Tel	*Sky & Telescope*
SoPh	*Solar Physics*
SpSciRev	*Space Science Reviews*
SvA	*Soviet Astronomy*
SSRv	*Space Science Reviews*
ZeAp	*Zeitshrift für Astrophysik*

Index

ASTRONOMY AND ASTROPHYSICS LIBRARY

The Early Universe Facts and Fiction
4th Edition By G. Börner

The Design and Construction of Large Optical Telescopes By P. Y. Bely

The Solar System 4th Edition
By T. Encrenaz, J.-P. Bibring, M. Blanc, M. A. Barucci, F. Roques, Ph. Zarka

General Relativity, Astrophysics, and Cosmology By A. K. Raychaudhuri, S. Banerji, and A. Banerjee

Stellar Interiors Physical Principles, Structure, and Evolution 2nd Edition
By C. J. Hansen, S. D. Kawaler, and V. Trimble

Asymptotic Giant Branch Stars
By H. J. Habing and H. Olofsson

Series homepage – http://www.springer.de/phys/books/aal